Measurements from Maps

Principles and Methods of Cartometry

Related Pergamon titles

LISLE	Geological Structures and Maps
GHOSH	Analytical Photogrammetry, 2nd edition
ROBERTS	Introduction to Geological Maps and Structures
ALLUM	Photogeology and Regional Mapping
GORSHKOV	World Ocean Atlas Volume 1: Pacific Ocean Volume 2: Atlantic and Indian Volume 3: The Arctic Ocean

Journals (*free sample copies gladly sent on request*)

Space Technology

Journal of Structural Geology

Journal of South American Earth Sciences

Journal of South East Asian Earth Sciences

Measurements from Maps

Principles and Methods of Cartometry

by

D. H. MALING

Formerly University of Wales

PERGAMON PRESS

OXFORD · NEW YORK · BEIJING · FRANKFURT
SÃO PAULO · SYDNEY · TOKYO · TORONTO

U.K.	Pergamon Press plc, Headington Hill Hall, Oxford OX3 0BW, England
U.S.A.	Pergamon Press, Inc., Maxwell House, Fairview Park, Elmsford, New York 10523, U.S.A.
PEOPLE'S REPUBLIC OF CHINA	Pergamon Press, Room 4037, Qianmen Hotel, Beijing, People's Republic of China
FEDERAL REPUBLIC OF GERMANY	Pergamon Press GmbH, Hammerweg 6, D-6242 Kronberg, Federal Republic of Germany
BRAZIL	Pergamon Editora Ltda, Rua Eça de Queiros, 346, CEP 04011, Paraiso, São Paulo, Brazil
AUSTRALIA	Pergamon Press Australia Pty Ltd., P.O. Box 544, Potts Point, N.S.W. 2011, Australia
JAPAN	Pergamon Press, 5th Floor, Matsuoka Central Building, 1-7-1 Nishishinjuku, Shinjuku-ku, Tokyo 160, Japan
CANADA	Pergamon Press Canada Ltd. Suite No. 271, 253 College Street, Toronto, Ontario, Canada M5T 1R5

First edition 1989

Library of Congress Cataloging in Publication Data

Maling, D. H.
Measurements from maps.
Bibliography: p.
Includes index.
1. Cartometry. I. Title.
GA23.M35 1988 912'.01'48 88-15252

British Library Cataloguing in Publication Data

Maling, D. H.
Measurements from maps: principles and
methods of cartometry.
1. Maps. Areas. Measurement
I. Title
912'·0287

ISBN 0-08-030290-4 (Hardcover)
ISBN 0-08-030289-0 (Flexicover)

Phototypesetting by Thomson Press (India) Ltd., New Delhi
Printed in Great Britain by A. Wheaton & Co. Ltd., Exeter

When you can measure what you are speaking about and express it in numbers, you know something about it; but when you cannot measure it, when you cannot express it in numbers, your knowledge is of a meagre, unsatisfactory kind.

Lord Kelvin

Preface

This book describes and analyses the accuracy of a variety of measurement techniques which may be used in many fields of science, administration and travel. The potential uses are summarised in Chapter 1. The content of that chapter and, perhaps even more, the subject range of the references quoted in the Bibliography on pages 552–564 indicates that no single discipline has monopolised the development of the subject.

It is usual that a group working in a specialised scientific field knows about the methods used by colleagues having similar interests. With improved communication it has also become common for them to have some acquaintance with similar work proceeding in other countries. However, this usually has linguistic limitations. In the field of cartometry, the Russian contribution to the subject has been especially important and, not surprisingly, nearly all of this has been published in Russian and other East European languages. Although a feature of this book is that the author has been able to explore these sources, this only fills one linguistic gap. Because the author has no knowledge of Chinese or Japanese, he has not been able to make much use of any work done there.

Often knowledge about the technique used is more strongly confined by the artificial boundaries of an academic discipline so that there may be even less communication of ideas between one subject and another than between one country and another. In cartometry, for example, there are three major groups of literature about measurement of area. The first is that appearing in the textbooks of surveying, cartography and geography. The second group comprises foresters. The third major group are the scientists working with photomicrographs rather than maps, who have developed techniques of linear and area measurement which correspond closely to those used by others on maps and aerial photographs. They call it stereology, which further obscures the connection. Users of maps and photographs belonging to any of these groups are probably likely to know about the contributions made to their own subject, but are significantly ill-informed about equivalent work in the other two. Consequently there has been a considerable duplication of effort in making experimental tests. As will be seen in some of the following chapters, most of these tests have been designed to study a few familiar ideas, such as the comparative accuracy of two different instruments, or the relationship between accuracy of area measurement and shapes of parcels. An example of the incompleteness of the experimental record is the study of the influence of paper deformation in Chapter 10. Here we have to make use of four different experimental studies to build up the picture of what happens to measurements made on paper maps. It compares the results of measurements obtained by a French professor, a Russian professor, a serving officer in the Royal Air Force and a Japanese lady working in a Swiss university who was funded by a U.S. Army research grant. Not surprisingly, none of the experimenters were aware of what the others had done but fortunately many of the questions which we need to ask have been answered by one or other of them.

There is also considerable lack of awareness of what has been done by our forefathers – indeed there always has been. Not infrequently we encounter the claim that the author has discovered a new method of measurement, only to find that this is already a well-established procedure in other disciplines, or was so in earlier years. Time after time we find that writers describe their own pet methods of measurement as being new inventions, and they were quite unaware that the method had been known and used earlier. It is sufficient to mention only one example here. This is the periodic rediscovery of the method of area measurement by weighting portions of the map which have been cut into the parcels to be measured. This method was known at least as early as the late sixteenth century, for there is reference to the method in *Methodus Geometrica* published in Nürnberg in 1585 to *a curious method of arriving at the area of a field ... cut out with scissors ... and balanced on goldsmith's scales.* Its use for measuring the areas of countries is frequently attributed to the astronomer, Edmund Halley, who, a century after *Methodus Geometrica*, described his techniques in a letter to John Houghton (1693) which I have quoted in full on page 208. During the next century it was used by at least three people, Scherer in 1710, Long in 1742 and Oeder in 1777. Certainly the last of these believed that *he* had invented the method. Two centuries later Proudfoot (1946) dryly commented that an U.S. Government employee had recently invented the weighing method for determining area, and in 1956 it was recommended as a suitable method to be used by the New Zealand Forestry Service. Even in 1967, the late C. J. McKay, who at that time was the Superintendent in charge of the Areas Section of the Large Scales Division of the Ordnance Survey, informed me that:

> This week a London newspaper 'phoned me to say they were recently very interested in an article in the American 'Life' magazine which described with enthusiasm a 'new' measurement technique whereby aerial photographs were cut up and each irregular area weighed in a precise balance.

An important consideration in the preparation of a book which is likely to be read by many different categories of map-user is the assumed level of mathematical knowledge needed to understand the text. There are many people working in the natural sciences who do not claim to be expert mathematicians; indeed this is probably why they originally found themselves studying geography, geology or biology rather than physics and chemistry.

The author has therefore assumed that the reader has comparatively little mathematical knowledge at the outset so that some very elementary concepts, such as the use of the simple scale conversion expressions and the determination of the arithmetic mean are explained in the early chapters. Clearly a book of this size and subject cannot provide a comprehensive, step-by-step development of all the mathematical arguments, so that there must be some significant leaps in the mathematical standard needed for a full understanding of the second half of the book. Some of the necessary background is supplied through the medium of studying the statistical theory or errors and sampling methods in Chapters 6, 7 and 8.

A typical example of the approach which has been adopted occurs in Chapter 11, where it is necessary to derive the formulae to calculate the area of a portion of the curved surface of a sphere or spheroid which is bounded by two parallels and two meridians. The simpler derivation of the spherical area is explained in order to demonstrate how it is done, but not the corresponding proof for the spheroid, which would require two or three additional pages of algebra.

Some younger readers, who have grown up in an environment which is so dominated by digital computing and data processing, may well glance at this book and dismiss it as being quite out of date; that this was how measurements may have been made in olden times – which they judge to have been before 1980. Since the hardware exists to scan or manually digitise maps and photographs, it may be argued by them that dividers, planimeters and steel scales are all things of the past. The digital revolution in North America and Western Europe has put microcomputers into the hands of virtually everyone who needs one, but their availability has not yet really altered the opportunities of making measurements. There are many limitations upon their use in this field. For example, much of the available hardware is only suitable for small-format imagery. Again, most data processing equipment and the peripherals require connection to a reliable supply of electricity (or frequent access to fully charged batteries), so that the equipment may not be continuously available in some parts of the world. A third point is that digital methods for determining areas or volumes, distances or angles, still need a suitable file of machine-readable data from which to work, or the facilities to carry out digitising. Chapter 21 contains some pertinent observations concerning the availability of such data and the cost of it. Moreover, where digital methods of cartometry are already available, the author will show that although it is easy enough to obtain acceptable results for the measured length of a coastline, the limitations imposed by the organisation of the system, may be such that the digital method of "rubber-banding" is possibly three to five times more time-consuming than doing the same work on a paper map with dividers at possibly one-thousandth of the cost. I think that the manual methods, requiring no more digital aid than a pocket calculator, will certainly outlive me – and probably the critical youngsters to boot.

The author would like to acknowledge the help of Guy Lewis, who has drawn the text figures, and of John Wilson, who made many of the Yorkshire coast measurements when we were 20 years younger. Dr Lucy Starr, formerly of the Department of Physiology, University College, London, effectively put the cat amongst the pigeons by introducing me to the subject of stereology when the original draft of this book was already far advanced. Dr Chris Wooldridge and his colleagues in the Department of Maritime Studies at UWIST introduced me to the Draft Convention on the Law of the Sea, with all its cartometric requirements and absurdities. My sincere thanks to all of them.

DEREK MALING
Defynnog, Powys

Contents

Index to Symbols and Characters Used in the Text

The author has attempted to retain certain letters to represent particular quantities, such as A for area, L and l to represent length; d to mean the separation of the points of a pair of dividers. However, in quoting so many different authorities writing on so many different subjects, there is inevitably some overlap in the meanings of symbols. It is not always possible, or desirable, to convert all of these to common usage and consequently the following list is required to show the range of symbol usage and help to ease the intellectual leaps which have to be made each time A, b, k, n or p appear in different guises. The following pages list the common usage, referring to the page or equation where they first appear. The convention adopted is that reference to an [equation] is given in square brackets; and (page) is in parentheses.

The following letters in the list indicate common usage in certain subjects:

[*C*] cartography, including the study of map projections
[*G*] geodesy
[*M*] mathematics
[*N*] navigation
[*P*] photogrammetry
[*S*] statistics.

To economise in space, certain well-known mathematical symbols, for example π, x and y are not included in this list, except where they are used in a special, and unfamiliar, form.

Distinction between different quantities of length, area, etc. is usually indicated by a suffix, such as A_0 or l_1 etc. This is used so commonly that no attempt is made to list all of their occurrences.

A Generally the italic A refers to area, the upright Roman A is retained for other geometrical uses such as:
Steinhaus' term for an arc (3.17)
Geodetic expansion coefficient [*G*] [11.36]
The event when Buffon's Needle intersects a family of lines, *a* (15.3)

A	Area of a parcel (4.7)
A_M	Area of a map (19.4)
A_B	Area of a parcel (19.4)
A'	Area of a geometrical figure (16.4)
A_0	True area of a parcel (11.6)
	Area of a circular parcel defined by the planimeter calibration jig [17.56]
	Constant ($= 56{\cdot}77\ \mathrm{cm}^2$) used by Montigel (17.38)
a	Major semiaxis of ellipsoid [*C*] [*G*] [*M*] (2.6)
	Separation of circle centres on Perkal's square pattern longimeter [3.14]
	Area of a single square cell [4.01]
	Constant [*M*] (6.15)
	Maximum particular scale at a point [*C*] (10.6)
	Side of a geometrical figure (15.3)
	Coefficient [*M*] (17.26)
a'	Area of part of a single square cell [4.02]
δA	Planimetric displacement of a contour line (9.10)
B	Geodetic expansion coefficient [*G*] [11.37]
	The event when Buffon's needle intersects a family of lines, *b* (15.3)
	Coefficient used by Yuill in the study of area measurement by point counting [19.21]
b	Minor semiaxis of ellipsoid [*C*] [*G*] [*M*] (2.6)
	Separation of circle centres in Perkal's triangular pattern longimeter [3.15] (3.22)
	Minimum particular scale at a point [*C*] (10.6)
	Side of a geometrical figure (15.3)
	Coefficient [*M*] (14.18)
	Coefficient used by Bonnor in the study of area measurement by point counting [18.22]
C	Binomial coefficient [*S*] (6.21)
	Geodetic expansion coefficient [*G*] [11.38]
	Constant ($= 75456{\cdot}835$) in Balandin's formula (11.20)
	Planimeter constant (17.14)
	Coefficient used by Yuill in the study of area measurement by point counting [19.12]
	Integration constant [*M*] (22.6)
c	Bessel's correction [*S*] (7.6)
	Vertical interval between contours [*P*] (9.11)
	Side of a geometrical figure (16.4)
	Number of intersections of a long Buffon's needle with the side of a grid (15.5)

	Coefficient [*M*] (17.26)
	Coefficient used by Baer in the study of planimeter movements (17.34)
	Coefficient used by Bonnor in the study of area measurement by point counting [18.22]
D	Geodetic expansion coefficient [*G*] [11.39]
	Richardson's term for dividers separation (d) expressed in the same ground units as the length of a feature (14.10)
	Percentage difference between areas measured by counting and by planimeter, as used by Yuill [19.12]
	Difference in meridional parts [*N*] (22.11)
	Arc distance between two points on the spheroid [*G*] [23.05]
d	Separation of dividers points (3.6)
	Separation of equidistant parallel straight lines [3.09]
	Separation of points on an overlay (4.6)
	Width of a strip on an overlay (4.8)
	Parameter ($=\delta/s_0$) used to determine the critical difference between two means [*S*] [17.23]
	Limit of resolution of a scale etc. (17.29)
	Distance between points (21.28)
d_1	Separation of grid lines on a square pattern overlay (4.6)
d_2	Separation of grid lines on a triangular (hexagonal) pattern overlay (4.7)
$d_1, d_2,$	Different settings of dividers, associated with the determination of the limiting distance along a line measured by dividers (5.16)
d'	Separation of dividers points in the final step of a measured line ($d' < d$) (3.9)
d_{MAX}	Maximum deviation of an observation from the mean used in Chauvenet's criterion [*S*] (6.17)
E	Allowable error [*S*] (7.11)
	Maximum allowable error [*S*] (18.19)
	Easting coordinate [*C*, *G*]
	Easting coordinate of a point on the ground (9.05)
	Maslov's shape factor (17.31)
E'	Mapped easting coordinate (9.05)
e	Eccentricity of spheroid [*C*] [*G*] [*M*] (2.6)
	Base of the expontential function (2·718282...) [*M*] (6.14)
	Maximum permissible error [*S*] (8.18)
	Percentage error derived by Bonnor from the comparison of area measurement by counting with planimeter measurements [18.22]

F	Håkonson's shoreline development (19.24)
f	Ellipticity or flattening of spheroid [C] [G] [M] (2.6)
	Relative error [S] (6.1)
	Sampling fraction [S] (7.7)
	Relative root mean square error expressed as a percentage (18.13)
	Frolov's correction for area measurement on Mercator's projection (22.23)
f	Focal length of camera lens [P] (2.16)
G	Mandelbrot's term corresponding to the length of a dividers' step (d or D) (14.12)
g	centre of gravity of a parcel, used as the suffix φ_g, p_g etc. (22.25)
H	Flying height of aircraft [P] (2.12)
	Ground height (9.5)
H'	Mapped ground height (9.5)
H_v	Horizontal scale factor on Landsat M.S.S. imagery (21.28)
h	Particular scale in the direction of the meridian [C] (10.6)
	Ground height [P] (13.5)
	Relative humidity (10.17)
	The height of a triangle [M] (16.4)
	The combination of the movements of the measuring wheel of a planimeter (17.13)
i	Generally a counter such as A_i = the area of parcel number i. Number of a strip (4.8)
K	Coefficient in determination of reduced length (5.17)
	Volkov's coefficient used to determine reduced length (5.17)
	The shape of a triangle (16.9)
	Coefficient expressing parcel shape (18.13)
	Reciprocal of the area scale ($1/p$) (22.16)
	Chord distance between two points on the surface of a spheroid [G] [23.04]
K'	The shape of a rectangle (17.31)
k	Number of trials or applications (3.18)
	Volkov's coefficient for determination of limiting distance (5.16)
	Constant [6.13]
	Particular scale in the direction of the parallel [C] (10.6)
	Coefficient used by Malovichko for correcting linear measurements (14.17)
k	Amount of reduction in scale from one document to another in map production (9.20)
	Conversion factor used with planimeters (17.10)

k_0	A pair of parallels on the normal aspect Mercator's Projection which are lines of zero distortion (22.7)
k_1, k_2, k_3	Coefficients used to determine the influence of shape, etc., upon the precision of area measurements by point counting methods (18.12)
L	Total length of line (3.10)
	Finite length of a straight line [3.05]
	Limiting distance of a line [5.01]
	True length of a line (6.1)
	The length of two sides of a square of area A (17.30)
	Cumulative length (18.11)
	Total length of all lines measured on a map containing a parcel to be measured [20.01]
L_{red}	Reduced length (5.17)
L_1, L_2	Lengths of the same lines on different scale maps, these lines having been corrected to their respective limiting distances (5.17)
L'	Theoretical length of a line measured by stream digitising (21.19)
l	Length of an arc element (3.19)
	Length of Buffon's needle [3.09] (3.19)
	Measured length of a line (6.1)
	Distance recorded by the measuring wheel of a planimeter (17.4)
	Undistorted length of a straight line on a truly vertical aerial photograph (13.15)
	Length of the protion of the parcel perimeter with a marginal grid cell (18.10)
	Total length of lines sampled within a parcel to be measured [20.01]
l	Length of the side of an unit rectangle (18.04)
l'	Measured length of a line l on a tilted aerial photograph (13.15)
l_1, l_2	Lengths of a line measured by dividers set to d_1 and d_2 respectively (5.16)
$l_1, l_2, l_3 \ldots$	Independent measurements of the length of the same line (6.1)
δl	Absolute linear error (6.1)
log	logarithm to base 10 [M]
ln	logarithm to base e [M]
M	Mandelbrot constant (14.12)
	Meridional or Mercatorial parts [N] (22.11)
m	Limiting value of k in Steinhaus' order of measurement (3.21)
	Constant (= 1·004285) in Carpenter's formula (11.19)
m	Side of a rectangular grid cell (18.03)

N	Total number [S]
	Number of separate independent variables [6.17]
	Size of sample [S] (9.30)
	Northing coordinate [C] (9.05)
N'	Mapped Northing coordinate (9.05)
NF	Normalised shore development (Håkonson) (14.24)
n	Number of equidistant steps along a line measured by dividers [3.01] (3.10)
	Number of successes (3.18)
	Number of observations [S] (7.10)
	$(a-b/a+b)$ [G] (11.17)
n'	Number of marginal cells surrounding a parcel (4.5)
p	Percentage value of variable (8.13) [S] [K] (6.16)
	Area scale [C] (10.6)
p_i	Expected proportion [8.02]
Q	Qualitative map accuracy as defined in acceptance sampling (9.30)
	Volkov's correction for area measurement on Mercator's projection (22.20)
Q_A, Q_B	Frolov's corrections for area measurement on Mercator's projection (22.23)
q	$(=1-p)$ or inverse probability [S] (6.16)
	Linear deformation on a map (10.20)
	Perimeter ratio (Zöhrer) (18.19)
R	Radius of sphere [C, G, M] (2.6)
	Radius of the (spherical) earth [C] (10.3)
	Length of the tracing arm of a planimeter (17.4)
R'	Length of the pole arm of a planimeter (17.2)
	Radii of the gears of a polar disc planimeter (17.5)
R	Scale reading on the variable tracing arm of a planimeter (17.22)
r	residual error [S](6.11)
	radius of a globe [C] (10.3)
	radius vector of polar coordinates [M] (10.4)
	correction coefficient [S] (10.20)
	radial distance to a point from the nadir homologue of an aerial photograph [P] (13.6)
S	Map scale [2.01]
	Scale of an aerial photograph [P] (2.12)
	Average expected error (Yuill) (18.19)
S'	Photo scale (13.5)
$S\%$	Percentage standard error (18.18)

S_E, S_N	Root mean square of δE and δN respectively [9.04, 9.05] (9.7)
S_H	Horizontal scale factor in Landsat M.S.S. imagery [*P*] (21.28)
S_K	Skew correction factor in Landsat M.S.S. imagery (21.28)
S_V	Root mean square error of a vector (9.9)
	Vertical scale factor in Landsat M.S.S. imagery (21.28)
s	Standard error (7.7) or root mean square error [*S*]
	Rhumb line distance between two points (22.9)
s_M	Standard error of the mean (7.7)
s_S	Standard error of the standard deviation (7.7)
s	Track followed by the measuring wheel of a planimeter (17.4)
t	Student's distribution [*S*] (7.6)
	The length of the arc of the circumference of the measuring wheel corresponding to 1 vernier unit on the scale (17.14)
	Chernyaeva coefficient for reduction of linear measurement (14.19)
	Frolov coefficient (d_2/d_1) for reduction of linear measurement (14.20)
	Acceptance tolerance adopted by OS for measurement of area [20.01]
U	The length of the perimeter of a parcel (17.35)
u	Reduced latitude [*G*] (23.22)
V	Vectorial distance between ground and mapped positions [9.03]
v	Planimeter scale reading in vernier units (17.4)
W	Perkal's Index of variation (15.15)
X	Number of allowable errors in acceptance sampling (9.30)
	An event or trial using Buffon's needle (15.1)
x	Coefficient ($\approx 0{\cdot}5$) used by Maslov in the method of squares [18.17]
Y	Expected range of error (Yuill) (18.16)
y	Coefficient ($\approx 0{\cdot}3$) used by Maslov (17.31)
z	Angular distance on sphere [*C*] [*N*] (3.16)
	Independent variable [*M*] (6.19)
	Probability level (9.29)
	Observed proportion (n/N) of points counted in a parcel on a map [19.05]
α	Limits of a rectangular distribution (7.8)
	Ground slope (9.10)
	Coefficient of (thermal) linear expansion (10.16)
	Producers risk in acceptance sampling (9.30)
	Angular movement of the pole arm of a planimeter (17.5)
	Direction (course) of a rhumb line [*N*] (22.9)

β	Consumers risk in acceptance sampling (9.30)
	Coefficient of (hygrometric) linear expansion (10.17)
	Angular movement of the tracing arm of a planimeter (1z7.9)
	Difference in ground height expressed as a vertical angle (13.17)
	Coefficient used by Köppke in the study of the r.m.s.e. of area measurement by point counting [18.20]
γ	Coefficient used by Beckett ($\approx$ 0·014) in the study of variation in distance with scale (14.11)
δ	Finite difference [M] (3.15)
	Difference between two means [7.21] (7.16)
	Relative error or the sum of the squared deviations [8.02] (8.19)
δA	Planimetric displacement of a contour line (9.10)
δE	Difference in eastings [C] [G]
δ_E, δ_N	Random and independent errors in eastings and northings (9.08)
	Differences in eastings and northings between a mapped point and its true position
δ_i	Difference between the results in area measurement obtained by a single application of a dot grid and the mean of several such applications [18.22]
δl	Absolute linear error (6.1)
$\delta\lambda$	Difference in longitude [C] [G] (3.16)
δN	Difference in northings [C] [G]
$\delta\varphi$	Difference in latitude [C] [G] (3.16)
δ_V	Difference between two planimeter vernier readings (17.4)
$\delta x, \delta y$	Finite coordinate differences [M] (3.15)
δ_1	Discrepancy in the position of the tracing index of a planimeter before and after tracing a parcel (17.29)
Δ	Frolov correction for dividers spacing (14.20)
Δ_H, Δ_V	The number of pixels in the horizontal (along) and vertical (perpendicular) to the scan lines of Landsat M.S.S. imagery between two points (21.28)
$\Delta x_0, \Delta y_0$	Displacement of a grid line corrected for paper distortion (11.6)
ε	Perkal's order of length (3.22)
η	Ratio of Perkal's order of length to the measured length of a line (18.12)
θ	Vectorial angle of polar coordinates [M] (10.04)
	The angle between a parallel and a meridian on a map (10.6)
	Aerial camera tilt [P] (13.7)
	The angle between Buffon's needle and a line (15.2)
	Angle at the pivot between planimeter arms (17.29)
λ	Longitude [C] (3.16)
	An angle (17.11)

μ	Population mean [S] (6.22)
	Particular scale [C] (10.04)
μ_0	Principal Scale [C] (10.03)
ν	Transverse radius of curvature of ellipsoid [C] [G] (2.7)
	Number of degrees of freedom [S] (7.13)
ρ	Meridional radius of curvature of ellipsoid [C] [G] (2.7)
	Conversion from radians into degrees (11.20)
	Radius of the zero circle of a planimeter (17.16)
$\rho\sigma$	Standard deviation [S]
σ_E, σ_N	Standard errors (OS) (9.07)
σ_C	Circular standard error (9.09)
φ	Latitude (specifically geodetic latitude) [C] [G] (2.7)
	An angle (17.11)
φ_M	Mean latitude [22.22]
φ'	Middle latitude (22.9)
ω	Maximum angular distortion [C] (10.06)

1

The Nature of Cartometry

It may be argued... that, rather than pursue the chimera of measurement, we should abandon the idea altogether. There is indeed some justification for adopting such advice. But the interesting thing is that in order to measure effectively we require deep analytical understanding and considerable thought regarding the controls necessary and the errors naturally incurred. Measurement may not be a satisfactory end in itself, but we can be sure that in pursuing this end we will turn up problems and difficulties, the solution of which will provide major advances to our understanding. This, however, depends entirely on a sound understanding of the nature and principles of measurement. Without such an understanding we are plainly lost.

(D. Harvey, *Explanation in Geography*, 1969)

Cartometry has been defined by the International Cartographic Association (I.C.A., 1973) as – "*Measurement and calculation of numerical values from maps...*"

Four kinds of measurement may be regarded as being basic techniques of cartometry:

- *distance measurement,*
- *area measurement,*
- *measurement of direction,*
- *counting the numbers of objects shown on maps.*

Other quantities or indices may be derived from combinations of these with other data. For example, *density* (of population, livestock, etc.) is the combination of number with area; *volume* is derived from areas measured in a particular way (of features on maps enclosed by contours or isobaths) in order to introduce the third dimension; *slope* or *gradient* is derived from distance measured between points of known height. Two of the operations listed, distance measurement and area measurement, are of importance to a wide range of users of maps having different interests in the end-product. Counting may arise in both of these activities so that this form of map use can be dealt with in describing the other techniques. Measurement of direction is, however, of much less practical importance to the majority of map users. Those specialists who need to measure direction from maps and charts, such as surveyors, navigators and gunners, already have extensive literature which deals with this subject. Thus measurement of direction is treated superficially in this book compared with the other subjects.

From the definition of the subject we may recognise two of the three stages which comprise the work. First, although this is implied and not defined, the limits of the object to be measured must be identified. Obviously we must know *what* we want to measure before we can start. Secondly, the measurements themselves are made. Third, the raw data being the results of these measurements must be reduced to a numerical form which can be compared with other data collected from other sources.

The English-language version of the definition of cartometry continues: "*... together with their graphic presentation*" – but these words are deemed to be unnecessary to the definitions in the other languages used in the *Multilingual Dictionary of Technical Terms in Cartography.* The author does not regard this to be an essential fourth stage in the work, because quantitative results are frequently used and compared without ever resorting to any kind of illustration. Consequently the formal description of the methods of graphic presentation is avoided in this book, although some of the examples used to illustrate particular points provide adequate description of the methods without having to labour the point.

The I.C.A. definition of cartometry is extended here to include sources other than maps and charts, such as measurements made on photographs and other kinds of graphic output from the methods of remote sensing; for example the imagery created by infra-red and radar scanning systems, electron microscopy and the computer-generated graphics recorded by sensors mounted in unmanned space vehicles, such as the *Landsat* imagery. Provided that we know about the geometrical properties and constraints of such images, just as we have to know about the geometrical properties of maps and charts, objects may be measured directly from these media without subjecting them to an intermediate mapping process.

THE USES OF CARTOMETRY

There are three major fields of human activity and organisation which are associated with map use and the need to make measurements upon maps:

- *scientific uses,*
- *route-finding uses,*
- *administrative uses.*

Cartometry in Science

Cartometry is an important technique of the natural sciences and in some branches of engineering, for the map is an important tool in geography, geology and certain branches of the biological sciences, together with the applications of these disciplines in agriculture, forestry and mining. The map, plan and section are vital documents to the civil engineer. In the submission made by the Royal Society to the Ordance Survey Review Committee (often called the Serpell Committee) in 1978, eighteen different subject headings were listed to identify the principal users of spatial data. Although it is in these subjects that the greatest use is made of measurements taken from maps, the use of cartometric techniques on photographs, recording charts and other kinds of graphic display extends the applications into the realms of pure laboratory investigation. Of particular interest in laboratory studies is the development of the methods of *stereology* during the last quarter-century, as described, for example, by Briarty (1975).

Notwithstanding the evident importance of cartometry as a scientific tool, it is remarkable how little has been written about the principles and methods of making measurements in any book which is comprehensive and easily accessible. It is quite difficult for the scientist to find any reference to the accuracy, repeatability and consistency of different methods of measuring area or distance, or how best to reduce the measurements to their corresponding ground dimensions. For example, the entire subject of measuring the length of an irregular line on a map is described in the best-known of all English textbooks on geographical cartography in the following eleven lines:

> It is frequently necessary to measure the length of some irregular line on a map, such as a road, railway or river. If the line is not too irregular, a number of short straight portions can be stepped off successively with dividers and summated. Alternatively, the end of a piece of fine thread is placed at the starting-point and then laid along the line, carefully following each curve. Again, a small toothed wheel fitted with a recording dial, known as an *opisometer*, can be run carefully along the line, the total length given on the dial read off in inches or centimetres, and this is converted into actual length by applying a scale factor. In each of these cases, it is well to measure the line twice, once from each end, and calculate the mean of the two results....
>
> (Monkhouse and Wilkinson, 1952)

It seems, therefore, that important scientific conclusions may be founded on little more than the mean of two measurements which have been multiplied by a scale factor to convert them into the appropriate units. The potential user of such data might be forgiven for supposing that this represents incontrovertible evidence, for there is nothing in this description to disabuse him of this belief, or sow any seeds of scepticism about accepting such measurements as gospel.

Moreover, the different methods of measuring distance are described in this quotation as if all were of equal precision and equally applicable to the measurement of any kind of line, irrespective of its length or the degree of its irregularity. The same criticism applies equally to the descriptions of area measurement given in practically every textbook which deals with the subject in the English language. There is no indication of the suitability of the methods for a particular job, nor their dependence upon the size of a parcel or the irregularity of its outline. Anyone who has tried to measure the area of a large parcel by counting points or squares on a finely-ruled grid placed over the map (Chapter 18, p. 412) has soon come to the conclusion that there must be better ways of occupying his or her time.

Most cartometric measurements made for scientific purposes are intended for comparative use. For example, we might wish to compare the lengths of streams of similar hierarchical order, or derive drainage density values for a collection of different catchment areas using techniques described by

Doornkamp and King (1971) and in Chorley (ed.) (1972). In each case it is not the single measured quantity which is required, but a collection of quantities relating to different objects which need to be compared. The result of a single measurement, however carefully executed, can only be stated baldly as a particular number of millimetres on a map or kilometres on the ground. It has little practical value until it is compared with other measurements.

Cartometry and Stereology

Weibel (1979, 1980) has described stereology as *the knowledge of space*, and defines it more elaborately as follows:

Stereology
is
a body of mathematical methods
relating
three-dimensional parameters defining the structure
to
two-dimensional measurements obtainable in sections of the structure.

The term *quantitative microscopy*, used, for example, by De Hoff and Rhines (1968) is regarded as being synonymous with stereology. The equivalent term in petrology is *modal analysis* as used, for example, by Chayes (1956). The word *morphometry* is used by Aherne and Dunnill (1982) with respect to biological and medical microscopy to mean the same thing. However, the last word is confusing to use in a work which is likely to be read by geographers and geologists, and to whom morphometry has an entirely different meaning.

Cartometry represents that part of the spectrum of measurement techniques which seeks to understand different phenomena from the information provided by images whose scales are generally much smaller than the feature being studied. Another part of this spectrum is occupied by the quantitative methods of microscopy, where the objects to be studied and measured have been enlarged many times from their natural sizes.

In principle there is no real difference between the measurements made on a map and made on a micrograph. Measurement of distance, area, direction and number can be carried out using the same techniques and instruments on both. In optical microscopy, indeed, there are additional techniques available which make use of the special facilities offered by the *graticule eyepiece* and the movements of a mechanical stage which can be recorded by means of adjustments to a micrometer. Techniques which may be used with an image analyser are intended for use with micrographs, rather than maps which are generally of too large a format to fit the equipment.

An important distinction must be made between the intended use of the measurements. The object of cartometric measurements is to discover the quantitative relationships between objects shown on maps and similar graphics. These are represented upon a two-dimensional surface, and it is not possible to measure also those conditions occurring at other levels in the

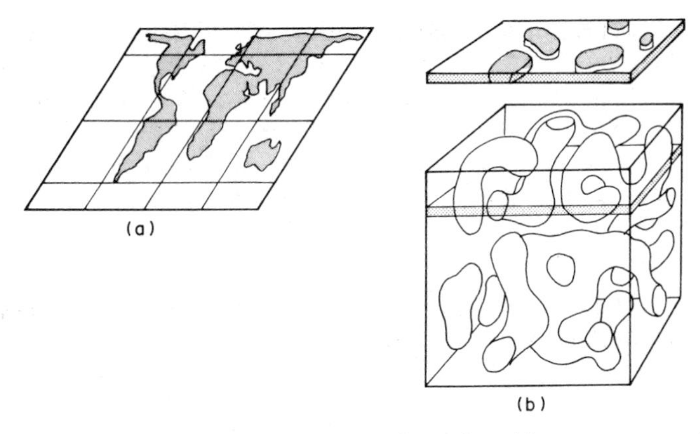

FIG. 1.1. The difference between cartometry and stereology: (a) cartometry comprises measurements made upon a plane map; (b) stereology comprises similar measurements made upon a thin section taken from a three-dimensional body being a specimen of an organ, rock or piece of metal.

atmosphere, beneath the ocean or the land surface solely from measurements made upon one map.

Nearly all studies in quantitative microscopy make use of sections or slices cut through a three-dimensional specimen, whether this be an organ or a piece of rock or metal. Here we have the possibility of cutting several slices to provide a series of thin sections representing different levels in the original specimen. The cartographic analogy would be a series of maps, each depicting a different level or horizon within the atmosphere or within a geological structure. But these maps could only be compiled if we already had sufficient information about the conditions occurring at each level or horizon, which would render unnecessary the sampling of a variable in the third dimension. Figure 1.1 illustrates the relationship between the cartometric example of making measurements on a single map and the stereological example of measuring a sample of sections.

Resulting from this preoccupation in stereology of recovering the third dimension through sampling, many of the actual techniques which are used to measure area and distance are based upon probabilistic, or Monte Carlo, methods. This is also an important aspect of the methods of cartometry, and it is in this field that both subjects can learn a great deal from one another. It is some measure of the utility, convenience and precision of certain probabilistic methods that they have been found to be equally useful for measuring the features shown upon a map or upon a micrograph.

Cartometry in Resources Inventories

An important use of area measurement which overlaps into both scientific studies and administration is the establishment of various kinds of resource inventories and, having established these in the first instance, the repetition of them periodically in order to detect changes. Obviously these are most important in the studies of land cover, land use and crop evaluation in agriculture and forestry. Such studies are not new, and have been attempted using maps since the late nineteenth century and using conventional aerial photography from the 1920s onwards. However, it is with the development of other kinds of remote sensing, and especially the availability of small-scale imagery of large parts of the earth's surface acquired from artificial satellites such as the Landsat series, that studies of this sort have burgeoned and suitable cartometric methods of doing the work have been developed. To begin with the majority of investigations were associated with mapping projects so that the features of land cover were mapped as a matter of course, and measurements of the area of woodland, for example, were carried out after the maps had been prepared. Later it became commonplace to sample the aerial photograph or the satellite image directly, and these areas could be measured without having to locate and plot the boundaries between different kinds of land cover, between different crops or different stands of timber. This is normally done by placing an overlay showing a network of points over the photograph and listing the categories of land cover, land use or other variable which coincide with each point on the overlay. The methods of area measurement by counting points, are obviously well suited for evaluating such lists of occurrences. We shall quote a variety of examples of such work ranging from the use of large-scale aerial photography for purposes of measuring urban land use to the use of Landsat imagery for studies of land cover. The examination of agricultural crops over large areas in the LACIE investigations (Large Area Crop Inventory Experiment), described, for example, by Macdonald and Hall (1980) has been a particularly important and continuing study carried out by various research teams in the U.S.A. We shall find that the work by some of the LACIE teams has advanced the subjects of two-dimensional statistical sampling and the measurement of area by sampling.

MEASUREMENT FOR ROUTE-FINDING

This is the least clearly defined branch of cartometry because it encompasses such a huge range of activities – from elementary map reading on a country walk to the sophisticated methods of navigating a modern oil tanker or bulk carrier. Yet the techniques of measurement do not differ much in outline. Compare the action of the walker who measures track and distance on a map before attempting to cross a comparatively featureless moorland with that of the navigator of a ship or aircraft who uses both of these measures in the conduct of graphical *dead-reckoning* (D.R.) navigation on a chart. Both measure direction in order to set a suitable course which will follow the required track between the starting point and destination. The only real difference in procedure is that the walker will only use his compass to the nearest 2° whereas the navigator works to the nearest $\frac{1}{2}$°. The walker hopes to discover, from measurement of the length of his journey and an estimate of his average walking speed, how long it will take to reach the destination on the other side of the moor. The navigator calculates this in the same way to obtain an *estimated time of arrival* (E.T.A.). However, the navigator can usually obtain a more accurate estimate of speed so that, once again, he works to closer or more precise limits.

For many kinds of cross-country navigation it is sufficient to make quite crude measurements because an error as large as 10% will not influence the outcome. For example, if it has been calculated that an isolated road junction will be reached by car in 22 minutes, it probably does not matter if the point is reached in 20 or 24 minutes. This is because there is additional and unique visual information available (it was described as being an *isolated* junction) and the E.T.A. is only confirmation that the place has been reached after approximately the correct interval in time from starting from the place where the calculations had been made. It is only considered to be unreliable if there are other road junctions situated fairly close to that sought. If so, more precise measurement of distance and more careful maintenance of a constant speed would be needed to meet the E.T.A. accurately.

The principle is also true for cross-country flying during daylight and in good weather, and for coastal navigation in good visibility. In both instances visual recognition of mapped landmarks is more important than the methods of D.R. navigation, for there should be an abundance of information confirming the location of the craft. Contrasted with this are the navigation problems which arise when the craft is out of sight of land, whether this be owing to distance, darkness or bad weather. An accurate D.R. plot then becomes essential and the two operations of measuring distance and direction on the chart have to be done with greater care.

The execution and reduction of the measurements made by motorists, cyclists and walkers as part of their map-reading skills may be described colloquially as "Boy Scout methods", with some justification, for it was through such activities that many boys first learned how to use a map. Figure 1.2 illustrates an extract from a scouting booklet which dates from the late 1920s or early 1930s.

The Boy Scout methods are adequate for many purposes because it is rare for the ordinary map user to have to depend upon accurate dead-reckoning alone. However it should be observed that, when an element of competition enters recreational activities, the navigation problems are made more difficult and greater precision in map-work becomes essential. Thus the sports of orienteering, competitive long-distance walking and motor rally navigation have created the need for greater expertise in making measurements of distance and direction. The map-reading and measurement skills which are needed in these sports are akin to those for military operations in which the

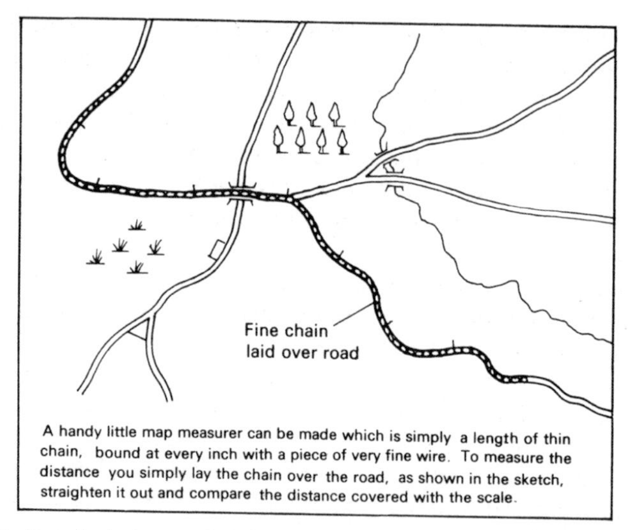

FIG. 1.2. A method of measuring distance as it was described and illustrated in *The Scout's Book of Gadgets and Dodges*, published in the early 1930s.

problems of cross-country route finding are tempered with the need for speed, stealth and surprise.

Nevertheless it is difficult to detect much difference between the described methods of measuring distance on a map, whether this be for the benefit of scouts, soldiers or scientists. The quotation from *Maps and Diagrams*, which was given on page 3, is, if anything, even less informative about the method of using a thread to make a measurement than is the use of a jeweller's chain illustrated in Fig. 1.2. At least *The Scout's Book of Gadgets and Dodges* explained what has to be done with the thread *after* it has been laid along a sinuous line on the map.

In the B.B.C. television documentary programme series about the Palace of Westminster, which was first broadcast in November and December 1983, there was a revealing glimpse of a member of the staff of the Fees Office checking the validity of members' claims for travelling expenses, using the Ordance Survey 1/625,000 Administrative Areas map and stepping off distances by dividers (an expensive pair befitting the importance of this work) set to a separation of one-half inch or more. The commentary described this as being done "with meticulous care". Perhaps this book, and especially Table 5.2 on page 68, may be read one day by a Member of Parliament who can draw the appropriate conclusions.

CARTOMETRY IN ADMINISTRATION

The third major application of cartometry arises in government and administration. Some departments of the administration, such as those concerned with Planning, Health, Highways and even Finance (for travelling expenses) have continuing need for cartometric measurements to be used in economic and engineering applications which are identical to the scientific uses. In addition, area measurement has considerable fiscal importance because much data collection and taxation is based upon land areas. In a properly governed country there must be agreement between landlord and tenant, landowner and administration, about the area of every parcel of land which is subject to rent, rates or land tax; to the payment of agricultural subsides and grants which are related to area. Because of this need for objective and authoritative records of area there is usually a statutory obligation for one or other department of the administration to undertake such work, with the corollary that the results of this work should be regarded as being legally binding both by the administration and the private citizen. In many countries it is the national survey department which is responsible for making measurements of area. For example, in Britain the Ordnance Survey do this for each parcel of land identifiable on the basic mapping of the country. Many countries also maintain a cadastral survey department to record and register titles to land. These departments almost certainly measure the areas of all parcels which are registered.

In an important study of the role of national mapping in government and administration, Smith (1979, 1980) lists the various responsibilities of a government and, in a detailed breakdown of the information content of all Ordnance Survey maps he lists "Areas" as being relevant to the following responsibilities:

General administration	Property boundaries
Planning	Rating
Water resources	Agriculture and forestry

Control and relief measures resulting from flood and earth movements.

CARTOMETRY AND THE LAW OF THE SEA

Another potentially important application of cartometry in administration arises from modern attempts to modify international law as this relates to national rights and sovereignty in coastal waters and further offshore. It is generally agreed that the navigable waters of the world may be subdivided into four areas, each part of the sea having a different legal status. These are (Fig. 1.3):

- *internal waters,*
- *territorial waters,*
- *contiguous zone,*
- *the high seas.*

The intricacies of how national sovereignty or jurisdiction applies to the

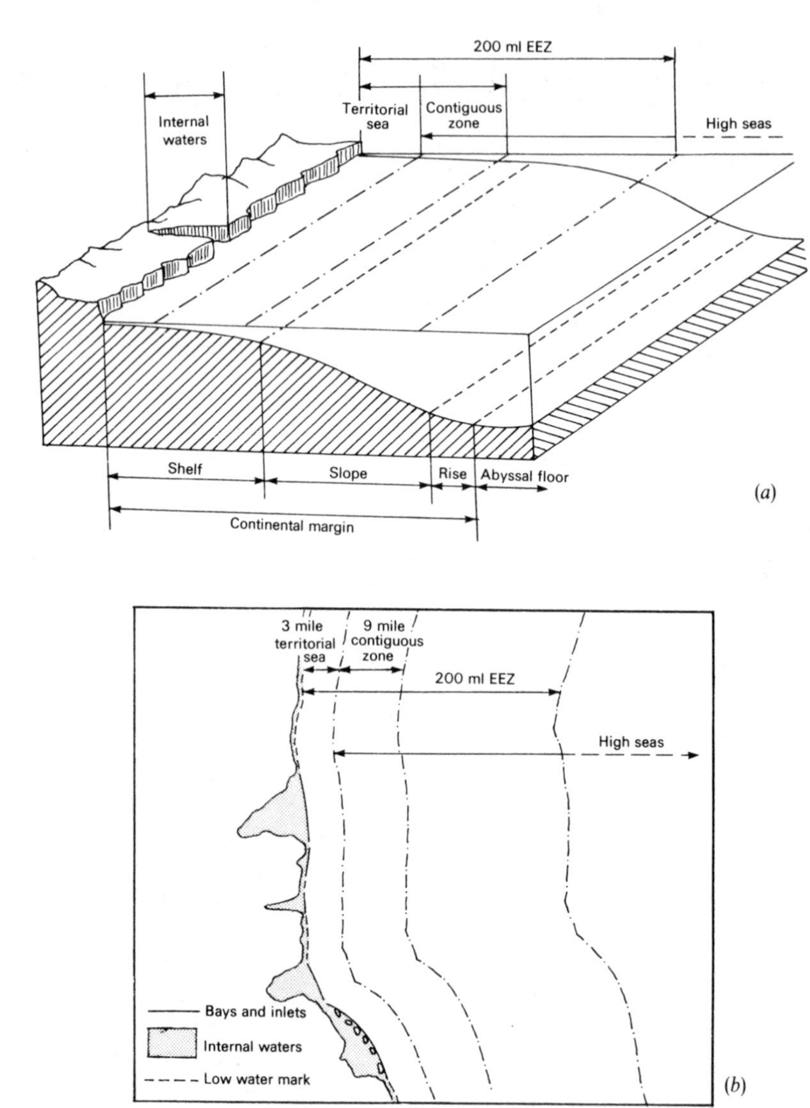

FIG. 1.3. Definitions (a) and cartographic representation (b) of the principal subdivisions of the sea for legal and political purposes. (Source: Couper, 1978).

different zones need not concern us here. The reader is referred to Couper (1978), Open University (1978) and Langeraar (1984) for details. The need for defining the boundaries of the various sea areas is especially well introduced in the second of these references as follows:

For the television newsreader and the newspaper reporter it may be sufficient to say that a particular country has, or has claimed, a territorial sea having a width of x miles. But things are not really quite as simple as that; on the contrary, they are remarkably complicated. Where is the x miles measured from? The obvious answer is 'outwards from the coastline'; but where does that mean in practice? If all coastlines comprised gentle curves there would be no problem. But nature is seldom like that. Real coasts often have bays, gulfs, estuaries, promontories, peninsulas, fjords, offshore islands, man-made earthworks, reefs and 'islands' that are only just above water at low tide.... When it comes to deciding where the *baseline* for measuring the territorial sea should be, all such features must be examined closely and rules laid down to accommodate them. This is far from a trivial matter. Unless the position is made clear, disputes may arise, wars may be fought and lives may be lost.

(The Open University, Oceanography: *Law of Sea*, 1978)

Every maritime nation needs reliable nautical charting of all its claimed waters if the full potential of its offshore national assets is to be realised. Moreover it is necessary for maritime nations to ensure the safety of shipping in their waters.

The way in which cartometry enters these deliberations depends upon the methods used to describe the different zones which have been listed, and where the baselines should be located. Fundamental to the methods of delimitation is the concept of a series of baselines which correspond, in general, to the low-water tidemark with a series of rules relating to the ways in which the baselines may be drawn of complicated coastlines. Some of these are described in Chapter 12, pp. 228–246, but we reserve the most detailed consideration to Chapter 23, pp. 525–551. Distance measurements come into the location of the baselines because there are certain maximum distances between headlands, or between the baseline and an offshore island which determines how certain features should be defined. Area measurement may be necessary, too, in the application of the *semicircular rule* (page 237) used to define a true bay and distinguish this from a minor indentation of the coastline. When the baselines have been established for a country, the boundaries between the offshore sea areas, as well as those for the *Exclusive Economic Zone* (E.E.Z.) on the continental shelf may be determined by measuring the appropriate distances at right angles to the baselines. It is at this stage that cartometric errors may occur, and these may have extremely important consequences. Before the exploitation of oil and natural gas deposits in sea areas bounded by different countries, an error of a few hundred metres in the position of the line serving as the boundary between national interests was unimportant. In any case, before the 1960s it was difficult to locate position out of sight of land with an accuracy greater than about 100 metres even in those areas best served with navigation aids. Today the economic value of a few square metres of the sea bed within the Frigg or between the Murchison and Statfjord oil fields of the northern North Sea is such that precise determination of the boundary between the British and Norwegian sectors is essential. We return to this aspect of the subject in Chapter 23.

UNCLOS III has allowed for the establishment of an *International Sea-Bed Authority* which is intended to supervise and control certain activities in the continental shelf areas as well as in or under the high seas. One of its intended roles will be to supervise the *Basic Conditions of Prospecting, Exploration and*

Exploitation as these were laid down in Annex III to the draft *Convention on the Law of the Sea.* This was adopted on 30 April 1982 at the eleventh session of the Third United Nations Conference on the Law of the Sea. One aspect of this supervision will be to make important and far-reaching decisions based amongst other factors upon area measurements of various parts of the oceans of the world.

At present, of course, the Draft Convention has not been ratified, the International Sea-Bed Authority does not exist and the measurements still have to be made. For different reasons, the U.S.A., U.K., Federal German Republic, Turkey, Israel and Venezuela all failed to sign the Convention before the deadline set for 10 December 1984. Although, at this stage, it seems that a stalemate has been reached in the proceedings, this may well be only a temporary delay. The Convention will come into force one year after 60 states have ratified the Convention, although by June 1985 only 18 had done so.

Should the Law of the Sea as represented by this Convention ever become a reality, this will herald a massive increase in the amount of hydrographic surveying, chart production and of related cartometric work needed to convert the legal definitions into lines on charts which can actually be agreed as representing valid maritime boundaries. Moreover, a considerable amount of the initial work of such an Authority will be to clarify the validity of, and resolve conflicts concerning, some boundaries which are already claimed by some nations, Since many of the states which will be affected most by these developments are least equipped to provide the information from their own resources, it is a matter of some concern that the measurements which need to be made will conform to acceptable standards. At this stage, however, few people seem to have considered what these standards should be in the light of what can be achieved in practice. It is for this reason that the present book is strongly orientated in certain chapters, and in many of the examples, to the consideration of aspects of the delimitation of maritime boundaries.

2

The Nature of the Medium

Maps are the final distillate of geographical knowledge. Maps are the principal aid to geography. Maps are indispensible tools of geographical science. Maps are the basis of geography. Maps are the philosopher's stone of geography. Maps are the eyes of geography.

(Max Eckert, *Die Kartenwissenschaft*, 1921)

THE MAP AS THE MEDIUM

By definition cartometric measurements are made on maps and charts and are intended to determine dimensions on the ground. However, the map or chart is not an exact replica of the ground. Therefore we must study the ways in which the picture of the ground is transformed during the map-making processes, for the transformations surely affect the measurements. We therefore start from one possible definition of a map:

A map is a representation, normally to scale and on a flat medium, of a selection of material or abstract features on, or in relation to, the surface of the earth or a celestial body.... A chart is a map designed primarily for navigation.

(I.C.A., 1973)

Although exceptions may always be found, the following factors are common to the great majority of maps which are likely to be used and the way in which they have been made.

- The map has been printed by offset lithography in two or more colours on a sheet of paper.
- In order to print in more than one colour, each sheet of paper must pass through the printing press more than once. Moreover a separate printing plate must be used for each colour printed. This, in turn, means that a separate manuscript must be prepared to make each printing plate.
- Fair drawing of the map manuscript is now usually done by *scribing* or cutting away parts of the surface layer of a specially coated sheet of plastic or glass rather than by conventional drawing on paper in ink. The letterpress, some of the map symbols and ornament or repetitive symbols used to denote vegetation and land use, are put on the manuscript by attaching thin plastic film bearing a photographic image of the information.
- A map is only a selection of the objects which occur on the ground. Selectivity is necessary in order to achieve legibility at small scales. The

process of selection and simplification which must take place is known as *cartographic generalisation.*

- The *map detail,* i.e. the lines and symbols depicting various ground features, has either been plotted from aerial photography using the methods of *photogrammetry* or it is derived from existing maps. Ultimately the information shown on these source maps has also been plotted by photogrammetric methods unless the sources are so old that the detail was surveyed on the ground. It is exceptional, nowadays, for a new map to be plotted entirely from ground surveys.
- The photogrammetric methods of plotting detail relay, in turn, upon the availability of *ground control.* This represents the positions and heights of certain identifiable points on the ground which have been located by field survey methods.

Figure 2.1 illustrates the succession of operations which go into the production of a typical map in the order in which they are carried out.

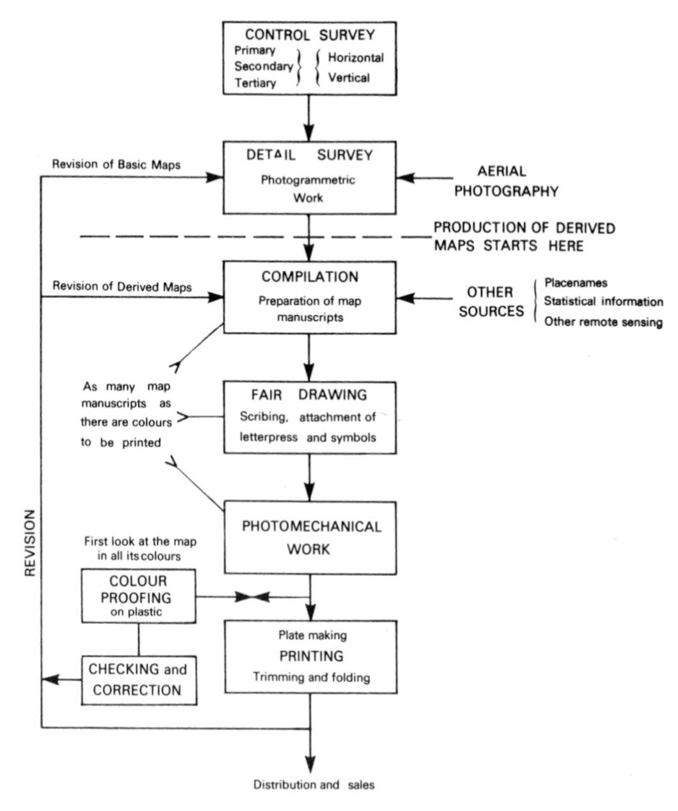

FIG. 2.1. The main stages in the production and revision of maps from the original survey through final printing and distribution.

Scale

The scale of a map is its most important mathematical property. This is often defined as:

> The ratio of the distance on the map to the actual distance that it represents on the ground.
> (I.C.A., 1973)

If two points A and B on the ground are represented by the corresponding points a and b on the map, the scale of that map is

$$1/S = ab/AB \tag{2.1}$$

The fraction $1/S$ is correctly called the *representative fraction* by many map users. Map makers tend to use the work "scale" without any qualification to mean the same thing. Equation (2.1) represents the commonest calculation which occurs in cartometric work, for it is used every time that it is necessary to convert from the measured distance (ab) into the corresponding ground distance (AB) with equivalent transformations to area measurements. Thus, if the measured length of a line on a map of scale 1/100,000 is 22·0 mm, the corresponding ground distance is 22·0 × 100,000 = 2,200,000 mm, more suitably expressed as 2,200 metres or 2·2 kilometres.

Classification of Maps by Scale

Because the concept of scale is so important to both map making and map use, it is not surprising that it is commonly employed as a primary means of classification for different categories of map. Thus the distinction may be made between:

Large-scale maps and charts
  Scale larger than 1/12,500
Medium-scale maps and charts
  Scale between 1/13,000 and 1/126,720
Small-scale maps and charts
  Scale between 1/130,000 and 1/1,000,000
Very small- or atlas-scale maps and charts
  Scale smaller than 1/1,000,000.

However, as has been demonstrated by Steward (1974) and other writers, there is no general agreement how these terms ought to be applied. A map which would be considered to be "small-scale" in one country may be of larger scale than any of the maps which are available in another. The numerical values listed above are the limits which have been adopted in this book for the purpose of definition.

The survey of a country is normally plotted as one or more *series* of maps at the largest scale which is convenient for the majority of administrative, economic or defence requirements. This is known as the *basic mapping* and all smaller-scale or special-purpose maps of the country are derived from this. The basic scale is not necessarily constant for a whole country. It would be extravagant to publish maps of sparsely populated mountain areas at the same scale which is needed to map urban areas. Thus the basic scale for a substantial part of Britain is 1/2,500 with less extensive cover of the major towns at 1/1,250. However the largest scale at which the whole of Britain is completely mapped at a single scale is 1/10,560, or Six-inches to One-mile (now replaced by maps at 1/10,000) for this is the largest scale at which the mountain and moorland areas have been mapped. All the other, smaller-scale map series produced by the Ordnance Survey are classified as *derived maps.*

Metric and Non-metric Maps

Another useful form of classification which is particularly important to cartometry has appeared in official U.S. publications during recent years, (e.g. Ellis, 1978). This distinguishes between metric and non-metric qualities. Many maps which are readily available to the user are intended for general information only. This category includes road maps, pictorial maps and even some atlas maps. They can be distinguished by having no kind of mathematical framework, and information about scale may be absent or imprecise. They may be termed non-metric maps to be distinguished from those showing a grid or graticule and reliable scale information.

It may be taken as an invariable rule that if a map gives no indication of its mathematical framework, then it is not good enough for making measurements.

The mathematical framework of a map comprises a *grid*, a *graticule* or both. A grid comprises a square network of equidistant parallel straight lines representing particular linear distances on the ground. Thus, depending on map scale, grid lines may be shown at every 100 metres, 1 km, 10 km, or even 100 km. The smallest scale at which grids are commonly shown is 1/500,000 or thereabouts. A graticule is a network of *parallels* and *meridians* for equal intervals of latitude and longitude on the earth's surface. The lines forming a graticule may also create a square or rectangular network but, depending upon the *map projection* employed, it may comprise two families of curves having variable separation between them. These subjects are treated in detail by Maling (1973) and Maling in I.C.A. (1984), as well as many other books.

We shall make use of the grid or graticule information to create essential geometrical control over any measurements which have to be made. This is because it is always possible to calculate the dimensions of a grid square or a graticule *quadrangle* from other data, namely a knowledge of the earth's shape and size, of map scale, the spacing of the grid lines or the parallels and meridians, and knowledge of the equations defining the particular map projection in use. By relating the theoretical dimensions, calculated from these data, to those measured on the map we may determine the extent to which paper deformation has affected a particular map sheet, or estimate a correction to be applied to the simple scale formula of (2.1) to allow for the changes in scale which are inherent in the projection used for a map.

On metric maps distances and areas can be measured and corrected precisely, either on the single map sheet or working over adjoining sheets which are based upon the same reference system and datum. Any attempt to continue measurement from one non-metric map to another is fraught with difficulty.

Throughout this book it is assumed that the user has access to metric maps for all cartometric purposes, and that he has the good sense to avoid attempting to make measurements on maps of non-metric quality.

THE GEOMETRY OF THE MEDIUM TO BE MEASURED

We have already suggested that there are certain differences between the geometry of the map and that of the photograph. Moreover, the various scanning devices employed in remote sensing also differ from one another, as well as from both the map and the photograph. Here we introduce some preliminary remarks about the nature of these differences, though detailed treatment of this subject is deferred until later chapters. There is an important reason why the geometrical differences between these media should be distinguished. A vertical aerial photograph and a Landsat image both *look* like a map. For many purposes they may be used just as if they were maps. However, the photograph is a perspective representation of the ground surface, whereas neither the Landsat image nor a map depicts the ground in this way. Therefore the positions of points appearing on a photograph differ from those of the corresponding points on the map. One consequence of this difference is that it is seldom possible to make a map from a photograph simply by tracing. More complicated methods have to be used, even if it is only to transfer a few points or lines from the photograph to the map. Although measurement of distance, area or direction may similarly be made on a photograph, the conversion of these into reliable ground units may similarly be difficult because of the ways in which the photographic images are displaced or deformed.

The Figure of the Earth

The earth is a solid body which approximates in shape to a sphere, but from the point of view of surveying and mapping of large portions of the planet corresponding to the continents and some of the bigger countries, the surface which needs to be determined is that of the *geoid*. This corresponds approximately to mean sea level in the open ocean without any tides, waves or swell to complicate it. It is only since 1970 or thereabouts that we have known enough about the configuration of the geoid to be able to appreciate the size and frequency of these complications over the whole world. Because of minor variations in gravity this surface is so irregular that it cannot be used for computation of position, distance, angle or area. Therefore we must make certain assumptions about the shape of the geoid. The most accurate in common use is that the earth corresponds in shape to an *ellipsoid of rotation* or *spheroid*. Less accurate is the assumption that the earth is truly spherical. Because plane and spherical geometry are both such simpler than spheroidal geometry, there are considerable practical advantages in using simpler solutions wherever possible. However, there are some applications which we shall show require a numerical solution which is based upon a spheroid. Therefore we have to introduce some of the equations which define the spheroidal earth.

Figure 2.2 illustrates the principal differences between a sphere and an ellipsoid of rotaion. The sphere has constant radius, R, in all directions. Therefore any plane section passing through the centre of the sphere may be represented by means of a circle with radius R. This is known as a *great circle*.

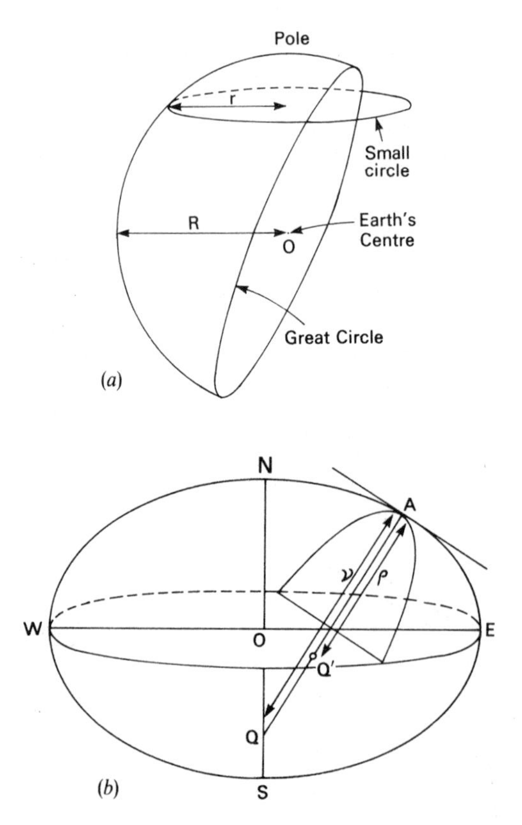

FIG. 2.2. The geometry of the earth regarded as a sphere (a) and as an ellipsoid of rotation, or oblate spheroid (b).

The arc of the great circle passing through two points on the spherical surface is the shortest distance between them.

An ellipsoid of rotation is elliptical in section, defined by its major and minor semiaxes, a and b respectively. From this it is possible to determine the additional parameters of

(1) *Ellipticity*, or *flattening*

$$f = (a - b)/a \tag{2.2}$$

which is usually expressed by means of the fraction $1/f$.

(2) *Eccentricity*

$$e^2 = (a^2 - b^2)/a^2 \tag{2.3}$$

These quantities vary with the adopted *Figure of the Earth*, defined primarily by the lengths of the two semiaxes. The choice of a particular figure for a given country has often been for historical or political reasons, and it may not necessarily be the best approximation to the geoid in the country of use. However, in recent years there has been a gradual trend towards figures for which $a = 6{,}378{,}150$ metres and $1/f = 1/298{\cdot}25$. The last three listed in Table 2.1 have been determined from the behaviour of the orbits of artificial satellites, and have to be used with different satellite navigation systems.

There are two radii of curvature at any point on the surface of a spheroid having *geodetic latitude* φ. The first is the *meridional radius of curvature*,

$$\rho = a(1 - e^2)/(1 - e^2 \cdot \sin^2 \varphi)^{3/2} \tag{2.4}$$

The second is the *transverse radius of curvature* lying in a plane at right angles to the meridian through the point

$$\nu = a/(1 - e^2 \cdot \sin^2 \varphi)^{1/2} \tag{2.5}$$

From a knowledge of e, ρ and ν it is possible to calculate the lengths of

TABLE 2.1. *Some of the Figures of the Earth in current use*

Name	Date	a (metres)	f^{-1}
Airy	1830	6,377,563	299·325
Bessel	1841	6,377,397·2	299·153
Everest	1830	6,377,276·3	300·80
Clarke	1866	6,378,206·4	294·98
Clarke	1880	6,378,249·2	293·47
International	1924	6,378,388	297·0
Krasovsky	1940	6,378,245	298·3
International Astronomical Union	1968	6,378,160	298·25
World Geodetic System	1972	6,378,135	298·26
Naval Weapons Laboratory NWL-9D		6,378,145	298·25
GRS-80	1980	6,378,137	298·26

geodesics which are the curves on the surface of an ellipsoid corresponding to a great circle on a sphere, the bearings or azimuths of such lines, the areas of *spheroidal quadrangles* and similar variables.

It is always preferable to use the simplest assumption which is compatible with the accuracy of measurement. Thus comparativly crude surveying techniques, using a tape or chain to measure distance and a compass to measure angles within a comparatively small area – for example in the survey of a small woodland or farm – need only be referred to a plane surface (or a flat earth) because any alterations in position resulting from more sophisticated treatment of the data in referring them to the curved surface of the earth are smaller than the errors in measurement arising from the instruments and methods used to make the map in the first place. At the other extreme, where very high accuracy is required to fix positions of points or to map boundaries, especially at long distances away from the origin of the survey, not only must the survey instruments and methods be more precise, it is necessary to compute positions with respect to a more reliable model of the earth. An important example is the need to locate maritime boundaries with extreme precision in offshore oilfields where the values of licence blocks are sometimes extremely high. We shall see later (Chapter 23) that in order to locate their boundaries with sufficient accuracy, it is necessary to use the methods of geodetic surveying to obtain the field observations and to use an acceptable reference spheroid as the basis for computing positions.

In cartometric work, area measurements made on a large- or medium-scale topographical map do not have to be referred to the spherical or spheroidal surface because the effects of earth curvature are smaller than the errors of the measurement processes. On the other hand, measurement of the area of a large country or oceanic feature upon a small-scale map or chart will require the use of control measurements (pp. 217–227) of spherical or spheroidal quadrangles whose areas have been calculated separately.

The Geometry of the Map

The map is a plane surface. In order to represent the curved surface of the earth upon this plane it is necessary to transform the positions of places which are to be mapped. There are an infinity of different transformations which are theoretically possible, each of which represents a unique *map projection* but few of them have practical importance in cartography. The transformations which take place are mathematically equivalent to stretching or tearing the spherical surface. Stretching causes variation in scale from one part of the map to another, leading to an evident contradiction of the simple definition of map scale given in equation (2.1). When we use this expression we assume that the ratio represented by $1/S$ is constant for a particular map, and that the value of the representative fraction does not vary with the length of the line measured, the position of the line on the map or the orientation of the line. Moreover, it is assumed that AB is a straight line on the ground and that the representation of it, ab, is also a straight line on the map. We shall see in Chapter 10 that none of these assumptions is true, for scale varies from place to place and in different directions on a map; a straight line on the ground is represented by means of a curve on the map. For the present, and in many practical applications, however, we may suppose that equation (2.1) is correct. This is a sufficiently accurate approximation for most of the everyday work which is carried out on large- or medium-scale maps and plans showing very small portions of the earth's surface on the single sheet of paper. This is because the variations in scale, though present, are too small to identify or measure on the map. Measurable scale changes within the single map sheet only occur when a sufficiently large portion of the earth's surface is shown in one map. Hence the problem of allowing for the map projection really only arises in the use of very small-scale or atlas-scale maps. Since both the nature and magnitude of the scale variations and other deformations depend upon the properties of the projection used as the base for the map, the usual remedy which has long been used in cartometry is to select that map which offers the smallest amount of deformation for the particular kind of measurement to be made. However,

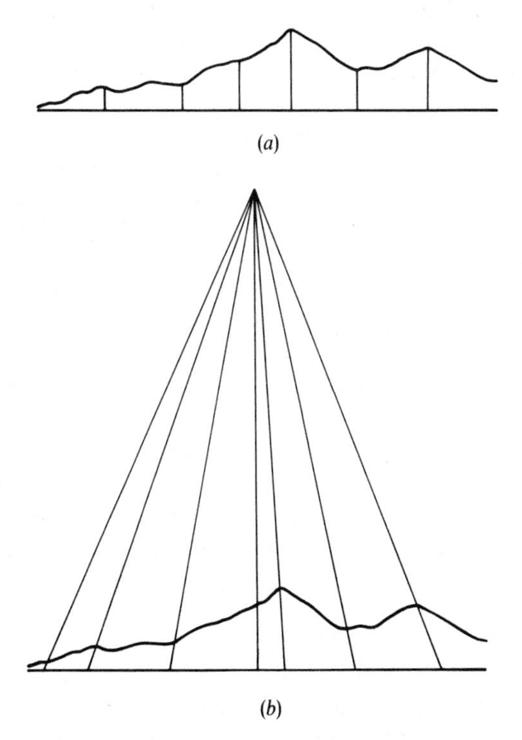

FIG. 2.3. Comparison of the orthogonal projection of surface relief to (a) the datum plane as carried out in surveying and mapping, with (b) the perspective central projection to the negative plane which is inherent in photography.

with the development of digital processing of cartometric measurements has emerged the possibility of applying numerous small corrections, these taking into consideration the scale changes and deformations which are inherent to any given map projection and which may be expressed numerically. This procedure should become standard practice one day.

The second important feature of the geometry of the map is the transformation which is needed to take into consideration the influence of surface relief. Its magnitude is small compared with the curvature of the earth's surface, but it must be treated systematically in order to remove any influence of it upon map scale. This is achieved simply by converting all distances measured in surveying into horizontal distances reduced to a surface which is parallel to mean sea level or whatever survey datum is employed. Consequently the map becomes the *orthogonal projection* of the topographical irregularities to the datum plane, as illustrated by Fig. 2.3a. As a corollary it follows that if the map user needs to know the slope distance between two points, it is first necessary to measure the horizontal distance from the map, estimate the heights of the points from the contours and calculate the slope distance from these data.

The Aerial Photograph as the Medium

If we use the aerial photograph, or any other kind of sensed image, for measurement, we are treating the photograph as a substitute for the map. We therefore avoid the complication of having to carry out a series of photogrammetric and cartographic processes which would otherwise be needed to transform the information into a conventional map.

Ideally measurements made on a photograph should be done by setting the stereoscopic pair of aerial photographs in a photogrammetric plotter and using the instrument scales to make the measurements. This is almost always the ideal solution, as, indeed has been shown in several studies for which plotters were used, such as the work described by Tomasegovič (1968). However, availability of a photogrammetric plotter to make measurements of distance or area is a luxury to which most users are unaccustomed. Usually the work has to be done on a paper print using the conventional instruments and methods of cartometry.

Just because a photograph has been taken from an aircraft, this does not necessarily endow it with magical qualities which make it suitable for all kinds of measurement or for mapping. We need to emphasise this point because intending users often expect that indifferent photography of doubtful origin can be used quantitatively – see Goodier (1971) for some useful discussion on this subject. If photographs are to be used for making measurements there are certain qualities which must be satisfied by the equipment used and the way in which the photographs have been taken. When we refer to aerial photography, we here imply that this has *survey quality* such as is normally taken for mapping purposes. We assume that this is the kind of photography to be used for cartometric work just as we have assumed that measurements are to be made on metric maps. Survey-quality photography has the following distinctive features:

- The photographs are of sufficiently large format for the work of interpretation and mapping to be done on prints or diapositives which have been reproduced by contact printing processes, thereby avoiding any optical complications and loss of definition which may result from having to enlarge each photograph from a small negative.
- Because of the large film size, great care must be taken to keep it flat during exposure so that the negative image can be regarded as representing the focal plane. This is provided by the special methods of film flattening used in survey cameras.
- A metric camera has a lens system which is sensibly free from distortion.

FIG. 2.4. The acquisition of vertical aerial photography showing how a pair of overlapping photographs of the ground are obtained from an aircraft which is flying truly straight and level. (Source: Wild–Heerbrugg).

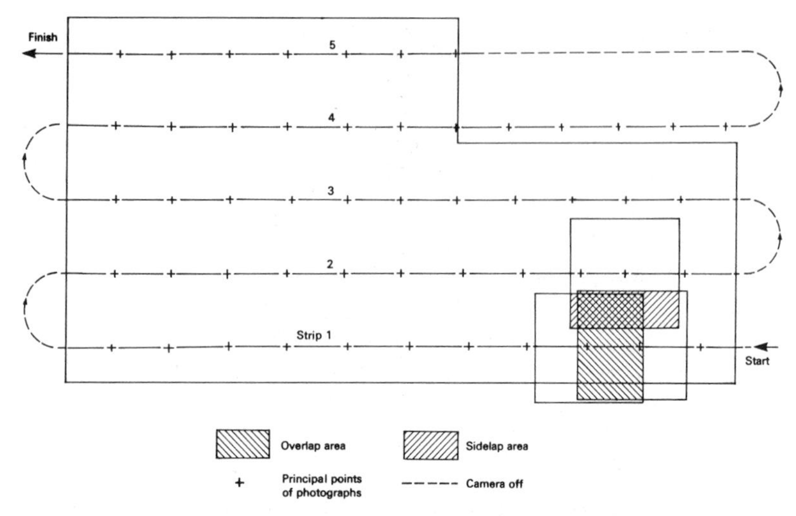

FIG. 2.5. An example of a photographic sortie comprising five strips and 52 photographs. The overlap between three neighbouring photographs has been shaded.

- The camera has been carefully calibrated after assembly so that the focal length of the lens is accurately known and the image of the optic axis on a photograph (the *principal point*) can always be recovered by reference to *fiducial marks* reproduced on every photograph taken through the camera.
- The majority of photographs taken and used for survey purposes have been taken from an aircraft flying straight and level with the camera pointing downwards. This is distinguished as being *vertical aerial photography*, in contrast to *oblique* photographs which result from using a strongly tilted camera. For reasons which will be described elsewhere, it is rare for oblique aerial photography to be used in modern mapping.
- Vertical photographs procured for survey purposes form a series of *strips* of photographs which are taken as the aircraft flies along a specified track. In order to obtain stereoscopic cover of the ground, which is necessary for measurement of height and plotting contours, as well as facilitating viewing and interpretation, the photographs have been taken at intervals which allow for 60% or more *fore-and-aft overlap* between successive photographs. In order to cover more than one strip of country the aircraft is flown along a succession of parallel tracks, each yielding a similar strip of photography. Each strip is flown to allow for a *lateral overlap* of 10–30% with its neighbours.

The Geometry of the Photograph

It has been stated that a photograph is a perspective projection of the view upon the focal plane of the camera. This is simply explained by means of a

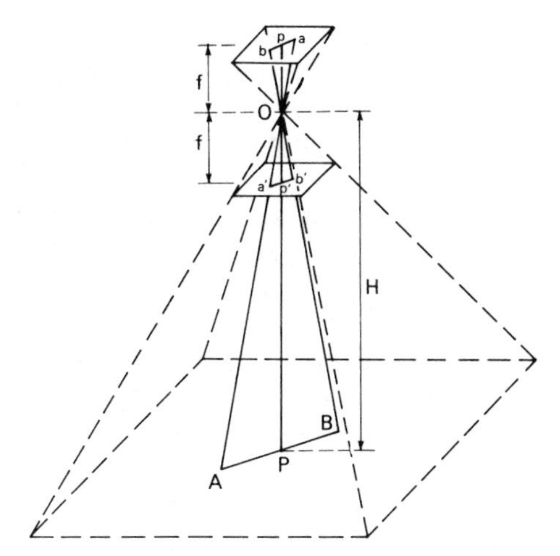

FIG. 2.6. The simple geometry of the vertical aerial photograph. The plane containing the points a, p and b is that of the focal plane of the camera and represented by the *negative*. A similar plane containing the points a', p' and b' shows these points in the same relationship as their ground homologues, A, P and B. This corresponds geometrically to the *positive* copy of the photograph.

diagram as in Fig. 2.6. The camera lens is here regarded as being a thin convex lens of focal length, f. A ray of light reflected from the point A passes along a straight line, entering the camera at the O and continues as the same straight line within the camera to create the image point a in the focal plane. Similarly a point B is represented by b on the straight line BOb. The point O is the perspective centre through which we imagine all light rays to pass. In the camera, the optic axis, pO is perpendicular to the focal plane at the *principal point*, p. Therefore bpO and apO are both right angles.

If A and B both lie on flat ground at the same height above survey datum, and the camera is pointed vertically downwards, the optic axis meets the ground at P and the angles APO and BPO are also right angles. These are the conditions of perfect vertical aerial photography of flat land. If they are satisfied we have four right-angled similar triangles, apO and APO, bpO and BPO, which give us the simple scale relationships

$$ap/AP = bp/BP = ab/AB = 1/S = f/H \tag{2.6}$$

where $1/S$ is the scale of the photograph and H is the flying height expressed in the same units as f. For example if $f = 152{\cdot}4$ mm (6 inches) and $H = 3048$ m (10,000 feet), the scale of the photograph may be determined from

$1/S = 0{\cdot}1524/3048 = 1/20{,}000$ where both f and H are expressed in metres,
$1/S = 0{\cdot}5/10{,}000 = 1/20{,}000$ where both f and H are expressed in feet,
$1/S = 6/120{,}000 = 1/20{,}000$ where both f and H are expressed in inches.

There is, however, a major difficulty which severely restricts the value of the single photograph as the medium for measurement. This may be summarised by the statement that *the scale of a photograph is not constant.*

There are two reasons for variation. The first is the effect of the perspective representation of objects which lie at different distances from the camera; that the images of closer objects are larger than similar-sized objects which lie further away. This can be seen on any photograph, and it corresponds to the perspective detected by the human eye. In the special case of vertical aerial photography of hilly ground, the foreground is occupied by the hill-tops, which are therefore represented at larger scale than the more distant valley floors. In aerial photographs, moreover, differences in height are represented by image displacement which increases radially outwards from the centre of the photograph. This can be seen on a vertical aerial photograph which shows a factory chimney or a tall building, because the position of the top of such a feature is offset from the position of its base. The same principle applies to natural objects but this cannot be recognised simply by inspection of the photograph; it is indicated diagrammatically in Fig. 2.3(b).

The second reason why scale may be variable is caused by camera tilt. In Fig. 2.7 this would be indicated by the angles APO and BPO being other than right angles. This may be intentional, as, for example, where the camera is mounted in the aircraft to take an oblique view of the ground and may therefore be inclined at an angle of 15–45° from the vertical. On the other hand, tilt may be accidental, resulting from the difficulty of flying an aircraft truly straight and level. In this case the amount of tilt ought to be small, diverging not more than 1–3° from the vertical. In this respect it should be remarked that the truly vertical aerial photograph is so rare that it is almost a geometrical abstraction. Small amounts of tilt are nearly always present, and

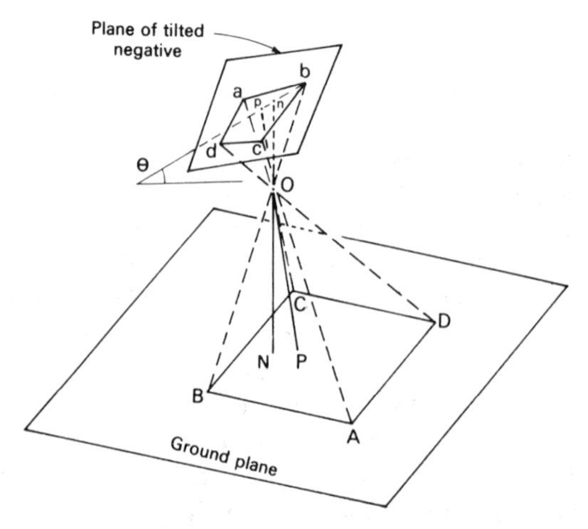

FIG. 2.7. The geometry of the tilted aerial photograph.

they seldom remain constant from one photograph to another. Strictly speaking, therefore, nearly all aerial photographs are obliques. However, we do not adhere rigidly to the definition of verticality in describing them. We refer to photographs being vertical by intention rather than by result. In either case the effect of tilt is to alter the scale of the photograph in a variable way so that a simple rectangular feature, such as a football pitch, becomes deformed into an irregular quadrilateral on the photograph. We treat with the deformations resulting from both tilt and relief in some detail in Chapter 13.

The combination of both variable scale resulting from surface relief and tilt creates images on the photograph which do not fit the corresponding features of a map. Since the deformation of the images increases with the amount of tilt, the vertical aerial photograph is now always used in preference to the oblique photograph for mapping. Obviously the vertical aerial photograph is also more convenient for cartometric use.

Although the geometrical deformations of the aerial photograph may appear to be similar to those caused by the projection of the map, we must not proceed too far with this analogy. We have seen that if we work with sufficiently large-scale maps the distortions caused by the map projection are small enough to be ignored. On an aerial photograph, however, the effects of relief and tilt have the opposite effect; namely that the distortions become greater on large-scale photographs.

The Geometry of Scanning Systems

The methods of remote sensing using a scanning detector vary according to which part of the electromagnetic spectrum is being used, and the nature of the sensor required to obtain a response at particular wavelengths. Consequently it is not practicable to review the possibilities in detail here. Colwell (1983) provide a comprehensive description of the geometry of most of the methods used to obtain images ranging from *sideways-looking airborne radar* (S.L.A.R.) within the microwavelengths through sensing of the infra-red and visible wavelengths, the last being exemplified by the *multispectral scanner* (M.S.S) employed by the Landsat satellites.

In these scanning systems the images are created in a succession of narrow parallel strips, each of which represents an individual sweep of the detector. It follows that the geometry of the resulting picture differs considerably from the simple perspective representation obtained by a conventional frame camera. Although a picture built up from SLAR images looks like an oblique aerial photograph it would be quite wrong to treat it as such. Generally the picture obtained approximates to an orthogonal representation of the ground, and therefore has more or less constant scale, but there are still some minor deformations which have to be corrected. Figure 2.8 shows diagrammatically how the scanned image differs from those of the vertical and oblique aerial photographs.

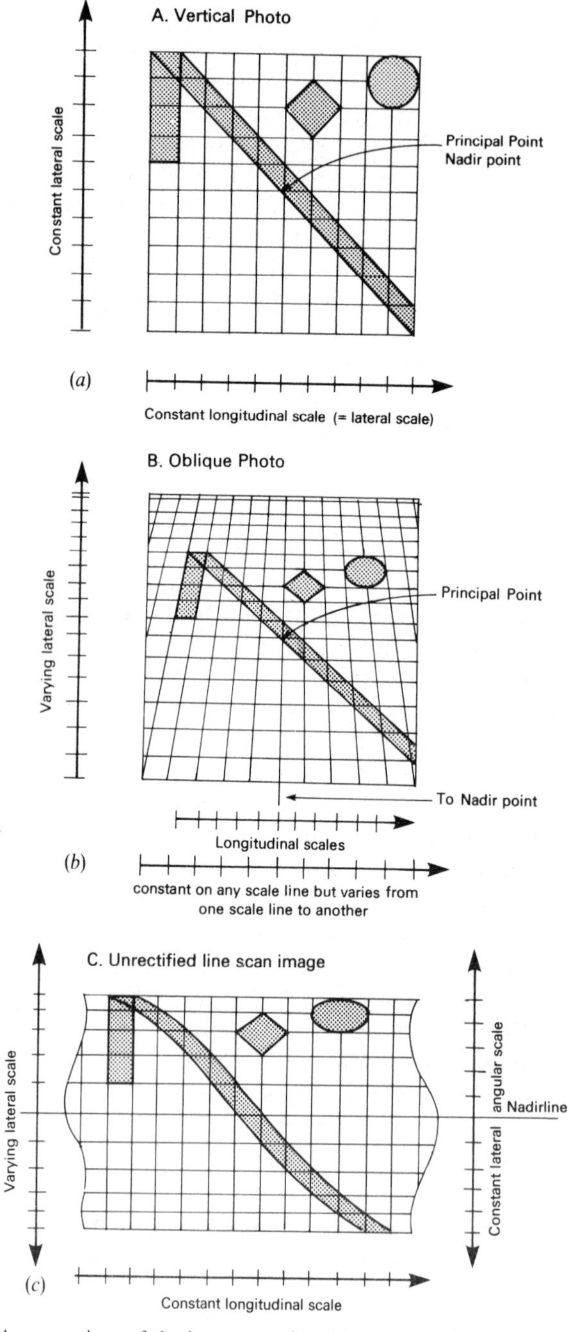

FIG. 2.8. A comparison of the images produced by (a) a vertical aerial photograph, (b) an oblique aerial photograph and (c) an unrectified line scan image created by scanning athwart the line of flight. (Source: Taylor and Stingelin, 1969).

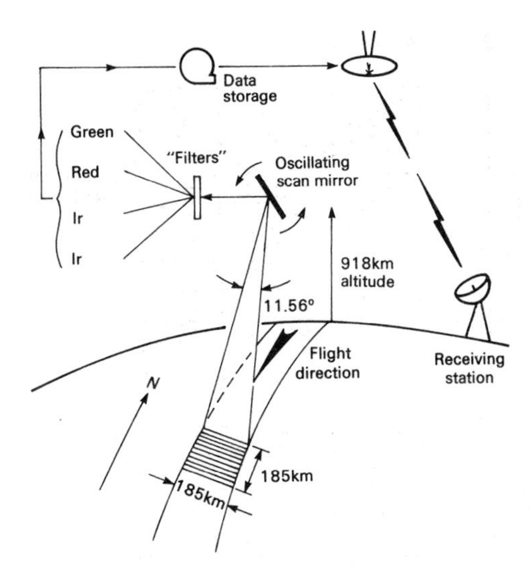

FIG. 2.9. The fundamentals of multispectral scanning as employed by the different versions of Landsat.

An additional factor arises in the use of scanned images obtained from space vehicles. Because of the increased distance of the craft from the earth's surface (the Landsat orbits vary between 900 and 950 km from earth), the scale of the images is small. Consequently there is need for map projection transformations similar to those to be found in using maps of similar scale. This is a problem which hardly arises in the use of survey photography because nearly all of this is still taken from flying heights which are less than 10 km. In a scanning system using electronic signals the transformations are introduced by computer during bulk processing of the pictures. The example of the M.S.S. used in Landsat is pertinent. During processing these images are subjected to 16 different kinds of rectification or correction for the known deformations affecting the system. These are illustrated in Fig. 13.13, p. 274. When they have been applied, the picture almost corresponds to an orthogonal representation of a portion of the earth's surface. However the fit is not exact and strictly speaking the Landsat picture must be based upon a *space cylindrical strip perspective projection.*

3

The Methods of Measuring Distance

There are nine and sixty ways of constructing tribal lays,
And-every-single-one-of-them-is-right!

(Rudyard Kipling, *In the Neolithic Age*, 1895)

Table 3.1 lists the methods which are mentioned in this book. It distinguishes between two kinds of measurement:

- The *classical* methods, usually employing a special instrument which makes a complete measurement of an object.
- The *probabilistic* methods of measurement, which are based upon the principles and methods of statistical sampling. These often do not require anything more complicated than a transparent overlay covered with regular patterns of squares, circles, dots or lines, and the ability of count.

Using sampling we work in terms of estimates, not exact measures, but by repeating the counting or measuring processes we may make the estimates as accurate as we wish. In their introduction to the methods of biological morphometry, Aherne and Dunnill (1982) have written:

> The methods of morphometry are indirect and essentially probabilistic ways of quantifying biological structures compared to the limited number of direct and deterministic methods available... they are rapid, flexible and versatile. But the beginner commonly has difficulty in trusting methods based upon probability; to him, at first, they appear to leave too much "to chance". This is not a logical difficulty; it is purely a psychological hurdle.

One major topic of this book is to emphasise the value of these probabilistic methods. A second preoccupation is emphasis that not all the methods listed are equally useful or convenient for all kinds of measurement. We shall see that each of the methods described is suited to the measurement of certain kinds of features on certain kinds of map, but no single technique is equally suitable for the measurement of any outline at any scale.

MEASUREMENT OF DISTANCE

Consider the simple task of measuring the length of a line about 150 mm, which is a straight line drawn on a sheet of paper. Using conventional drawing

TABLE 3.1. *The methods of measuring distance and area on maps, plans, charts and photographs*

Classical methods	Probabilistic methods
Measurement of distance	
Distance measured by scale	Inverse solution for Buffon's needle
Distance measured by dividers	
Equal steps	Steinhaus' longimeter
Unequal steps	Perkal's longimeter
The War Office, or card method	Håkonson's longimeter
Line following by measuring wheel	Matern's nethod
Comparison of a line with a thread or similar object	
Distance calculated from coordinate differences	
Distance measured by digitiser	
Measurement of area	
Measurement and calculation of simple geometrical figures	
Calculation of simple geometrical figures from coordinate differences	
Measurement by comparison with standard figures of known size	
	Point-counting methods
	Direct measurement by point-counting
The grid count method	Indirect measurement by point-counting
The method of squares	
Area determined from linear measurement	
The method of strips	
Scale-and-trace method	Linear or transect sampling
	Direct and indirect methods
Area measurement by tracing the parcel perimeter	
Polar planimeter	
Polar disc planimeter	
Rolling disc planimeter	
Hatchet planimeter	
System Gueissaz	
Miscellaneous methods of area measurement	
Cut-and-weigh	
Densitometric methods	
	Image analyser methods

and measuring instruments this work can be done in the following ways:

1 A scale is laid between the terminal points of the line. The distances between them may be determined from the difference between the two scale readings made against the points. This distance is measured in the units which are engraved on the scale, such as inches or millimetres.

2 A pair of dividers are opened until the two points coincide with the

terminal points of the line. The dividers are transferred to the scale and the distance is measured by comparing the positions of the dividers' points with the scale graduations.

3 A pair of dividers are set to some separation such as 10 mm or 20 mm. This distance is *stepped-off* or *walked along* the line in such a way that most of the line is measured by a succession of equidistant steps. Usually the length of the line is not a simple multiple of the initial setting of the dividers. Therefore the last step is measured separately.

If the piece of paper represents a map, plan or chart, the next stage is to convert the measurement into ground units, using equation (2.1) as in the example given on page 15. Therefore each of the methods which have been described comprises first, the *measurement process* and secondly, *reduction of the measurements* by calculation.

There are three further ways in which the measurement may be made and reduced without calculation.

4 This is a special case of method (1) which is commonly used by surveyors, architects and engineers, who use large-scale maps and plans. A special ruler, graduated in feet or metres at the appropriate map scale, is employed so that ground distance is measured directly. A straight edge can only carry two or four such scales, because a different ruling edge must be used for each scale. A scale which is triangular in section can have six different scales. Therefore this method is restricted to working at only the few map scales which are commonly used in these professions. These are in the range 1/48 to 1/10,560.

5 This method corresponds to (2) but the dividers are compared with one of the graduated scales, described under (4) or, more commonly, with the graphic representation of the scale of yards, miles, metres or kilometres which has been printed on the map.

6 This corresponds to (3) but the dividers are walked back again along the graphic scale on the map to find the distance in units of ground measure.

In methods (5) and (6) there are two measurement stage. The fourth method is unique in having only one measurement stage with no calculations.

Usually, however, the line is more complicated, and we must also consider the methods which have to be used to measure the length of a meandering river or a complicated coastal outline. For this we must use one of the classical methods described next.

Classical Methods of Distance Measurement

Measurement by Dividers

Table 3.1 indicates that many of the classical methods of measurement are based upon the comparison of the line with a suitable scale through the

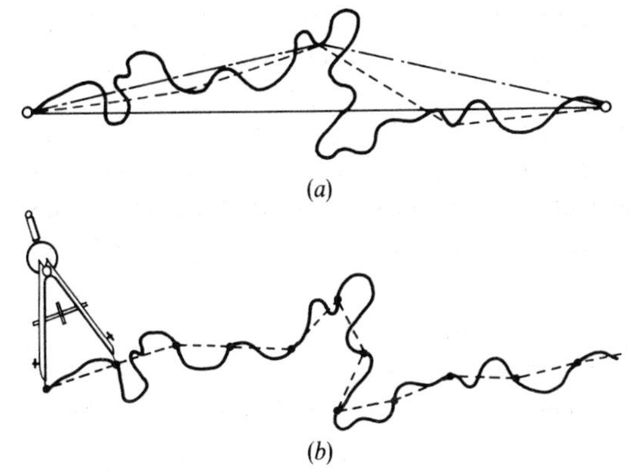

FIG. 3.1. An irregular line measured in (a) one (—), two (—·—) and four (----) equidistant steps compared with the same line measured in twelve equidistant steps (b) as would be done using dividers at a constant setting.

medium of a pair of dividers, a piece of card or a length of thread. Using dividers or card we imagine that the line has been resolved into a series of straight lines, and that the length of it is the sum of these rectilinear elements. It is obvious that such a procedure gives rise to results which are *always* shorter than they ought to be. Figure 3.1 shows that, if an irregular line AB is measured in two parts, its length is greater than the straight line distance AB because the two measured elements and two sides of the triangle in which AB is the third side. Similarly measurement of the same line divided into four rectilinear elements is longer than those measured in only one or two steps. Since an increase in the number of rectilinear elements is equivalent to a reduction in the separation of the dividers' points, we may state the invariable rule that *measured length increases as the span of the dividers is made smaller.*

This is seen in Fig. 3.1(b) where the rectilinear elements are now only one-twelfth of the length of the line. Nevertheless their total length is appreciably less than that of the line which is to be measured, for we can still see that some corners are cut and the broken line defined by the dividers misses some of the major irregularities. Not surprisingly, many map users regard this to be the fatal objection to using dividers, strips of card, or even digitised coordinates for measuring the length of an irregular line. For those who steadfastly adhere to the belief that dividers are inaccurate because they always underestimate the true distance, the only alternative is to make the measurements by chartwheel or by comparison with a piece of thread.

However comparison of Figs 3.1a and b demonstrates, albeit crudely, that as the separation of the dividers' points, d, is made smaller, so the outline measured by dividers approximates to that appearing on the map. Applying the reasoning of the calculus to this observation, we may state that as d is made indefinitely small, the line stepped off by dividers corresponds more closely to the original line, and that stepped off by dividers becomes closer, and at the limit when $d=0$ the two lines become identical. This is an important conclusion which underpins the whole theory of the treatment of measurements made by dividers to determine a *limiting distance* being, in effect, the length of the line which might be obtained if we could measure it with dividers having zero separation. This approach has been developed in Russia over more than a century, with significant contributions made by U. M. Shokal'sky, at the end of the nineteenth century and by Volkov (1950b). It is fundamental to Russian cartometric practice that only measurements made by dividers are sufficiently accurate for most scientific purposes. The other classical methods using a chartwheel or thread are usually rejected as being too crude.

The following types of dividers may be used in cartometry:

BEAM COMPASS. Figure 3.2 illustrates a typical instrument. The typical beam compass comprises a metal or wooden rod, or *beam*, to which are clamped two devices which hold the plotting or drawing instruments at right angles to the beam. The clamps allow the holders to be moved along the beam in order to make a rough setting of the distance between the points. A slow-motion screw on one of the clamps is used to make the final adjustment. This procedure is normally used to set a required distance against a graduated scale and then transfer it to the plotting sheet. Since the points are always set at right angles to the beam, the points should also meet the map, scale or plotting sheet at right angles. This allows for accurate location of the terminal points of a line, which is an important factor in the use of all kinds of dividers and compasses. Figure 3.3 illustrates how an instrument which has points set at right angles to the paper makes more satisfactory contact than dividers whose points are inclined to the plane of the paper.

The maximum distance which can be effectively measured in one step by beam compass is usually 800–1,000 mm. The limit is usually imposed by the flexibility of the beam, which must be light enough to handle easily. In this form it is used primarily for map construction, especially for plotting an accurate framework which will serve as the grid, graticule or the neat lines of a map. Maling, in I.C.A. (1984) describes its use for this purpose. Used with a

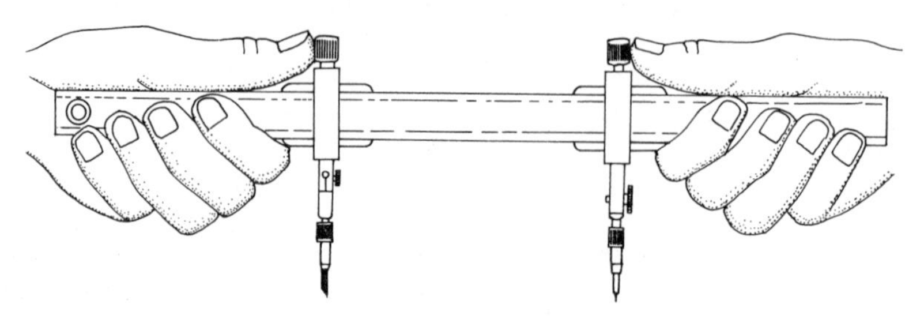

FIG. 3.2. Beam compass.

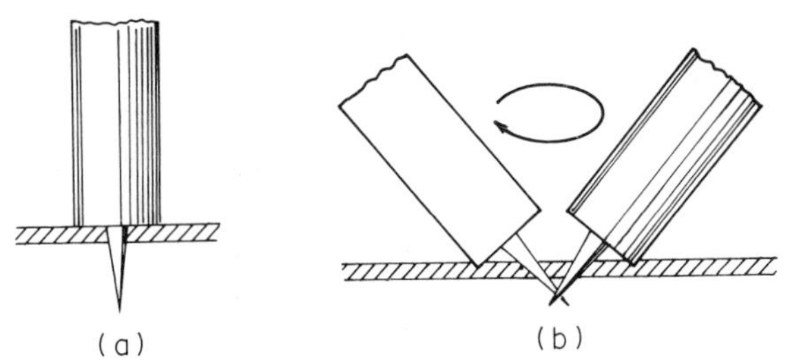

FIG. 3.3. The effect of vertical (a) and inclined (b) dividers points, showing how rotation of an inclined point makes a hole in the map.

short beam, which is easier to handle for cartometric work, a beam compass may be used for any of the applications for which a half-set or precision dividers are also suitable. Indeed because the points are held perpendicular to the plane of the map, a beam compass is a more precise instrument than any of the other types of dividers. We examine this statement on page 279.

HALF-SET (Fig. 3.4). This figure illustrates the typical instrument to be found in most boxed sets of engineering drawing instruments, usually

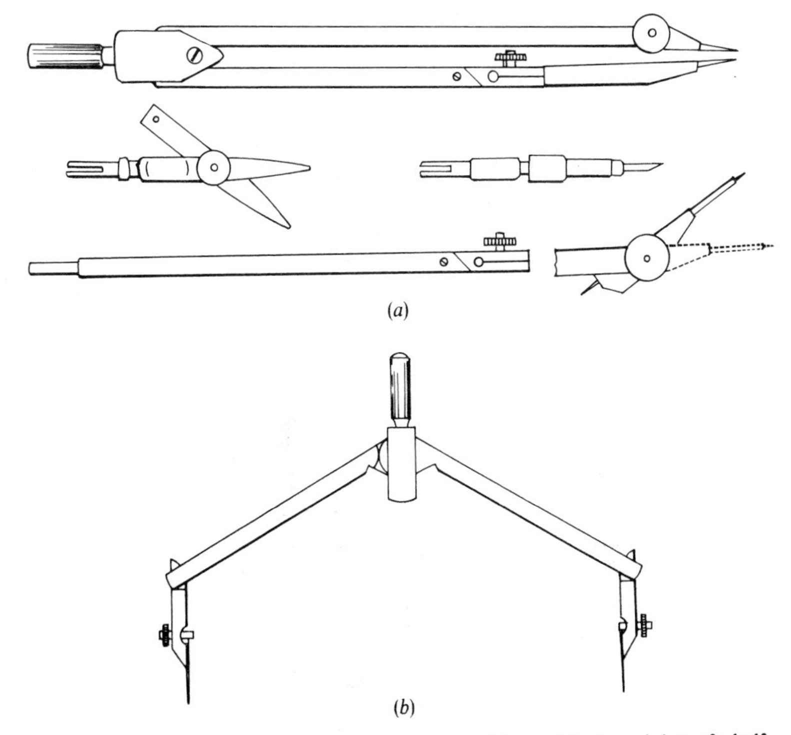

FIG. 3.4. (a) Half-set showing separate components; (b) use of the knee-joints of a half-set to make the points vertical.

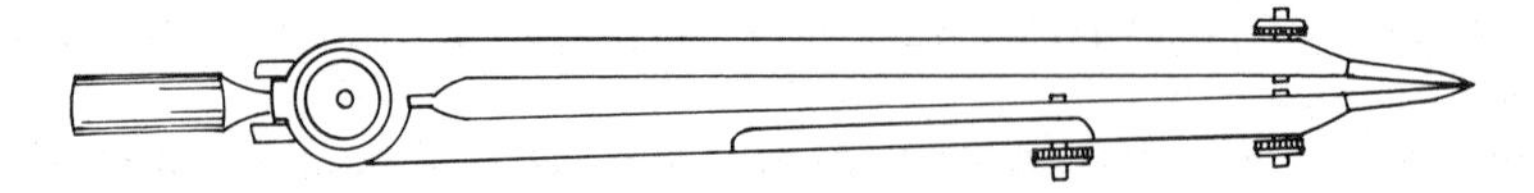

FIG. 3.5. Precision dividers.

supported by a number of interchangeable components (some of which never seem to be required).

For measurement the important parts include the extension bar, which allows the effective span of the instrument to be considerably increased. The knee-joints on the legs set the points perpendicular to the plane of the map, as shown in Fig. 3.4(b). At maximum extension a half-set has an effective spread of 350–500 mm. Without the extension bar, and using the instrument like conventional dividers, the effective spread is only about 150 mm, but this depends upon the length of the legs. Although a half-set is remarkably versatile in having so many possible settings, the various movable joints and hinges can cause accidental alteration of the setting of the instrument as it is moved from the map to the scale or vice-versa, thereby reducing confidence in the use of it for cartometric work.

PRECISION DIVIDERS (Fig. 3.5). These are usually smaller than the typical half-set. There is an adjustable screw on one leg which alters the width of a split in that leg. The purpose of this screw is to introduce a small change the distance between the points by bending that leg. This allows fine measurements and settings to be made against a scale. The maximum spread of the instrument is usually of the order of 150 mm, but because the points are fixed

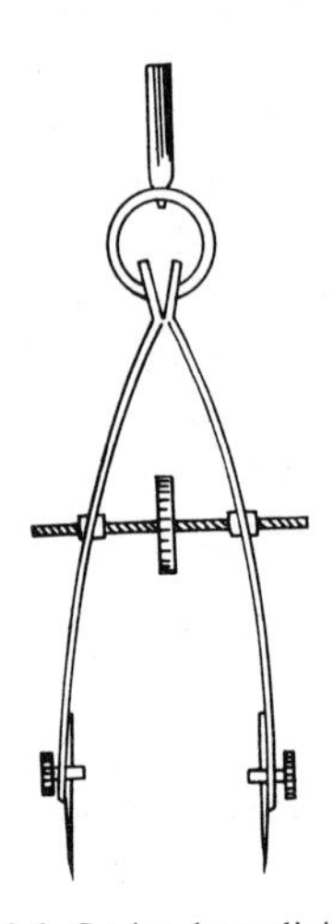

FIG. 3.6. Spring-bow dividers.

parts of the legs, they meet the plane of the map at an acute angle if the legs are too widely separated.

Dividers are used in navigation, primarily for D.R. plotting and associated chartwork. There is a special technique to be used for measuring distance on nautical charts, which are invariably based upon Mercator's projection. This technique is illustrated in Fig. 22.1, p. 500. For measuring lines of several tens or even hundred of millimetres in length on charts, it is necessary to use large dividers having legs of length 150–180 mm with an optimum span of about the same distance. Navigation dividers do not have the split leg adjustment but in all other respects resemble precision dividers.

SPRING-BOW DIVIDERS (Fig. 3.6). These are the instruments most commonly used for measuring distance by stepping-off short lengths along a line. The separation of the legs is held under tension by means of a circular spring operating against a screw which passes through each leg about mid-way along them. Adjustment of the milled wheel on this screw allows precise adjustments to be made in setting the dividers against a scale. The spring ensures that the separation of the points is maintained. The legs of a spring-bow are often only about 50 mm long so that the effective spread of the points is between 20 and 50 mm. Some manufacturers offer larger instruments having legs of length 120 mm or greater, but these are difficult to use for small separations of 10 mm or less.

CARTOMETRIC DIVIDERS. In order to achieve perpendicular setting of the points over a line, and to have an instrument in which the separation of the points cannot be altered accidentally during measurement, Shokal'sky developed the special instrument illustrated in Fig. 3.7. This has been used by Russian scientists since the 1920s, or even earlier, but appears to be virtually unknown elsewhere. A separate instrument is required for every value of d which is to be used. Generally the range of these is from $d = 1$ mm up to $d = 15$ mm.

The effective spread of each kind of divider has been noted in order to indicate the effective range of each for measuring the length of a straight line in a single span. If the line is too long to measure in one step, or if the line is too irregular to obtain satisfactory measurements by using dividers with a wide span, it is necessary to measure the length of the line in stages. This can be done in two ways; in equal steps or unequal steps.

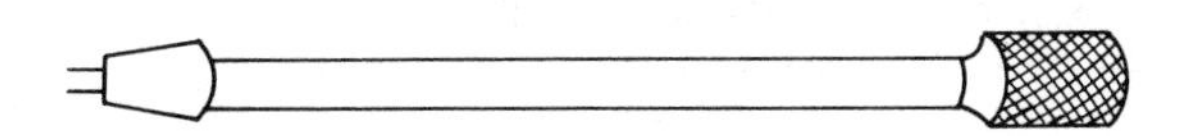

FIG. 3.7. Cartometric dividers.

MEASUREMENT BY DIVIDERS IN EQUAL STEPS

The dividers are set to some separation, d. A precise value for d is needed and this has to be done by calibration of the dividers, as described in Chapter 5. As the dividers are "walked along" the line to be measured the number of steps is counted. It is unusual for the length of the line to correspond exactly with an integer number of steps. Usually there is a final step of length d' which has to be measured separately. Obviously $d' < d$; otherwise it would have been possible to make another step of length d along the line.

It follows that if n steps have been counted, the total length of the line is

$$L = n \cdot d + d' \tag{3.1}$$

MEASUREMENT BY DIVIDERS IN UNEQUAL STEPS

This time a line is made up of a succession of rectilinear elements $d_i (i = 1, 2, 3, \ldots, n)$, which are not necessarily equal. Consequently each setting of the dividers ought to be compared against a scale and the distance recorded. The sum of all the parts is the required length of line, or

$$L = \sum_{1}^{n} d_i \tag{3.2}$$

The task of measuring the length of each element d_i individually makes the method much slower than stepping-off with dividers set to a single value of d. One way of overcoming this inconvenience is to manipulate the dividers in

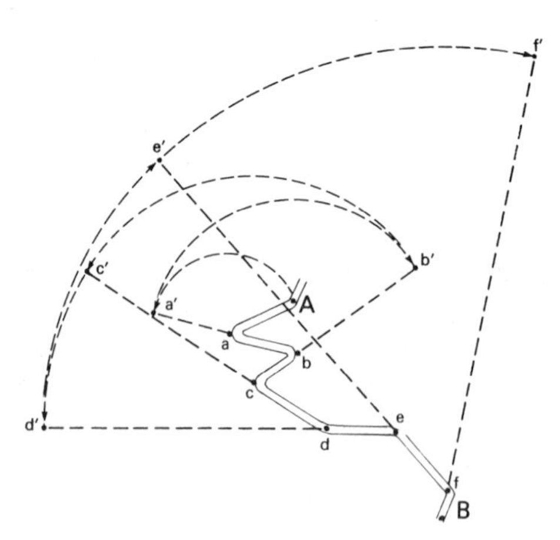

FIG. 3.8. Measurement of the length of the irregular line AB using dividers which are opened progressively. (i) With one point on A open the dividers to the first bend in the road, indicated by a. (ii) Swing the dividers about the point a to form the next straight line element $a'ab$. (iii) Open the dividers to the distance $a'b$. (iv) Swing the dividers about the point b to align the next straight line element $b'bc$. (v) Continue this procedure until the final setting of the dividers, $f'B = AB$.

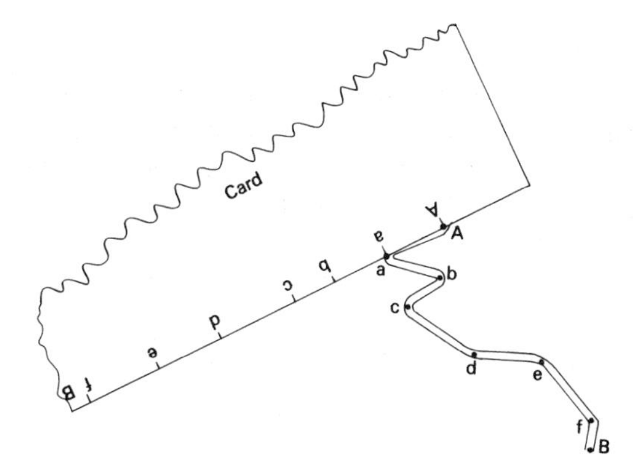

FIG. 3.9. The War Office or card method of measurement of the same line AB which is illustrated in Fig. 3.8.

such a way that they are progressively opened by the length of each new element as this is measured. Consequently the spread of the dividers represents the cumulative distance along the line and, finally, the measured length of the whole line. This technique is illustrated in Fig. 3.8.

The War Office, or Card Method

It is probably easier to appreciate the last technique if the comparison is made against a piece of folded paper or card rather than by manipulating dividers. Figure 3.9 illustrates this method as it is taught to British army recruits as an elementary map-reading technique, as in War Office (1957).

The edge of a piece of paper or card is placed over the starting point, A, and aligned along the first part of the line to be measured. The point A is marked on the card by sharp pencil. Where the line to be followed begins to deviate appreciably from the straight edge of the card, both card and map are marked at the point a. The card is rotated about this point until the next relatively straight element ab of the line is identified. The position of b is marked on both map and card, which is now aligned in the direction of bc. The procedure is repeated until the other end of the required line, B, has been reached. The card now has a series of pencil marks along its edge indicating each of the turning points, as well as a record of the total length on the map between A and B. It is placed over the graphic scale of the map to obtain the ground distance between these points, as well as that to any of the turning points en route.

MEASUREMENT OF DISTANCE BY DIGITISER

A digital method which is closely related to measurement of variable rectilinear elements by dividers or card is used to compute the length of an

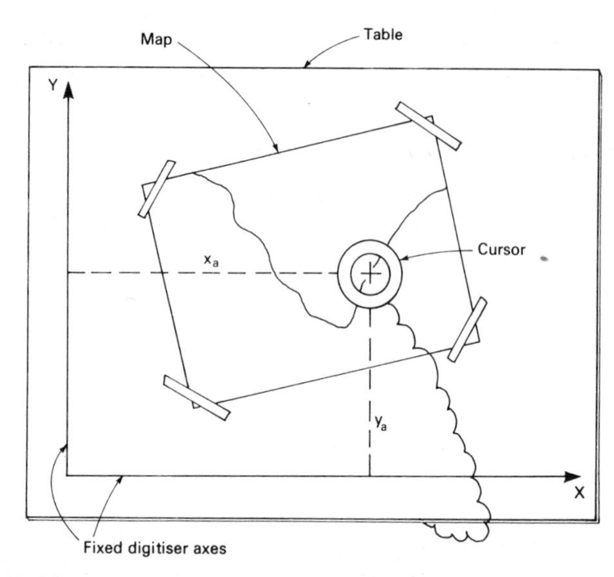

FIG. 3.10. Diagrammatic representation of a digitiser used to measure and output the plane Cartesian coordinates of the position occupied by the measuring mark on the cursor.

irregular line from the coordinate differences between adjacent points along it. The necessary measurements are made by *digitiser* operating in either the *point mode* or, preferably, *stream mode*. A digitiser is a specially wired table on which the (x, y) plane Cartesian coordinates of points may be measured. These points are selected by means of a cursor which is placed over the map. As the cursor is moved from place to place on the table, the coordinates corresponding to the position of the index mark may be output to magnetic tape or disc, or processed in real time in the computer. Point mode processing involves operating a key to record the coordinates of each point to be digitised separately. Stream mode processing does the keying automatically at prearranged intervals of time, so that the operator can concentrate wholly upon following the line with the cursor. In a sophisticated digitiser the timing interval may be set to read any interval between measurement from one point every 0·01 seconds through one point every 10 seconds. The speed with which a line can be followed by cursor is a function of the complexity of the line. If the line is irregular, the operator needs to trace it more slowly in order to keep the cursor index on the line. Therefore the points at which the coordinates are read are closer together than would happen in tracing a smooth curve along which the cursor may be traced more quickly.

If equation (3.5) (p. 44) is used to calculate the straight line distance between successive coordinated points, each straight line element corresponds to a distance d_i and the total length of the line is obtained from equation (3.2).

It should be noted that circular arcs, parabolae and other curves may also be fitted to clusters of three or more adjacent points.

Classical Methods of Continuous Measurement

The alternative to making comparison between map and scale by means of rectilinear elements is to measure distance by means of a piece of thread, or by chartwheel.

DISTANCE MEASUREMENT BY COMPARISON WITH A PIECE OF THREAD

This method has already been described and illustrated in Fig. 1.2, p. 8. Two further comments about the method are desirable. The first is that it requires considerable manual dexterity to make a thread or jeweller's chain coincide with a complicated line to be measured. There is an annoying tendency for a short length of thread to untwist or tie itself into knots if it is not held firmly in position. The author uses a number of small pieces of adhesive drafting tape, to stick down each part of the thread as this is fitted to the line. Some people recommend holding the line with a series of fine pins, but when the thread is taut this is equivalent to creating a series of rectilinear elements like those made with dividers.

The second weakness of the method is the difficulty in identification of the terminal points of the line on the piece of thread, so that these may be recovered accurately when the thread is transferred to the scale. An obvious choice is to mark the end of the line in ink. However, this is unsatisfactory because the ink tends to spread along the fibres of the thread and therefore the

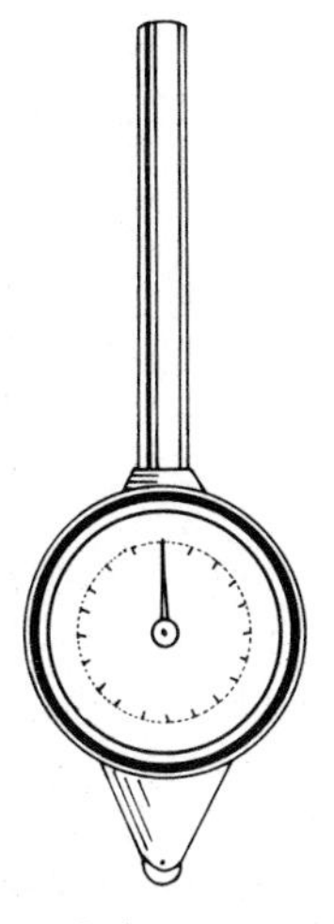

FIG. 3.11. Opisometer or chartwheel.

size of each mark is increased. Consequently the position of the terminal points cannot be recovered with an accuracy approaching 0·2 mm, which is the average diameter of prick marks made on paper with dividers. It follows that other potential sources of error, such as stretching of the thread when this is placed against the scale, are trivial in comparison with the major uncertainty of not knowing the whereabouts of the ends of the line. Despite these objections it seems that the method can produce acceptable results, as was shown by Kishimoto (1968). But my fingers are not as delicate or sensitive as hers.

MEASUREMENT BY CHARTWHEEL

The chartwheel, curvimeter or opisometer (Fig. 3.11) is the only instrument in common use which has been designed specifically to measure distance along irregular lines. Most of them resemble pocket watches; hence the continental usage of the terms *Instrument de Mésure Derby* or *Kurvenmesser Derby*, these originating from the old term of "Derby" for a pocket watch. The small wheel projecting beneath the instrument is the device whose rotations are geared to the pointer which moves round the graduated scale. Those instruments which have a measuring wheel of diameter 10–12 mm usually also have some graduations engraved round the circumference of it. For example, the Keuffel and Esser Map Measure 62 0320, which is graduated in British Standard units, reads a distance of 2 inches in one rotation of the wheel. This is therefore marked at intervals corresponding to a map distance of $\frac{1}{2}$ inch with further subdivisions corresponding to $\frac{1}{16}$ inch, this representing the shortest distance which can be recorded with this instrument. A wheel as large as this is quite difficult to manoeuvre round sharp bends, so such an instrument is better suited for work on large-scale plans than on small-scale maps.

Those instruments having smaller wheels, as illustrated by Fig. 3.11, seldom have any engraved subdivisions; not even a datum mark to be set against the

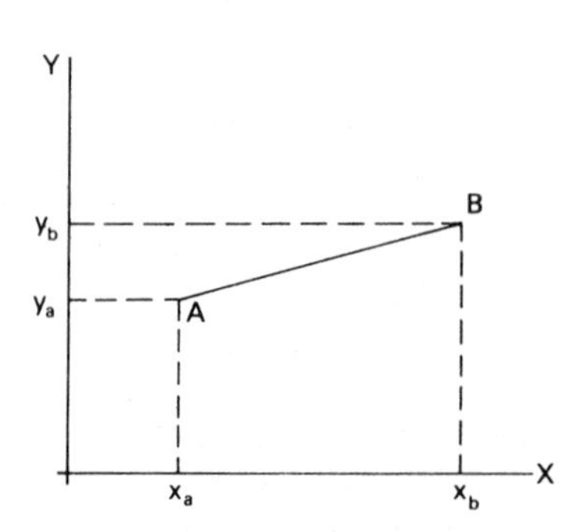

FIG. 3.12. Calculation of the straight line distance *AB* from the Cartesian coordinates of these points.

starting point on the map. It is therefore necessary for the user to make a small but legible mark on the side of the wheel to be used for this purpose. Unfortunately there are now very few good-quality instruments available. The majority of those offered for sale, even by suppliers of drawing-office equipment, have been designed and manufactured as plastic toys or gadgets for motorists. The value of these for reliable cartometric work must always be questioned. Often the scales are irregularly divided and crudely printed, or there is excessive backlash between the movements of the wheel and pointer. Indeed it is surprising that some of them work at all.

Two versions of the curvimeter developed in Russia since the middle 1950s have electronic readout of the scale as a dial reading or a digital display. The instruments have been described by Senkov (1958), whose instrument has been much used in hydrological work; the other has been described by Kopylov (1980). The author has no knowledge of any examples of these *semi-automatic curvimeters* in use in any western countries.

Another kind of chartwheel was first manufactured by the firm A. Ott in Germany before World War II. It was described and tested at that time by Werkmeister (1935). The instrument is a circular flat object, not unlike the kind of cursor now used with a digitiser, having a transparent window with a datum mark which is used to follow the line. The instrument sits upon a wheel having graduations and a vernier reading system similar to those on a polar planimeter. As the datum mark is moved along a line, rotation of this wheel records the distance travelled. Like other instruments with rather large measuring wheels, the curvimeter is evidently more satisfactory for use on large-scale plans with regular lines and smooth curves. The author has no knowledge of how it operates on irregular lines. This instrument was still being advertised in the Ott catalogues up to the early 1970s, but it seems to have been withdrawn from their range of products at about that time.

THE CALCULATED LENGTH OF A RECTIFIABLE CURVE. The word *rectification* relates to the ability to determine the arc length of a curve by calculation. We may therefore distinguish between those curves which are rectifiable and *empirical curves* which are not. The first are lines having definable geometrical properties; the second by irregular lines such as those to be found in nature and on maps. Let us therefore consider how the length of a simple curve may be determined by calculation.

THE CALCULATED LENGTH OF A STRAIGHT LINE. The simplest case is that of a straight line on a plane. Euclidean geometry describes the length as being the shortest distance between the two points A and B which therefore serve as the ends of the line. If the plane Cartesian coordinates of the points are

$A = (x_a, y_a)$ and $B = (x_b, y_b)$, the coordinate differences between them are:

$$\delta x = x_a - x_b \tag{3.3}$$

$$\delta y = y_a - y_b \tag{3.4}$$

and solving the right-angled triangle ABC having δx and δy as its two known sides, the required third side is

$$L = \sqrt{(\delta x^2 + \delta y^2)} \tag{3.5}$$

THE CALCULATED LENGTH OF A GREAT CIRCLE ARC ON A SPHERE. The second example of determining distance by calculation takes us from the plane to the curved surface of the sphere. This is the solution of the problem of how to calculate the distance between two points on the earth's surface which are too far apart to allow us to assume that the earth is a plane. We do not attempt to explain here what we mean by "too far apart", nor to justify the assumption that the earth is truly spherical. These subjects are treated in some detail in later chapters where we discuss use of the method in the location of offshore boundaries. At this stage it is sufficient to know that the required distance is the shorter arc of the great circle passing through the points A and B in Fig. 3.13.

The equation (3.6) which is presented here without proof is known as the *fundamental formula of spherical trigonometry*, because it can be derived from first principles by plane trigonometry, and it is from this formula that most of the theory of spherical trigonometry may be argued. In the example illustrated in Fig. 3.13 we already know the *geographical coordinates*, or latitude, φ, and longitude λ, for each point. Thus $A = (\varphi_a, \lambda_a)$ and $B = (\varphi_b, \lambda_b)$. Moreover we

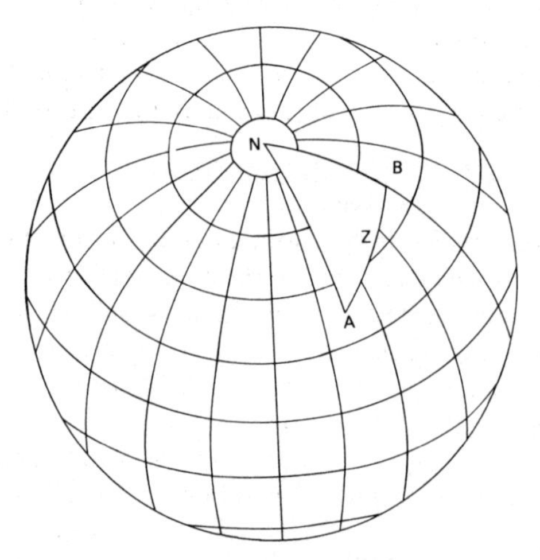

FIG. 3.13. Calculation of the great circle distance $AB = z$, from the geographical coordinates of A and B.

may put $\delta\lambda = \lambda_a - \lambda_b$, this being the difference in longitude between the two points. The required arc length is z. Then

$$\cos z = \sin \varphi_a \cdot \sin \varphi_b + \cos \varphi_a \cdot \cos \varphi_b \cdot \cos \delta\lambda \tag{3.6}$$

Because the sides of a spherical triangle are measured by angles at the centre of the sphere, the solution to equation (3.6) is a trigonometric function of the angle z. Therefore it is still necessary to convert z into linear measure. For an answer in kilometres,

$$AB = R \cdot z = 6371 \cdot 11z \tag{3.7}$$

where $R = 6371{\cdot}11$ km, which is one of the acceptable values for the radius of the earth regarded as a perfect sphere, and z is expressed in radians.

We have already seen that a better approximation is that the earth's shape corresponds to an ellipsoid of rotation or spheroid. There are about a dozen different ways of obtaining the length of the geodesic, which are described in the standard works on geodesy, for example, Bomford (1950). One method of making the calculation is described in Chapter 23, p. 548. The important point to make at this stage is that both the determination of arc length along a great circle on a sphere or along a geodesic on a spheroid are mathematically soluble and these are rectifiable curves.

THE LENGTH OF ANY RECTIFIABLE ARC ON A PLANE. Returning again to the calculation of arc length on a plane, we introduce an expression which is well known from elementary textbooks on calculus. We assume that there is a functional relationship between x and y which is satisfied continuously over the recognised length of a curve. Then we may imagine that the length of the curve is the sum of the small segments inscribed within it, as illustrated in Fig. 3.14, and is the chord CD of the curve as if this had been measured by dividers. If the rectilinear elements are all regarded as being indefinitely short, it can be shown that the length of the arc AB may be expressed in the form

$$AB = \int_A^B [1 + \mathrm{d}y/\mathrm{d}x)^2]^{1/2} \cdot \mathrm{d}x \tag{3.8}$$

The derivation of this expression may be found in many elementary textbooks on calculus. The formula may be applied to the conic sections, for example, to obtain the arc length of an ellipse or a parabola as well as to higher-order curves. However, none of these relates directly to the lines which appear on maps, apart from the grid and graticule lines which by definition are intended to satisfy certain mathematical expressions. It may also be assumed that certain linear detail, such as railways, roads and some field boundaries shown on large-scale maps and plans, represent good approximations to rectifiable curves. Indeed, many of these have been located on the ground by *setting-out* of arcs of mathematically identifiable curves by the surveyor or engineer who did this essential preliminary work in their construction.

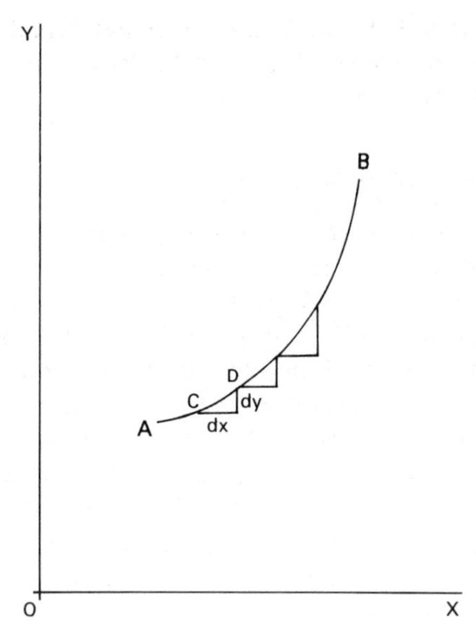

FIG. 3.14. The calculation of the length of any regular curve AB expressed in Cartesian coordinates.

Most other lines on maps may be classed as being empirical curves. As such their lengths cannot be calculated because we cannot devise expressions which describe such lines mathematically so that equation (3.8) cannot be applied. If such lines are measured in different ways, particularly at different map scales, the resulting lengths may differ substantially. Chapter 5 provides some examples of how even carefully designed and executed experiments illustrate this phenomenon.

THE PARADOX OF LENGTH. There is an important theoretical obstacle to the possibility of ever obtaining a "true" length of an empirical curve. This may be described as the *paradox of length*, a term first used by Steinhaus (1954), although earlier writers had evidently been close to understanding the phenomenon without calling it by this name. Steinhaus wrote:

> Length is a discontinuous functional. This means that we can trace in the vicinity of any rectifiable arc A another arc A' whose length exceeds an arbitrary, previously described limit, or even is infinite. This fact is something more than a mathematical curiosity: it has practical consequences...
>
> When measuring the left bank of the Vistula on a school map of Poland, we get a length which is appreciably smaller than that read from a map of 1:200 000... the left bank of the Vistula, when measured with increased precision would furnish lengths ten, a hundred and even a thousand times as great as the length read off a school map...

We shall return to the implications of the paradox of length elsewhere in this book; for the time being it is sufficient to be aware of the difficulties which it presents. According to both Steinhaus (1954) and Perkal (1958), one of the means of overcoming some of these difficulties is to use the so-called probabilistic methods of linear measurement.

Probabilistic Methods of Measurement

The methods which have been described, and which are listed in Table 3.1 as Steinhaus', Perkal's and Håkonson's longimeters and Mátern's method, owe much of their origin to an eighteenth-century gambling game which was investigated by Buffon in 1777 and which has therefore become known as *Buffon's needle*. The inverse solution to Buffon's needle does, in fact, provide the most elementary method of measurement using the probabilistic ideas which are known as *Monte Carlo methods*. This method is still extensively used in stereology.

Buffon's Needle Applied to Measurement of Length

The theory of Buffon's needle and later modifications of it will be examined in Chapter 15. It is only necessary to comment briefly here about how the technique may be applied in cartometry.

Imagine a wooden floor in which all the floorboards are d units wide. A stick which is less than d units in length is thrown to the floor. If the stick lies across the junction between two boards a "success" is counted; if it fails to intersect any line this constitutes a "failure". The object of the game, as known in the eighteenth century, was to wager upon the successful outcome of such an event.

Obviously this procedure can be scaled down to the study of the probability that a short straight line (the needle) intersects one of a family of equidistant parallel straight lines of separation d plotted on a sheet of paper.

The relationship established by Buffon was that if the needle intersects the lines n times in k trials, the probability of a success, or n/k, may be expressed as

$$n/k = 2 \cdot l/(\pi \cdot d) \tag{3.9}$$

where l is the length of the needle.

In the mathematical literature of the next two centuries it has been usual to demonstrate the validity of equation (3.9), or the efficiency of the sampling method, by calculating π from the results of a series of observations. Some examples of these determinations are given in Table 15.1 on page 305. It is easy enough to rewrite the equation so that this can be used to calculate the length of the needle. Thus

$$l = (\pi \cdot d \cdot n)/2k \tag{3.10}$$

Table 15.1, page 305, also gives the calculated lengths of the needle used for each of the published experiments.

This leads to two important conclusions. First, measurement of distance can be carried out by using probabilistic methods of assessing the chance that a random event may occur. Secondly, such measures are derived by repetition of the measurement process. In this case the trial is repeated k times and the count of the number of times that the needle intersects the line is an essential feature of the measurement process. Although it is easy enough to demonstrate that the method works for measuring the length of a short straight line, we still have to demonstrate that it can also be applied to the measurement of irregular lines.

Midway between the work of Buffon, just before the French Revolution, and that of Steinhaus in the middle of the twentieth century, lies that of Morgan Crofton, who investigated the branch of mathematics which is now called *geometrical probability*. In 1868 he proposed what is now known as *Crofton's theorem*, which is quoted on page 309.

In effect this provides the justification for assuming that the outline of a closed parcel may be regarded as being composed of a series of short rectilinear elements each equivalent to the single needle, so that if we count all the intersections made by all the needles in a single application of a grid over it (Fig. 3.15), then the sum of all the points counted may be used to calculate the length of the parcel perimeter.

We therefore have at our disposal a counting method which may produce acceptable results. However, it is obvious that we should know how many applications of the grid over the map are needed to obtain a satisfactory result. From this point of view the results obtained from the traditional approach to

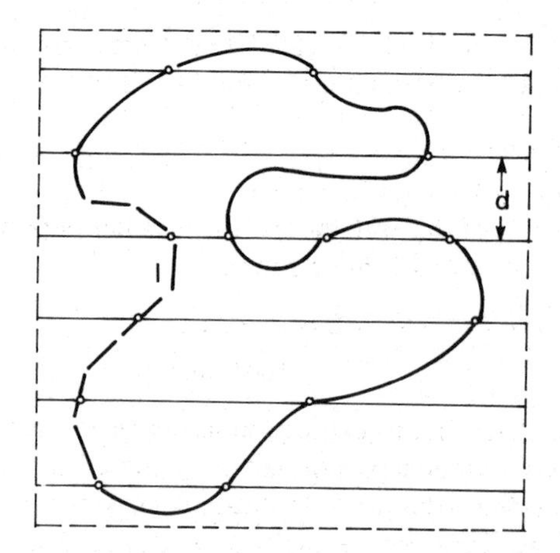

FIG. 3.15. The use of Buffon's needle to measure the length of a closed curve. The length of the needle, l, is represented by the series of straight line elements shown on the left-hand side of the parcel. Note that $l < d$. The same illustration may be regarded as representing a single application of Steinhaus' longimeter.

Buffon's needle are not much help. Examination of Table 15.1, p. 305, indicates that one experiment to determine π was based on more than 5,000 separate trials, and that many of the others used more than 1,000 throws. Because of the time needed to count all the points of intersection in each application to find the length of the perimeter of a parcel, even a sample as large as $k = 100$, would make the method unwieldy in practical cartometry.

Håkonson's Longimeter

An important addition to Buffon's theory was made by Laplace in 1812, who proposed that the single family of parallel straight lines should be replaced by a square or rectangular grid comprising two such families of lines. The significance of the modifications are discussed in Chapter 15 and need not be reviewed here.

Although the use of such a square grid is well enough known for area measurement, the application of it to distance measurement seems to have escaped attention until as recently as 1978, when its use was first described by the Swedish limnologist, Lars Håkonson. He called it the *C.T.P.* or *checkered transparent paper* method, which is an unfortunate choice because the term might be employed equally to describe the various methods of area measurement which can also be done using graph paper or a similar grid. In conformity with the terminology used by, and with respect to the work of, the Polish mathematicians, Steinhaus and Perkal, the present author proposes that the method under review should be called *Håkonson's longimeter.*

The transparent overlay simply comprises a square grid such as might be used for measuring area by the method of squares. It is placed over a closed line and the number of intersections which the line makes with the grid is counted. Thus in Fig. 3.16 there are 20 points where the parcel perimeter intersects a grid line. Then

$$L = d \cdot n \tag{3.11}$$

where d is the spacing of the grid and n is the number of points counted. If, as indicated in Fig. 3.16, $d = 0{\cdot}5$ cm or 5 mm,

$$\begin{aligned} L &= 20 \times 5 \\ &= 100 \text{ mm.} \end{aligned}$$

Håkonson argues that d in this expression has the same geometrical meaning as the use of d to express the separation of the points of dividers. Consequently small values for d will give better "statistical resolution" than a grid composed of large squares, but if the squares are too small it becomes difficult to count the number of intersections quickly or accurately. He therefore arrived at the empirical conclusion that the optimum separation of the grid lines should be about 5 mm. The dilemma of being unable to count very large numbers of grid

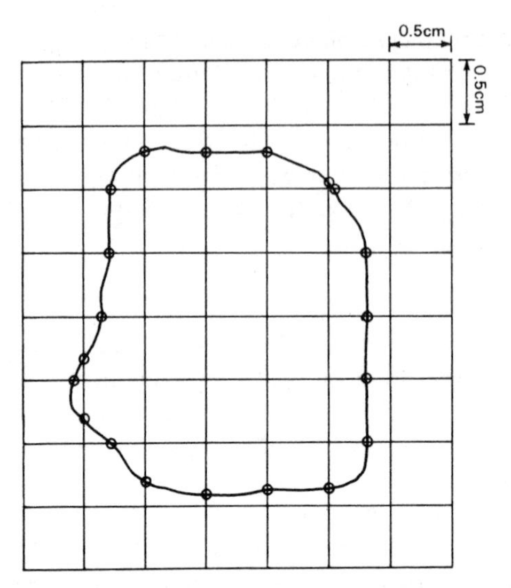

FIG. 3.16. Håkonson's longimeter used to measure the length of a closed curve. The 20 points where the parcel perimeter intersects the square grid are indicated by circles.

intersections characterises the methods of area measurement based upon counting squares.

Each application of the grid and the count of n_i intersections represents one sample measurement. The precision of the measurement process is improved by repeating the procedure for a series of k random applications of the grid to obtain a mean value $\bar{n}$ for the number of intersections counted. An example of Håkonson's own measurements, made of the shore of Lake Skagern in southern Sweden on a map of scale 1/500,000 is illustrated in Fig. 74, p. 111, where it is used to demonstrate some fundamental principles concerning statistical sampling. We therefore defer investigation of the desirable size of sample until later.

Steinhaus' Longimeter

Steinhaus' longimeter is a tranparent overlay showing a family of equidistant parallel straight lines d units apart. It follows that Fig. 3.15 illustrates this application as well as that for Buffon's needle. This overlay is placed over the boundary to be measured, and the number of points where it cuts a grid line is counted. Thus, for the first application of the grid we obtain a count of n_1 points. Turning the overlay through some angle $\pi{\cdot}k/m$ (where $k = 1, 2, \ldots, m$), we obtain n_k intersections for each of k applications of the grid. It is terminated at the number $k = m$, where m is defined as the *order* of measurement. The grand total of the points counted is

$$N = \sum n_k \qquad (3.12)$$

and the length of the measured line to order m is

$$L_m = (N \cdot d \cdot \pi)/2 \cdot m \tag{3.13}$$

Steinhaus derived empirical values for d and m. He suggested that $m = 6$ and $d = 2\,\text{mm}$ "give sufficient accuracy", but does not attempt to qualify the last statement so that we do not know for what use he considered it to be acceptable. It can be seen that equation (3.13) is practically identical to (3.10)

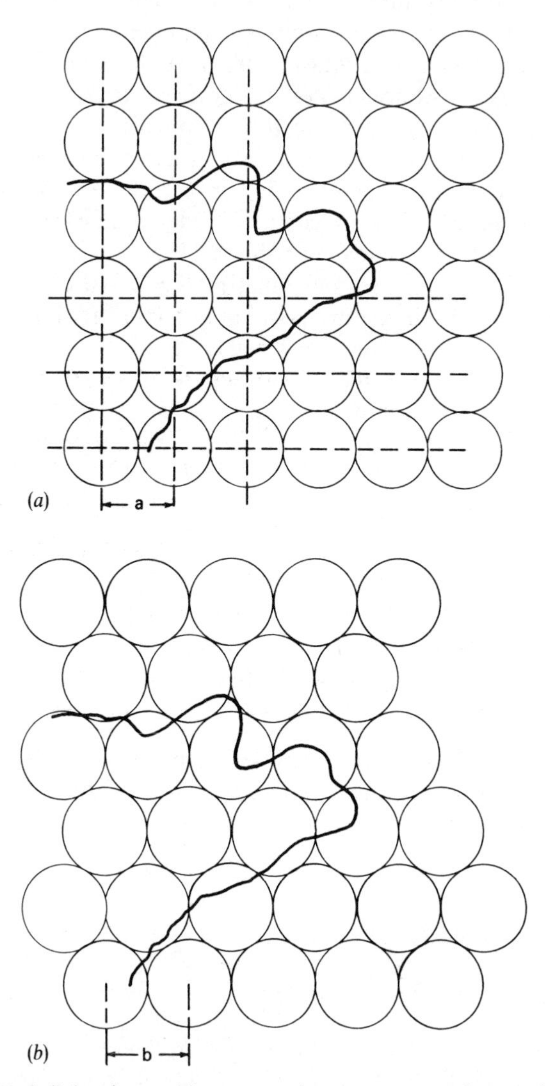

FIG. 3.17. Perkal's longimeter: (a) represents the square pattern of tangential circles at the constant separation of a units; (b) represents the triangular pattern of tangential circles which are b units apart.

but contains the restriction that the measurement procedure is terminated after m applications.

Perkal's Longimeter

Steinhaus' conclusions were criticised by Perkal (1958), who argued that there was a weak connection between length of order m and the classical length, measured in customary units of inches or millimetres. Consequently the concept of length of a particular order is unfamiliar to those working in the fields of study for which such measurements would be useful ("naturalists", according to Perkal). Although Perkal's criticism of Steinhaus now seems somewhat unconvincing, it led him to develop his own concept of order, expressed by the quantity ε, which may be interpreted as being the diameter of a small circle which has several interesting properties described in Chapter 15. For purposes of measurement of length "of order ε" an overlay comprising a regular pattern of identical circles of radius $\varepsilon/2$ is used. The pattern formed by the circles is either square, as illustrated in Fig. 3.17a, or triangular, as in Fig. 3.17b. In either case it is convenient but not necessary for the circles to be tangential to one another along rows and columns.

The overlay is placed over the boundary to be measured, and we count the total number of intersections of this line with the circles. In k such applications, a total of N circles have been counted. Then the length of the line, of order epsilon is

$$L_{(\varepsilon)} = a^2 N/2k - 1/2\pi\varepsilon \tag{3.14}$$

for the square pattern in which the centres of the circles are a units apart (or the diameter of each circle is equal to a units, so that $a = \varepsilon$).

For the triangular pattern

$$L_{(\varepsilon)} = b^2 \cdot \sqrt{3}/4k - 1/2\pi\varepsilon \tag{3.15}$$

where b is the distance between the rows of circles. Employing values of a or b expressed in millimetres, the number of applications, k, is so contrived that the result of measurement, though expressed to order epsilon, also provides a length expressed in millimetres.

Mátern's Method

This represents an application of Buffon's needle. Matern (1960) used it to demonstrate the principles of systematic transect sampling and autocorrelation, described in Chapter 8, pp. 139–140. In the particular example provided by Mátern, a map of the Stockholm Archipelago was overlain by 12 equidistant parallel straight lines, each of which was sampled at regular intervals. The presence of land at a point was denoted 1; of sea by 0. He used this to determine the length of the coastline shown on each map.

4

Measurement of Area

Area and volume are different things from length, and can no more be produced by squaring and cubing length than growth and happiness can be produced by squaring and cubing wealth.
(H. W. B. Joseph, *Lectures on the Philosophy of Leibnitz*, 1949)

It is convenient to start with two definitions:

Parcel

A parcel is any enclosed feature, the areas of which is known or required.

The word has the cadastral meaning of *A piece of land being part of an estate and having separate identity*, as used, for example, by Dale (1976). It has been used in this context since mediaeval times (*Rolls of Parliament I*, 387/1 of 1321) and is still commonly employed by those who are concerned with transactions in real estate, though less so in the scientific study of the earth's surface. In other disciplines different words may be encountered, and are used to mean the same thing. We have already given the example in Chapter 1 of the use in stereology of the word "profile" to describe any sharply outlined flat trace in the plane of a thin section, such as a single cell or organism viewed through a microscope or shown on a micrograph. In metallurgy the word which is most often encountered is *phase*. In recent years the word *polygon* has gained currency in North America with reference to remote sensing and inventory analysis. In this literature a polygon invariably means a piece of country having distinctive land use or land cover which can be classified, mapped and shown as an irregular parcel.

Give-and-Take Line

A Give-and-Take is a straight line which is chosen to represent the irregular boundary of part of a parcel, and has been chosen in such a way that the area contained within the triangle on one side of the straight line is equal to that enclosed on the other side of it.

Figure 4.1a shows an irregular line AB which is to be matched by a single straight line. It is necessary to choose two points A' and B' in such a position that the two triangular areas $AA'C$ and $BB'C$ are equal but on opposite sides of AB. If the line AB is extremely irregular, as shown in Fig. 4.1b, the give-and-

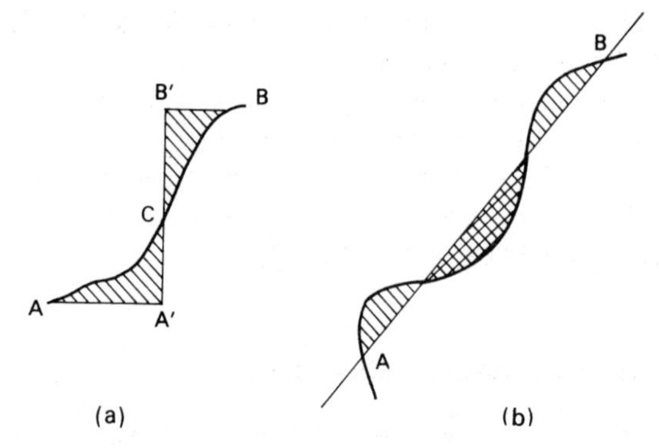

FIG. 4.1. The nature of the "give-and-take line,": (a) is an example which intersects the line AB once only; (b) is an example which intersects AB at four different points.

take line may intersect it several times. In this example the position of the give-and-take line is determined in such a way that the sum of all the small areas on one side should equal the sum of them on the other.

AREA MEASUREMENT BY CALCULATION

Measurements Based upon Simple Geometrical Figures

The method is based upon measurement of the sides of plane geometrical figures which correspond in shape to the feature to be measured. Figure 4.2 illustrates an irregular parcel which has been subdivided into a series of simple geometrical figures, the side of which have been measured. This may be done by scaling distances from the map or plan, or these may have been measurements made on the ground. Since the oldest and simplest method of

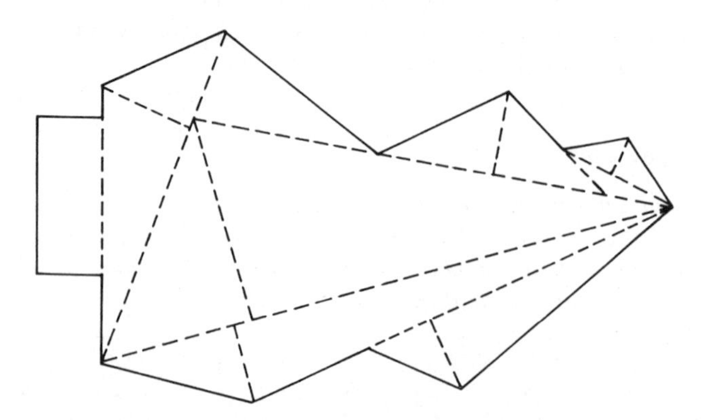

FIG. 4.2. An irregular parcel subdivided into triangles and rectangles for the purpose of measurement. Original illustration dates from 1618.

surveying a field or an estate is by chain survey, the geometrical method of measuring area is the natural concomitant to such a survey. To demonstrate the antiquity of this method, we show, in Fig. 4.2, how the parcel known as "Frith-wood" was subdivided for measuring the area of it from the component geometrical figures. This example first appeared as a woodcut illustration in John Norden's book *Surveiors Dialogue*, published in 1618, to show how the area of a parcel may be determined.

Areas Derived from Coordinate Differences

The calculation of parcel areas from the differences between the coordinates of a succession of points located along the perimeter is a later extension of the simple geometrical methods. Because the method was rather slow and clumsy to compute by logarithms, or using a hand-operated calculator, it was originally restricted to calculation of the areas of comparatively simple figures having only a few points on the perimeter, such as the area enclosed within a closed traverse. With the advent of automatic data processing by digital computer this method of area measurement became more important, for it can be used to measure parcels with extremely irregular outlines, the coordinates of which can be measured directly from the map by digitiser. This method is so efficient and fast that it is now the method most commonly used by any organisation which has access to a suitable digitiser. The geometrical and coordinate difference methods of area measurement are described in detail in Chapter 16, pp. 325–350.

AREA MEASUREMENT BY COMPARISON WITH STANDARD FIGURES

This comprises the use of one or more figures of standard shape and size as gauges, either to decide which of several standards corresponds best in size to a particular parcel or to estimate the number of times that a standard figure can be fitted within the parcel. Usually the standard is a square, rectangle or circle reproduced on a transparent overlay. Weibel (1979) has emphasised that the human eye is extraordinarily adept in matching two areas of equal size, even if one is a circle and the other a complicated jagged figure. He has described several devices showing graduated circles which are especially useful in certain applications of stereology, for example in histology, because many of the objects studied are approximately circular in outline. He recommends the kind of stencil used by draughtsmen to draw circles, which have a range of ten to fifteen circles of different diameter on the single template. The main disadvantage of using such a method is that it is only suitable for grouping objects into categories or classes according to size. The areas of parcels of intermediate size cannot be measured but only estimated by the method.

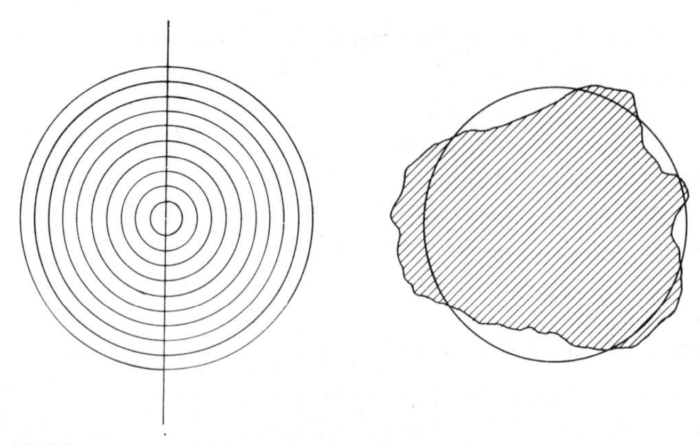

FIG. 4.3. Measurement of the area of an irregular parcel by comparison with a circle of known area selected from a set of concentric circles constructed as a transparent overlay. Note the use of give-and-take principles applied to the circumference of the appropriate circle.

The Grid Count Method

An important variation of the method is that described as the *grid count method* by Grattan-Bellew *et al.* (1978); but it is sometimes called "the method of squares", thereby causing some confusion between this and the method to be described next. In the stereological application the intention is to measure the areas of a large number of small objects which occur on a micrograph or slide. In the cartometric application it is to determine the areas of land use, land cover or similar variable shown on an aerial photograph or map. A fairly coarse square grid is placed over the photograph, and the proportion of each grid cell which is occupied by the feature to be measured is estimated. In cartometric work this might arise if the map grid itself was used as a substitute for a separate overlay. If the unit cells are smaller than the features to be measured, then each object occupies more than one grid square and the measurement procedure becomes a matter of counting the number of squares and part squares. This is the true method of squares.

THE METHOD OF SQUARES

A square grid is placed over the parcel to be measured. We assume that the unit area of the grid cell is much smaller than the parcel. The separation between the grid lines is d units in both directions, the area of the single cell is d^2 square units. If a parcel is occupied by n such cells, the area is

$$a = n \cdot d^2 \tag{4.1}$$

However, this is only an approximate value because no allowance has been made for the fact that the perimeter of the parcel passes through a succession of

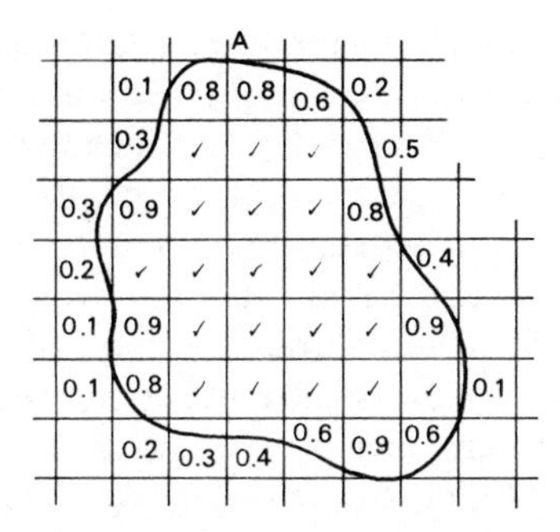

FIG. 4.4. The method of squares. The parcel contains 20 whole squares. The numbers in each of the marginal squares indicate the proportion of each lying within the parcel. The sum of these is 11·8 squares, so that the total area of the parcel is represented by 31·8 squares.

cells enclosing a variable fraction within each of them. It follows that a count of only those squares which lie wholly within the parcel is an underestimate of the area. There are three other ways in which we may treat with the marginal squares:

- These are all included in the count as if each lay wholly within the parcel. This obviously overestimates the area of the parcel.
- An estimate is made of the fraction of each marginal square which lies inside the parcel. The sum of these is added to the approximate area obtained in (4.1). Thus, if the sum of the marginal squares is a' square units, the total area of the parcel is

$$A = a + a' \tag{4.2}$$

- The marginal cells are counted but, instead of estimating the fractions, it is assumed that the average proportion of these is one-half. Thus if n' marginal cells have been counted the additional area contained within them is reckoned to be

$$a' = \tfrac{1}{2}n' \cdot d^2 \tag{4.3}$$

Obviously the second is the most complete method of determining the area of the parcel, and it is this which distinguishes the method of squares from the other point- or cell-counting techniques. If, however, the fraction of each marginal cell is estimated separately, as is indicated in Fig. 4.4, it takes appreciably longer to complete the measurement. This is why some users adopt the third method and determine a'.

POINT-COUNTING METHODS OF AREA MEASUREMENT

There are two distinct approaches to making measurements by counting points. These are described here as the *direct* and *indirect* methods. Just as some confusion has arisen about the true method of squares, so confusion exists about the two approaches to point-counting. The direct method employs a regular point pattern overlay on which the size of the unit cell has been determined from the separation of the points. The number of points contained within a parcel is counted, and this number is multiplied by the area of the unit cell. The indirect method is used to determine the area of a map occupied by a particular feature from the ratio of the number of points contained within it compared with the number of points lying outside it. This provides a proportional area which may be converted into the customary units by multiplying this fraction by the area of map or the district which has been determined by calculation, or from some other source. Because the indirect method does not depend upon the arrangement of points on the overlay it may be statistically more acceptable. For example it may be used with a random point pattern if so desired. It is also more flexible in making measurements of a variable which occurs in numerous small parcels; for example, the distribution of woodland or some other category of land use on a map. Using the indirect method it is possible to measure the total extent of woodland on the map, whereas using the direct method it would be necessary to measure each parcel of woodland separately.

The Direct Method of Area Measurement by Point-counting

Like the method of squares, this method of point-counting also makes use of an overlay – sometimes, indeed, showing a square grid, but either the lines have been replaced by a regular point pattern, each point enlarged to a legible dot to facilitate counting, or the grid lines are ignored and only the points formed at their intersection are regarded as the significant part of the overlay. Each dot in a point pattern is regarded as lying at the geometrical centre of a small cell of unit area, this being determined by the pattern and separation of

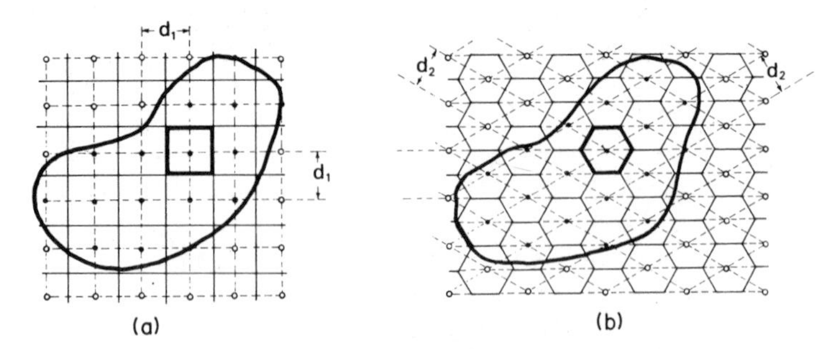

FIG. 4.5. Point-counting methods in area measurement. (a) This represents a square pattern of points in which the separation of them is d_1 units; each dot may be regarded as lying at the centre of the cell indicated by the thick lines. (b) This represents the triangular or hexagonal pattern formed by two families of lines of separation d_2 units. The dot may be regarded as lying at the centre of the unit hexagon indicated by the thick line.

the points. Figure 4.5(a) illustrates a square pattern of points in which all the dots are arranged in rows and columns of distance d_1 units apart. It follows that the individual cells are bounded on the overlay by lines which lie at $d_1/2$ units from each dot. These are indicated by the broken lines in Fig. 4.5, although these, of course, do not appear on the overlay. It follows that the area of one square cell is d_1^2 square units, so that the area of a parcel containing n such points is

$$a = n \cdot d_1^2 \tag{4.4}$$

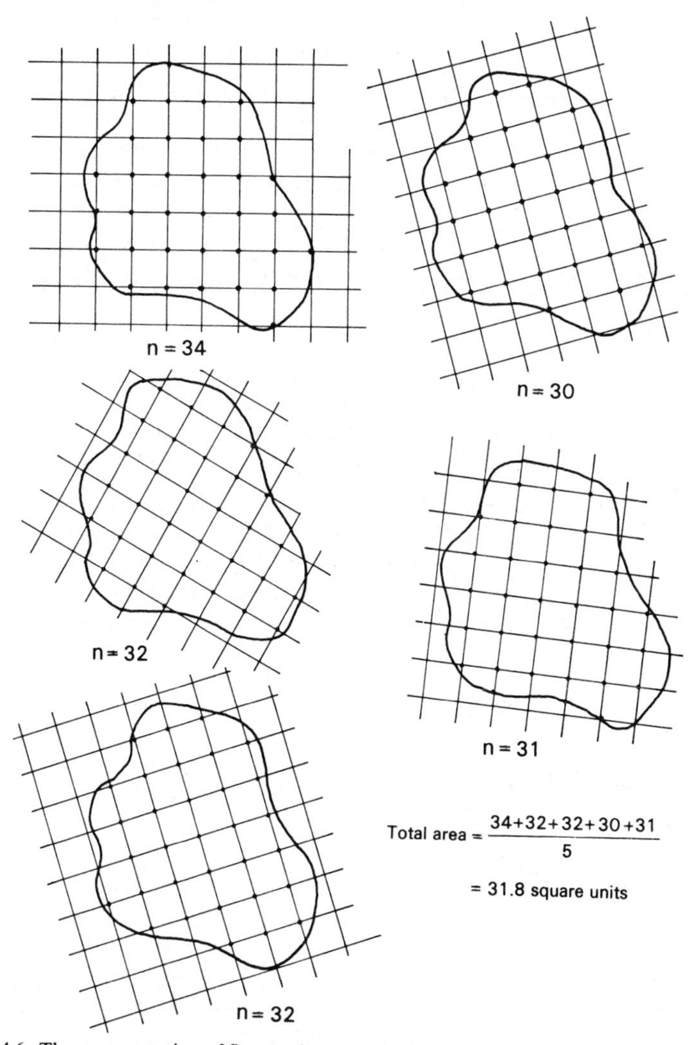

FIG. 4.6. The representation of five random applications of a square point pattern to the same parcel illustrated in Fig. 4.4. Note that the mean of the five separate counts is the same as the measurement obtained by the method of squares.

We do not estimate a' for the marginal cells, for there are no lines on the overlay to allow us to do this. In this respect, therefore, the method is identical to the simplest application of the method of squares as described in equation (4.1).

In order to overcome the obvious defects of that method several separate measurements are made, changing the position of the overlay between counts. The result is a series of separate counts, $n_1, n_2, n_3 \ldots n_N$ of the number of points lying in the parcel in each of N separate applications of the overlay. The mean count, $\bar{n}$, may be calculated and finally the area of the parcel is

$$A = \bar{n} \cdot d_1^2 \qquad (4.5)$$

for a square point pattern. If the triangular point pattern is used, the area is obtained from

$$A = \bar{n} \cdot \sqrt{3} \cdot d_2^2 \qquad (4.6)$$

where d_2 has the meaning illustrated in Fig. 4.5b.

These techniques represent *direct measurement* by point-counting. They are illustrated by Fig. 4.6, in which a series of different positions of the overlay is indicated.

THE INDIRECT METHODS OF AREA MEASUREMENT BY POINT-COUNTING

The overlay is placed over a portion of the map the area of which is already known from other sources. All the points which lie within this designated area, N_1 are counted. All the points which lie within the parcel or variable to be measured, N_2, are also counted. The proportion of the points within the parcel is N_1/N_2, so that if the area of the whole map covered by the overlay is A_M, the area of the parcel is

$$A = A_M \cdot (N_1/N_2) \qquad (4.7)$$

MEASUREMENT OF AREA BY LINEAR MEASUREMENT

The Method of Strips

A different kind of overlay pattern shows a single family of equidistant parallel straight lines, as used for Buffon's needle experiments and Steinhaus' longimeter, described in Chapter 3. If such an overlay is placed over a map it can be seen that a strip of width d units is formed by any pair of adjacent lines. Superimposed upon a parcel to be measured, the strip is terminated by give-and-take lines drawn with respect to the parcel perimeter where this intersects the strip. The distance l is measured between the give-and-take lines. It follows that the strip i is a rectangle having area $d \cdot l_i$ square units. Consequently for a parcel containing n strips, the area is the sum of the areas of these strips, or

$$A = \sum(d \cdot l_i) \tag{4.8}$$

Figure 4.7 illustrates the method and indicates how the give-and-take lines are selected in this application. The method is also illustrated on page 439, where the *scale and trace method* of area measurement is described. This crude form of automation, employing a wooden *computing scale* and transparent overlay, or *trace*, was introduced into production work at the Ordnance Survey before 1850. It continued to be used there as the sole method of measuring area on large-scale maps until it was finally superseded in 1966. It is described in detail in Chapter 20.

TRANSECT SAMPLING AS A METHOD OF AREA MEASUREMENT

Just as point-counting is a modification of the method of squares converting a classical form of measurement into probabilistic methods, so transect sampling is the equivalent modification of the method of strips. The overlay is placed in an arbitrary position over the parcel to be measured and the length l_i of each line i is measured. Figure 4.7 shows that the measurement corresponds, in effect, to that of a strip which extends $d/2$ either side of each line and that, irrespective of the configuration of the perimeter, the give-and-take lines are imagined to occur where each line meets the perimeter.

The area of the whole parcel may be obtained from an equation which is the same as (4.5) provided that it is understood that, whereas in that expression the term l_i represented the measured length of that strip, we now take it to be the length of that line. Since the method again involves sampling, a series of separate measurements should be made for each parcel with the overlay in different random positions.

As in point-counting, this represents the direct measurement procedure. The indirect method comprises measuring the lengths of a succession of lines which extend right across the map as well as the portions of the same lines which

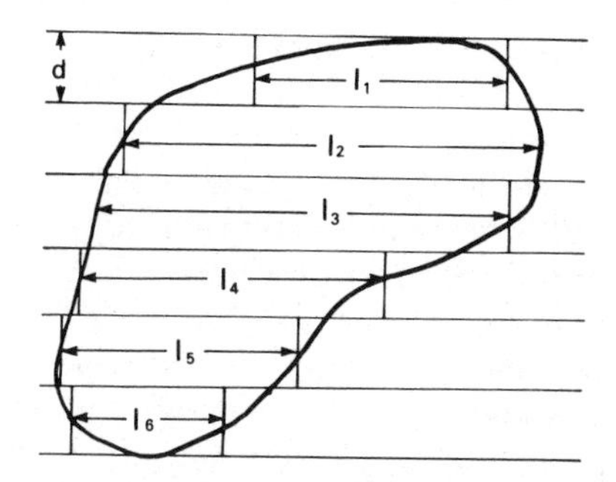

FIG. 4.7. Measurement of the area of a parcel by the method of strips, showing six strips of width d and lengths l_1–l_6 units.

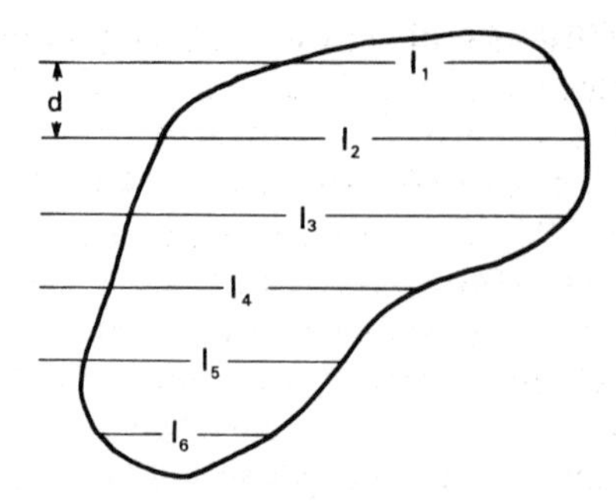

FIG. 4.8. Measurement of the area of a parcel by transect sampling. In this case the measurements l_1 through l_6 are made along the lines themselves, rather than between the give-and-take lines shown in Fig. 4.7.

occur within the parcel or distribution to be measured. The method is to be used, for example, if the statistical model requires that the lines to be measured are randomly positioned and orientated on the overlay, or accessibility dictates that only certain lines can be followed on the ground in the execution of a vegetation survey or a forest inventory. Under such circumstances there is no constant value for d which can be used to make direct calculation of the areas of strips. Therefore it is necessary to determine the proportion of each line passing through the parcel to be measured to that crossing the whole map. If there are i lines and if on each line the distance l_2 measures across the map or country, the distance l_1 being that across the parcel, then

$$A = A_M \sum (l_1/l_2)_i \tag{4.9}$$

These methods are further studied in Chapter 20, pp. 442–451.

MEASUREMENT OF AREA BY TRACING PARCEL PERIMETERS

These are the methods of area measurement which employ a specially designed instrument known as a planimeter. That used for cartometric work is usually a polar planimeter or polar-disc planimeter.

The polar planimeter, illustrated in Fig. 17.1, p. 352, comprises two hinged arms, one anchored at a fixed point, or pole, the other having a pointer or tracing lens for tracing. As the perimeter of the parcel is traced, the movements of the instrument about both the pole and the hinge are converted into linear measurements made by rotation of a small wheel containing the measuring drum, which travels over the surface of the map. The scale of the measuring drum is read when the tracing pointer is situated at the same point on the perimeter of the parcel before and after tracing. The required area is the difference of the two readings, multiplied by some factor to convert from the vernier units as the planimeter reading is often called, into the customary units of areas measurement.

Since the theory of the polar planimeter and detailed operating instructions are provided in Chapter 17, pp. 351–393, we do not repeat these here. However it is desirable to comment briefly about some of the other instruments which are available, because many different variations have been used.

Many of these have been intended for specialised and sometimes esoteric kinds of measurement such as the areas of skins or pieces of leather. Another group of instruments are used for the measurement of traces drawn on elongated or circular charts used with scientific instruments which automatically record variables on paper. Of particular importance in this field is the rolling planimeter, and the more sophisticated rolling-disc planimeter, in which the principal movement is the rolling of a special carriage to and from during tracing. This is more suitable for measurement on elongated strips of paper than the polar instruments, which remain firmly anchored to one part of the map.

In this context we should also mention the *hatchet planimeter*, a curious device which is remarkable for its simplicity. This instrument (Fig. 4.9) comprises nothing more than a metal rod which has been bent through two right angles. One end of the rod is regarded as the tracing point and has been suitably shaped; the other has been flattened and engraved with an index mark. The instrument is operated by tracing round the perimeter of a parcel, but unlike the other planimeters, tracing begins and ends at a point near the centre of the parcel. Area is determined by marking the positions of the index mark before and after tracing, and measuring the distance between these two points. This value is multiplied by the length of the bar to obtain the required

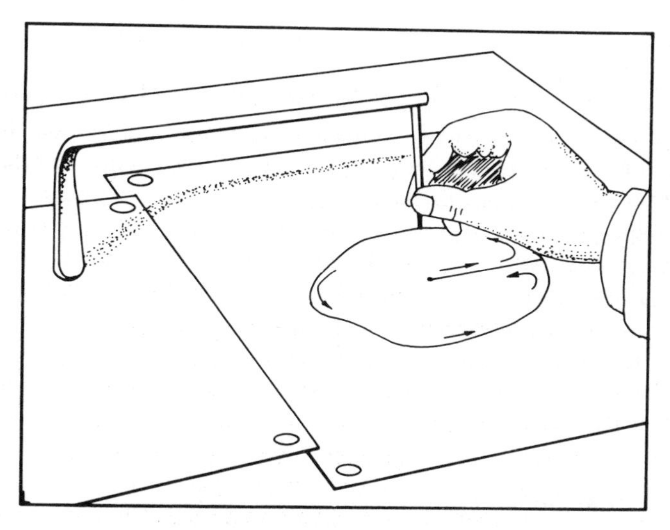

FIG. 4.9. The hatchet planimeter, indicating the method of measurement.

area. A variation of this instrument was introduced in the 1950s as the *System Gueissaz*. It comprised two components which could be mounted on a typical half-set, one of which served as the tracing mark; the other as the datum.

MISCELLANEOUS METHODS OF AREA MEASUREMENT

Cut-and-weigh is the best known of the traditional methods of area measurement which does not fit into any of the categories listed above. It has been periodically rediscovered through the last 400 years. It seems likely that the first description of it appeared in *Methodus Geometrica* published in Nürnburg in 1585. During the late seventeenth century it was used by Sir Edmund Halley to measure the areas of the counties of England and Wales, the results of which were published by John Houghton in January 1692/3 (p. 208). One of the most recent accounts of its use is by Naylor (1956), who was enthusiastic about the application of it to the work of the New Zealand Forestry Service. Even later, Chernyaeva and Molchanova (1963) made comparative tests of measurement by weighing compared with use of a planimeter for measurement of river catchment basins. The technique simply comprises cutting the required parcel from the map and weighing this upon a delicate balance. Usually the comparison is made against a series of other pieces of the same map sheet which correspond to known areas. This is considered to be better than weighing the piece of paper representing the parcel in grams, because the result depends to no small measure upon the thickness, composition and especially the moisture content of the paper. In 1585 a pair of goldsmith's scales were recommended, but today the measurements would be made on a delicate chemical balance. Obviously the method destroys the map which is being measured.

A *densitometric method* has been described by Nash (1949) using photoelectric cells and a galvanometer to measure the amount of light which can pass through the parcels when these are subjected to even illumination. Investigation of these possibilities was also made by Coppock and Johnson (1962) and Latham (1963). Because the ordinary printed map is translucent, it is usually necessary to increase the opacity of the parcels to be measured. Nash describes mounting the map on card and then cutting out the parcels to be measured. An alternative would be to reproduce the map by photomechanical methods using presensitised strip-mask material in such a way that the parcels to be measured are converted into opaque patches. The time and expense involved in either of these methods makes the so-called "photoelectric planimeter" slow, clumsy and expensive compared with any of the other methods described.

Image Analyser Methods

Much more important nowadays are the various kinds of image analyser, such as the Cambridge Instruments *Quantimat*, which are available for a variety of scientific applications. Given copy of suitable format, usually a positive image of a photograph or micrograph, it is possible to isolate those images having a selected range of grey tones. If a series of similar objects having similar tonal signatures can be isolated, they can be counted. Therefore point-counting may be employed as a fully automatic process, the unit corresponding to the unit cell being the *pixel*, the term used nowadays to describe the smallest area unit which may be used in remote sensing and computer technology to describe the unit representing the limit of resolution of the system in use. Linear sampling is also used in some forms of scanning systems.

Writing about the large number of semi-automatic instruments which are now available commercially, Henderson, in Aherne and Dunnill (1982), has described these as being "basically electronic planimeters". Most of them are difficult to use for cartometric purposes because the instruments only take photographic material of quite small format. They may be used with some kinds of remotely sensed images, but even the standard format survey-quality aerial photography, which in most English-speaking countries measures 9 × 9 in. (228 × 228 mm) is too large to fit some of these instruments without trimming. If these machines were big enough to accommodate maps, plans and charts, much greater emphasis would be placed upon their actual and potential use. As it is, those who are working in quantitative microscopy have the technological edge upon those who work with maps and charts, for this single reason.

5

The Variability of Cartometric Measurements

This shows how much easier it is to be critical than to be correct.
(Benjamin Disraeli, House of Commons Debate, 24 January 1860)

Early in the analysis of cartometric measurements we have to appreciate that there may be large variations in the recorded dimensions of the same object, notwithstanding the fact that all have apparently been made with similar skill, care and integrity. It will require the greater part of this book to account for some of these variations. It is therefore useful to present some examples of cartometric work which has been repeated in various ways, and which therefore illustrates the magnitude of the discrepancies.

Without having a special or detailed knowledge of the variability of cartometric measurements the typical map user might list several factors which possibly influence the results. Different results may be caused by:

- the scales of the maps on which the measurements were made;
- the way in which the feature to be measured has been defined;
- the instruments and methods used to make the measurements;
- the way in which the actual measurements have been reduced or converted into the appropriate ground units;
- variation in the accuracy of the maps used.

There are also two factors which might not immediately occur to the beginner or the casual map user as being significant. However, both of these may be important under certain conditions.

- The influence of the geometry of representation (or the projection) of the objects shown on the map.
- The possibility that the paper on which the map has been printed may have been deformed by variation in temperature and humidity which has occurred during storage and use of the document.

In the following examples we concentrate upon the single cartometric problem of measuring the length of a coastline. These all present similar problems of definition and handling maps or charts of widely different scales.

A major source of variation is the commonly observed phenomenon that *the*

TABLE 5.1. *The length of the coastline of Istria from Salvore to Cabo Promontore*

Map scale	Distance in km	Map scale	Distance in km
1/15,000,000	105	1/750,000	199·5
1/3,700,000	132	1/300,000	190·6
1/1,500,000	157·6	1/75,000	223·81

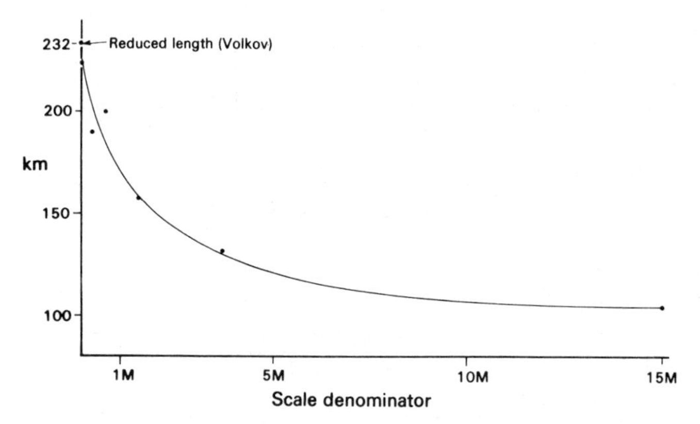

FIG. 5.1. Variation in the measured length of the coastline of Istria carried out at different map scales. The reduced length of 232 km is that derived by Volkov (1950) from fitting a parabola to the data in Table 5.1 according to equations (5.3) and (5.4).

larger is the scale of the map, the greater is the length of the measured line. In his classic work on geomorphology, Albrecht Penck (1894) listed the results of some measurements of the length of the coastline of Istria (northern part of the Adriatic Sea) which he had made on maps of different scales. The results of his measurements are summarised in Table 5.1, and their relationship to one another is illustrated in Fig. 5.1. From the table it can be seen that the measured distance at the largest scale (1/75,000) is more than double that measured at the smallest scale (1/15,000,000). When these data are plotted as a graph, as in Fig. 5.1, most of the points lie close to a smooth curve which has the appearance of a parabola.

Penck noted that, even at the scale 1/75,000, the numerous small inlets along the coast were not shown in the map and therefore these were not included in the measured length. We shall see that the same kind of relationship exists between measurements made at much larger scales.

EXAMPLE 1: THE LENGTH OF THE COASTLINE OF YORKSHIRE

Table 5.2 summarises the results of an experiment comprising more than 100 separate measurements of the same feature made with dividers on maps at nine different scales, carried out by J. S. Wilson and the present writer in 1968 and

TABLE 5.2. *The length of the Yorkshire coast, in kilometres, measured by dividers having constant separation, d. The numbers of measurements are given in parentheses*

Map scale	$d = 2$ mm	$d = 3$ mm	$d = 4$ mm	$d = 5$ mm	$d = 15$ mm
1/25,000	190·56 (2)	–	186·37 (2)	–	–
1/63,360	188·89 (2)	188·30 (2)	186·12 (2)	188·77 (2)	179·41 (4)
1/250,000	179·58 (2)	178·41 (2)	176·69 (2)	175·25 (4)	170·26 (2)
1/500,000 (A)	175·99 (2)	173·42 (2)	173·62(2)	172·22(2)	–
1/625,000	173·05 (2)	170·88 (2)	170·64 (2)	170·29 (2)	160·67 (2)
1/1,000,000 (I)	170·51 (2)	169·54 (6)	167·52 (2)	166·09 (4)	155·91 (2)
1/1,000,000 (A)	172·58 (2)	169·17 (2)	165·75 (2)	163·60 (2)	–
1/2,000,000 (A)	165·04 (2)	164·37 (2)	164·85 (2)	162·15 (2)	–
1/2, 500,000 (A)	161·99 (2)	156·41 (2)	157·00 (2)	159·00 (2)	–
1/6,000,000 (A)	158·18 (2)	157·48 (2)	154·50 (2)	156·75 (2)	–
Range (1/63,360–1/6,000,000)	30·71	30·82	31·62	28·02	

1969. The feature to be measured was the coastline of Yorkshire between the northern extremity at Teesmouth lighthouse and the southern extremity at the tip of Spurn Head, and represented by the continuous line representing high-water mark as shown on the larger-scale maps used. Since those days Yorkshire has ceased to exist as a separate administrative unit. Strictly speaking, therefore, this coast ought now to be referred to as that of North Yorkshire, with parts of Cleveland and Humberside, but for reasons of convenience, as well as priority, we continue to call this the *Yorkshire Coast Experiment*.

The maps used were those published by the Ordnance Survey within the scale range 1/25,000 through 1/1,000,000, the last being the sheet prepared by the Ordnance Survey in 1964 according to the revised specification for the *International Map of the World* (I.M.W.). A second series of measurements were made on maps contained in the *Atlas of Britain* (1963) at scales 1/500,000, 1/1,000,000 and smaller. These maps are marked (A) in Table 5.2. The measurements were made using spring-bow dividers. Each measurement was repeated at least once; a pair of measurements comprising one which started from the northern end, followed by another which started from the southern end of the coastline. The number of measurements made for each combination of dividers and map is given in parentheses after the mean length of each line expressed in kilometres. Each setting of the dividers was calibrated twice; once before and once after measurement of a line. This was done by walking the dividers along a straight line through a large number of steps, e.g. $n = 100$, and usually longer than the line to be measured. The distance traversed by the dividers was then measured independently by steel scale or coordinatograph. The calibrated value for d is obviously the length of one step, or L/n where L is the length of the straight line measured independently. Although the values of d are given in Table 5.2 as integer values, those determined by calibration of the dividers for each setting hardly ever correspond exactly, but differ from the nominal value by as much as 0·2 mm. The lengths of the lines have, of course, been determined by using the calibrated values for d in equation (3.1). We stress the value of making an independent calibration check after measuring a line, for this ensures that faulty measurements resulting from accidental change of the setting of the dividers may be detected. However, this procedure means that the final step, corresponding to d' in equation (3.1), cannot be measured until calibration has been completed. One way round this is to keep a separate pair of spring-bow dividers available for making such measurements.

Table 5.3 gives a series of 57 measurements made of the same line by chart wheel on some of the Ordnance Survey maps used for the previous experiment. These measurements were made by four different operators, all of whom were students taking the postgraduate Diploma in Cartography at the University College of Swansea in the early 1970s. As before, the measurements were made in pairs, alternately starting the measurements at Teesmouth lighthouse and at Spurn Head. The total numbers of measurements are recorded in parentheses after the mean length determined from them.

Table 5.4 summarises a series of 69 measurements made by the present writer using two different digitisers operating in stream mode. The first few measurements were made on the early d-mac Line Follower which was being manufactured in the late 1960s. The more extensive series were measured on a Ferranti Freescan Digitiser which was installed in the Geography Department, University College of Swansea, in 1975. The lengths derived from these

TABLE 5.3. *The length of the Yorkshire Coast, in kilometres, measured by chartwheel. The numbers of measurements are given in parentheses*

Map scale	Operator				Mean length (km)
	1	2	3	4	
1/63,360	194·9 (4)	186·4 (4)	–	–	190·7
1/250,000	191·0 (4)	180·0 (4)	176·9 (6)	176·6 (4)	181·2
1/625,000	176·4 (4)	174·2 (4)	–	–	175·3
1/1,000,000	168·3 (7)	172·5 (4)	169·6 (6)	166·7 (6)	169·3

TABLE 5.4. *The length of the Yorkshire coast, in kilometres, calculated from digitised coordinates measured in stream mode. The numbers of measurements made at each scale are given in parentheses*

Map scale	d-mac Line Follower	Ferranti Freescan [Digitiser]	BBC Domesday System
1/50,000	–	–	182·45(1)
1/63,350	188·437(2)	–	
1/250,000	180·48(2)	182·48(23)	
1/625,000	175·35(4)	–	174·6(1)
1/1,000,000	171·54(4)	175·95(35)	

data were calculated according to the assumption that each pair of adjacent coordinates was connected by a straight line, as in equation (3.5). For the methods used with the BBC Domesday Geographical Information System to measure length, see pages 495–497.

EXAMPLE 2: THE LENGTHS OF THE COASTLINES OF THE WORLD

We now describe two independent sets of measurements of many of the coastlines of the world. These measurements were made by two different departments of the U.S. administration within about 15 years of one another.

U.S. Coast and Geodetic Survey Measurements

These are contained in the paper entitled *World Coastline Measurements* (Karo, 1956). According to Karo, the measurements of all countries outside the U.S.A. were made on small-scale maps and charts, described by him as being "the best available maps of large regional components". The measurements were made by dividers. Mainland coasts of countries other than the U.S.A. were measured in steps of 50 statute miles. Islands which were less than 100 miles in circumference were measured in units of either 10 or 20 miles, and the lengths of the shorelines of small islands were estimated. Consequently the separation of the dividers depended upon the scale of the map or chart used. Some of the values of d, expressed in millimetres, which correspond to some typical map scales, are given in Table 5.5. Only the last two entries correspond even approximately with the range of values for d used for the Yorkshire Coast Experiment. Karo does not state how many measurements were made of each coastline. However he claims that the "mainland measurements as well as those of larger islands are considered reasonably accurate within a margin of error ranging to not more than 5 to 10 percent". The results are rounded off. The lengths of the coasts of the coterminous United States (i.e. the 48 states, excluding Alaska and Hawaii) are recorded to the nearest statute mile, but for most other countries, including those of Western Europe, the measurements are rounded to the nearest 5, 10 or even 100 statute miles. Obviously there is no point in recording results to some spurious degree of accuracy if the quality of the measurements or the maps used does not warrant this, but at that time most the coasts of Western Europe

TABLE 5.5. *Separation of dividers, d, corresponding to one step of 50 statute miles at different map scales*

Map scale	d (mm)	Map scale	d (mm)
1/500,000	161	1/5,000,000	16
1/1,000,000	80	1/10,000,000	8
1/2,000,000	40		

were better mapped than those of many parts of the U.S.A., and one might expect to find rather more detailed results on this account. Since the use of 50 mile steps naturally produces crude results irrespective of the quality of the maps, this evidently explains the rounding of these data. We shall see later that different procedures were used to measure the coastline of the U.S.A., and evidently this has been considered good enough to record these measurements to the nearest statute mile.

U.S. Department of State Measurements

Geographical Bulletin No. 3, entitled *Sovereignty of the Sea*, was produced in 1969 by the Geographer in the Office of Strategic and Functional Research of the U.S. Department of State. This publication includes, in Table II, the measurements of the coastlines of the "world's major political entities".

These were measured by dividers using the single map scale of 1/1,000,000 for all coastlines. The separation of the dividers is recorded as being "ten miles". Consequently the value of *d* is either 16 mm or 18·5 mm, depending upon how this statement is interpreted. The reason for the uncertainty is that all the distances are tabulated in nautical miles, but it is not clear from the text whether the measurements were made with dividers set to 10 statute miles and the results subsequently converted into nautical miles, or were measured directly in nautical miles from the maps. If the second, and more likely, interpretation correct, the larger value for *d* was that used. This, too, is greater than any of the dividers settings used for the Yorkshire Coast Experiment.

The 10-*mile rule* was used to determine which islands should be regarded as forming part of the mainland and which inlets should be considered to be part of the internal waters of a country. In other words, as illustrated in Fig. 5.2, p. 74, measurements of estuaries and other inlets are made as far inland as the place where the estuary first narrows to a width of 10 nautical miles. This and similar rules are discussed further in Chapters 12 and 23. The State Department measurements are all recorded to the nearest nautical mile without any rounding. There is no indication about the number of times each coastline was measured, nor is there any estimate of the order of accuracy claimed for the results.

These measurements have been much used in conjunction with recent studies of the Law of the Sea, for example, by Gamble (1974), and by Gamble and Pontecorvo (1974), with repetition of the factor analyses based upon these data elsewhere. By contrast, the measurements made by the U.S. Coast and Geodetic Survey seem to be little known outside hydrographic circles.

In making a comparison between these two sets of data it is desirable to ensure that they are compatible to the extent that the terminal points of each coastal element are the same. What happens to the coastline length between these points is a different matter, which will be examined in detail later. It is

TABLE 5.6. *Percentage differences in coastal lengths measured by the U.S. Department of State and the U.S. Coast and Geodetic Survey. Positive differences indicate that the State Department measurements are the longer (Total 56)*

Percentage difference	No. of cases	Percentage difference	No. of cases
40 to 49·9	1	0 to − 9·9	50
30 to 39·9	4	−10 to −19·9	30
20 to 29·9	8	−20 to −29·9	9
10 to 19·9	12	−30 to −39·9	6
0 to 9·9	35	−40 to −49·9	3
		−50 to −59·9	2
		−60 to −69·9	0
		−70 to −79·9	1

reasonable to suppose that the end-points of each coast are the same if no boundary changes took place between the times at which the measurements were made. During the period which elapsed between publication of the two sets of data, a series of changes in sovereignty occurred, especially in Africa and Asia, when many colonial territories achieved independence. Some of these changes, especially those in West Africa, involved the creation of new boundaries and some realignment of the old ones. Consequently it is not possible to match every coastline listed in the two sources. If we exclude those examples where boundary changes have occurred, there still remain 156 coastal elements which can be compared. These are summarised in Table 5.6.

The average difference is − 3·2%, corresponding to a discrepancy of 3·2 km on a coastline of length 100 km or 32 km on a coast of 1,000 km. It should be noted that only half of the results lie within the range + 10% to − 10%, which Karo took as the margin of error of the U.S.C.&G.S. results.

EXAMPLE 3: THE LENGTH OF THE COASTLINE OF THE U.S.A.

Five different parts of the coastline of the U.S.A. may be compared in both of the world lists already described. These parts correspond to the principal sea areas surrounding the U.S.A., together with the coasts of Alaska and Hawaii. In addition to the world measurements there are three more official measurements carried out by N.O.A.A. (National Ocean and Atmospheric Administration) in 1975 and the U.S. Army in 1971. These comprise three sets of figures which have been listed, state by state, in Ringold and Clark (1980). Here they appear in Table 5.7, listed by sea area. Three different definitions have been proposed in these publications.

1 *General coastline.* According to Karo, the general coastline was measured in units of 30 minutes of latitude on charts as near the scale 1/1,200,000 as possible. Since 1 minute of latitude corresponds to 1·85 km, this is equivalent, at the stated scale, to $d = 46$ mm. It follows, moreover, that if

this separation is used for measurement, it is also logical to apply the related "30-mile rule" to define the internal waters of the U.S.A.

2 *Tidal shoreline.* This definition of the coast is intended to overcome the obvious generalisation of the coastal outline measured with dividers set to 30 nautical miles. Karo distinguishes between two different measurements under this heading:
   (a) *The general tidal shoreline* was measured by the U.S.C.&G.S. in units of 3 statute miles on maps or charts of scale 1/200,000 to 1/4,000,000, which corresponds to using $d = 24$ mm and $d = 1$ mm at the two limiting scales. The "3-mile rule" was used to define the landward extent of tidal waters.
   (b) *The detailed tidal shoreline* was, according to Karo, measured on the largest-scale maps and charts which were available in 1940. The coastline is defined as extending as far inland as the place where tidal waters measure only 100 feet wide. It is believed that this definition corresponds to the term *tidal shoreline* used by Ringold and Clark.

3 *National shoreline.* This was the coastline measured by the Corps of Engineers, U.S. Army, and extends as far inland as the tidal limits of all estuaries and creeks.

These definitions indicate markedly different amounts of estuarine and inshore waters which are to be interpreted as lying to one side or the other of the coastline. This is demonstrated in Fig. 5.2, which distinguishes these differences in interpretation. The line $ABCD$ is that satisfying the 30-mile rule. The line $AB'C'D$ is that corresponding to the 10-mile rule as used by the State Department and $AB''C''D$ is that corresponding to the 3-mile rule. The point X indicates the place where the tidal waters narrow to only 100 feet. Finally the tidal limit may occur some distance further inland. Obviously this defines a coastline which is significantly longer than those defined by the other rules. It

TABLE 5.7. *Different measurements of the length of the coastline of the U.S.A. (distances in kilometres)*

Coast	General coastline		Dept. of	Tidal shoreline	National shoreline
	N.O.A.A.	Karo	state	(N.O.A.A.)	(U.S. Army)
Atlantic	3,330	3,038	2,987	46,145	30,080
Gulf	2,625	2,670	3,111	27,586	14,466
Pacific	3,288	2,081	2,131	64,853	85,130
Coterminous U.S.A.	8,035	7,789	8,229	86,853	52,053
Alaska	10,686	10,686	12,127	54,563	76,122
Hawaii	1,207	1,247	1,232	1,693	1,497
Total U.S.A.	21,634	19,722	21,590	146,698	129,676

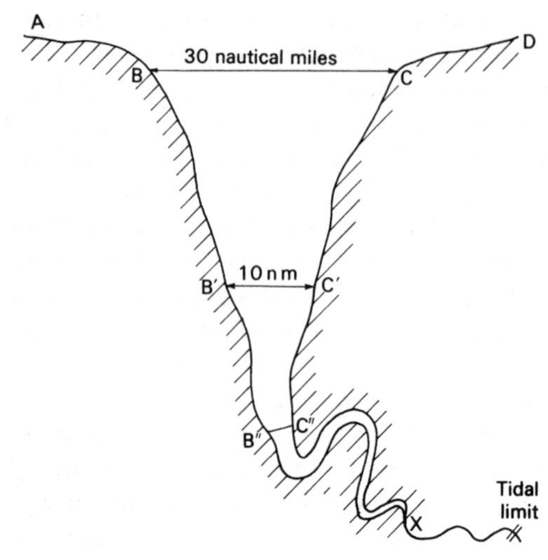

FIG. 5.2. The five different definitions of an estuary employed in official U.S. publications.

is demonstrated in Table 5.7, where the final two columns show a ten-fold or twenty-fold increase in the measured lengths compared with those to the left of them.

Since N.O.A.A. is the department which now includes within its responsibilities the work of the former U.S. Coast and Geodetic Survey, it follows that the general coastline of 1975 ought to agree reasonably well with the values listed in Karo (1956). We see that the figures for Alaska are identical, but generally the 1975 values are slightly longer than those measured earlier. We also see that there are rather large differences between the lengths of the tidal shoreline and the national shoreline, both of which more or less eliminate the simplifications resulting from truncation of the upper reaches of inlets and estuaries. Clearly, however, there must be other reasons for the large discrepancies between these two columns.

EXAMPLE 4: THE LENGTH OF THE COASTLINE OF THE U.S.S.R.

This example compares the two American measurements with the results of an exhaustive study by Vorob'ev (1959). In the Russian work the techniques of measurement pioneered by Shokal'sky (1928) and the reduction of the measurements proposed by Volkov (1950) have been followed. Vorob'ev's measurements were all made by dividers, using at least two different settings for d on maps at two different scales. Thus each recorded distance is based upon at least four independent measurements. The smaller value for d is the range 1–1·25 mm; the larger is $d = 4$ or 5 mm.

By virtue of the great variability in the quality of maps of the U.S.S.R. which still existed in the 1950s, it was impossible to make the measurements at uniform scales everywhere. Vorob'ev's work evidently antedates the completion of the 1/100,000 scale topographical mapping of the U.S.S.R., but that map series may not have been available to him.

Vorob'ev lists the map scales used for the different parts of the coast. These range from a larger scale of 1/150,000 for the Black Sea and 1/200,000 for the Baltic and parts of the Pacific coasts, down to 1/700,000 for some Arctic islands such as Severnaya Zemlya and Zemlya Frantsa Iosifa, for which no

TABLE 5.8. *Comparison of the length of the coastline of the U.S.S.R. measured by Vorob'ev (1959) with corresponding measurements made by the U.S.C. & G.S. (Karo, 1956) and the U.S. Department of State (1969). All distances in kilometres. Totals marked with an asterisk do not necessarily represent the sum of the entries tabulated here, because only those islands which are separately identified in the American lists are included in the table*

				Differences (%)	
Coastline	Vorob'ev	U.S.C.&G.S.	U.S. Dept. of state	Col 2 – Col 3	Col 2 – Col 4
(1)	(2)	(3)	(4)	(5)	(6)
Baltic Sea	7,190	2,189	3,163	228	127
Black Sea	2,928	–	1,607	–	82
Black Sea and Sea of Azov	4,731	3,098	–	53	–
Caspian Sea	7,183	4,345	–	65	–
Pacific Ocean					
Mainland	17,871	11,426	11,258	56	59
Islands					
Karaginskie	333	241	–	38	–
Kommandorski	437	265	319	65	37
Sakhalin	3,006	2,358	2,481	27	21
Shantarski	495	257	–	92	–
Kurilski	2,921	1,207	1,740	142	68
Total Pacific*	26,752	15,754	15,798	67	68
Arctic Ocean					
Mainland	27,699	17,461	15,133	59	83
Islands					
Vrangela	460	381	358	21	28
Novo Sibirskie	3,222	1,569	–	105	–
Severnaya Zemlya	3,498	1,529	–	129	–
Zemlya Frantsa Iosifa	4,425	1,127	–	293	–
Novaya Zemlya and Vaigach	8,213	3,001	2,113	171	289
Kolguev	246	217	–	13	–
Oleniy	177	137	–	29	–
Sibiryakova	144	121	–	19	–
Beliy	223	161	–	39	–
Total Arctic*	46,967	25,704	23,569	83	99
Total U.S.S.R.					
Mainland	64,674	38,519	31,161	63	101
Islands	27,740	12,571	12,796	121	114
Grand total	92,414	51,090	44,137	77	105

TABLE 5.9. *The length of the coastline of the Black Sea. Measurements by Vorob'ev (1959), in kilometres*

Map scale	$d = 1–1{\cdot}25$ mm	$d = 4–5$ mm	Limiting distance
1/150,000	2,540	2,430	2,650
1/600,000	2,350	2,215	2,485
1/2,000,000	2,030	1,780	2,280
1/5,000,000	1,830	1,640	2,020
1/10,000,000	1,770	1,580	1,960
1/30,000,000	1,575	1,320	1,830
1/50,000,000	1,530	1,280	1,780

larger-scale maps exist. The smaller-scale maps range from 1/750,000 for the Black Sea, down to 1/2,000,000 for many of the less accessible parts of the Soviet Arctic. In the Black Sea sector a large number of additional experimental measurements were made at different scales to demonstrate the variation in length with both dividers spacing and map scale. These results are given in Table 5.9. It is interesting to compare these with the results of the Yorkshire Coast Experiment in Table 5.2, because the length of the Black Sea coast is much greater, it is more difficult to define, and because many smaller scale maps were used.

DISCUSSION OF EXAMPLES 1–4

We now return to the list of factors at the beginning of this chapter to seek which of them may have played a significant part in causing the variations. However, this will only be a preliminary survey of these; more exhaustive analyses must come later.

A cursory glance through the various tables indicates that there is a considerable difference between the magnitute of the discrepancies obtained from the Yorkshire coast measurements and those for the other examples. A variety of reasons might be suggested to account for this; for example the larger number of measurements which were made for every combination of instrument, operator and map scale; the larger scales of the maps used and the comparative simplicity of this coast, which lacks estuaries or offshore islands to complicate definition and measurement. However, it is likely that the principal factor is that the measurements of the Yorkshire coast were all supervised by the author, who maintained a consistent interpretation of how the coastline should be defined. Moreover, the individual operators could be made aware of some possible sources of gross error, such as the risk of duplicating the measurements of part of the coastline where two map sheets overlap. For example, on one pair of 1/25,000 sheets (TA23 and TA32) there is overlap of rather more than 1 kilometre of almost featureless coast, where the extent of one of the sheets has been increased to cover more than the standard 10 × 10 km format of the Provisional Edition of the 1/25,000 map. Such an overlap is so unusual in this map series that it provides an obvious trap for the unwary.

Examination of the other examples, and the much larger discrepancies found between them, indicates how difficult it may be to reconcile the measurements if there is no comprehensive account of how they were done. The nature of many of the discrepancies in examples 2–4 suggests that by far the most important causes of variation are:

The Way in which the Feature has been Defined

From the point of view of the world coastline measurements we have already made the obvious comment that discrepancies will arise if the ends of the lines to be compared are not identical, and that we have eliminated all those elements of the world's coasts where boundary changes may have altered the terminal points. However, there are three more ways in which the definition of the coastline may give rise to variation in the measured lengths. Each of them is equally effective in making comparison a waste of time, although the role of each is perhaps not as obvious as those discrepancies which arise from boundary changes.

1. Some coastlines which lie within enclosed waterways have been excluded from the national total because these waters have been interpreted as being part of the internal waters of a country. A typical example is the southern part of Chile, where the coast is composed of an intricate labyrinth of narrow inlets and channels. Do the coasts of these inlets and their associated islands represent the coastline of the country as a whole?
2. Large islands or island groups may be regarded as having coasts which should be included in the length of the national coastline, even if they lie a considerable distance from the mainland. For example, Hawaii is a state of the U.S.A. and as such its coastline is duly recorded in Table 5.7. However, the Hawaiian Islands lie about 3,500 km from the nearest mainland coast of the U.S.A.
3. The coastline may be interpreted variously as extending as far inland as the tidal limit, or it may be arbitrarily truncated by the application of some geometrical standard such as the 10-mile rule.

Table 5.10 lists these coasts for which there are large differences between the world coastline measurements. A large difference is here regarded as having a difference between the recorded lengths of 35% or greater.

We may list the examples of Chile, Mexico and U.S.S.R. as clearly being examples where definition of the extent of internal waters is the main cause for the disagreement. We know that the State Department measurement of the coast of Chile is restricted to the oceanic coastline because the original table lists that as "excluding passages within archipelago". Indeed, the application of the 10-mile rule to the archipelago of southern Chile excludes virtually all

TABLE 5.10. *Countries for which there are particularly large (>35%) differences between the coastline lengths measured by the U.S. Department of State and the U.S. Coast and Geodetic Survey (distances are in kilometres)*

Country	U.S.C.&G.S. measurement	U.S. Dept. of state measurement	Difference	
			km	Percentage
Chile	12,466	5,341	7,125	133
Brazil	9,687	6,842	2,845	42
Mexico (Pacific)	7,242	4,844	2,398	50
Norway	4,989	3,058	1,931	63
Greece (Mainland and Aegean Is.)	4,345	2,418	1,927	80
U.S.S.R. (Black Sea)	3,098	1,607	1,491	93
Venezuela	2,873	2,003	870	43
France (Atlantic)	2,696	1,635	1,061	65
Finland	2,353	1,362	991	73
Ecuador	2,057	849	1,208	142
Yugoslavia	1,448	789	659	84
Dominican Republic	966	602	364	60

the channels and inlets, together with the Straits of Magellan. In the example of Mexico, the application of the 10-mile rule excludes half of the Gulf of California because the northern part is separated from the open sea by a series of small islands reducing the width of the Gulf to a series of channels of width 6–9 nautical miles. In the example of the Black Sea, the State Department interpretation of the Sea of Azov is that this lies wholly within the internal waters of the U.S.S.R. (as, too, is their opinion about the Caspian Sea, which does not appear in their table under either the Soviet Union or Iran). The U.S.C.&G.S., on the other hand, include the coast of the Sea of Azov in their measurements for the Black Sea.

The second kind of variable is exemplified by Ecuador and, in part, by Venezuela. In the case of Ecuador we know from Karo that measurement of the national coastline included those for the Galapagos Islands, whereas that of the State Department is for the mainland coastline only. The Isla de Margarita and some smaller islands of the Lesser Antilles, which are part of Venezuela, are excluded from the State Department measurement but are evidently included in the other set.

The third major variation in the definition of coastlines is most clearly demonstrated in Table 5.6 by the difference between the general coastline and the tidal shoreline or the national shoreline. We have already noted that the difference amounts to a ten- to twenty-fold increase in the length of the coastline where the tidal limit is taken as the landward extent of the coast compared with some arbitrary truncation rule. Even in the case of an estuary which is comparatively small by world standards, such as that of the River Thames, the tidal limit is at Teddington Weir which is 103 km upstream of the conventional mouth of the river (taken as a straight line across the estuary which passes through the Nore Light vessel). The choice of the tidal limit in this case effectively increases the length of the coastline by more than 200 km. Alternatively, this choice of boundary between land and sea effectively reduces the measured length of the River Thames by about 30%. The tidal limit of the Amazon is situated at Obidos, which is about 1,000 km upstream of the position of a straight line drawn from Cabo Norte to Cabo Maguarinho which appears, at the scales of atlas maps, to separate the mouth of the Amazon from the Atlantic Ocean. In this example the use of the tidal limit has the effect of increasing the length of the coast of Brazil by about 2,000 km, or reducing the total length of the river by about 16%. It will be observed that these differences are of similar order of magnitude to the discrepancy in the lengths of the coast of Brazil recorded in Table 5.10.

Instruments and Methods Used

Only the Yorkshire Coast Experiment provides any information about the influence of different methods, for virtually all the other measurements discussed here have been made using dividers.

The first impression to be obtained from comparison of Tables 5.2, 5.3 and 5.4 is that there is comparatively little variation between the results obtained at any given scale. Table 5.3 indicates that there is a tendency for the chartwheel measurements to be longer than those obtained by dividers. This was a preliminary notion first advanced in Chapter 3, p. 33. However the chartwheel measurements have considerable variability amongst themselves, shown in Table 5.11 by the range between the longest and shortest measurements obtained at each scale.

Using dividers set to a separation d, it is reasonable to suppose that the greatest discrepancy between a pair of measurements ought to be less than one whole step of length d. This is borne out by Table 5.11, where indeed the majority of the differences relating to the dividers measurements are less than $d/2$. In terms of distance, however, this statement disguises the fact that the discrepancy may be quite large for a wide separation of the dividers. Thus at a scale 1/1,000,000 and $d = 15$ mm, the range between the pair of measurements

TABLE 5.11. *Range (in kilometres) between the longest and shortest distances measured for the Yorkshire coast using different spacings of dividers and chartwheel (numbers of measurements are given in parentheses)*

Map scale	$d = 2$ mm	$d = 3$ mm	$d = 4$ mm	$d = 15$ mm	d = 15 mm	Chart-wheel
1/25,000	0·28 (2)	–	0·19 (2)	–	–	–
1/63,360	0·15 (2)	1·20 (4)	0·08 (2)	0·55 (2)	0·54 (2)	8·5 (8)
1/250,000	2·26 (2)	2·26 (4)	0·11 (2)	1·13 (4)	1·09 (2)	14·4 (18)
1/625,000	0·03 (2)	0·70 (4)	0·10 (2)	1·57 (4)	4·45 (2)	5·3 (8)
1/1,000,000 (I)	0.06 (2)	2·30 (6)	0·54 (2)	2·58 (4)	7·05 (2)	5·8 (24)

is over 7 mm, which is slightly greater than the range of chartwheel measurements made on that map. Does this imply that a chartwheel has similar precision to dividers set to a wide spacing of 10–15 mm? We return to consideration of this possibility in Chapter 14.

Both Tables 5.2 and 5.8 illustrate the phenomenon which has already been stressed, namely that the length of a line increases with decreasing separation of the dividers and with increasing map scale. The first of these trends is satisfied by all except five of the series of Yorkshire coast measurements, but with only one exception these anomalies only amount to a few tenths of a kilometre in magnitude. Although it is evident that all the American measurements have been made with d set to much greater values than used by Vorob'ev, we should not jump to the immediate conclusion that the huge differences between the measured lengths of different parts of the coastline of the U.S.S.R. listed in Table 5.7 are due only to the setting of the dividers, for the discrepancies also result from the methods used to reduce the Russian measurements in an attempt to correct for both the setting of the dividers and map scale.

The application of a correction for the spacing of the dividers given in the measurements of the Black Sea coast made by Vorob'ev is illustrated in Table 5.9. The fourth column of this table is headed "*Limiting distance*", and the values given in it are always greater than either of the measured values obtained for each scale of map. The limiting distance has been calculated from an equation proposed by Volkov (1950b), which determines the theoretical value for length of line measured by dividers with zero separation of the points. This may be expressed as

$$L = l_1 + k(l_1 - l_2) \tag{5.1}$$

where

$$k = \sqrt{d_1}/(\sqrt{d_1} - \sqrt{d_2}) \tag{5.2}$$

In these equations L is the limiting distance to be calculated from two sets of measurements made on the *same* map using dividers set to d_1, giving the distance l_1 and to d_2 giving l_2. We assume that $d_1 < d_2$ so that $l_1 > l_2$. Equation (5.1) represents a parabola which can be uniquely defined by the four variables d_1, l_1, d_2 and l_2. Hence no other measurements are required for this solution. However it should be stressed that it is not the only way in which the limiting distance may be calculated. Alternative formulae have been proposed by Soviet scientists working in this field. These are described in Chapter 14.

Variability Owing to the Scales of the Maps Used

This variation characterises the paradox of length mentioned in Chapter 3. This is clearly illustrated by Tables 5.1 and 5.2, in which distance always increases with increasing map scale. There is only one example where this condition is not satisfied and, even here the difference is small (0·7% of the length of the line). The importance of map scale as a factor influencing the variability of the measurements is clearly evident from the range of values in each column of Tables 5.1 and 5.8. For the Yorkshire coast the range between the measurements made at 1/63,360 and 1/6,000,000 (i.e. a scale change of nearly one hundred-fold) is a regularly about 30 km or approximately 16% of the length of the line.

Figure 5.1, p. 67, shows that the relationship between measured distance and map scale is not linear. Consequently it is not wise to treat these figures as being invariable for all map scales or for all natural features. However, they do indicate the kind of variation which may be experienced through using maps of widely different scales for comparative purposes.

Although the easiest apparent solution may be to make all the measurements at the same scale, as suggested by Penck (1894), and as was done by the U.S. Department of State for their set of world coastal measurements, this solution is possible only for features having worldwide or continental distribution if measurements made on small-scale or atlas-scale maps are acceptable. The 1/1,000,000 scale is the largest for which there is uniform map or aeronautical chart cover of the land areas of the world, in the form of the *International Map of the World* and the *World Aeronautical Chart*. There are several other possibilities at still smaller scales. Attempts to work at any larger scales may well be frustrated by the absence of any suitable mapping in some areas, or by the lack of availability of the larger-scale maps to the unofficial, civilian map user. There are plenty of countries in the world, not exclusively Eastern Bloc countries, where all topographical maps, and even general maps of scale larger than 1/1,000,000, have some sort of security classification.

Because working at an uniform scale may not be a feasible solution, we turn to the procedure originally proposed by Volkov (1950) and which has been followed, for example, by Vorob'ev, in the reduction of the measurements of the coastline of the U.S.S.R. This is to interpret Fig. 5.1 as a parabola, so that in equation (5.2) we extract the roots of the independent variable, which in this case is map scale.

If the measurements have been made on two maps of scales $1/S_1$ and $1/S_2$ and the limiting distances, determined from equation (5.1), are L_1 and L_2 respectively, then the *reduced length*, L_{red}, may be expressed by

$$L_{red} = L_1 + K(L_1 - L_2) \tag{5.3}$$

where

$$K = \sqrt{S_1}/(\sqrt{S_2} - \sqrt{S_1}) \tag{5.4}$$

We assume that $1/S_1$ is the larger scale so that $L_1 > L_2$. Hence the reduced length can be determined from a minimum of four independent measurements made on two different maps with dividers at two different settings. The geometrical significance of the reduced length is that it represents the point where the curve intersects the ordinate, corresponding to an hypothetical distance which might be measured on a map of scale 1/1. A comparison is

TABLE 5.12. *The length of the coastline of the U.S.S.R., in kilometres, measured by Vorob'ev. A comparison of the actual measured values with the reduced lengths recorded in Table 5.7*

Coast	Length on the larger-scale map used		Length on the smaller-scale map used		Reduced length derived from these four measurements
	$d =$ 1–1·25 mm	$d = 4$–5 mm	$d =$ 1–1·25 mm	$d = 4$–5 mm	
Baltic Sea	5,148	4,353	3,745	3,083	7,190
Black Sea	2,648	2,533	2,424	2,290	2,928
Sea of Azov	1,677	1,597	1,615	1,504	1,803
Caspian Sea	6,377	6,026	5,820	5,305	7,183
Pacific Ocean	23,924	22,208	21,708	19,353	26,752
Arctic Ocean	48,957	42,468	43,549	36,395	62,491
Total U.S.S.R.	88,731	79,185	78,861	67,930	108,347

made between the actual measurements and the reduced lengths of different sections of the coastline of the U.S.S.R. in Table 5.12. The reduced length is given in the sixth column of this table. It will be seen that these are the same values as given in Table 5.8, and that they are appreciably longer than any of the original measurements obtained on the smaller-scale map with the larger value for *d* are much closer to the American measurements than was indicated by the differences in Table 5.8. For example, the measurement of the length of the Black Sea and Sea of Azov given in the fifth column of Table 5.12 is only about 22% greater than the result obtained by Karo (1956). It would be reasonable to suppose that much of the remaining discrepancy can be accounted for by variation in spacing of the dividers and in the use of maps of different scales.

The conclusion to be drawn from the comparison of these results is that the Russian measurements given in Table 5.8 cannot be compared with the American values. This is obvious enough when the reasons for the differences are explained, but when the figures are presented in a gazetteer or a statistical abstract without any detailed explanation, it is difficult to accept them at other than face value, and it is more or less guesswork to decide which should be used for a particular purpose. Clearly, therefore *consistency in the treatment of data is extremely important.*

Notwithstanding this conclusion, there still may be occasions when variations seem to be unaccountable. A good example of this is to be found in Table 5.2, where there are two sets of data for the Yorkshire coast which ought to be identical, for they represent measurements made by the same people using identical methods and definitions, at the same map scale, though on different maps. These are the dividers measurements made at scale 1/1,000,000 on the I.M.W. map compared with those made on page 60 of the *Atlas of Britain*. The differences between the measurements are given in Table 5.13. We

TABLE 5.13. *Comparison of the measurements made of the Yorkshire coast by dividers at the scale of 1/1,000,000 by Wilson and Maling*

d (mm)	I.M.W. (km)	Atlas of Britain (km)	Difference	
			km	Percentage
2	170·51	172·38	−1·9	−1·1
3	169·54	169·71	−0·17	−0·1
4	167·52	165·75	+1·8	+1·0
5	166·09	163·60	+2·5	+1·5

see that the difference between the results is not consistently positive or negative, and therefore it is not what we shall call in later chapters a *systematic error*. The differences fluctuate, so that a difference of − 1·9 km is more or less balanced by another of + 1·8 km. On paper these distances correspond to − 1·9 mm and + 1·8 mm. In view of the fact that these measurements were all made under the controlled conditions, which were described at the beginning of this chapter, such discrepancies seem difficult to explain.

Since we cannot attribute them to definition, method or difference in map scale, it is necessary to enquire whether any of the other factors, such as map accuracy, map projection or paper deformation may give rise to variations of this order of magnitude. First, however, it is desirable to examine the theory of errors of measurement and place the study of their variability upon sounder statistical foundations.

6

Errors of Measurement and Their Analysis

Measure thrice before you cut once.
(Italian proverb)

THE NATURE OF AN ERROR

Consider the measurement of a straight line by the methods described in Chapter 3, pp. 31–32. If the true distance between its terminal points is denoted by L and the measured distance read from the scale is l, the *absolute error* of the measurement, δl, is the difference

$$\delta l = L - l \tag{6.1}$$

This may also be written in the form of the *relative error*

$$f = \delta l / L \tag{6.2}$$

In practice, however, the true value of L is unknown. It is impossible to determine the true length of a line or the true area of a parcel because every measurement which is made, irrespective of the method used and the care taken, is encumbered with errors. Even if we were to magnify our view of the line so that we could read increasingly fine subdivisions of the scale, errors would still occur. Ultimately there must be a theoretical limit at which the individual molecules comprising the sheet of paper, divider points and scale all become visible. These would all be seen to be in constant Brownian motion so that the terminal points of the line are no longer fixed points on the paper and the engraved graduations of the scale are moving too.

It follows that each measurement, l, can only be made with a certain degree of accuracy which is governed by a variety of factors which are continually varying. Consequently each time the measurement of the line is repeated, a different value for l will be obtained. It follows that an approximation to the true length of the line may be obtained from the combination of several independent measurements, such as $l_1, l_2, l_3, \ldots l_n$.

We shall see later that the *arithmetic mean* of these measurements may be considered to be the *most probable value* of the measurements which have been made. Analysis of the spread of the measurements through the *standard deviation* (or the *standard error* of sampling theory) will provide a measure of their *precision*. The errors which occur in each measurement are subject to certain mathematical laws, and although it is not possible to determine their absolute magnitude, it is possible to evaluate the statistical *probability* of whether errors of a certain magnitude will occur.

THE CAUSES OF ERRORS

The causes of errors arising in laboratory measurements have been summarised by Lyon (1970) under the following headings:

Measurement stage
(1) Personal
(2) Instrumental
(3) Environmental
(4) Natural fluctuations of the quantity itself

Reduction stage
(5) Computational
(6) Errors in auxiliary data
(7) Errors in theoretical assumptions.

It is convenient to split the analysis of these causes into the two stages already noted in Chapter 3. The first is the *measurement stage*, which involves the actual manipulation of the instruments. This is followed by the *reduction stage*, in which the results of the measuring process are converted into their ground dimensions and any other kind of numerical reduction and corrections are applied to take into account the geometrical properties of the map or photograph used for measurement. Some of the possible causes of error arising in the simplest case of measuring the length of a line may be listed as follows.

Measurement Stage

Personal errors include mistakes in reading the scale or in manipulating the dividers. They also include the *personal equation* of the observer, which is the way in which a reading lying between the closest subdivisions of the scale is estimated visually.

Instrumental errors include:

1 Small defects in the dividers and the scale, such as excessive flexibility in the legs of the dividers, slackness of the joint at which they are hinged, small errors in the positions of the engraved subdivisions of the scale.
2 Variation in the way in which the dividers are held with respect to the plane of the paper or the scale and variation in the depth of penetration of the points as these pierce the paper or rest on the surface of the scale.
3 Technique of measurement – for example the difference between spanning the whole distance by a single setting of the dividers compared with making the same measurement in a succession of smaller steps.

Environmental errors which comprise the changes resulting from variations in temperature, humidity and illumination at the time of measurement. Since the majority of cartometric measurements for scientific or administrative purposes are likely to be made in the laboratory or drawing office, rather than in the field, we do not imply comparison of measurements made at midday in the Sahara with those made in Antarctica during winter. We must consider under this heading the local but nevertheless significant changes in temperature and humidity caused by attempting to illuminate the work with a powerful reading lamp placed too close to the map and the instruments. In cartometry these causes are closely related to the:

Natural fluctuations of the quantity itself. It may seem that a straight line drawn on paper is a quantity which does not fluctuate like an electric current or the pressure of a gas may vary during a physical experiment. However, the dimensions of the paper may be changed by variations in temperature and humidity during measurement. If a sheet of paper is placed over a light table in order to facilitate illumination of it, the combination of heating and drying of the paper causes it to change in shape and size, and consequently the representation of the line changes, too.

Reduction Stage

Computational errors. We do not mean mistakes in the calculations, for we must assume that the observer has got the sums right. However, the errors which arise in rounding off, and from using an unsuitable conversion factor to transform from one system of units to another may give rise to significant discrepancies in the results.

Auxiliary data errors. How accurate is the map upon which the measurements have been made? Obviously errors in the positions of points or lines on the map will be transmitted to any measurements which are made between these points or along these lines. We investigate this subject in Chapter 9, where we will find that no single, simple statement that "this map is inaccurate" or "this map is accurate" can suffice to describe such a complex document.

Theoretical assumptions and resulting errors. A typical source of error of this type results from any neglect of the deformations which are inherent in the representation of the curved surface of the earth as a plane map. The study of this deformation is part of the theory of map projections. The main influence of this upon linear measurement is that the scale of the map varies continuously along the length of a line, and that the use of a constant scale factor, the representative fraction, may be a gross simplification of the calculations needed to convert from measured distance into ground distance. We return to this subject in Chapter 10.

Because of the progressive simplification in outline which accompanies any decrease in map scale, the length of a feature measured on a map at one scale differs from that measured at a different scale. We have already seen how important this may be from the examination of Penck's measurements of the coastline of Istria, the present writer's Yorkshire coast measurements and those of the coastline of the U.S.S.R. by Vorob'ev. Much of the analysis of linear measurements of natural objects depends upon the theoretical assumptions which are made about the relationship between measured length and map scale, and therefore upon the mathematical model used to determine the reduced length of a feature.

From this brief summary we can see that the causes of error in the measurement stage are more or less common to all kinds of distance measurement, and do not vary according to the nature of the line, the purpose of the measurement or the quality of the maps used. These are, in effect, the *technical errors of cartometry*. By contrast, those errors which arise in the reduction stage correspond broadly to *cartographic errors*, because they have varying significance according to the nature and purpose of the measurements and upon the scale and accuracy of the maps on which the work has been done.

We do not have to take all the causes of error into consideration if we believe some of them to be insignificantly small. For example, we may wholly ignore the influence of the projection upon which the map is based if we are only measuring short distances on large- or medium-scale maps. On the other hand the properties of the projection are extremely important if we wish to measure a long line on a very small-scale map or chart. Similarly there is no point in determining the reduced length of a short straight line by repeating the measurements on maps of different scales, to walk or drive along a road between two points, but this procedure is vitally important in determining a satisfactory value for the length of a major river or the complicated coastline of a country.

In order to treat systematically with the different types of error which may arise, it is desirable to define their nature in some detail, together with the statistical assumptions which have been made about them. Many of these theoretical principles have not been properly verified in their special application to measurements made on maps, but they have been exhaustively studied in cognate disciplines, such as surveying and engineering. We find that on the face of it there is not much difference between the errors arising in cartometry and in other branches of science. Therefore we may safely assume that the statistical theory of errors applies to measurements made upon pieces of paper.

A CLASSIFICATION OF ERRORS

Some of the causes of errors have been listed under seven different headings. The nature of the errors themselves are usually classified under three headings:

1 *Gross errors*, or *Mistakes*

2 *Cumulative errors* or *Systematic errors*
3 *Accidental errors* or *Random errors.*

Gross Errors

Gross errors arise from inattention or carelessness of the observer in handling instruments, reading scales or booking the results. For example, in using dividers, clumsy handling may cause alteration of the separation of the points as the instrument is transferred from the map to a scale. In reading a scale, gross errors usually arise because the observer is concentrating upon estimating and correctly recording the decimals of the unit of measurement but disregards the larger multiples of these units. Thus a gross error is often 10, 100 or even 1,000 times greater than the order of accuracy which can be obtained with the equipment in use. For example, if the length of a line which has been measured is 165·7 mm, the observer who is concentrating upon correct estimation of 0·7 mm might read the larger units of the scale incorrectly and record 265·7 mm, 175·7 mm or 164·7 mm. Another common mistake is the *transposition error* affecting the order of the numerals when these are written down or copied, such as 156·7 mm or 167·5 mm, this despite the fact that the scale had been read correctly. Each of these mistakes can be detected if the measurement is repeated, particularly if the second measurement is made quite independently in order to reduce the risk of making the same mistake twice. This might be done by using different parts of the same scale, or making one measurement in British Standard, the other in metric units, and then converting into the same units. Gross errors are nearly always detected if the measurements are made independently by two different observers.

Simple repetition of the measurement is usually sufficient to demonstrate the presence of a gross error so that an incorrect observation may be discarded from the data. Occasionally one of them succeeds in slipping past undetected to influence the later analysis. We comment later upon a check which can be made at that stage. However, it is fundamental to the statistical treatment of errors of measurement that gross errors should have been eliminated before proceeding to further analysis of the data. If an undetected gross error is only recognised at a later stage it is necessary to discard that measurement and start again at the beginning of any statistical calculations which included it. An example is given on page 100.

Cumulative Errors

These are often subdivided into two categories: *constant errors* and *systematic errors*. Constant errors are considered to be those arising from some defect which will remain constant during the lifetime of an instrument. Faulty graduation of a scale is commonly cited as an example of this error. Systematic errors arise at the time of measurement and only affect the measurement process temporarily. The small discrepancy in setting a pair of dividers from the intended positions of the points gives rise to a systematic error which will be present only for as long as the setting of the dividers remains unchanged. Usually it is difficult to separate these two components, for their influence is the same. The actual error may be small; indeed it might even be regarded as negligible, but if the measurement process is repeated many times in the same fashion, the generation of the same error many times will allow it to grow until it cannot be disregarded. The classic example of cumulative error, described in nearly every textbook on surveying, is that propagated in the process of measuring a straight line distance on the ground using a tape which is longer or shorter than its supposed length. If the constant or systematic error in the length of the tape is $+e$ units and the distance measured comprises n tape lengths, the cumulative error in the measured distance is $+n \cdot e$ units.

The comparable example in cartometry is that resulting from measuring the length of a line by stepping-off equidistant short steps using spring-bow dividers. For example, a line of length 165·7 mm might be plotted by making ten steps of length 16·0 mm with a final measurement of 5·7 mm. If, however, the separation of the dividers is really 16·05 mm, and even if no other errors are introduced to the construction, the effect of the systematic errors upon the construction of this line is to plot the line of length $160{\cdot}5 + 5{\cdot}7 = 166{\cdot}2$ mm. Alternatively, if the observer is trying to find the distance between two points by this method, the measured length will comprise ten steps of length 16·05 mm each, plus a final measurement of 6·2 mm, giving the total of 165·7 mm. However, the observer thinks that the separation of the points is 16·0 mm, so that the measured distance is recorded as $160{\cdot}0 + 5{\cdot}2 = 165{\cdot}2$ mm. Although the systematic error amounts to only $+0{\cdot}05$ mm, which cannot be detected by naked eye, it has grown to $+0{\cdot}5$ mm by virtue of the technique used. This kind of error can be avoided by adhering to the principle of *working from the whole to the part*. Thus in plotting a map, or measuring straight line distances from it, the error can be avoided by measuring each line as a complete unit. See I.C.A. (1984) for a graphic description of this technique. For measurement of irregular lines by dividers, the technique of stepping-off cannot be avoided. Since this method violates the principles of working from the whole to the part, it is most important to reduce the influence of cumulative errors in some other way. This may be done by careful calibration of the dividers. These are set to the intended separation and then walked along a straight line through a sufficiently large number of steps for any cumulative error to become appreciable. The length of this line is now measured independently, preferably by laying a steel scale along it. Division of this length by the number of steps taken with dividers provides a more reliable measure of the separation of the dividers than can be made by visual inspection of the dividers against a scale.

Scale Resolution and the Personal Equation

Suppose that a steel scale is subdivided in millimetres, and that the observer can reliably estimate to the nearest one-fifth of the space between adjacent divisions. The scale is being read by estimation to the nearest 0·2 mm and we call this the *resolution* of the scale. It follows that each estimated fraction must serve for all readings which lie within 0·1 mm of the actual scale reading. For example, if the length of the line is 165·7 mm, a series of measurements recorded by the name person should theoretically by subdivided between the estimate of 165·6 mm and 165·8 mm. Assuming that no other errors are present to modify such estimates, and that 10 such measurements of the line have been made, five readings would be estimated as being 165·6 mm and five more to be 165·8 mm. Shown diagrammatically, as in Fig. 6.1, the frequency of the scale readings would have a *rectangular distribution.*

The reader may be surprised to find that the author has suggested reliable estimation to 0·2 mm, whereas most people believe that this should be done to the nearest 0·1 mm; if only because the result is written as a decimal. However, the author questions the justification for the assumption because he knows that he is unable to estimate decimals consistently and he suspects that others cannot, but are unaware of their inability.

Table 6.1 records the results of more than 1,200 individual estimations of decimals carried out by the authors some 20 years ago, using a scale having graduations at 1 mm intervals. The table also gives the results of some measurements made by Yule (1927). To facilitate comparison, the values are all recorded as percentage frequencies. Both sets of data are *continuous*, in the sense that any intermediate scale reading is theoretically possible. Moreover, each scale reading has been made independently of those which precede and follow it. Because the two samples are large, it is reasonable to suppose that all

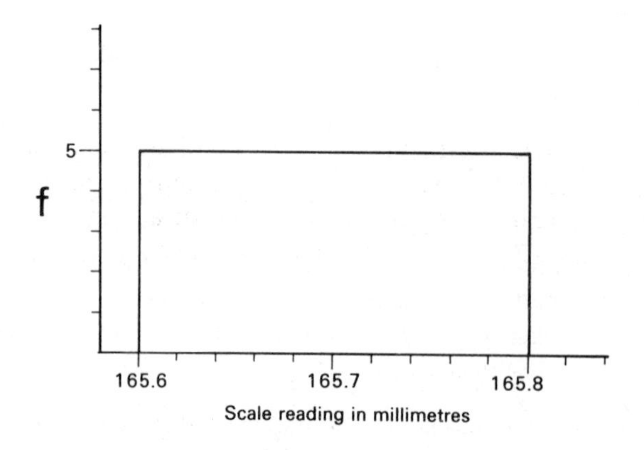

FIG. 6.1. Rectangular distribution.

TABLE 6.1. *Comparison of the personal equations of two observers in their ability to estimate tenths (percentage frequencies)*

Digit	Yule	Maling	Digit	Yule	Maling
·1	16	7	·5	7	4
·1	10	9	·6	9	7
·2	13	23	·7	6	9
·3	7	9	·8	13	17
·4	8	4	·9	13	10

decimal values would have equal chances of being recorded. If this be true we should expect that every decimal digit ought to occur with equal frequency, giving a rectangular distribution in which the percentage frequency in each class is 10. The table clearly indicates the reluctance of the author to accept that any measurement coincides exactly with an engraved subdivision of the scale, or that it may lie exactly half way between the two marks. The very high frequencies for ·2 and ·8 are painfully obvious, and Yule dryly commented "I seem to have a special dislike for 7".

Does knowledge about an observer's idiocyncrasies in estimating tenths influence subsequent readings? Turnbull and Ellis (1952) believed that the personal equation remained constant, even after an observer had been informed about these preferences. The present author believes that the observer who knows about them will tend to correct and possibly overcompensate for them in later work, so that the personal equation changes. Perhaps, therefore, it is better to live in blissful ignorance of one's peculiarities or defects, but the obvious moral to be drawn here is that some people cannot do better than make mental subdivision of a scale into quarters or fifths of the scale division. In such cases it is rather a waste of time to record all estimates to the nearest decimal.

The Random Event

The idea that each decimal digit ought to have had an equal chance of being recorded introduces us to the concept of randomness which is central to statistical theory and to the methods of statistical sampling which are discussed in Chapter 7 and 8. It is therefore necessary to define what we mean by the word random:

> *A process of selection applied to a set of objects is said to be random if it gives to each object an equal change of being chosen.*

In the special case of measurements this implies that each measurement which is made is wholly independent of those which have already been made, as well as those made subsequently. In order to satisfy the condition of independence it seems sufficient to make the deliberate effort to measure an object in a different way. For example, dividers might be closed and reopened between measurements, or a scale is shifted laterally through an unspecified amount so that a different part of it is read at each measurement. An example of how such procedural variations may be employed in actual measurement is given on pages 107–108.

Accidental Errors

The rectangular error which is primarily caused by scale resolution, though modified by the observer's personal equation, may be further augmented by small errors arising from other causes. These include uncertainty in the recovery of the exact positions of the terminal points of a line (especially when using dividers after these points have become enlarged through repeated application of the dividers), failure to hold the dividers upright, variation in the penetration of the dividers' points, variations in illumination, etc. These errors have the effect of modifying the rectangular distribution, both by extending the range of acceptable measurements and accumulating many errors which are smaller than the resolution of the scale. Experience has shown that for a prolonged series of measurements of the same object, made under apparently similar conditions, these errors have the following properties:

1 Small errors are more frequent than large errors.
2 Positive and negative errors are equally frequent.
3 Very large errors do not occur.

They are known as *accidental errors* or *random errors.* The frequency distribution of these errors is characterised by the symmetrical bell-shaped curve illustrated in Fig. 6.2. This is known as the *normal curve* and it depicts the *normal distribution* of statistical theory. This curve indicates that zero errors are the most frequent, and that the frequency with which large errors occur decreases rapidly in conformity with property (3) above and symmetrically in conformity with property (2). However it should be appreciated that the distribution is unbounded so that it extends to $-\infty$ on the left-hand side, and

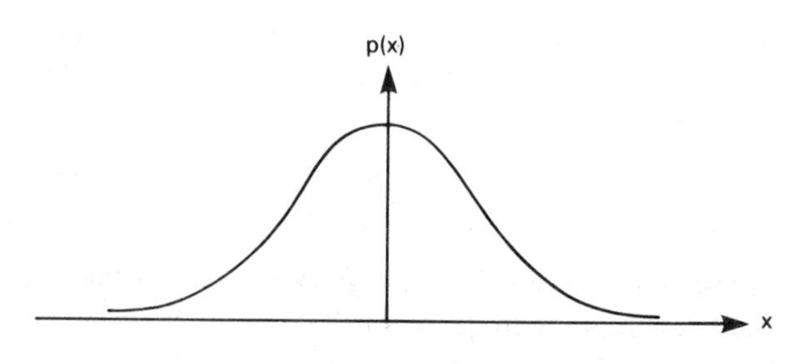

FIG. 6.2. Normal distribution.

to $+\infty$ on the right-hand side, of the curve. Theoretically, therefore, there is always the very small probability of a large error occurring by chance. It follows that the normal distribution cannot provide unequivocal proof of the occurrence of an event which gives rise to a specified error, but only a measure of probability concerning the likelihood of the occurrence. It is a matter of personal judgement how rare the occurrence of an event may be for this to be deemed insignificant. It follows that the whole basis of the analysis of accidental errors (which should, of course, be the only errors requiring analysis) is *statistical inference* which, in turn, is based upon probability.

The "normal curve" illustrated in Fig. 6.2 should strictly be called the *normal probability density curve* because this represents the total probability of the occurrence of the continuous random variable, x. Since probability is reckoned on a decimal scale from 0 (zero probability) to 1 (absolute certainty), the total area of the distribution, i.e. the area beneath the curve, is equal to 1.

THE NORMAL DISTRIBUTION

Two parameters are required to define this distribution. These are the arithmetic mean and the standard deviation. For a succession of values of the variable

$$x_i = x_1, x_2, \ldots, x_n \tag{6.3}$$

we obtain the mean from

$$\bar{x} = 1/n \cdot \sum_{i=1}^{i=n} = \sum x/n \tag{6.4}$$

In these equations the symbol $\sum$ denotes summation of the variable within the indicated limits of x_i from $x_i = 1$ to $x_i = n$. It is usual to omit these limits from the expression where they are self-evident, as in the right-hand expression of (6.4).

In such a series of observations, each value of x_i will differ from the arithmetic mean. The difference $x_i - \bar{x}$ is called the *residual error* or *residual*, denoted by r_i. Thus for a series of measurements

$$\begin{aligned} x_1 - \bar{x} &= r_1 \\ x_2 - \bar{x} &= r_2 \\ x_3 - \bar{x} &= r_3 \\ & \\ x_n - \bar{x} &= r_n \end{aligned} \tag{6.5}$$

If the distribution is symmetrical about the arithmetic mean, positive and negative residuals are equally frequent so that

$$\sum r_i = 0 \tag{6.6}$$

If this be true, the arithmetic mean may be regarded as being the most probable value of the variable which has been measured. It can be shown, for example by Greenwalt and Schultz (1962) as well as in many textbooks on statistics, that if any value of the variate other than the arithmetic mean is selected as the value from which a set of residuals r'_i are determined, the sum of the squares of these residuals, $\sum_i'^2$ is greater than $\sum r_i^2$, thereby demonstrating the special property of the arithmetic mean. We emphasise that the term "most probable value" relates only to those measurements which have been made and used to calculate the mean. Everyday usage suggests that if we wish to find the most probable value of the variate x it is simply necessary to make a succession of observations of x and calculate the mean. However this procedure merely identifies one value in what we describe later as the *sampling distribution.* Nevertheless, knowledge about the next parameter enables us to gauge the probability whether the true value of x lies within a certain range of the arithmetic mean.

The determination of the standard deviation from the data summarised in the equations (6.5) follows from the need to eliminate the distinction between positive and negative residuals. This is most conveniently accomplished by squaring them. If, therefore, we determine

$$\sigma^2 = \sum_{i=1}^{i=n} r_1^2/n \tag{6.7}$$

We have obtained the mean value of the squared residuals. This is known as the *variance.* The square root of the variance is known as the standard deviation.

In the special case of studying the errors of measurement, a succession of independent measurements x_i, having mean $\bar{x}$, have residuals $(x_i - \bar{x})$ representing errors measured from $\bar{x}$. Therefore the parameter corresponding

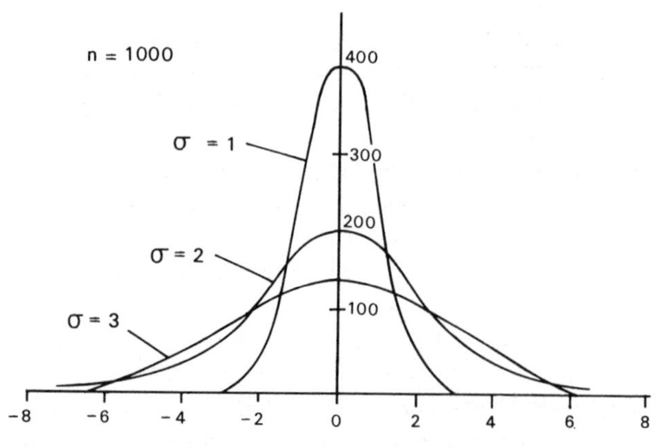

FIG. 6.3. Three normal curves representing the same size of population ($n = 1{,}000$) and mean 0, but having standard deviations $\sigma_1 = 1$, $\sigma_2 = 2$ and $\sigma_3 = 3$ units respectively.

to the standard deviation is often called the *root mean square error* (r.m.s.e.). Sometimes this is shortened to the *mean square error* (m.s.e) although this is clearly a more appropriate description of the variance than the standard deviation. Some care is needed to interpret usage of these terms by some writers, but usually the term "mean square error" is to be regarded as synonymous with standard deviation.

A small numerical value for the standard deviation indicates that the measured variables cluster more closely round the mean than they do for a distribution comprising the same number of observations but having a larger standard deviation. This is illustrated in Fig. 6.3 in which the three distributions each have the same mean, but σ equal to 1, 2 and 3 units respectively, are superimposed upon one another.

The Distinction Between Accuracy and Precision

It is now possible to make a distinction between the two words "accuracy" and "precision" which are much used in this and subsequent chapters. The *accuracy* of a measurement process describes the degree of *conformity to the truth* of measurements generated by repeated applications of the process under fixed conditions. The *precision* of a measurement process only describes the degree of conformity of the measurements *among themselves*, and hence to the degree of conformity to the average irrespective of whether the average value is or is not the "true" value.

We may therefore represent four contrasting examples of accuracy and precision by means of the four parts of Fig. 6.4. In each of these it is assumed that the true value of the variate, T, is known. The two left-hand diagrams (Fig. 6.4a and c) indicate that the mean of the measurements, $\bar{x}$, coincides with T. It follows that in each of these examples of measurements may be considered to be accurate. On the pair of right-hand diagrams (Fig. 6.4b and d) the two curves show *bias*, or lateral displacement, so that x no longer coincides with T. These represent inaccurate measurements.

The pair of curves in the upper part of the diagram (Fig. 6.4a and b) are narrower than those in the lower pair. This, as we have just seen, corresponds to the difference between a small numerical value and a larger value for the standard deviation. From Fig. 6.3 we infer that the upper pair of curves represent precise data, having comparatively small spread either side of the mean and therefore smaller numerical values for the standard deviation (or root mean square error). By contrast, the lower diagrams (Fig. 6.4c and d) illustrate less precise data, having wider spread corresponding to larger numerical values for σ.

Since we can never know the value of T, the *first desirable aim in making repeated measurements is to produce a precise set of data.* This is achieved by careful handling of the instruments and reading scales so that the accidental errors are kept as small as possible and, we hope, will produce results approximating to the upper pair of diagrams of Fig. 6.4. In order to approach

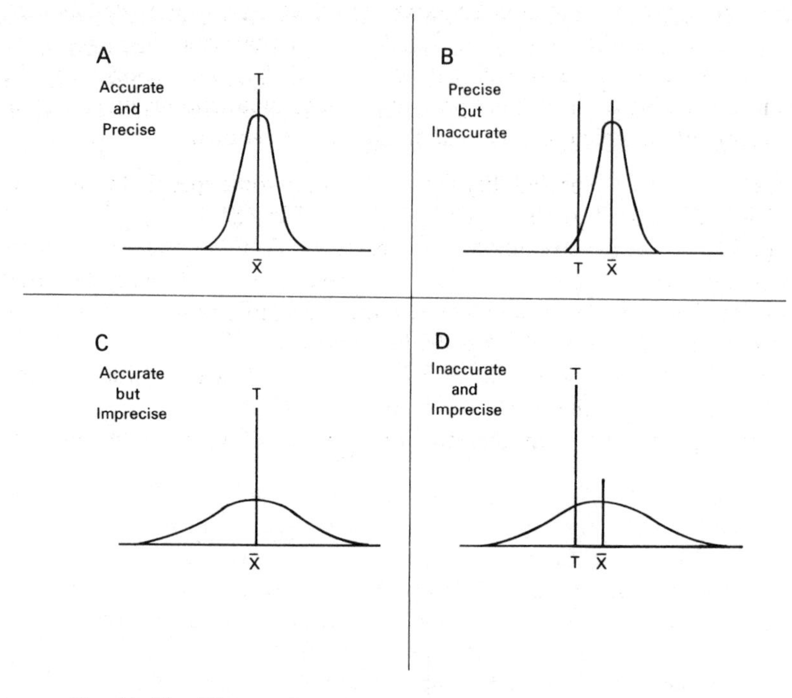

FIG. 6.4. The differences between accuracy and precision shown graphically.

the ideal result, which is, of course, illustrated by Fig. 6.4a, it is necessary to eliminate any causes of bias in the measurements. *The bias clearly represents undetected systematic error* which may be reduced by careful standardisation or calibration of the instrument and by using measurement techniques which do not encourage the accumulation of systematic errors.

The Normal Distribution and Probability

The equation defining the *normal probability density function*, $p(x)$ may be written in the form

$$p(x) = \frac{1}{\sigma \cdot \sqrt{2\pi}} \cdot e^{(-r_i^2/2\sigma^2)} \tag{6.8}$$

In this equation, x_i is any randomly selected measurement of the unknown quantity x, whose mean is $\bar{x}$ and standard deviation is σ. The terms $\sqrt{2\pi}$ and e are constants.

$$P(x) = \int_{-\infty}^{x} p(x) \cdot \mathrm{d}x \tag{6.9}$$

$$= \int_{-\infty}^{x} \frac{1}{\sigma \cdot \sqrt{2\pi}} \cdot e^{-(r^2/2\sigma^2)} \cdot \mathrm{d}x \tag{6.10}$$

If we integrate this function between the limits $-\infty$ and x we have the *normal probability distribution function* $P(x)$, As x approaches its upper limit, $P(x)$ approaches 1; as x approaches its lower limit, $P(x)$ approaches zero. The graphical meaning of the normal probability density function is illustrated in Fig. 6.5(a)–(d), in which four typical conditions are shown:

1 $p(x < +X)$, i.e. the probability that x is less than some specified value $+X$. This is illustrated by the shaded portion of Fig. 6.5(a).
2 $p(x > +X)$, i.e. the probability that x is greater than the value $+X$. This is equal to $1 - p(x < +X)$ because we have already defined the total probability, or the total area beneath the curve, to be equal to unity. The condition is illustrated by the shaded part of Fig. 6.5(b).
3 $p(x < -X)$, i.e. the probability is greater than the specified value $-X$ and is represented by the shaded portion of Fig. 6.5(c).
4 By integrating between the two limits X_1 and X_2 we may find the

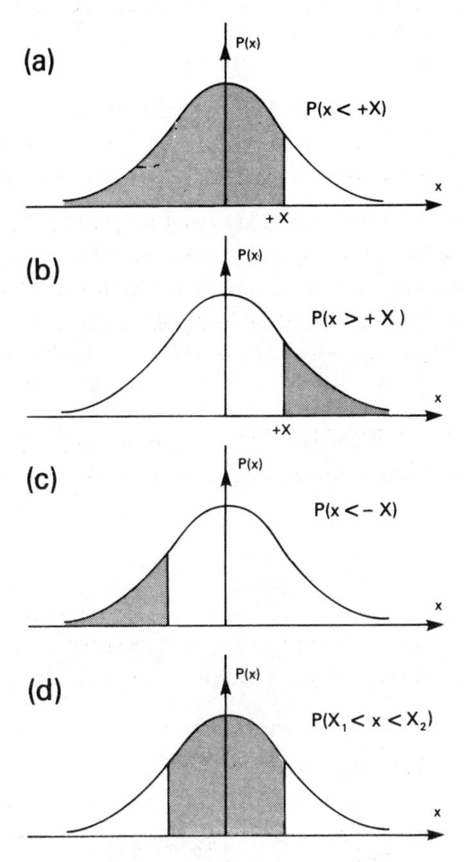

FIG. 6.5. Normal probability distribution functions.

probability that x lies between X_1 and X_2, i.e.:

$$p(X_1 < x < X_2) = p(x < X_2) - p(x < X_1)$$

is illustrated by the shaded part of Fig. 6.5(d).

The last of these conditions is of particular importance if we specify that X_1 and X_2 coincide with specific multiples of the standard deviation, i.e. $X_1 = -a \cdot \sigma$ and $X_2 = +a \cdot \sigma$.

Where $a = 1$, the solution of the integral is $P(x) = 0.6827$; for $a = 2$, it is 0·9544 and for $a = 3$, it is 0·9973. Since these decimals also correspond to the areas enclosed under the curve, these results may be rephrased in the form

68·27% of all the measurements will lie between $+\sigma$ and $-\sigma$
96·45% of all the measurements will lie between $+2\sigma$ and -2σ
99·75% of all the measurements will lie between $+3\sigma$ and -3σ

They have the graphic meaning illustrated in Fig. 6.6.

Converting these ideas into a practical example, we may draw the following inferences from a particular set of measurements. Suppose that we have measured the same line many times and have determined the mean length to be 166·69 mm with a root mean square error $\pm 0{\cdot}09$ mm. We may infer that

68·27% of the measurements will lie between 165·60 mm and 165·8 mm,
96·45% of the measurements will lie between 165·51 mm and 165·87 mm,
99·73% of the measurements will lie between 165·42 mm and 165·96 mm.

We have already seen that the normal curve is asymptotic to the abscissa and can be extended indefinitely to $\pm\infty$. This indicates that it is unrealistic to talk about 100% probability, or absolute certainty, in the field of measurement. For most practical purposes the limits of the distribution may be regarded as lying somewhere between $\pm 2{\cdot}5\sigma$ and $\pm 4{\cdot}0\sigma$. The actual limits which may be chosen depend upon convention, or upon the judgement of the particular individual, whether one error in 100 exceeds the acceptable odds or that one is only willing to accept the chance that one measurement in 10,000 is significant. In many aspects of inferential statistics, an acceptable risk is often taken to

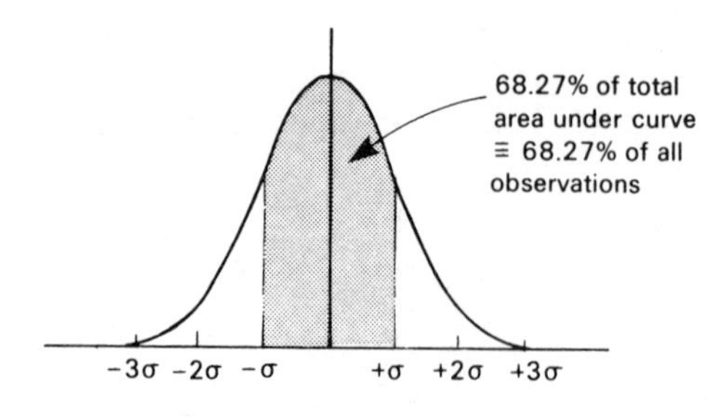

FIG. 6.6. The geometrical meaning of $\pm\sigma$.

TABLE 6.2. *The normal distribution. Percentage of the observations lying within the range $+a\sigma$ to $-a\sigma$; of the probability, $q(=1-p)$ or $1/k$ that an observation may exceed these limits by chance*

a	Percentage of observations	$1-p$	$1/k$
0·6745	50·0	0·5	1/2
1·0000	68·27	0·333	1/3
1·5	86·64	0·143	1/7
1·6449	90·0	0·1	(or 10%) 1/10
1·96	95·0	0·05	(or 5%) 1/20
2·0	95·45	0·045	1/22
2·5	98·76	0·02	(or 2%) 1/50
2·576	99·0	0·01	(or 1%) 1/100
2·807	99·5	0·005	(or 0·05%) 1/200
3·0	99·73	0·003	1/333
3·29	99·9	0·001	(or 0·1%) 1/1000
3·5	99·95	0·0002	1/2000
4·0	99·99	0·0001	1/10000

correspond to a probability of 0·05 or 0·01. In other words we are willing to accept the likelihood that a value will lie between specified limits on 95% or 99% of all occasions, or that these limits are only likely to be exceeded, by chance, on 5% or 1% of occasions.

Table 6.2 indicates the probabilities to be associated with different multiples of the standard deviation within the normal distribution. In addition to tabulating probability as a decimal, we give the percentage and reciprocal values for easy comparison with the variations in usage which are commonly quoted.

Rejection of Gross Errors by Chauvenet's Criterion

An immediate application of these conclusions is to use them to eliminate any gross errors which have escaped detection, and which have therefore been used to calculate the mean and standard deviation of a group of measurements. If any of the measurements differs by more than $\pm 3\sigma$ there is, from Table 6.2, a probability of only 0·003 (or 0·3%) that this has arisen by chance. In the example quoted above this would be exemplified by any measurement which is less than 165·4 mm or greater than 166·0 mm. Since the probability that this has occurred by chance is so small, it is reasonable to suppose that a gross error has been made, and that the measurement should be rejected. Having done this, it is necessary to calculate the mean and standard deviation again, because the presence of the "wild" measurement affected the values of the parameters calculated in the first instance. However, simple use of Table 6.2 is based upon the assumption that the number of measurements made plays no part in evaluation of the criterion for rejection. Since the total number of

TABLE 6.3. *Chauvenet's criterion for rejecting a measurement*

Number observations (n)	Ratio of maximum acceptable deviation to standard deviation (d_{max}/σ)	Number of observations (n)	Ratio of maximum acceptable deviation to standard deviation (d_{max}/σ)
2	1·15	10	1·96
3	1·38	15	2·13
4	1·54	25	2·33
5	1·65	50	2·57
7	1·80	100	2·81

individual measurements made of a variable, the *size of the sample*, to be described in Chapter 7, is usually quite small, a more restrictive test should be applied. This is known as *Chauvenet's criterion.* It specifies that a reading should be rejected if the probability of obtaining the particular deviation from the mean is less than $1/(2n)$ in a collection of n measurements. Table 6.3 lists values of the ratio of the maximum deviation, d_{max} to the standard deviation, using all the data to determine σ. In applying Chauvenet's criterion to eliminate doubtful observations the first stage is to calculate the mean and standard deviation using all the data. The deviations of individual observations are the compared with the calculated standard deviation. If any of these ratios exceed those given in Table 6.3 for a given size of sample, the observation is rejected.

An example of the use of the method may be taken from the measurements of the length of the Yorkshire coast made by one of the operators using a chartwheel on the map of scale 1/1,000,000. This was described in Table 5.3, p. 69. Eight separate measurements had been made by this operator, who obtained the mean length 169·75 ± 4·66 mm. The largest deviation from this mean was one measurement of 180·0 mm. Does this represent a gross error, or is it within the acceptable range for chartwheel measurements?

The difference from the mean d_{max} is 180·0 − 169·75 = 10·25 mm. Therefore 10·25/4·66 = 2·20. From Table 6.3 the limiting ratio lies between 1·80 and 1·96 for a sample of eight measurements. Since the calculated ratio is greater than these values, this observation ought to be rejected. It follows that the revised value for the mean of the chartwheel measurements given in Table 5.3 is 168·3 ± 2·8 mm, based upon seven measurements.

Combination of Mean Square Errors

The results of certain measurement processes may contain errors which are attributable to a variety of different causes or different stages in the work. If these can be determined separately it becomes necessary to combine the various values for the root mean square error on order the derive and overall estimate of precision.

The general expression, which is derived in many of the standard textbooks on statistics, describes the relationship between a dependent variable, y, which is a function of several independent variables $x_1, x_2, \ldots, x_n$. This may be written

$$y = f\{x_1, x_2, \ldots, x_n\} \tag{6.11}$$

Then the root mean square error of y, or σ_y, may be expressed in the form

$$\sigma_y = \pm\left\{\left(\frac{\partial f}{\partial x_1}\cdot\sigma_{x_1}^2\right)^2 + \left(\frac{\partial f}{\partial x_2}\cdot\sigma_{x_2}^2\right)^2 + \left(\frac{\partial f}{\partial x_n}\cdot\sigma_{x_n}^2\right)^2\right\}^{1/2} \tag{6.12}$$

The general equation simplifies considerably for many of the common functions which arise in the study of the errors of measurement and map accuracy. We give some of the commonest of these.

(A) Multiplication by a Constant

This is the case where

$$y = k\cdot x \tag{6.13}$$

k being a constant. Then

$$\sigma_y = k\cdot\sigma_x \tag{6.14}$$

A typical example of this occurs where k is the scale factor relating a measurement made in millimetres on the map (x) to the distance in metres or kilometres on the ground (y). Indeed Barford (1967) uses this typical cartometric problem to derive equation (6.14) from first principles.

(B) The Root Mean Square Error of a Sum or Difference

If

$$y = x \pm z \tag{6.15}$$

$$\sigma_y = \pm\sqrt{(\sigma_x^2 + \sigma_z^2)} \tag{6.16}$$

This may be extended to include as many independent variables as occur in the function. It is demonstrated by the following example.

In Chapter 17, reference is made to the problem of measurement of a large parcel of land shown on a map which is too large to measure by polar planimeter as a single unit using the "pole-inside" mode of operation. Consequently the parcel has to be measured in parts, each of which gives rise to a different root mean square error. It is required to determine the root mean square error for the whole parcel. The results of the partial measurements are given in Table 6.4.

From equation (6.16) the root mean square error for the whole parcel is

TABLE 6.4. *Area and root mean square error (in square centimetres) of a large parcel which has been measured by planimeter in four separate parts*

Subdivision	Area	Root mean square error
1	32·17	±0·153
2	37·47	0·162
3	38·25	0·163
4	30·18	0·150
Sum	138·07	0·628

determined from

$$\sigma_y = \sqrt{(0{\cdot}153^2 + 0{\cdot}162^2 + 0{\cdot}163^2 + 0{\cdot}150^2)}$$
$$= 0{\cdot}314\,\text{cm}^2$$

whereas the simple sum of the four values is twice as large.

(C) The Root Means Square Error of a Sum or Difference Where These are Equal

This is a special case of (6.16) where $\sigma_y = \sigma_z = \sigma$. Then, for N separate independent variables,

$$\sigma_y = \sqrt{N}\cdot\sigma \qquad (6.17)$$

This particular calculation appears repeatedly in cartometry. For example, in Chapter 7, the data given in Table 7.1 represent the results of measuring the same line 100 times using a plastic scale and reading the scale at each terminal point. In this we have calculated the root mean square error of the length of the line to be $\sigma_y = \pm 0{\cdot}008$ mm. Since each measured length comprises two scale readings, what is the root mean square error of finding the length of a line by scale? From (6.17)

$$0{\cdot}088 = \sqrt{2}\cdot\sigma_1$$

so that the root mean square error of the determination of length is

$$\sigma_1 = 0{\cdot}062\,\text{mm}$$

In the example in Table 6.4 the root mean square errors are all about $\pm 0{\cdot}16$. If we accept this as a representative value for all the component parcels, then $\sigma_y = \sqrt{4} \times 0{\cdot}16 = \pm 0{\cdot}32$ compared with $\pm 0{\cdot}31$.

The Statistics of Discrete Data

Thus far the statistical treatment of errors of measurement has been based upon the assumption that the measurement process gives rise to continuous data; that is to say, a measurement may assume any value between two others. In measuring distance, for example, the resolution of the scale allows recognition of measurements which differ from one another by intervals of 0·1 or 0·2 mm, depending upon the ability of the observer to estimate correctly between the scale divisions. Notwithstanding the limiting resolution of the scale, the measured distance might have an intermediate value verifiable by a more refined reading process.

There is a different category of variable which only advances in discrete steps. For example, human or animal populations are discrete, because we cannot have one-half of a person or one-quarter of a cow. In evaluating land-use maps at particular points, such as the point of intersection of two grid lines, the observed category of land use may be grassland or it may be woodland. It should not be recorded as being one-half grassland and one-half woodland unless the mapping specification has recognised this particular category of mixed land use, that it has been interpreted in this light during the mapping process and the map shows such mixed categories. Such variables are called *discrete variables*, and they possess special statistical properties. In cartometric work such distributions arise when thematic maps are used to sample the attributes of such data as land use, soil types and vegetation types, using point-counting methods and linear sampling, which are also used to determine their areas, using the techniques described in Chapters 19 and 20. These data frequently appear in the form of ratios or percentages.

The Binomial Distribution

If p is the probability that an event will happen in any single trial and $q(=1-1)$ is the probability that it will not happen in any single trial, then the probability that the event will happen exactly X times in N trials is given by

$$p(X) = {}_NC_X \cdot p^X \cdot q^{N-X} \tag{6.18}$$

$$= \frac{N!}{X!(N-X)!} \cdot p^X \cdot q^{N-X} \tag{6.19}$$

where $X = 0, 1, 2, \ldots, N$ and $N! = N(N-1)(N-2)\ldots 1$.

The discrete probability distribution (6.18) is known as the *binomial distribution*, since for $X = 0, 1, 2, \ldots, N$, it corresponds to successive terms in the binomial expression

$$(q+p)^N = q^N + {}_NC_1 \cdot q^{N-1} p + {}_NC_2^2 + \cdots + p^N \tag{6.20}$$

where ${}_NC_1$, ${}_NC_2 \cdots$ are *binomial coefficients*.

The binomial distribution has been recognised since the seventeenth century because of its relevance to gambling and, in particular, to tossing coins and such games as that known as Buffon's needle. Hence in the common usage of the terminology of probability, p is often categorised as a "success" whereas q is a "failure".

To indicate its application in the indirect methods of area measurement we imagine that a land-use map has been examined and a collection of identifiable points upon it have been used to extract information about a category of land use. The result of such an investigation is a series of tabulated frequencies, one column for each of the categories of land use shown on the map. If we wish to examine the category labelled as woodland, and, incidentally, measure the areas of woodland on the map (p. 141), we might assign to those points representing woodland on the map the value p. It follows that q represents all the points which have been counted which do not represent woodland, i.e. all the other land-use categories added together. It follows that because every point examined on the map must be either p or q,

$$p + q = 1 \tag{6.21}$$

or the number of successes (in finding woodland) plus the number of failures (or not finding woodland) represents all the points which have been counted.

It can be shown that for the binomial distribution the following parameters apply:

Mean: $$\mu = \Sigma X \cdot p(X) = Np \tag{6.22}$$

Variance: $$\sigma^2 = \Sigma(x - \mu)^2 \cdot p(X) = Npq \tag{6.23}$$

Standard deviation: $$\sigma = \sqrt{(Npq)} \tag{6.24}$$

For proportions we have

Mean: $$\mu_p = p \tag{6.25}$$

Standard deviation: $$\sigma_p = \sqrt{(pq/N)} = \sqrt{[p(1 - p)/N]} \tag{6.26}$$

7

Statistical Sampling and Cartometry

If we have a hunch that most of the cats in Cumbria are black, it is no good trying to test the idea by deliberately concentrating the sampling in those areas where we have reason to suppose black cats are especially cherished by maiden aunts.

(M. W. Holdgate, *A Perspective of Environmental Pollution*, 1979)

In statistical usage the word *population* is applied to any collection of individuals or the values of a variable. The population has two essential characteristics. First, the individuals forming a population are of the same kind. They may be people or they may be sheep, but they are not a mixture of people and sheep unless they belong to a much greater population, such as that of all mammals. Secondly, the individuals of a population differ in respect of one typical feature or attribute, known as the *variate*. Two kinds of population may be recognised:

1 *Finite populations*, as the name suggests, are composed of a finite number of objects, units or individuals, such as people or sheep. A complete census enumeration of these is possible, although it may be too time-consuming or expensive to execute without the backing provided by a government and its administrative services.
2 *Infinite populations* are, obviously, infinitely large. Most physical measurements belong to this category. For example the population of the measured lengths of a line is the infinity of separate measurements which would result if the measurement process could be and were repeated indefinitely under similar conditions.

In order to describe the characteristics of a group of objects, whether these belong to finite or infinite populations, it is generally either impracticable or impossible to select or measure every item. Instead of attempting to examine the entire population, only a *sample* of them is studied. A sample may therefore be defined as:

Part of a population, or a subset from a set of units, usually obtained by deliberate selection, which is intended to represent the whole population.

The emphasis in this definition lies in the word represent.

All sampling models assume at the outset that entities exist whose proportions or sizes can be represented with varying degrees of confidence by the subset drawn from the parent population. The parameters corresponding to the mean and standard deviation calculated from a sample of measurements of the length of a line may be regarded as being *estimates* of the mean and standard deviation of the parent population. Perfect agreement between the sample estimates and the true population values is unlikely because the estimates may be presumed to contain errors. The absolute magnitude of these errors cannot be established if the true mean and standard deviation of the parent population cannot be known, as in the study of the errors of measurement. Therefore it is only possible to estimate the magnitude of them in terms of probability, according to the characteristics of the *sampling distribution.* This is the distribution which would be created by drawing a large number of separate samples for the same population and treating the mean of each sample as if it were an individual observation of measurement. Whether or not a sample provides results which are representative of the population depends upon keeping the sample unbiased and the sampling errors small. If bias is to be avoided, the selection of samples must be determined by some process which is not influenced by the qualities of the objects studied and which is free from any element of choice on the part of the observer. An efficient sample has a relatively small spread of the errors of estimation, and therefore it corresponds to the ealier description of a precise set of measurements. The prime requirement of any sampling model is accuracy which, as was demonstrated in Chapter 5, represents absence of bias. The two aims ought to be achieved through the use of a suitable *sampling strategy* which involves the choice of suitable sampling methods and sample size.

THE PROPERTIES OF A SAMPLE

A statistically valid sample has two important properties, irrespective of the nature of the population or the design of the sampling model.

1 Every individual population should have a known chance of being selected. This condition can only be fulfilled if the population has been divided into equal units, such as areas of certain size, which can serve as the sampling units. These units should cover the whole of the population but should not overlap, in the sense that every element of the population belongs to one and only one unit.
2 The method of selection must be consistent with the first condition – that the chances of being selected are equal. This can *only* be rigorously satisfied by some method of random selection of the sampling units.

The formulae which provide the estimates of the population parameters and express sampling precision are based upon these premises. If random selection is not satisfied, the estimates may no longer be considered reliable because it is no longer possible to assign definite values or limits of the probability that an event may have occurred by chance. We shall see that problems of this sort arise in the use of some kinds of *systematic sampling* in which the results are treated as if the units had been selected at random.

In addition to the aims that a sample should be unbiased and should yield an estimate of its own error, there is an additional requirement that a sample should be *representative*; that it should have nearly the same characteristics as the parent population. This is not inconsistent with the need for random choice, but it often means, especially in spatial sampling, that purely random sampling is less satisfactory than some other form, such as *stratified random sampling*. We return to this subject for detailed consideration later.

An Experiment of Sampling in Measurement

The theoretical principles which have been outlined above apply to sampling of any kind of variable. Those which apply to samples of measurements and the treatment of their errors represent only a subset of statistical sampling theory, but since these are of particular importance in the context of this book it is helpful to explain much of the theory in terms of errors of measurement using a simple experiment which is easy to reproduce.

A straight line is drawn on a sheet of draughting film and the two terminal points, *A* and *B*, are chosen to define the distance to be measured. The length of the line should be approximately one-half of the length of the scale which is to be used for the experiment so that an appreciable part of the reading edge of the scale may be used. For example, if the scale is a ruler which is about 310 mm in length (the typical foot ruler in British Standard measure), the line ought to be about 150–170 mm in length. The terminal points are marked by pricking the plastic with a fine needle, and the position of each point is identified by a pencil ring drawn round it. The observer now places a scale along the line, deliberately *not* setting zero on the scale against either point. The scale is read at both points so that the measured distance is always the difference between two scale readings. The procedure is repeated many times, using a different part of the scale to make each measurement. The order of reading should be changed. For example, the first few measurements might be made by reading the scale first at the point *A* followed by a reading at *B*; the next few measurements by reading *B* first, and so on. One sample may be measured by placing the scale along one side of the scale, the next by placing the scale on the other side of it so that the direction of the ruler is periodically reversed. The measurements made of the same line by two or more observers may also be combined.

Strictly speaking any deliberate variation in method should be subject to *randomisation* so that the position of the scale, the order of reading etc. is also derived by chance, using random numbers to control the choice. Chatfield (1970) describes the method. These variations in procedure are intended to introduce a measure of independence into the readings so that it is unlikely

TABLE 7.1. *The results of 100 separate measurements of the length of a straight line by plastic scale (distances in millimetres)*

	165·9	165·7	165·7	165·7	165·7	165·8	165·7
	165·7	165·8	165·5	165·7	165·6	165·5	165·7
	165·6	165·6	165·7	165·6	165·7	165·6	165·6
	165·9	165·7	165·7	165·7	165·7	165·7	165·8
	165·7	165·7	165·5	165·7	165·7	165·6	165·8
Mean:	165·76	165·70	165·62	165·68	165·68	165·64	165·72
s.e.s.o.:	±0·13	±0·07	±0·11	±0·05	+0·05	±0·11	±0·08
	165·7	165·7	165·7	165·6	165·6	165·7	165·8
	165·5	165·6	165·7	165·7	165·6	165·6	165·7
	165·6	165·8	165·6	165·7	165·7	165·7	165·7
	165·8	165·8	165·8	165·6	165·7	165·7	165·7
	165·7	165·6	165·9	165·9	165·7	165·7	165·7
Mean:	165·66	165·70	165·74	165·66	165·68	165·68	165·72
s.e.s.o.:	±0·11	±0·10	±0·11	±0·06	±0·08	±0·05	±0·05
	165·7	165·6	165·5	165·7	165·7	165·7	
	165·8	165·5	165·7	165·6	165·7	165·8	
	165·7	165·8	165·6	165·6	165·6	165·7	
	165·7	165·8	165·7	165·8	165·7	165·7	
	165·6	165·6	165·7	165·6	165·6	165·8	
Mean:	165·70	165·66	165·66	165·66	165·66	165·74	
s.e.s.o.:	±0·07	±0·13	±0·11	±0·09	±0·06	±0·06	

that any one measurement has any influence upon another. The measurements are repeated N times, where N is a large integer. In the example described here, the length of the line is evidently about 165·7 mm and 100 separate measurements have been made. The results are tabulated in Table 7.1 in the order in which the measurements were made. They are grouped into 20 separate samples of $n = 5$ measurements each. The frequency of the measurements, listed in classes of 0·2 mm, are given in Table 7.2 and illustrated in Fig. 7.1.

The mean and standard deviation calculated from all 100 measurements are

$$\bar{x} = 165{\cdot}69 \text{ mm}$$

$$\sigma = \pm 0{\cdot}09 \text{ mm}$$

TABLE 7.2. *Frequency of measurements listed in Table 7.1*

Length (mm)	Frequency	Length (mm)	Frequency
165·4	0	165·8	17
165·5	6	165·9	3
165·6	25	166·0	0
165·7	49		

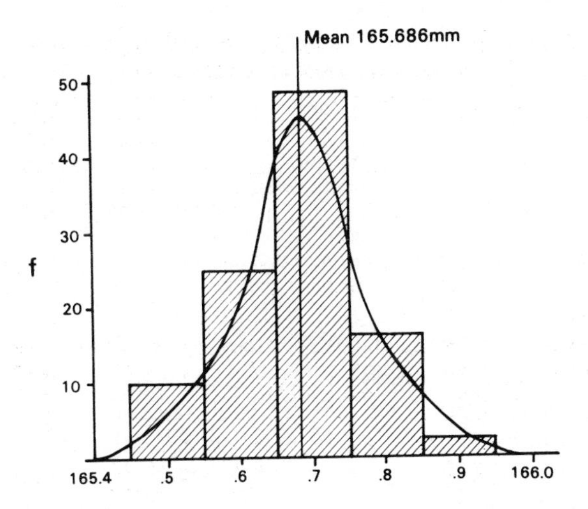

FIG. 7.1. Histogram representing the frequency distribution of the 100 measurements of the length of a straight line recorded in Table 7.1. A normal curve has been fitted to these data.

Figure 7.1 indicates a satisfactory approximation to the normal distribution which we have supposed to be the characteristic of the random or accidental errors of such measurements. The smooth curve illustrated in the figure is a normal curve calculated from x and σ. By comparison, Fig. 7.2 illustrates the frequency of the five measurements comprising the first sample (at the top of the left-hand column) in Table 7.1. This shows no apparent resemblance to the shape of a normal curve. We may now calculate the mean values of each of the 20 sub-samples into which the measurements have been grouped. The sampling distribution formed by the 20 mean values is illustrated in Fig. 7.3. This shows some approximation towards a symmetrical distribution, although the total of only 20 results is not really large enough to give more than a rough demonstration of this property.

In order to show that the same principles apply to measurements of other kinds of lines, made in other ways, we illustrate in Fig. 7.4, the results of some of Håkonson's measurements of the length of the shoreline of Lake Skagern in southern Sweden, made at a scale of 1/500,000 using the method which we

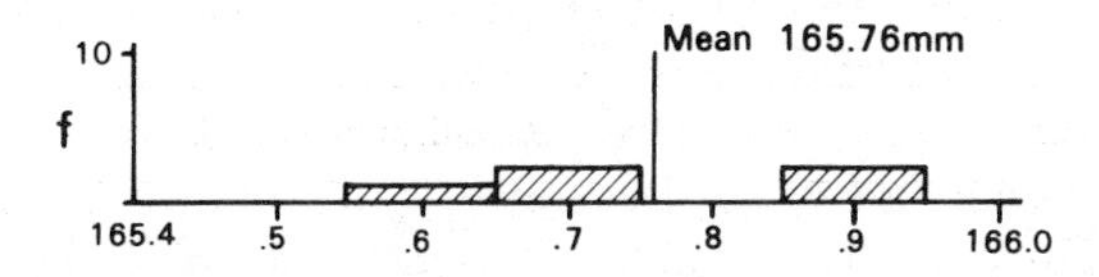

FIG. 7.2. Histogram representing the frequency distribution of the five measurements contained in the first sample recorded in Table 7.1.

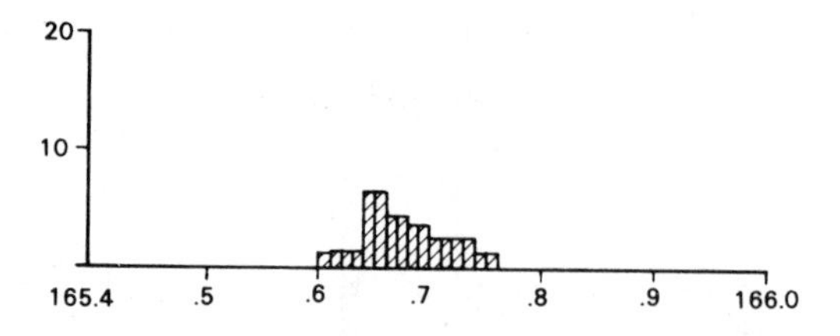

FIG. 7.3. Histogram representing the frequency distribution of the 20 mean values obtained from sub-samples of five measurements each, recorded in Table 7.1.

have called Håkonson's longimeter. The diagrams represent the frequency of measurements made in samples of 10, 20, 40 and 80 separate random applications of a square 5 mm grid to the map. These show increasing tendency towards the normal distribution as the size of the sample is increased. However, the mean count derived from each sampling distribution does not change much.

The tables and illustrations indicate that there may be some justification in making the following statements about the sampling procedure:

(1) If a series of samples are drawn from the parent population at random the means of these samples are normally distributed. This is in accordance with the *central limit theorem*, which states that, as sample size increases, sample means tend to be distributed normally, even if the parent population is not normal. This is not true of small samples, the distribution of which tends towards *Student's "t" distribution.*

(2) If we determine the arithmetic mean of a sample and from this derive the residuals, as in (6.4), we may find the parameter which is the measure of the spread of measurements within the sample. It is equivalent to the standard deviation (6.6) which tends to overestimate the precision of a sample of measurements. This may be overcome by applying *Bessel's correction*, namely

$$c = \sqrt{[n/(n-1)]} \tag{7.1}$$

which is applied in the form

$$s = c \cdot \sigma \tag{7.2}$$

$$= \sqrt{[\Sigma r^2/n]} \cdot \sqrt{[n/(n-1)]}$$

$$= \sqrt{[\Sigma r^2/(n-1)]} \tag{7.3}$$

where $r_i = x_i - \bar{x}$ as before and n is the number of measurements in the sample. The parameter s is known as the *standard error of the single observation*, abbreviated to s.e.s.o. in Table 6.1 and elsewhere. Often it is confused with the standard deviation or the root mean square error and is described in these terms. The denominator of the fraction, $n-1$, is known as the *number of degrees of freedom* in the sample. Obviously the difference between a denominator n and $n-1$ has a greater effect upon a small sample (e.g. $n = 5$) than upon a large sample (e.g. $n = 100$).

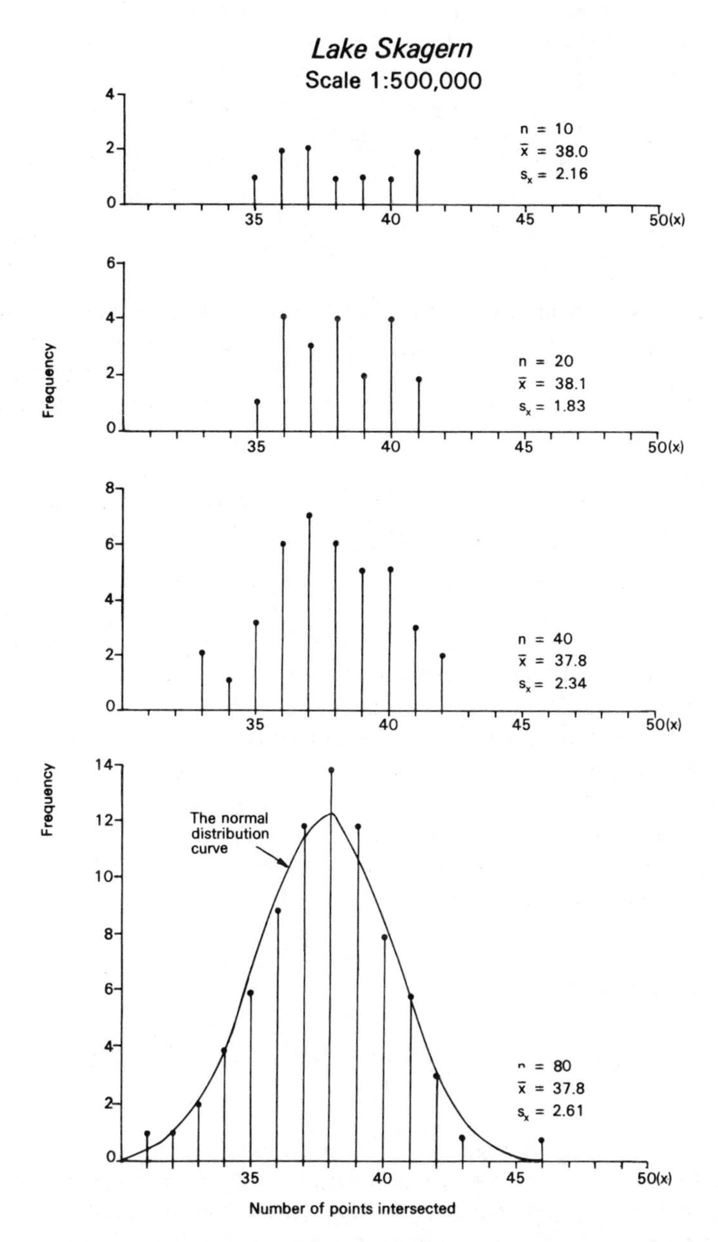

FIG. 7.4. Frequency distributions for the measurements of the length of the shoreline of Lake Skagern by Håkonson made on a map of scale 1/500,000. The diagrams illustrate increasing approximation towards the normal distribution for samples comprising 10, 20, 40 and 80 measurements each. Note that the mean values for the samples do not vary much.

It is desirable to explain the distinction between a standard deviation and a standard error. Both measure the dispersion of a probability or frequency distribution and both are square roots of the variance. In order to obtain the standard error of n observations empirically, it would be necessary to draw many samples, each of size n, calculate the mean of each sample and then obtain the standard deviation of all these sample means. The reader is invited to do this using the data tabulated in Table 7.1. Such an empirical determination is never done in practice because the standard error, s, may be more easily determined from

$$s = \{\Sigma x^2/(n-1) - (\Sigma x)^2/[n(n-1)]\}^{1/2} \tag{7.4}$$

(3) For a finite population, in which the number of individuals is known, a *sampling fraction*

$$f = n/N \tag{7.5}$$

should also be taken into consideration. Here N is the total size of the population. The sampling fraction has no effect upon calculation of the mean. It affects the determination of the standard error and enters (7.3) in the form

$$s = \pm(1-f)\{\Sigma r^2/(n-1)\}^{1/2} \tag{7.6}$$

If the parent population is infinitely large, as for physical measurements, the correction approximates to unity because f is infinitely small. Then equations (7.3) and (7.6) are taken as being equal.

(4) If the population has mean μ and variance σ^2, and we draw a large number of random samples from it, the means of these samples will be distributed with a variance of σ^2/n. Moreover the mean of the distribution will also be μ. We may estimate the mean of the population from the single sample of n measurements, and argue that this estimate lies as close to the population mean as is indicated by the *standard error of the mean* within the probability limits indicated in Table 6.2, p. 99. This parameter is, of course, the square root of the sample variance and may be expressed as

$$s_M = s/\sqrt{n} \tag{7.7}$$

Finally, since s_M is an estimate of the standard deviation of the population, it is subject to its own standard error which may be written

$$s_S = s_M/\sqrt{2n} \tag{7.8}$$

Equation (7.7) is a very important result. It shows how the precision of a sample varies with the size of the sample, or how the precision of a measurement may be improved by increasing the number of measurements which are made. This is very useful in practice when it is impossible to increase the precision of results further by any improvements in experimental technique. Thus, in the example of the measurement of the straight line, treating all 100 measurements as a single sample, the standard error of the single observation, $S = \pm 0{\cdot}09$ mm so that $s_M = \pm(0{\cdot}09/\sqrt{100}) = \pm 0{\cdot}01$ and we might refer to the results as

$$\bar{x} = 165{\cdot}70 \pm 0{\cdot}01 \text{ mm.}$$

On the other hand, if we only have at our disposal a sample of five measurements such as the sample in Table 7.1 at the foot of the left-hand column, for which $x = 165{\cdot}70$ and $s = \pm 0{\cdot}07$, the standard error of the mean is $s_M = \pm(0{\cdot}07/\sqrt{5}) = \pm 0{\cdot}03$ mm. Normally this would be the only result available to us, which would be written

$$\bar{x} = 165{\cdot}70 \pm 0{\cdot}03 \text{ mm.}$$

Obviously there is good agreement between the two values of the mean. Although there is a three-fold increase in the standard error of the mean as n is reduced from 100 to 5, we ought to remember that the variable represents millimetres read from the scale and decimals of 1 millimetre estimated from that scale using unaided vision. Therefore these standard errors represent much shorter distances than could be recognised when the scale is placed alongside the line. In practice, therefore, the variations to be analysed should be substantially larger than the minimum interval of the scale reading, for there is little to be gained by taking 100 readings if all of them are identical. Inspection of Table 7.1 suggests that the range of variation is determined by the resolution of the scale rather than by random factors. This, indeed, is one of the reasons why this particular experiment has been presented in detail. We shall see later, p. 279, that direct application of a scale to a straight line is the most precise method of measuring its length with simple equipment, because the errors are virtually confined to the reading errors caused by the resolution of the scale. It was demonstrated in Chapter 5 that a succession of such readings gives rise to a rectangular distribution. In this distribution the probability of all values within the interval, e.g. $x \pm \alpha$ is constant. Therefore $(px) = 1/2\alpha$ within the interval and is zero outside it. It can be shown that the variance of x is given by

$$\sigma^2 = \int_{-\alpha}^{+\alpha} x \cdot dx/2\alpha \tag{7.9}$$

$$= \alpha^2/3 \tag{7.10}$$

Consequently the standard deviation of a rectangular distribution is

$$\alpha = \alpha/\sqrt{3} \tag{7.11}$$

In the example illustrated in Fig. 6.1, p. 90, $\alpha = 0{\cdot}2$ and therefore $\sigma = \pm 0{\cdot}11$ mm.

In general, if other experimental errors have an essentially continuous distribution with variance σ^2, then the reading error will introduce an additional component to the total error. We have seen in equation (7.11) that if the reading interval of the scale is 2α and the corresponding error distribution is rectangular, the standard deviation of the reading error is $\alpha/\sqrt{3}$. Since the scale has been read with estimation to 0·1 mm, $\alpha = 0{\cdot}05$ and the standard deviation of the reading error is $\pm 0{\cdot}03$ mm. However every measurement of length comprises two such readings of the scale. Using equation (5.17) the total error attributable to this cause is

$$= \sqrt{2}\cdot(\alpha/\sqrt{3})$$
$$= \pm 0{\cdot}04 \text{ mm} \tag{7.12}$$

In the analysis of the errors which arise in reading a scale, Lyon (1970) examines three possible cases which are illustrated in Fig. 7.5.

If $\alpha < \sigma$, as in Fig. 7.5(a) the two components of error can be considered to be independent and hence the total variance

$$s_t^2 = \sigma^2 + \sigma_r^2 \tag{7.13}$$

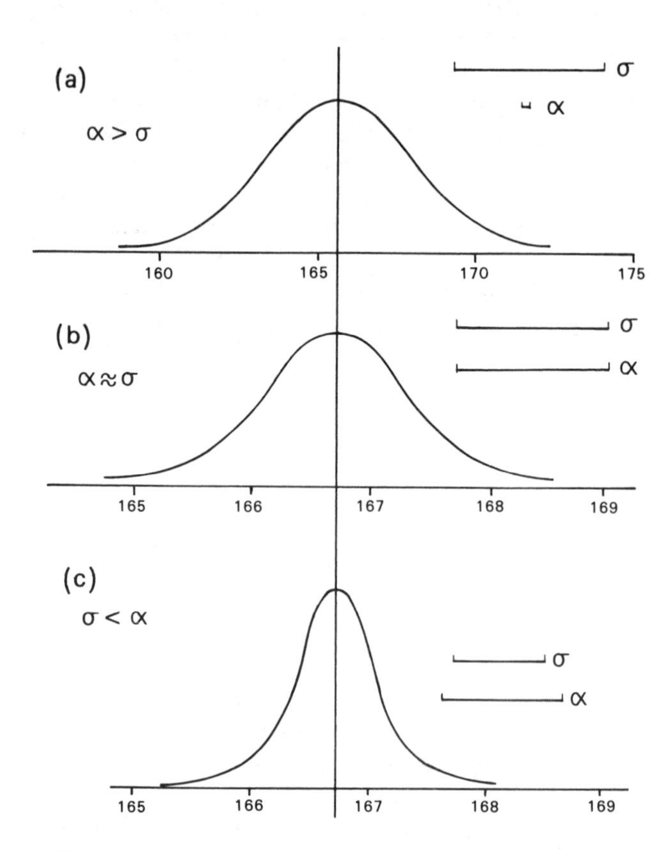

FIG. 7.5. The three possible relationships which exist between the reading interval of a scale and the standard error of a sample of measurements.

If the opposite holds good and $\sigma < \alpha$, as in Fig. 7.5(c), then the reading error will predominate and the two components cannot be considered to be independent. In the extreme case, where $\sigma = 0$ (i.e. the only uncertainty is due to reading error), repeated readings would always give the same value, but its uncertainty would be assessed as $\alpha/\sqrt{3}$ and therefore $\sigma_r = s_M$ and no improvement can be obtained by increasing the number of readings. We can see that this condition approximates to the results of the experiment described in this chapter, but inspection of Fig. 7.1 indicates that even this example demonstrates a normal rather than a rectangular distribution. The conclusion evidently represents a strong argument for restricting the number of measurements to only a small sample. For intermediate cases, such as is illustrated in Fig. 7.5(b), s_M is indeterminate and we can only say that it lies somewhere between $\sigma_t/\sqrt{n}$ and $\alpha/\sqrt{3}$. If it is necessary to give a single indicative value, one might select a value about midway between the two, but, as Lyon has noted, this is only as rough guide and it has no rigorous statistical basis.

The conclusion obtained for the case illustrated by Fig. 7.5(a) justifies the practice of making repeated measurements in order to improve upon the precision of the results. For example, it is the principle adopted by the field surveyor to obtain mean readings which are more precise than the reading resolution of the theodolite. There is a class of theodolites whose circles are graduated for reading the scales to the nearest 20 seconds of arc. By judicious repetition of the measuring routine used to observe horizontal angles, described in textbooks on surveying, it is possible both to remove the small systematic errors which exist within the instrument and to reduce the standard error of the mean of these angles to something appreciably more precise than is the least count of the scales. For example, if the standard error for an angle measured using one complete set of readings is of the order $\pm 40''$, by increasing the number of sets of readings to 16 the standard error of the mean is reduced to $40/\sqrt{16} = \pm 10''$. Nevertheless Weightman (1982) offers the following homily as a warning about too indiscriminate use of the argument:

> There is the old chestnut of the Eskimo standing outside his igloo who can tell the azimuth of a line to within 20 degrees by looking at it; one then asks 100 Eskimos and means the answer to an accuracy of 2 degrees – while 10,000 Eskimos will yield 12 minutes of arc and so on. Clearly there is a limit beyond which one cannot go.

SIZE OF SAMPLE

Thus far we have not indicated that the size of the sample has any influence upon theory. However, the principles which have been established apply generally to "large" samples, whereas we have repeatedly hinted that many cartometric results can only be regarded as being based upon "small" samples. It is often suggested that a sample size, $n = 30$, is approximately the borderline between what may be deemed to be large and what is small, but this cannot be taken as being more than an approximate guide. From a practical point of

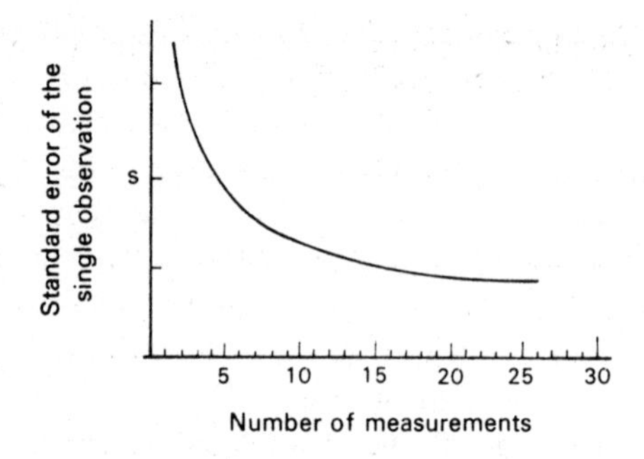

FIG. 7.6. The relationship between the standard error and the size of a small sample. (Source: Barry, 1964).

view, moreover, this is far too large. The relationship between the standard error of the single observation an sample size may be represented by an illustration from Barry (1964) (Fig. 7.6).

Figure 7.6 indicates that an increase in sample size is always accompanied by a decrease in the standard error for samples of similar precision, but that the rate of decrease in standard error becomes less for a sample of $n > 15$. Conversely the diagram emphasises a rapid increase in standard error as the size of the sample is reduced to less than 10 measurements. The interpretation to be placed upon this figure is that a sample ought to exceed 15 measurements.

Sampling involves a compromise between precision and economy of effort. A parameter can only be obtained with 100% accuracy by measuring or counting the whole population and, obviously, this is only possible for finite populations – usually small ones, to boot. The smaller is the sample and, to a lesser extent, the smaller the sampling fraction, $1 - f$, the wider is the margin of error of any estimate made from it. In order to avoid the extremes of drawing too few or too many samples it is desirable to adopt an analytical approach and estimate a desirable size of sample. We do this through a rearrangement of equation (7.7), which defines the standard error of the mean.

In assigning a standard error to a mean we are determining the range through which it might vary. If we wish to reduce this range it is, as we have already seen, necessary to make more observations. In order to reduce the standard error of the mean to one-half, four times the number of observations should be made. Since it is often necessary to determine a mean with a certain degree of accuracy, or to detect differences of a certain size, the number of observations necessary for this should be known.

We achieve this by estimating the standard deviation. This may be done by drawing a small pilot sample first, or it may be based upon past experience of working with a particular kind of variable. The standard deviation is used to determine the limits for an arbitrary number of observations at a required accuracy. We do this by substituting $E = s_M$ in equation (7.7) where we call E

the *allowable error* for the determination of the mean. Then

$$n = a^2 \cdot s^2 / E^2 \tag{7.14}$$

in which a is the probability integral tabulated in the first column of Table 6.2, p. 99. Most work is done at either the 95% probability level, for which $a = 1{\cdot}96$, or at the 99% probability level, for which $a = 2{\cdot}58$.

As an example, consider the size of sample needed to measure the length of a line approximately 200 mm in length by chartwheel. Suppose that we wish to estimate the length of line with an accuracy of ± 1 mm (which is really quite crude). From Table 5.3, page 69, the mean square error of a chartwheel is about $\pm 2{\cdot}2$ mm for a line of this length. Since

$$s_M = \pm 2{\cdot}2/\sqrt{n}$$

we are 95% certain that for a sample of n observations the true mean lies within $1{\cdot}96 \times 2{\cdot}2/\sqrt{n}$ of the estimate mean. Therefore, putting $1{\cdot}96 \times 2{\cdot}2/\sqrt{n} = 1{\cdot}0$, we may solve the expression for n, or

$$n = (1{\cdot}96 \times 2{\cdot}2/1{\cdot}0)^2 = 19 \text{ measurements.}$$

For 99% probability the corresponding value for n is 33 and for 90% probability it has fallen to 14 measurements.

A similar argument may be used to determine the size of sample to be drawn from a population of ratios or proportions having a binomial distribution. In this case the allowable error for 95% confidence probability is

$$E = 1{\cdot}96(pq/n)^{1/2} \tag{7.15}$$

Therefore the sample required to attain the required level of E is

$$n = [3{\cdot}842p(1 - p)]/E^2 \tag{7.16}$$

In these equations p and q have the significance outlined on page 104. Solution of equation (7.16) requires an advance estimate of p. Table 7.3 provides values for the product $p{\cdot}q$ from 0·1 through 0·9. The table shows that the product $p{\cdot}q$ is symmetrical about a maximum value of 0·25, where $p = q = 0{\cdot}5$. Moreover the value of the product only changes slowly between $p = 0{\cdot}3$ and $p = 0{\cdot}7$. Within that range of probability the precise value of p is

TABLE 7.3. *Values of* $p{\cdot}q = p(1 - p)$ *for* $0{\cdot}1 < p < 0.9$

p	$1 - p$	$p(1 - p)$	p	$1 - p$	$p(1 - p)$
0·1	0·9	0·09	0·6	0·4	0·24
0·2	0·8	0·16	0·7	0·3	0·21
0·3	0·7	0·21	0·8	0·2	0·16
0·4	0·6	0·24	0·9	0·1	0·09
0·5	0·5	0·25			

TABLE 7.4. *Size of sample, n, according to the specified probability and permissible sampling error, E, for a binomially distributed population*

Sampling error		Probability					
E	$E(\%)$	99%	95%	90%	85%	80%	70%
0·01	1	16,590	9,604	6,766	5,184	4,109	2,684
0·02	2	4,148	2,401	1,692	1,296	1,027	671
0·03	3	1,844	1,067	752	576	457	298
0·04	4	1,057	600	423	324	257	168
0·05	5	664	384	271	208	165	108
0·06	6	461	267	188	144	115	75
0·07	7	339	196	139	106	84	55
0·08	8	260	156	106	81	64	42
0·09	9	205	118	84	64	51	34
0·10	10	166	96	68	52	41	27

unimportant because it has little effect upon n. It is only where p is close to zero or unity (100%) that it may be significant, so that it is important to know the precise value of the term. In this respect, however, the principle of taking a pessimistic view of things enters into the argument. Because we wish to determine the size of the sample which is large enough to fulfil the condition imposed by the acceptable degree of sampling error, E, it is sufficient to take the maximum value of $p(1-p)$ and solve equation (7.16) using this. For the 95% confidence level, $3{\cdot}842 \times 0{\cdot}25 = 0{\cdot}961$ and only E varies. Table 7.4 illustrates the sizes of samples which correspond to values of $E = 0{\cdot}01$ through 0·1. Calculation of sample size from equation (7.16) is particularly important in the methods of indirect measurement of area by point-counting which are described in Chapter 19.

THE STATISTICS OF SMALL SAMPLES

Because many cartometric measurements are based upon what we have established as being small samples, it is necessary to comment about some of the statistical theory which relates specifically to small samples. The statistics of small samples departs from ordinary sampling theory in the following ways:

1 The distribution of individual measurements within a sample seldom shows any tendency towards any recognisable pattern. This, of course, is primarily because the sample *is* small, as has been illustrated by Figs. 7.2 and 7.4(a).

2 The principle that the sampling distribution of sample means tends towards the normal distribution is only true if the parent population itself is normally distributed.

3 The sampling distribution approximates more closely to ***Student's t distribution*** for the size of the sample being studied, Fig. 7.7 illustrates the t distribution for the values of $v = n - 1$, for three different levels of v,

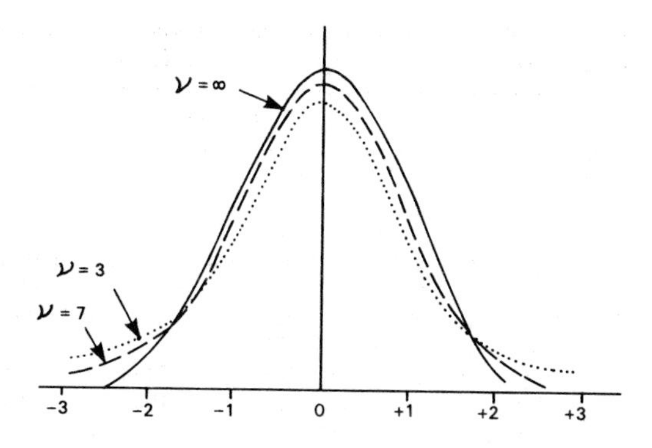

FIG. 7.7. The t-distribution for 3, 7 and ∞ degrees of freedom.

namely $v = 3, 7$ and ∞. The quantity v is known as the *number of degrees of freedom* of the sample. Clearly the overall appearance of the distribution resembles that of the standard normal distribution and, indeed, the curve for $v = \infty$ is a normal curve.

Since we are dealing with a different distribution pattern for each sample size, the relationship between the multiples of the standard error and probability, which were given for the normal distribution in Table 5.2, needs some modification. The usual form of the table for probabilities and the probability integral, or area beneath the curve is that given in Table 7.5. Student's t is a measure of the ratio of the mean of a sample to its standard error of the mean. Thus

$$t = (\bar{x} - \mu)/(s/\sqrt{n}) \tag{7.17}$$

where $\bar{x}$ and $s/\sqrt{n}$ are the mean and its standard error for a sample of size n.

The values of t corresponding to $v = \infty$ are, as we have seen, the same as those for the corresponding probability levels for the normal distribution as indicated in Table 6.2, p. 99. For small samples these values no longer apply. Thus the numerical values for t which corresponds to a probability of 95% and a sample of five measurements is 2·776, whereas it is 1·96 for the normal distribution.

We need to find limits for the assumed population mean μ so that, using a small sample, it will lead to a calculated ratio which is not significantly different at a particular probability level. If our choice is the 0·05 or 5% probability level, as before, there is a 5% probability that the value lies outside the confidence range, or a 95% probability that it lies within the range. Thus the confidence range for μ is that range of values of μ which, given a particular sample, will furnish a value of $|t|$ which is less than the value of t_1, this being the value of t for specified probability and sample size extracted from Table 7.5.

This requires that

$$|x - \mu| \cdot (\sqrt{n}/s) < t_1 \tag{7.18}$$

TABLE 7.5. *Student's t distribution*

Size of sample		Probability		Size of sample		Probability	
n	ν(d.f.)	90%	95%	n	ν(d.f.)	90%	95%
2	1	6·314	12·706	9	8	1·860	2·306
3	2	2·920	4·303	10	9	1·833	2·262
4	3	2·353	3·182	11	10	1·812	2·228
5	4	2·132	2·776	21	20	1·725	2·036
6	5	2·015	2·571	31	30	1·697	2·042
7	6	1·943	2·447	61	60	1·671	2·000
8	7	1·896	2·356	121	120	1·658	1·980
				∞	∞	1·645	1·960

so that

$$x - s \cdot t_1/\sqrt{n} < \mu < x + s \cdot t_1/\sqrt{n} \tag{7.19}$$

or

$$x - s_M \cdot t_1 < \mu < x + s_M t_1 \tag{7.20}$$

These extreme values may be regarded as being the *confidence limits* of the population mean within the probability value indicated by t_1.

In the example represented by the data in Table 7.1 we have measured the line AB in samples of five measurements each, and we have seen that s_M varies from 0·02 mm to 0·06 mm. From Table 7.5, for $n = 5$ and 95% probability, $t_1 = 2{\cdot}776$. Thus for the first sample

$$165{\cdot}76 - (2{\cdot}776 \times 0{\cdot}06) < \mu < 165{\cdot}76 + (2{\cdot}776 \times 0{\cdot}06)$$

or

$$165{\cdot}59 < \mu < 165{\cdot}93$$

and for the fourth sample

$$165{\cdot}68 - (2{\cdot}776 \times 0{\cdot}02) < \mu < 165{\cdot}68 + (2{\cdot}776 \times 0{\cdot}02)$$

or

$$165{\cdot}52 < \mu < 165{\cdot}73$$

If we assume that the value $\bar{x} = 165{\cdot}69$ mm determined from all 100 measurements corresponds more closely to the population mean, we see that our assumption has been satisfied in each of these small samples.

THE SIGNIFICANCE OF THE DIFFERENCE BETWEEN TWO MEANS

A particularly useful application of the t-distribution in cartometry is to investigate the hypothesis whether there is a statistically significant difference between the means of two samples. This may be to discover whether there is any significant difference between measurements of the same object made using different techniques, instruments or observers.

Suppose that we have determined the mean $\bar{x}_1$ and standard error s_{M1} from a sample of size n_1 and that we have also determined the mean $\bar{x}_2$, standard error s_{M2} from a second sample of size n_2. We require to know if any difference between $\bar{x}_1$ and $\bar{x}_2$ is significant or may just have arisen by chance. We denote the difference between the means as

$$\delta = |\bar{x}_1 - \bar{x}_2| \tag{7.21}$$

From equation (5.16), the standard error of this difference is

$$s_D = \sqrt{(s_{M1}^2 + s_{M2}^2)} \tag{7.22}$$

Putting

$$d = \delta / s_D \tag{7.23}$$

we determine the ratio of the difference between the means and the standard error of their difference. Entering the t-distribution for $v = n_1 + n_2 - 2$ degrees of freedom we find the probability whether the difference is significant. From Table 7.5 we find that for 95% probability and $v = 60$, the value of $t = 2{\cdot}0$.

This value for t has been used by Barry (1964) as a quick way of assessing significance without having to resort to tables or calculation on each occasion. His method was used extensively by Kishimoto (1968) in her comparative tests of different cartometric methods. Barry argues that for

$$\delta > 2{\cdot}s_D \tag{7.24}$$

the difference δ is significant at the 95% probability level. Strictly speaking this is only true for two samples totalling 62 measurements, and it should not be used with smaller samples. For example, in our use of two samples of five measurements each the appropriate value for t is 2·306. Therefore Barry's useful method is only a rough guide which should not be followed slavishly if there is only a small difference between δ and $2{\cdot}s_D$.

As an example of this technique we examine a question which arose in making the experimental measurements of the length of the coastline of Yorkshire, described in Chapter 5 and elsewhere in the book. Duplicate measurements were made for every combination of dividers setting and map scale. One measurement was made in the northward direction and this was always paired with another made in the southward direction. Does the direction of measurement have any effect upon the result?

TABLE 7.6. *Measured lengths of the Yorkshire coast from the International Map of the World at 1/1,000,000*

Direction of measurement	Mean distance (mm)	Standard error (mm)	sample size
Northwards	166·52	6·018	8
Southwards	167·15	3·623	8

We choose the measurements made on the *International Map of the World* at 1/1,000,000 scale, for which there are 16 measurements; eight made in the northward direction and eight in the southward direction. The results may be tabulated as shown in Table 7.6.

From equation (7.21) $\delta = 0{\cdot}63$ mm
From equation (7.22) $s_M = 7{\cdot}025$
Therefore $d = 0{\cdot}0897$

Since there are 14 degrees of freedom, the value for t for 95% probability is 2·145 and

$$2{\cdot}145 s_D = 15{\cdot}07$$

It follows that $\delta < 2{\cdot}145 s_D$ so that there is no significant difference between the two sets of measurements.

8

Spatial Sampling and Cartometry

Areal sampling is an especially suspect portion of statistics.

(William Bunge, *Theoretical Geography*, 1966)

In treating with the errors of measurement, discussion of the nature of the *sampling frame*, the *sampling unit* and the *sampling strategy* hardly appears as a subject for serious consideration. Provided that the physical limits of the object, such as the terminal points of a line or the perimeter of a parcel, are adequately defined, the samples are drawn from infinitely large populations without any attempt to make conscious or deliberate selection. The only kind of precautions which may be taken to avoid collecting a biased set of measurements is the attempt to make each measurement independent of the others. Usually it is considered sufficient to vary the ways in which the measurements are made systematically or randomly, as was described for the measurement of a straight line on page 107. We refer elsewhere to methods of achieving similar independence between other kinds of cartometric measurements, for example in the detailed description of the operation of the polar planimeter in Chapter 17.

Consideration of a suitable sampling strategy is important to most kinds of statistical analysis. For example, in demographic and sociological studies it is necessary to consider the statistical merits of different kinds of *list*, of households, voters and the like to form a sampling frame to meet particular purposes and requirements. Having chosen a suitable list, the sampling strategy is concerned with the way in which the sampling units are extracted from the list. Figure 8.1 illustrates some of the methods by which units may be drawn from such sources.

Many textbooks on statistics deal with this kind of sampling in detail. However the special needs for *two-dimensional, plane areal* or *spatial sampling* are less often considered. Two-dimensional sampling has direct cartometric applications in the following areas:

- The methods of area measurement, described in Chapters 19 and 20. The outcome of sampling a map or a collection of aerial photographs may be the production of a table of the frequencies of occurrence of the different variables studied, which are, or are easily converted into, proportions of

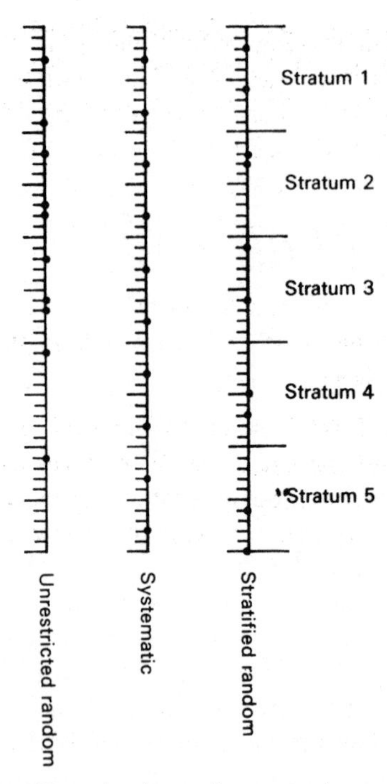

FIG. 8.1. The three one-dimensional sampling strategies envisaged as units (dots) drawn from a list, which is represented by a vertical scale. (Source: Quenouille, 1949).

the study area. For example, the outcome of sampling a land-use map might be a list of the percentages of arable, grassland, woodland and moorland. If the area of the site, region or the map is known from independent sources, it is a simple matter to determine the areas belonging to the different categories of land use from these percentages.

- Spatial sampling methods play an important part in the evaluation of the qualitative accuracy of maps and other graphics. This is done by sampling corresponding units of a map and on the ground; on a photograph and on the map; on a Landsat image and on the conventional photograph or map, to discover whether the category of variable represented by that unit has been correctly identified, interpreted and delineated on the document which is being tested.

SAMPLING FRAME, SAMPLING UNITS, SAMPLING STRATEGY

Sampling Frame

It is desirable to define clearly and unambiguously some kind of framework within which sampling is to be conducted. For example, in studying a human population using lists of households as the source of units to be sampled by interview or measurement, there must be available an accurate and up-to-date list of all the households comprising that population to serve as the *sampling frame*. In sampling spatial variables it is often a map which creates the frame both figuratively and geometrically. Even if the units to be sampled are subsequently examined or measured in the field, the selection of them, as well as actually finding them, has to be done with reference to a map or a set of aerial photographs. It follows that:

> *A sampling frame is a list, map or other specification of the units which constitute the available information relating to the population designated for a particular sampling scheme.*

This specification of the frame implicitly defines the geographical scope and extent of the survey and the categories of material covered. Since we are primarily concerned with sampling from a map or a collection of photographs, we use the word "frame" to mean some kind of graphic which covers the region of study.

Sampling Units

An aggregate of data must be divided into units for the purpose of sampling, each unit being regarded as individual and indivisible when selection is made. These are sampling units.

Division may be according to recognisable entities, such as one person, one household or one farm, but it may also be based upon an arbitrary or mathematical concept, for example by the grid coordinates which may be used to locate a point on map. Generally the sampling procedure, including estimation of population parameters and sampling errors, is simplest where the sampling units are of similar size and contain the same number of units.

Three different elementary types of sampling unit may be distinguished.

- points,
- transects or lines,
- areas.

These are illustrated in Figs. 8.2–8.10, where the sampling strategies are also indicated.

The point sampling unit is used to determine the presence or absence of some characteristic, or the position where a variable has been measured regarded as a point in space. This is probably the most commonly used sampling unit in cartometry, simply because the map or aerial photograph is a much-reduced image of the earth's surface. Then a house, or even a city, occupies such a small portion of the map that it may be regarded as only a point. We shall study an investigation of this form of spatial sampling by Smartt and Grainger (1974), who used selected points on a map to sample different categories of vegetation.

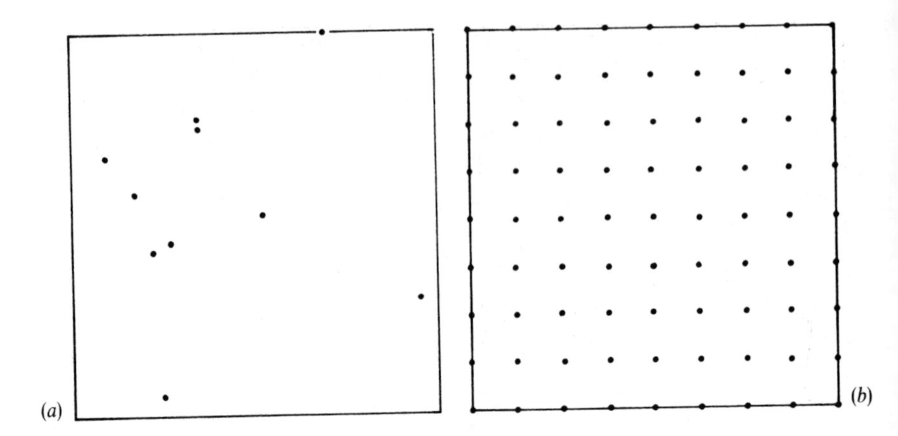

FIG. 8.2. Point sampling units: (a) unrestricted random points within the sampling frame; (b) systematically aligned points within the sampling frame.

The transect or linear sampling unit represents a line along which the variables are counted, listed or measured. Linear sampling was formerly especially important in forestry because, in the absence of adequate topographic mapping and since "timber cruising" was formerly carried out on the ground, it was easier to follow a compass bearing across country, sampling information along this line or along a narrow strip rather than locate a network of points or quadrats. Even where good maps are available, the method may be used to good effect. A classic cartometric example of this method of sampling was its use by Yates (1949) for the 1942 Census of Woodlands, which had to be carried out in wartime with extreme urgency and with very limited resources. The sampling method and its use for measurement of areas of woodland on the One-inch to One-mile map is described in Chapter 20, p. 445–447.

The area sampling unit can readily be employed in sampling continuous or discrete distributions, such as ground slope, vegetation cover, temperature or other surface characteristics. It can also be used to sample discrete data, such as plant or animal populations, crops, etc. Where the scale is small enough the area units may appear as if they occupied points on the ground, and may be treated as if they did so. The most typical application of area sampling units is in the method of quadrat sampling used in the biological sciences. With the increasing availability of good topographical mapping, and the use of vertical aerial photographs, area sampling has also increased in importance as a method of forest evaluation. We quote two examples in this field, both of which were designed to investigate and test the effects of sampling strategy and sampling size where the parent population is already known.

Loetsch and Haller (1965) created four "populations" of imaginary forest, comprising 400 square cells of area 0·1 ha each, in which the timber volumes

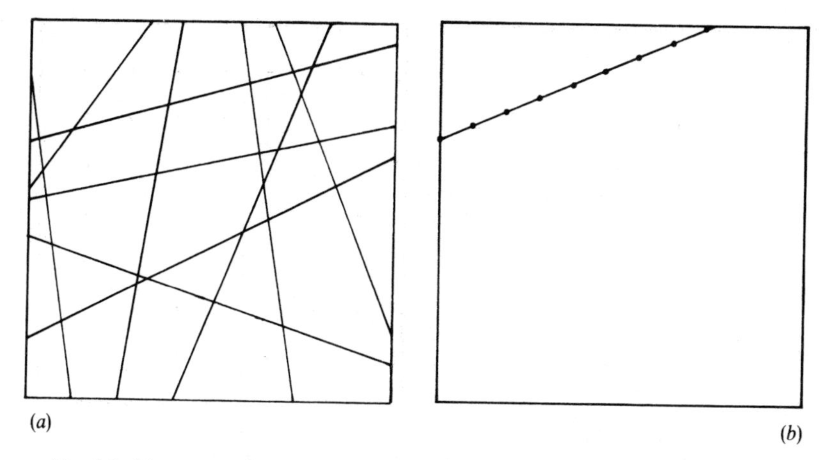

FIG. 8.3. Linear sampling units: (a) unrestricted random lines within the sampling frame; (b) systematic (equidistantly spaced) sampling points along a single line of random orientation.

had been recorded in units of cubic metres per hectare. The different populations were created by arranging these cells in different patterns to simulate the different kinds of plantation which might be found in the field. Since each population comprised the same data, it was possible to examine the relative merits of different sampling strategies in great detail. The leader seeking an exhaustive analysis of the methods is recommended to study their work. As a less ambitious example, based upon real data, we quote an investigation by Haggett (1963). He devised an exercise in sampling to be carried out by students, which was based upon the extent of woodland, expressed as the percentage area within squares of the National Grid for a block of country measuring 100 × 100 km of the borderland between England

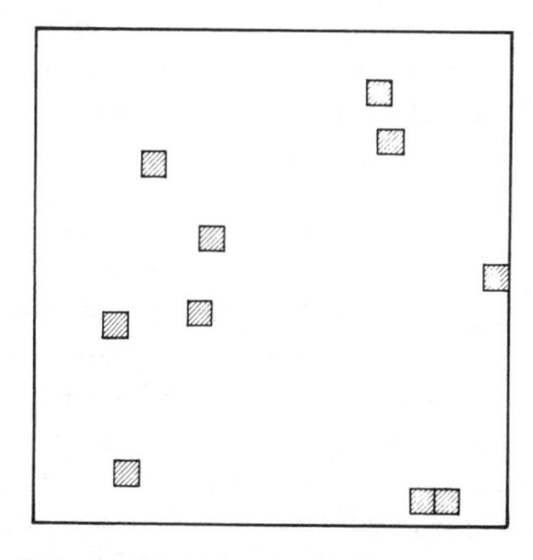

FIG. 8.4. Area sampling units. Unrestricted random quadrats of equal size within the sampling frame.

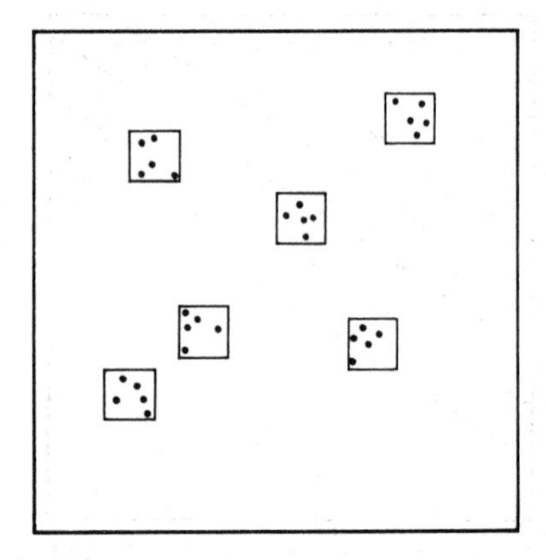

FIG. 8.5. Multi-stage sampling units. The primary units are the quadrats, or squares of equal size which are randomly distributed within the sampling frame. The second-stage units are the randomly distributed points occurring within each quadrat.

and Wales. For this investigation the sampling unit was the 5 × 5 km grid square. Variation in the sampling pattern and the size of samples was made by selecting different members of these units from the population of 400 such units.

The threefold subdivision of sampling units should be extended to include the concept of *cluster sampling*, which is normally associated with hierarchical *multi-stage sampling*. The term multi-stage means that the population is divided into a number of first-stage, or *primary units* which are sampled in the ordinary manner. These primary units are then subdivided into smaller second-stage units, which are also sampled. This is best illustrated in Fig. 8.5, where the primary units are squares of equal area and the second-stage units are the points lying within each square. Usually the method is selected for specific reasons. Occasionally, however, the choice of method is imposed by the nature of the source material, for example in the use of a block of vertical aerial photography to sample a variable which occurs throughout an extensive area.

Sampling Strategy

The main possibilities for the sampling scheme, and therefore the pattern of units to be adopted, may be summarised as follows: The *basic* or *primary strategy* is either *unrestricted random*, as in Fig. 8.2(a), or *two-dimensional systematic sampling* (Fig. 8.2(b)).

In unrestricted random sampling every point within the sampling frame has an equal chance of being selected. The sole criterion for the selection of points is the method used to draw a random sample from the population. Position on the map or photograph is usually defined by (x, y) Cartesian coordinates which

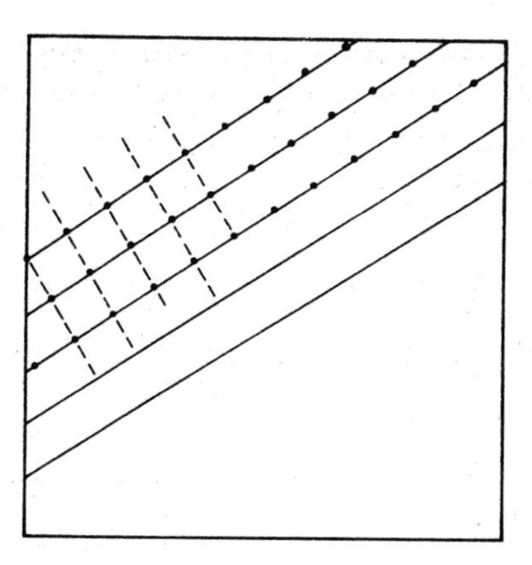

FIG. 8.6. Two-dimensional systematic pattern of points which result from sampling at equal intervals along parallel transects.

are chosen by a method of random selection, such as choosing suitable groups of random digits to form coordinate values. In the case of using point sampling units the coordinates obviously refer to the position of each point which has been chosen. For area units the coordinate pair may be used to refer to the point at the geometrical centre of the quadrat, or it may apply to a corner of the unit cell. There is no objection to the use of such arbitrary definitions to locate each cell, provided that the rule is applied consistently. Transect lines require coordinates for two points to define their ends. In two-dimensional systematic sampling the choice of the first coordinate pair, representing one point in a regular grid, may be drawn at random, but thereafter the location of every other point is determined on the basis of regular intervals, which is equivalent to placing a regular grid over the sampling frame. A random transect sample, Fig. 8.3(a), is a network of straight lines which intersect one another at any angle and do not form regular families of lines. An objection to its use is that it contravenes the statement on page 106 that sampling units should not overlap. It is possible to select a point at the intersection of two lines as being a member of two different samples.

A systematic transect sample, Fig. 8.3(b) is a straight line from which samples are drawn at equidistant intervals along it, or it is a family of equidistant parallel straight lines. The combination of the two is equivalent to a grid pattern of points as in Fig. 8.6. Restrictions of one kind or another may be incorporated into the basic strategy, and the resulting *derived strategy* becomes a two-stage model. The most common form of restriction is that of *stratification* of the population prior to making the actual location of the sampling units. This may be done in a variety of ways. The sampling frame may be regularly stratified by subdividing the map into blocks of grid squares as shown in Fig. 8.7(a). On the other hand the selection of strata may be made in the light of specialised knowledge or experience of the variables, attempting

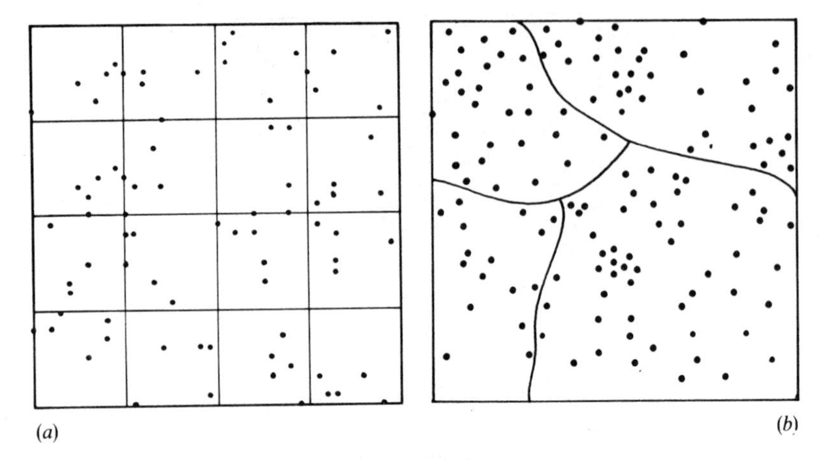

FIG. 8.7. Stratified random sampling: (a) of points within a regular and uniform grid; (b) of points randomly located within strata which have been identified on the basis of different criteria such as terrain, land use or political divisions.

to make some kind of regional subdivision of the mapped area. This is a practice much favoured by geographers who have been taught to subdivide a suitable area shown on a map into its physiographic or other regions, but the technique is used in other disciplines. For example, Loetsch and Haller indicate many different methods of stratifying forest areas on the bases of species, stocking density, terrain and other variables. (Fig. 8.7(b)). The second stage in stratified random sampling is to draw units from each stratum according to the various rules concerning the number and pattern of units drawn. Thus a *stratified random sample* selects points or areas at random within each stratum as this was defined in the primary stage. Figure 8.8

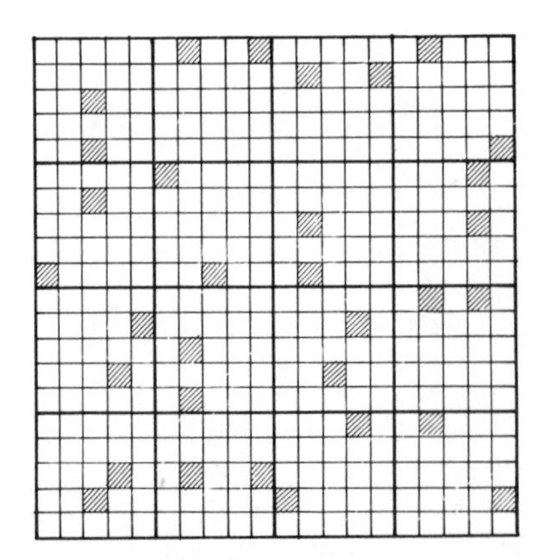

FIG. 8.8. Stratified random sampling of unit areas within strata formed from larger groups of the unit areas. The example illustrated is that described by Haggett in his measurement of the area of woodland within a block of Welsh borderland country.

illustrates the application of this to the methods used by Haggett to select a stratified random sample of two units from each of 16 strata of dimensions 25 × 25 km based upon the National Grid.

A systematic stratified sample may be *aligned*, as in a square grid, or it is *unaligned* as shown in Fig. 8.9 and described on page 135.

SPATIAL SAMPLING METHODS

Unrestricted Random Sampling

When the object of the analysis is estimation of population parameters and statistical inference in general, random sampling designs are essential.

The commonest method of drawing such a sample is to create some simple rules to convert random numbers into locations. The random numbers may be obtained from statistical tables, e.g. Fisher and Yates (1948), or by computer. Most versions of BASIC and similar programming languages used with microcomputers have subroutines to generate random numbers. The actual choice of the sampling unit may be done through the medium of counting, having first subdivided the area into a collection of small cells which are consecutively numbered, or through the use of the random numbers as grid references. Both of these methods need to be treated with caution, because the use of unit cells which are too large or grid references having too few significant figures may have the effect of converting the strategy into a form of systematic sampling.

In unrestricted random sampling the number of samples occurring in each category will be proportional to the area occupied by it on the map. The main advantage of unrestricted random sampling is that it provides an unbiased estimate of the population parameters, together with the precision of the standard errors from (6.3) (for an infinitely large population) or (6.4) (for a finite population). The sample is extendable. In other words, if the results of a survey indicate that more units should be drawn, additional sampling units can be drawn without upsetting the existing sampling pattern and the new data may be combined with the original sample. This cannot be done with some of the other methods of sampling.

Nevertheless the method has certain disadvantages. The most important is that, unless the sample size is extremely large, the cover of the region studied may be uneven with some parts oversampled whereas others have been undersampled. Estimates of the accuracy of the smaller classes are based on fewer observations, and are therefore less reliable. This is especially important if point-counting is used to measure areas of land use, for example, where certain minor categories may produce anomalous results. They may be wholly unrepresented by the sample or may be given apparently much exaggerated significance. Many people who have been confronted with such anomalies comment about them, some to the extent of having doubts about the whole philosophy of spatial sampling. However, the pragmatic view of Stobbs (1967), for example, that it is sufficient for the sampling to show that a minor category exists, is the most realistic way of treating with such anomalies. He has written

> In the extreme case of a very small unit [of land use] in which a given category occupies only a very small proportion this can produce a large percentage error.
>
> In practice we have found that the larger the error, the less important the unit for which results are being given, and the less significant the category involved. The overall picture of land use is not in consequence distorted in a way which hampers the planners for whom the work is primarily intended. Conversely to achieve a high standard of accuracy throughout would involve a disproportionate increase in the amount of work required. The point to emphasise is that a balance has been struck between objective and method.

In other words, the method does not take into account any spatial variations within the population. Consequently unrestricted random sampling does not usually lead to the most precise estimate possible for a given size of sample. As noted above, the method is not entirely devoid of bias which results from the finite spacing of the coordinates of the points sampled. This may be demonstrated with reference to sampling on the 1/50,000 Landranger maps of the Ordnance Survey, having the standard-format 40 × 40 km which, on the map, measures 80 × 80 cm. Suppose that it is considered that one pair of random digits is sufficient to define the sampling units. Given the sequence of numbers 22 17 19 36... the first point in the sample would be identified by the coordinates (2,2), the second by (1,7) and so on. This would be equivalent to dividing the map into ten lines in each direction. The space between each pair of lines would be 8 cm and only 100 points could be located on the map by means of these coordinates. If, however, the resolution of the grid were increased by using two pairs of digits, so that the coordinates of the first point are (2·2, 1·7), this would be equivalent to using a grid with 8 × 8 mm of the map and a possible total of 10,000 points might be identified uniquely. The use of even larger numbers of digits for the coordinates further increases the resolution of the grid and the number of available points.

Systematic Sampling

Systematic sampling depends upon the maximum dispersion of the sampling units at fixed distances for both its precision and objectivity. Thus in a systematic sample the units are measured at a predetermined distance from one another, the first unit to be measured being either a predetermined point or a point which has been selected at random. This may be called a *random start* as, for example, by Bellhouse (1981). The sampling units may be systematic in one direction only, as in the measurement of every *n*th tree along a transect defined by means of a compass bearing through a forest, or at equidistant intervals measured along such a line. Some methods of selecting which aerial photographs from a block should be used for sampling are described on pp. 425–429. Yates (1949) describes the use of linear sampling for measurement of areas of woodland. This was done by measuring the lengths of equidistantly spaced E–W lines on the One-inch to One-mile Ordnance Survey maps, comparing the length of line coinciding with woodland with the total length of line across the map.

The most important systematic sampling pattern is a square grid illustrated in Fig. 8.2(b). An important advantage of using this pattern is the ease with which this may be carried out on maps and photographs. A uniform dot grid, such as that described in Chapter 18 for use in area measurement, provides a sampling pattern which may be used as an overlay for map or photograph. However, there is often no need to use a separate overlay. The grid printed on large- and medium-scale maps will serve perfectly well for many purposes. Similarly the *réseau* on many survey-quality aerial photographs provides a network of small crosses at 1 cm intervals which serves very well as a basis for sampling.

The determination of the mean of a systematic sample is the same as that for the unrestricted random strategy and is the method of equation (5.4). Where known populations have been sampled the accuracy of the results is generally as good as may be obtained by unrestricted random sampling. There is abundant evidence to support this, for example from the work of Smartt and Grainger and that of Loetsch and Haller. However, a major disadvantage of systematic sampling is that *there is no wholly dependable method for estimating the variance of the mean*. This is because the sampling units have been selected by position rather than by chance, so that this fundamental tenet of inferential statistics has not been satisfied.

The use of a random start does not convert the system into a random sample. The argument that it may be considered to be random because there seems to be no relationship between the grid pattern and the distribution of the variables is a denial of the definition of random. Even more specious is the "pseudo-theoretical argument" as Milne (1959) describes it, which states that because a systematic arrangement of points, such as is illustrated in Fig. 8.2(b) *might* have arisen by chance, that it can therefore be regarded as if the points were randomly located. It is necessary to emphasise that the systematic pattern of points on a grid cannot, under any circumstances, be regarded as if it were a random pattern of points. As Greig-Smith (1964) succinctly put it:

> Consideration of a uniform distribution, such as that of trees in an orchard, shows that the probability of finding an individual is not uniform over the area but rises at the corners of an imaginary grid. Put another way, in a random distribution the presence of one individual does not raise or lower the probability of another occurring near by.
>
> In an uniform distribution the probability is lowered...

Consequently there are three schools of thought concerning the use and treatment of systematic sampling.

1 There are some, for example Greig-Smith, who consider that this inability

to determine sample variance is a fatal objection to the use of it, and would therefore prefer to use stratified random sampling methods rather than become involved in the uncertainties of testing systematic samples.

2 There are others to whom the problem is not unsurmountable and who have, indeed, offered possible methods of obtaining estimates of the variance of the mean. This position has been maintained by some of the leaders of thought in this branch of statistics. Thus Cochran (1946), Quenouille (1949) and Yates (1949) have all offered possible solutions. However, the methods of collecting the necessary data have generally involved repetition of the sampling process, or using techniques which are so closely allied to stratified sampling that it would seem better to use this strategy from the outset. The simplest case of all comprises systematic linear units and the variance estimate is based upon the differences between successive samples. We describe the procedure in detail in Chapter 20.

   One method of determining the variance of the mean of a systematic sample, which was proposed by Quenouille, appears elsewhere in this book in a different guise. He proposed the use of sets of systematic samples to be derived from placing a regular grid pattern in random positions with respect to one another. This is a well-known cartometric device, for it is identical to the direct method measuring area using a dot grid, described in Chapter 18.

3 The third school of thought is that it does not really matter if the variance calculated by conventional methods has no statistical validity, for the method is so useful this balances any theoretical objections. Indeed, many users of systematic sampling are probably unaware that the method suffers from such defects, and a considerable amount of work for environmental evaluation and for indirect measurement of area makes use of statistical inference in this context despite the impropriety of so doing. For example in order to make preliminary assessment of the sample size which is needed to attain a specified level of accuracy, using the arguments presented on pp. 115–118, the fundamental criterion which underpins the method is that each sampling unit has an equal chance of being selected. There are plenty of people who have used equation (7.16) to determine the size of the sample and follow this by drawing a systematic sample. Milne (1959) endeavoured to discover the effects of this statistical misdemeanour by investigating 50 different populations which had been collected for biological and agricultural purposes. By employing different sampling procedures with each, he was able to establish the merits of both random and systematic strategies. Milne concluded that, with proper caution, one will not go far wrong in treating most systematic samples as if they were random. However, he is at pains to emphasise that this should not be taken as licence for anyone to believe that the insistence of statisticians on random sampling can, in general, be ignored.

Stratified Sampling

The second major subdivision of plane sampling methods are the derived strategies which involve stratification of the population. The basic intention of stratification is to combine the advantages of both systematic and random sampling, namely their accuracy and their statistical validity respectively. This type of sampling is most often used when accuracy statements are to be derived for each thematic class represented on a map, as well as an accuracy statement for an entire map.

It has been noted that there are two basic criteria for defining the strata. One is simple geometrical dissection of the sampling frame into contiguous compartments, usually of equal size; the other is subdivision based upon prior knowledge of the study region, derived from field-work, photographic interpretation or simply from study of existing maps. Choice of the number of strata required is also a matter for judgement. The population variability and the purpose of the investigation are major considerations which influence this. We have already seen that there are two principal kinds of stratified sampling is common use.

STRATIFIED RANDOM SAMPLING

In this case, illustrated by Figs. 8.7 and 8.8, the sampling units are drawn at random within each compartment or stratum. Since stratification extends throughout the sampling frame, every unit comprising the population has the same chance of being drawn if every stratum is of the same size and contains the same number of units. Consequently this form of sampling strategy allows the exercise of statistical tests which are the same as those used for unrestricted random sampling.

Stratified Systematic Unaligned Sampling

This is a hybrid form of sampling strategy in which one coordinate of each unit is selected at random, but the other is arranged systematically. The intention, as before, is to combine the advantages of both random and systematic sampling and avoid the possible influence of bias owing to periodicity if this happens to coincide with the stratification.

Consider the simple stratified pattern illustrated in Fig. 8.9 in which the compartments are uniform squares corresponding to subdivisions of the map grid. Allocation of the sampling units within these regular strata is carried out as follows:

1 In the top left stratum the point A is chosen from random coordinates, $A = (x_a, y_a)$.
2 In the next compartment of the top row the point B is selected by

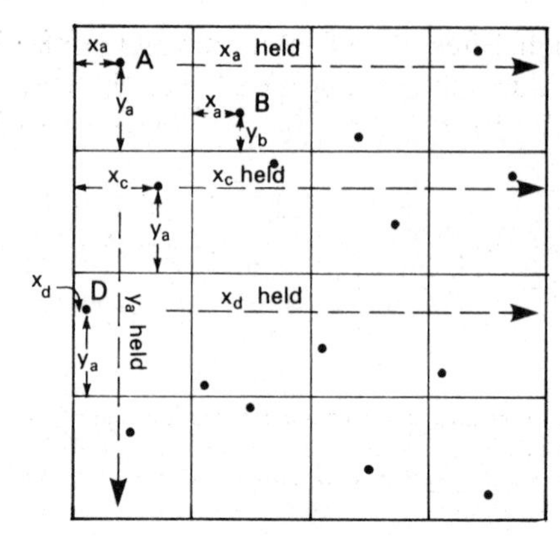

FIG. 8.9. The technique for selecting points within a stratified systematic unaligned sample.

combining the x_a abscissa, already determined, with a new ordinate y_b. Thus $B = (x_a, y_b)$. This procedure is continued along the top row.

3 In the second row the point C is chosen by means of a new random abscissa, x_c, but the original coordinate y_a is retained. Thus $C = (x_c, y_a)$ and since the same procedure is carried out for the left-hand cell of each row, the first column is labelled "y_a held".

4 The remainder of the points to be located in the second row have the same abscissa, x_c, together with a new random value for every coordinate.

5 In the third of strata the point D is defined by a new random abscissa x_d together with y_a. In other words, $D = (x_d, y_a)$. The other points in this row have coordinates determined in the same way as has been outlined for the first and second rows.

Completion of these instructions for every stratum within the sampling frame creates one sampling unit in each. Usually more units are required so that it is necessary to repeat the sequence until each stratum contains $2, 3, \ldots n$ sampling units.

COMPARISON OF SPATIAL SAMPLING METHODS

The work by Smartt and Grainger (1974) is an example of a carefully controlled test of the four principal methods of spatial sampling using point units. In was done on a single sheet of the Ordnance Survey Six-inches to One-mile (1/10,560) series, sampling the eight varieties of vegetation depicted on the provisional edition of the series. The area covered by the chosen sheet (SU 30 SE) represents part of the New Forest. The investigation was intended to test both sampling strategy and sampling size. Each method was applied with increasing size of sample, in steps of 25 units until a total 400 points had been sampled by each method. Data are published for the percentage frequencies of occurrence for the eight categories of vegetation and for the 16 sizes of sample. For simplicity, the values tabulated here only record the frequencies obtained for samples of size 25, 50, 100, 200, 300 and 400 points. These are given in Table 8.1.

The estimates for both unrestricted random sampling and systematic sampling generally lie within the specified error limits of the actual values, and there appears to be little to choose between the two parent strategies in terms of overall accuracy.

TABLE 8.1. *Spatial sampling study of Smartt and Grainger (1974). Percentage frequency values for different vegetation types shown on OS 1/10,560 sheet SU 30 SE as determined by different sampling strategies and sizes of sample*

	Number of points sampled					
	25	50	100	200	300	400
Unrestricted random sampling						
Parkland	44·0	30·0	28·0	30·0	33·0	35·5
Heathland	24·0	24·0	26·0	21·0	22·0	22·3
Deciduous woodland	12·0	16·0	18·0	21·0	20·0	18·3
Mixed woodland	16·0	16·0	15·0	12·5	9·7	10·0
Coniferous woodland	4·0	10·0	7·0	7·0	7·0	6·8
Orchard	0·0	0·0	3·0	5·5	5·0	5·0
Furze	0·0	0·0	1·0	2·0	1·7	1·3
Marsh	0·0	4·0	2·0	1·0	1·3	1·3
Systematic sampling						
Parkland	36·0	40·0	33·0	33·5	34·0	34·3
Heathland	20·0	16·0	19·0	21·0	21·0	21·0
Deciduous woodland	12·0	16·0	17·0	14·5	15·3	14·8
Mixed woodland	4·0	4·0	9·0	10·0	9·7	10·3
Coniferous woodland	12·0	12·0	9·0	10·5	9·7	9·3
Orchard	0·0	2·0	4·0	4·5	4·3	5·3
Furze	4·0	2·0	2·0	1·5	2·0	1·8
Marsh	12·0	6·0	5·0	3·0	2·7	2·0
Stratified random sampling						
Parkland	36·0	38·0	34·0	34·5	34·0	34·0
Heathland	20·0	18·0	23·0	23·0	21·7	20·7
Deciduous woodland	16·0	14·0	16·0	15·0	17·7	17·7
Mixed woodland	8·0	10·0	8·0	8·5	9·3	9·0
Coniferous woodland	12·0	12·0	10·0	10·0	8·7	8·8
Orchard	0·0	0·0	3·0	4·0	4·0	4·5
Furze	0·0	0·0	0·0	0·5	0·7	0·7
Marsh	4·0	4·0	2·0	2·0	2·3	2·5
Stratified systematic unaligned sampling						
Parkland	28·0	30·0	33·0	31·0	31·7	31·8
Heathland	24·0	24·0	25·0	24·0	24·7	24·8
Deciduous woodland	24·0	16·0	15·0	18·5	16·3	16·5
Mixed woodland	4·0	10·0	7·0	8·5	8·7	9·3
Coniferous woodland	4·0	10·0	10·0	9·5	8·7	8·3
Orchard	8·0	4·0	5·0	4·5	4·3	4·3
Furze	4·0	2·0	1·0	1·0	1·3	1·3
Marsh	4·0	2·0	2·0	1·5	1·7	1·5

In order to compare the merits of different sample sizes and patterns over all eight categories of vegetation, Smartt and Grainger calculated values for the *root mean sum of squares deviation*, or RSSD, which has the form

$$\mathrm{RSSD} = \sum_{i=1}^{n} (x_i - p)^2 \tag{8.1}$$

where x_i is the observed percentage in the ith replicate and p is the actual percentage determined independently for each category of vegetation. The number of replications for a given sampling intensity is n. On the basis of these calculated values, the values were ranked and the rank order for each vegetation type was summed. Using only the results for the four smallest sample sizes (25, 50, 75 and 100) the overall rating values for the four methods of sampling place them in the following order of merit:

1	Stratified systematic unaligned sampling	43·5 (best)
2	Stratified random sampling	81·5
3	Systematic sampling	90·5
4	Unrestricted random sampling	104·5 (worst)

This agrees broadly with the conclusions of Cochran (1953), Yates (1949) and Milne (1959). For estimating the area covered by forest or by water shown on a map, which is typical of the cartometric use which we shall make of spatial sampling, Mátern (1960) found that a square grid was superior to random sampling. Yates also suggested that in extensive area surveys, systematic line or grid sampling is to be preferred to random sampling. For ecological work, Greig-Smith (1964) considers that random sampling is the better strategy if overall information concerning the composition of vegetation is desired. On the other hand he agrees that systematic sampling is to be preferred if interest centres on variability within the area. Smartt and Grainger found greater variability exhibited by the estimates for random sampling compared with those for systematic sampling. The magnitude of this variation is between three and five times greater for random sampling. Moreover, in terms of the variability of the estimates, both appear to conform to the pattern set by their respective parent strategies, with stratified random sampling exhibiting greater variability that the stratified unaligned systematic method.

Systematic sampling in the form of a network with regular distances between the sampling units equalises the spread of information over the area. There is an inherent bias towards over- or under-sampling of units of different sizes when these are compared with the intersample distance. If this is substantially less than the average size of the larger units, but greater than that of the small units, then small scattered units tend to be less well represented than large parcels. This disadvantage may be overcome by using a closely ruled grid, corresponding to a large number of points to serve as the sampling units. However as in the methods of area measurement, either by the method of squares, or in the various point-counting methods, very high sampling intensities are impracticable because of the time needed to do the work. It is therefore important to recognise that the minimum size of parcel which can be recovered is determined by the intersample distance. Unrestricted random sampling, in contrast, has a more flexible sampling distribution, but may be subject to local clustering effects. This is most noticeable at low sampling intensities, for additional units tend to fill in the gaps and eventually with high sampling intensities a more or less even distribution of units is reached. Once again, the smallest size of parcel which is recoverable is set to the intersample distance which, as we found on page 132, is the resolution of the coordinate system employed to pick units at random.

Unrestricted random and stratified random samples are not constrained geographically. In other words, when the size of the sample is small, large parts of the map remain unsampled. In theory this feature is not a difficulty unless the surface being sampled exhibits marked *spatial autocorrelation.*

Spatial autocorrelation is an important statistical concept which is particularly applicable to the kind of spatial analysis required for the natural and environmental sciences, but which only impinges to a limited extent upon the main themes of this book. Indeed, one might argue that the study of spatial autocorrelation and related subjects begins more or less where this chapter ends. For our purposes the principal tenet of spatial autocorrelation has been encapsulated with breath-taking simplicity in the statement by Tobler (1970) that it is:

> the first law of geography: everything is related to everything else, but near things are more related than distant ones.

For a proper appreciation of the methods and implications which arise in consequence, the reader is referred to the work of Cliff *et al.* (1975), Cormack and Ord (1979), and Cliff and Ord (1981).

It follows that most geographical areas do exhibit spatial autocorrelation between the features being mapped. Indeed, if they did not, it could be argued that there would be little point in making a map of them in the first place.

The simplest case is when correlations are highest for adjacent pairs of points and decrease monotonically with increasing distance.† This may be exemplified by the form of curve known as a *correlogram*, which illustrates variation in the autocorrelation coefficient plotted against what is usually termed *lag*, but which here may be construed as the furthest point from the origin in sequence of points from which the autocorrelation coefficient, r_k, has been determined. The form of correlogram illustrated in Fig. 8.10, which is concave upwards, is especially common in the environmental sciences.

The example illustrated in Fig. 8.10 which has particular cartometric interest was produced by Mátern (1960), who investigated the example of the relationship between land and water shown on maps of scales 1/250,000 and 1/50,000 of part of the Stockholm archipelago by sampling along transects. He

†A sequence x_n is said to be monotonic decreasing if $X_n \geq X_{n+1}$ for all n.

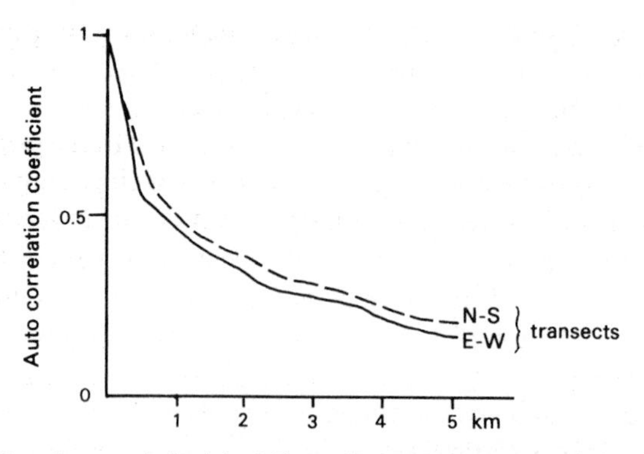

FIG. 8.10. Correlogram derived by Mátern from measurements of the relationship between land and water within the Stockholm archipelago by systematic sampling along parallel transect lines.

sampled the 1/250,000 scale map along twelve equidistantly spaced E–W lines at points every 1 mm along each line. If the point tested lay on land it was given the value 1; but if it lay in water it scored 0. Serial correlations were calculated along each line, giving a series of values for r_k for each distance sampled. As a result of this and other experiments derived from the use of Buffon's needle, Mátern was able to show that the method can be used to measure the respective areas of land and water on the map. Moreover, where sampling was done in random directions he was able to calculate the length of the shoreline in units of centimetres per square centimetre of map area. In an earlier paper, Mátern (1947) studied some other area distributions, such as total land and forest areas, the areas of sites occupied by various timber quality classes and variation in tree volume. All of these variables exhibited the same monotonic decrease with distance.

It can be shown that where the sample size is small, a systematic sample will be the more accurate in all cases where the correlogram is concave upwards. In this situation the stratified random sample loses more in precision when the distance interval is shortened below the constant interval. Moreover the systematic sample will also be unbiased if there is no periodicity in the data that can interact with the regular spacing of the grid. The influence of periodicity, which is the fortuitous coincidence between the sample units and a pattern on the ground, has often been quoted as being an objection to the use of two-dimensional systematic sampling. Although Yates and some other statisticians harbour some doubts about the possible effects of periodicity, Milne has claimed that its influence affords a negligible risk, and that it may be safely ignored. Smartt and Grainger reach the same conclusion. American geographers, on the other hand, are accustomed to dealing with rectilinear patterns in the landscape, whether these be the grid-iron patterns of the road network in the Middle West or the street patterns of the majority of North American towns. To them the threat of periodicity appears much greater. To provide against the event of a fortuitous coincidence of a square grid sampling pattern with such rectilinear distributions, the systematic unaligned sampling strategy seems to them to be the sovereign remedy.

Smartt and Grainger suggest that stratification of the frame prior to random sampling guards to some extent against any marked clustering of samples in the early stages. Local clustering within strata may occur, and the estimates for stratified random samples can still exhibit greater variability than the equivalent systematic model. However, compared with unrestricted random sampling, the more even spread of samples at low sampling intensities effects a general improvement in the representation of large vegetation units; but whilst smaller coordinate intervals limit to some extent any discrimination against the smaller units, there is small comparative loss of accuracy at this level. On the whole, however, systematic unaligned sampling allows a much greater flexibility in the distribution of sample units than pure systematic sampling. Although some sample clusters may be formed between neighbouring strata, the system is such that this is likely to be rare. The combination in one model of the elements of regularity and flexibility which provides good representation of both large and small units alike results in relative errors which are consistently lower than those of the parent strategy. These views are also held by Berry (1962), Berry and Baker (1968), Rosenfield *et al.* (1982), and Fitzpatrick-Lins (1981).

THE SIZE OF SPATIAL SAMPLES

It has been established that an increase in the size of a sample leads to an increase in the precision of the estimate of the population parameters. For a random sample from a binomially distributed population the size of sample may be calculated using equation (7.16) as follows:

Suppose that we wish to measure the area of woodland on a map by sampling randomly distributed points. At the end of the sampling process we have counted the number of points coinciding with woodland, p, and the number of points corresponding to land which is not woodland, q. The total area investigated is obviously $p + q$. Suppose, too, that we wish to determine the amount of woodland with a probability of 95%, and that this estimate may have a maximum permissible error $E = \pm 1\%$. In other words, we are willing to accept a variation in probability between 94% and 96%.

Substituting (7.16) and using the arguments presented on pp. 116–118

$$\begin{aligned} n &= (1{\cdot}96^2 \times 0{\cdot}25)/0{\cdot}01^2 \\ &= 9604 \end{aligned}$$

as in Table 7.4, p. 118. Note that the results have been obtained without any reference to the size of the area to be sampled, for the argument is wholly based upon the two-fold classification of attributes. In controlled experiments where the area of a variable has already been determined by other methods (e.g. by planimeter) and this is now used to test the accuracy of different sizes of sample, the graph relating percentage area to size of sample is characteristically L-shaped. In other words the results may be widely erratic for very small samples, but as the sample size increases it soon settles down to produce results which are not far removed from the "true" value determined by other means. Thus Haggett's work showed that a sample of 3% of the units measuring 5 × 5 km was large enough to produce a satisfactory estimate of the percentage of woodland in the area tested. For larger samples the values of the percentage of woodland oscillated within about 3% of the "true value".

We illustrate the more complicated conclusions obtained by Smartt and Grainger in their comparison of 16 different sizes of sample, eight categories of variable and four different sampling strategies. In order to make the comparisons, Smartt and Grainger made use of the *relative error* or *sum of the squared deviations*, δ, where

$$\delta = \sum_{i=1}^{8} (x_i - p_i)^2 \qquad (8.2)$$

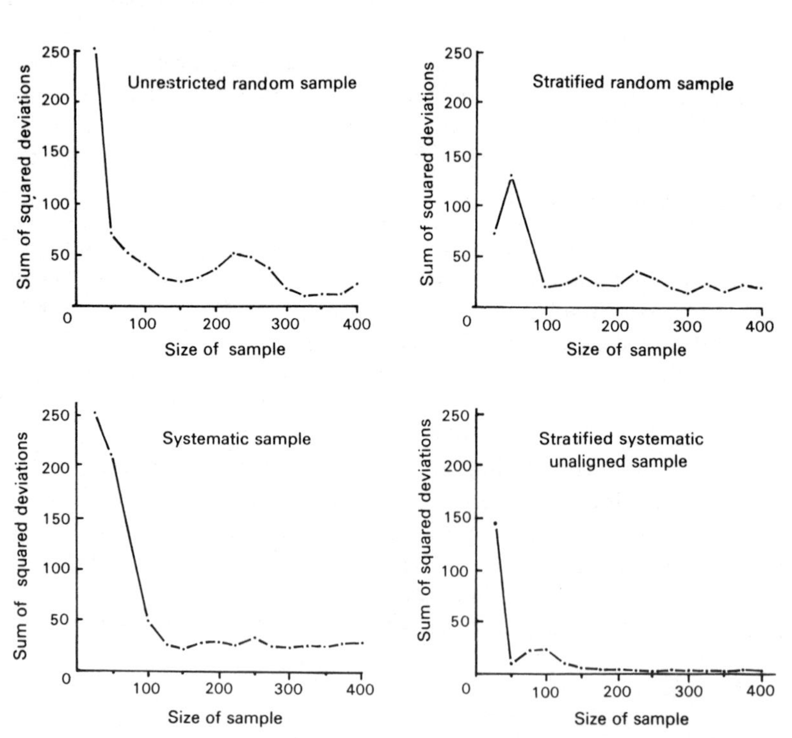

FIG. 8.11. Comparison of the relative efficiency of different sample sizes and different sampling strategies based upon the results of Smartt and Grainger. For each sampling strategy the plot of the sum of squared deviations (δ) against sample size indicates the characteristically rapid decrease in δ to a value which remains more or less constant, irrespective of sample size.

where x_i is the observed proportion of a vegetation category and p_i is the expected proportion. The four curves (one for each sampling strategy) which are based upon (7.2) are illustrated in Fig. 8.11.

The curves are L-shaped just as that described for the different sizes of sample drawn by Haggett, indicating how the relationship between δ and n quickly settles down to a nearly constant value. Figure 8.11 demonstrates yet again the superiority of stratified unaligned systematic sampling over the other methods tested.

9
The Concept of the Accurate Map

The greatest enemy of progress is ignorance and I think I am safe in calling maps the most used and least understood documents of modern civilization. The tendency of the uninformed to believe everything they see in print extends to maps and charts. These people have not been trained to examine printed matter with a critical eye, and they are therefore prone to accept without question the facts set forth either in a book or on a map, as gospel. They have little, if any, conception of the virtues and limitations of maps or how to use them.

(Lloyd A. Brown, *Surveying and Mapping*, 1953)

The word accuracy is frequently employed with respect to maps, both in making and using them. It is nearly always misused because an unqualified adjective "accurate" or "inaccurate" is inadequate to express the evaluation of a document as complex as a map. The problem has been summarised succinctly by Harley (1975) as follows:

Accuracy is relative rather than an absolute concept as far as cartography is concerned. It cannot be endowed with a rigid definition such as "exact conformity to truth" or "free from error or defect"; not only are maps deliberately generalisations of reality, "representative models of the real world", but all survey and map production processes inevitably introduce error at some stage. What the map user as well as the map maker should be concerned with is a systematic study of factors affecting error, and to seek to establish their cause and variability and the statistical parameters by which error is characterised.

The object of the present chapter is to make this study, but after investigation of the errors which may arise in different stages of map production, we are left with the firm conviction that map accuracy, as such, is not an important source of cartometric errors. However, it is not enough just to make this statement without any supporting evidence. It is desirable to treat with the subject of map accuracy in some detail and lay, once and for all, this ghost of an opinion that where the results of measurements are unreliable, this is invariably due to the fact that the maps are inaccurate. This may be true of some kinds of map or charts in some parts of the world, but, as we shall see, most metric-quality maps produced for developed countries are accurate enough for the majority of our cartometric needs.

THE COMPONENTS OF MAP ACCURACY

We have seen in Chapter 6 that, in the general statistical sense, accuracy represents absence of systematic error or bias and, where bias is absent, the mean of a variable corresponds to its true value. Applying this idea to the study of map accuracy, we must attempt to compare positions on two different media; a map and the ground, a photograph and the map, or on two different maps, by measuring the differences in position. In order to make suitable measurements, the positions of both representations of the same object must be related to the same reference system. Therefore a preliminary definition of map accuracy is *the closeness of location of points of map detail to their true ground positions, each measured with respect to the same grid or graticule.* It follows that if two points have been plotted on a map without error, the scaled distance between them (allowing for any corrections due to other causes) should correspond to the ground distance between the points. It follows that this kind of accuracy can be measured and therefore the errors in position may be expressed numerically and treated statistically. Therefore it is possible to compare a map with certain agreed standards of accuracy and consequently with other maps which have been tested in the same way. We shall call this *quantitative accuracy* or *positional accuracy.*

However, the meaning of the word accuracy must be extended to describe two other cartographic concepts. These are *qualitative accuracy* and *completeness.* Map detail is depicted by means of lines, symbols or ornament which distinguish different kinds of ground features. In addition, a map bears a considerable amount of letterpress information, such as place names, graticule and grid numbers, road classification numbers, mileages and heights. This provides information which cannot otherwise be shown but is an important aid to map use. The accuracy of representation of the qualitative information cannot be measured on an interval scale but it is possible to distinguish what is correct from that which is incorrect. Methods of analysing qualitative information of this sort are described later. However, there are also many instances where the choice of symbols, boundaries, ornament or letterpress is not simply a matter of correctness but involves some measure of judgement on the part of those who contributed to making the map. Typical examples of this need for judgement may arise in the selection of natural boundaries, for example those for forest, marsh or desert which are often not clearly defined on the ground.

The completeness of the map is the third factor which contributes to the accuracy of the finished product. This, too, may be difficult to express quantitatively or concisely. Completeness depends in part upon the quality of the original survey or the sources used in the compilation of the map, but it depends mainly upon the degree of generalisation of the ground detail which can be shown at a particular map scale. Cartographic generalisation comprises two distinct processes; first, smoothing of outlines and even deliberate displacement of some of the linework or symbols in order to improve legibility; secondly, reduction in the amount of information shown on the map. The first process of generalisation affects the quantitative accuracy of the map; the second obviously influences the completeness of it.

From the point of view of the general map user, the document will be considered to be accurate if it appears to be so in the qualitative sense and if it also seems to be complete. Thus, if the map shows the inn and the church and the path across the fields between them, this information will probably satisfy the user, who is more likely to criticise the map if the Inn proves to be a church, or if the path is not shown, than for the reason that the distance between the buildings measured on the ground does not agree with that scaled from the map. The requirement that the different features shown on the map ought to occur in their correct relative positions does, of course, impose some sort of positional accuracy upon the map as a whole, but this is not particularly precise.

From the point of view of using maps to make measurements for scientific purposes, or, indeed, for the more precise kinds of route-finding and navigation, the influence of the quantitative accuracy of a map upon the results is vitally important. It indicates that the map maker ought to have adopted standards of accuracy in the work which should satisfy the normal, reasonable requirements of specialised map use. Because of the practical need for reliable measurements made on topographical maps for military purposes and upon navigation charts, most of the published work on the subject refers to quantitative accuracy and to various kinds of topographical map and chart.

Statistical methods of describing qualitative accuracy have emerged only in recent years, and they have developed from the need to evaluate maps produced wholly from remote sensing. Since 1972, the use of Landsat imagery as the primary source of information has offered possibilities of producing detailed small-scale maps directly from enhanced versions of the images. The work on qualitative accuracy carried out by the U.S. Geological Survey has been supreme in this field. Evidently the methods which have been described to measure the qualitative accuracy of Landsat-based maps are also applicable to other kinds of cartographic evaluation. At this stage, however, comparatively little work seems to have been done in this field.

MAP ACCURACY STANDARDS

The typical specifications for the accuracy requirements of a topographical map are known as *map accuracy standards.* These may be used by government departments and private companies as means of monitoring the quality of new work. The original form of a standard was proposed in the U.S.A. in the early 1940s; see Marsden (1960) on this subject. The specific requirements laid down by the U.S. Bureau of the Budget for use with U.S. Geological Survey maps were stated in 1941 as follows:

With a view to the utmost economy and expedition in producing maps which fulfil not only the broad needs for standard or principal maps but also the reasonable particular needs of individual agencies, it is proposed that standards of accuracy be adopted as hereinafter defined.

1. Three publication scales are selected for reference in accuracy. These are 1:62,500 (1 inch

equals approximately 1 mile); 1:24,000 (1 inch equals 2,000 feet); 1:12,000 (1 inch equals 1,000 feet)...

2. *Horizontal Accuracy* for the first reference scale shall be such that not more than 10 percent of the points tested shall be in error by more than 1/50 inch; for the second scale 1/40 inch; and for the third scale 1/30 inch.

These limits all apply in all cases to positions for well-defined points only.... In general what is "well-defined" will also be determined by what is plottable on the scale of the map within 1/100 inch. Thus while the intersection of two roads or property lines meeting at right angles would come within a sensible interpretation, indentification of the intersection of two such lines meeting at an acute angle would obviously not be practicable within 1/100 inch. Similarly, features not identifiable on the ground within close limits are not to be considered as test points within the limits quoted, even though their positions may be scaled closely upon the map. In this class would come timberlines, soil boundaries, etc.....

3. *Vertical Accuracy*, as applied to contour maps on all scales shall be such that not more than 10 percent of the points tested shall be in error by more than one-half of the contour interval. In checking elevations taken from the map, the apparent vertical error may be decreased by assuming a horizontal displacement within the permissable horizontal error for a map of that scale.

The corresponding metric values are:

1/62,500, 0·51 mm corresponding to 31·9 metres on the ground;
1/24,000, 0·63 mm corresponding to 15·1 metres on the ground;
1/12,000, 0·85 mm corresponding to 10·2 metres on the ground.

In 1947 this statement was amended and the planimetric error was set at 0·02 inches ($\approx$ 0·5 mm) for all map scales smaller than 1/24,000. This gives rise to the much-quoted statement referring to the basic 1/24,000 scale mapping of the U.S.A. that 90% of the points tested ought to lie within 40 feet of the true position on the ground. An indication of the quality of this mapping is provided by Fig. 9.1, which refers to 110 mapping projects tested by means of 3,623 test points.

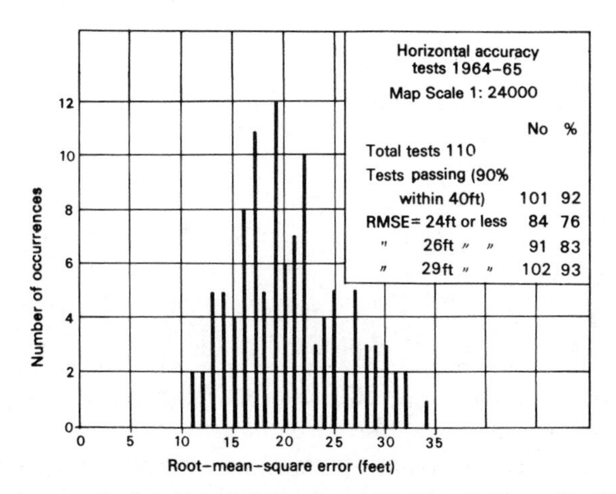

FIG. 9.1. An example of testing U.S.G.S. maps at 1/24,000 scale. The work relates to the years 1964–1965. (Source: Thompson and Rosenfield, 1971).

For more than 30 years, the marginal statement:

THIS MAP COMPLIES WITH NATIONAL MAP ACCURACY STANDARDS

has been added to those U.S.G.S. maps which satisfy this requirement.

Following this lead, similar standards were applied to military topographical maps elsewhere. In 1958, proposals were made by the Institut Géographique National to the 2nd Regional Cartographic Conference of the United Nations for Asia and the Far East, as part of a more extensive report entitled *Standard Cartographic Practices Recommended for International Use for Land Maps* (I.G.N., 1958). In general the proposed system follows that described above, the highest accuracy classification corresponding to American standards, with six lower grades. These standards for evaluation and description were subsequently adopted by NATO for military mapping as a system of classification of operational land maps according to accuracy.

Similarly worded specifications are now to define the horizontal accuracy requirements for nautical charts. For surface navigation charts (Harbour, Approach, Coastal and General Charts), the criteria are defined by Vogel (1981) as:

90 percent of all well-defined planimetric features, except those unavoidably displaced by symbolic exaggeration, are located within 2 mm of their geographic position with reference to a prescribed datum.

Thus a feature on a chart of scale 1/2,500 is required to lie within 5 metres of the true position, but on a chart of scale 1/600,000 this uncertainty increases to 1,200 metres. Many charts published before 1974 have not yet been evaluated for accuracy according to these criteria, and are unlikely to meet such standards.

TESTING QUANTITATIVE ACCURACY

Any scheme for testing the accuracy of a map ought to satisfy the three criteria that:

- The test procedure should have a low probability of accepting a map of low accuracy.
- Conversely, there should be a high probability of accepting a map of high accuracy.
- This should be done using the smallest sample required to satisfy these criteria at a specified level of probability and specified accuracy.

In quantitative accuracy evaluation we are attempting to compare the complicated three-dimensional surface of the ground in which the position of points are usually expressed by the Cartesian coordinates (E, N, H), with another surface mapped in the two dimensions (E', N') with height H' represented by means of spot heights and contour lines. The first of these surfaces is never exactly known in its entirety, for even the most careful measurement of it will contain gaps. Moreover the test measurements themselves can never be wholly error-free. The second surface is never completely defined. Apart from those simplifications introduced by generalisation, the third dimension is only approximately shown because the detailed configuration of the ground between adjacent contour lines has not been established. Following Imhof (1965) we may therefore ask the rhetorical question

How can we compare an incorrectly defined surface with an area which itself cannot be determined exactly?

The American map accuracy standard refers to the *number of points tested* and it is this approach which is the most widely used. In Fig. 9.2, $A = (E, N)$ represents a point on the ground which is shown on the map by the point $A' = (E', N')$. The coordinate differences are

$$\delta E = E - E' \tag{9.1}$$

$$\delta N = N - N' \tag{9.2}$$

and the vectorial distance corresponding to these is AA' or

$$V = \sqrt{(\delta E^2 + \delta N^2)} \tag{9.3}$$

The linear distances δE, δN and V are measures of the horizontal or planimetric displacement of the point A' on the map. If, therefore, the test procedure is repeated for other points, we obtain samples from which the means of the coordinate differences, the mean distances and their standard errors may be determined. Problems concerning the nature of the sampling strategy inevitably arise. Normally the map should be tested by selecting a sample of well-defined points as was specified in the original declaration of map accuracy standards. Following the procedures outlined in Chapter 8, in

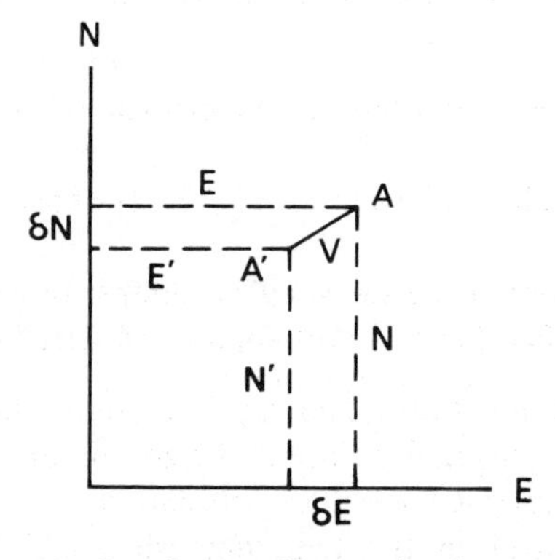

FIG. 9.2. The relationship of a point A on the ground to the corresponding point A' on the map.

which suitable samples may be drawn either by random coordinates or the random application of a systematic point pattern, the majority of those selected are unlikely to be well-defined, or even identifiable, because they are more likely to fall in a featureless area, such as the middle of a field, wood or lake than at the corner of a building or the perpendicular intersection of two roads. Consequently some rule must be devised to allow for selection of the nearest well-defined point to that indicated by the sampling procedure. However, such a rule destroys any illusion of randomness or systematic regularity throughout the sampling frame. Dozier and Strahler (1983) have argued that the preferred method of testing should be: first, to identify all the well-defined points on the map and list these; secondly to make the selection from this list using one of the acceptable methods of drawing a one-dimensional sample described in Chapter 7. This is an elegant solution to the difficulty, but it may be impracticable to execute. A map of an industrial or urban area in a developed country may contain tens or hundreds of thousands of such well-defined points, each of which is a member of the population from which the sample is drawn. The smaller the scale of the map, the larger is this population. It seems, therefore, that the identification and listing of them would become an excessively laborious preliminary stage in the work, even if it were statistically desirable. Such an approach may become practicable when suitable data banks have been created by digitising topographical maps, but implementation of the idea lies in the future.

There is now widespread agreement about the desirability of using the root mean square error and standard error as measures of quantitative map accuracy. Therefore unrestricted or stratified random sampling is to be preferred to the methods of systematic sampling.

The two parameters, root mean square error and standard error, may be used to mean different quantities, as is current Ordnance Survey practice, described by Harley (1975).

Consider a sample of measurements giving rise to the series of values δE_1, $\delta E_2 \ldots$, δE_n and δN_1, $\delta N_2 \ldots$, δN_n for n test points on a map. The root mean square error in each direction is

$$s_E = \pm \sqrt{(\Sigma \delta E^2/n - 1)} \tag{9.4}$$

and

$$s_N = \pm \sqrt{(\Sigma \delta N^2/n - 1)} \tag{9.5}$$

If these errors are random, without any systematic bias, then $\Sigma \delta E = 0$ and $\Sigma \delta N = 0$ and the consistency of the map is demonstrated by s_E and s_N. If, however, the sums δE and δN do not equal zero, some systematic error is present in the map. We determine this as an average discrepancy for the points comprising the sample from

$$\overline{\delta E} = \Sigma \delta E/n \tag{9.6}$$

$$\overline{\delta N} = \Sigma \delta N/n \tag{9.7}$$

The consistency of the map is now found by subtracting the systematic component from the root mean square error or the individual error at each point.

$$\sigma_E = (s_E^2 - \overline{\delta E}^2)^{1/2} \tag{9.8}$$

$$= [\Sigma(\delta E - \overline{\delta E})^2/n]^{1/2} \tag{9.9}$$

and

$$\sigma_N = (s_N^2 - \overline{\delta N}^2)^{1/2} \tag{9.10}$$

$$= [\Sigma(\delta N - \overline{\delta N})^2/n]^{1/2} \tag{9.11}$$

The parameters σ_E and σ_N are called the standard error by the Ordnance Survey. Other organisations are less specific in their use of the terms, and refer indiscriminately to root mean square error, standard deviation and standard error to mean the quantities s_E and s_N.

The most important criticism which may be levied against this treatment of samples of coordinate differences or vectorial distances as one-dimensional errors is that equations such as (9.3) take no account of the signs of δE or δN, and therefore do not indicate any variation or preferred directions in the orientation of the discrepancies. Although the vectorial angle might also be determined and treated statistically, this is uncommon. Usually the directions of the vectors at test points are treated diagrammatically, as in Fig. 9.3, without any further attempt to analyse the data. It is sufficient for these vectors to be short and to show no preferred orientation for the results to be acceptable. Of course the presence of such preferred direction indicates

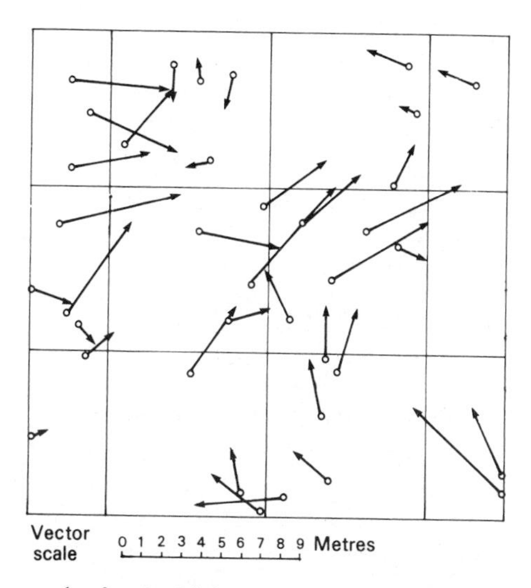

FIG. 9.3. An example of a plot of the vectors, V, as defined in Fig. 9.2, in which the lengths of the vectors have been drawn on a conventional scale. The position of the points tested on the maps are indicated by the small circles.

systematic errors in a map which would become apparent in the use of equations (9.6) and (9.7).

CIRCULAR PROBABILITY

There is a second school of thought concerning the statistical treatment of planimetric error at points. This derives from dissatisfaction with the treatment of them as being one-dimensional errors, and is therefore based upon the combination of linear probabilities in two dimensions. The study of bivariate distributions is well enough known in other fields. It has been studied from the cartographic point of view by Greenwalt and Schultz (1962).

It can be shown that two random and independent errors such as δ_E and δ_N give rise to a two-dimensional probability function having the form of the equation for an ellipse. Thus we may imagine that for some point A' on the map there may be superimposed a concentric series of ellipses whose axes represent different levels of probability. Then the likelihood that the true position lies within one of these ellipses is measured by the probability corresponding to each figure. However, there is no *a priori* reason why s_E should differ from s_N. If they are equal, the probability ellipses become circles, which are easier to manipulate mathematically. It follows that we may imagine a series of concentric circles superimposed upon the point A' on the map, the radius of each circle corresponding to a different value of the probability

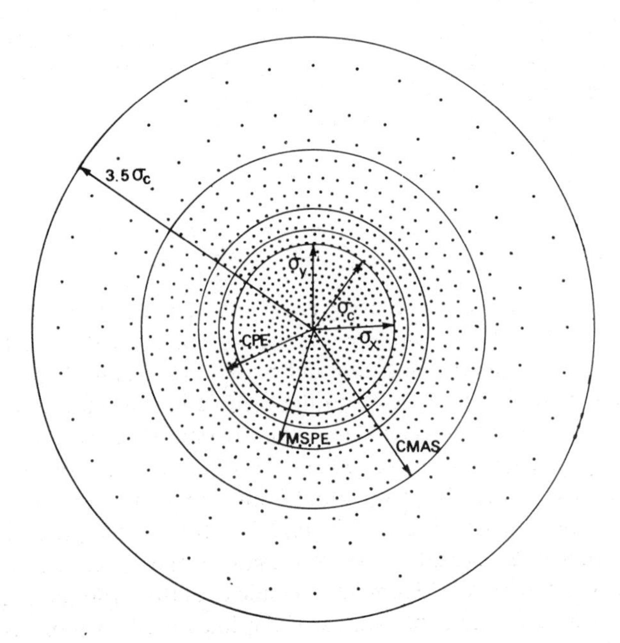

FIG. 9.4. Circular precision indices. (Source: Greenwalt and Schultz, 1962).

TABLE 9.1. *The Definition and Derivation of Circular Precision Indices*

Name	Symbol	P_c	Derivation
Circular standard error	σ_c	0·3935	$1{\cdot}0\sigma_c$
Circular probable error	CPE or CEP	0·5	$1{\cdot}1774\sigma_c$
Circular mean square positional error	MSPE	0·6321	$1{\cdot}4142\sigma_c$
Circular map accuracy standard	CMAS	0·9	$2{\cdot}1460\sigma_c$
Three-five sigma error	3.5σ	0·9978	$3{\cdot}5\sigma_c$

distribution function. In other words, the pattern of these circles provides a series of *circular precision indices.*

Greenwalt and Schultz treat with five of these corresponding to different values of the probability distribution function, which are illustrated in Fig. 9.4 and specified in Table 9.1. It is easy to show that where $s_E = s_N$ and the root mean square error of the vector, V, is

$$s_V = \sqrt{(s_E^2 + s_N^2)} \tag{9.12}$$

so that

$$\sigma_C = s_V/\sqrt{2} \tag{9.13}$$

The most commonly used index in Table 9.1 is circular map accuracy standard, because it relates to the 90% criterion used in map accuracy standards. From its tabulated derivation it follows that, knowing the r.m.s.e. of vectors such as AA', we may convert to CMAS through

$$\text{CMAS} = 2{\cdot}146 s_V/\sqrt{2} \tag{9.14}$$

HEIGHT ERRORS

The evaluation of planimetric accuracy is usually kept separate from that of altimetric accuracy. One reason why this is so stems from the fact that the measurements of planimetric position and height are usually kept separate throughout the mapping process up to the completion of photogrammetric plotting of detail. However a second, and far more important, consideration is that of the scale change between ground and map. Features shown on a map of scale $1/S$ have been reduced by this amount from their ground dimensions. If a check reveals a planimetric discrepancy of 2·5 metres on the ground, this is represented on the map of scale 1/5,000 by a distance of 0·5 mm, at 1/10,000 by 0·25 mm and at 1/20,000 by 0·13 mm. Discrepancies of this order of magnitude are hardly any greater than the widths of the lines used to represent such features as field boundaries, road edges or river banks. By contrast, an error of

2·5 metres in the height recorded at a spot height remains an error of this magnitude irrespective of the scale of the map on which it is depicted, for this information is read from the map and has not been measured on it.

Errors in height may be determined by remeasuring spot heights by independent field survey or photogrammetric methods. If the spot height shown on the map has height H', but an independent check indicates that it is H, there is a height error

$$\delta H = H' - H \tag{9.15}$$

at that point. A similar procedure may be used to determine the height error at any identifiable point lying on a contour line. The method is frequently used by mapping organisations as the means of evaluating contour accuracy. However, there is a relationship between height error and planimetric displacement which we have already seen may be allowed for in the U.S. map accuracy standard. Figure 9.5 illustrates the relationship between a correctly located contour, A, and the incorrect version of it at A'. The discrepancy may be regarded as the planimetric displacement δA or as the height error δH. The two variables are related through the ground slope α, for

$$\tan\alpha = \delta H/\delta A \tag{9.16}$$

It follows that $\delta H = \delta A$ where $\alpha = 45°$, but this represents an exceedingly steep slope which is only likely to be encountered in high mountain country. Invariably, therefore $\delta H < \delta A$. The relationship between the root mean square error in height, planimetric displacement and ground slope may be expressed by means of a form of equation proposed by Koppé (1902), still known as *Koppé's formula*. If the root mean square error in height is represented by s_H and the corresponding r.m.s.e., of planimetric displacement is a_A, we have

$$s_H = \pm(a + b\tan\alpha) \tag{9.17}$$

or

$$s_A = \pm(b + a\cot\alpha) \tag{9.18}$$

Table 9.02 indicates some typical values for the expressions.

The U.S. map accuracy standard states that 90% of the contours tested shall lie within one-half of the contour interval. In Fig. 9.6, A and B are two

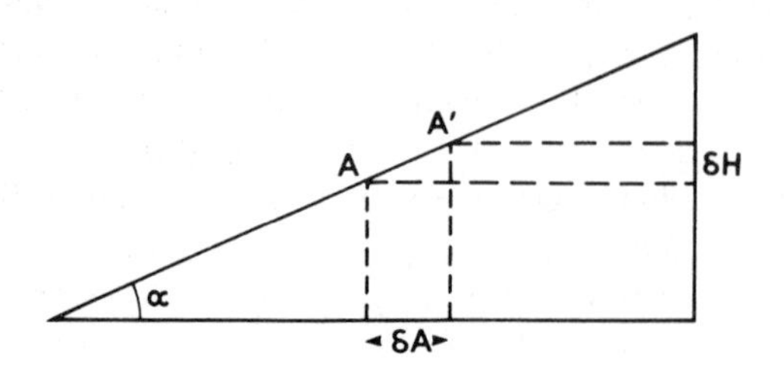

FIG. 9.5. The relationship between the position on a slope of a correctly located contour, A, and an incorrect version of it at A'.

TABLE 9.2. *Relationship between root mean square error in height and in planimetric displacement according to Koppé's formula*

Map scale	Contour interval (m)	Root mean square error in height s_H(m)	Root mean square error in planimetry s_A(m)
1/1,000	1	$\pm(0{\cdot}1+0{\cdot}3\tan\alpha)$	$\pm(0{\cdot}1\cot\alpha+0{\cdot}3)$
1/5,000	5	$\pm(0{\cdot}4+3\tan\alpha)$	$\pm(0{\cdot}4\cot\alpha+3)$
1/10,000	10	$\pm(1+5\tan\alpha)$	$\pm(\cot\alpha+5)$
1/25,000	10	$\pm(1+7\tan\alpha)$	$\pm(\cot\alpha+7)$
1/50,000	20	$\pm(1{\cdot}5+10\tan\alpha)$	$\pm(1{\cdot}5\cot\alpha+10)$

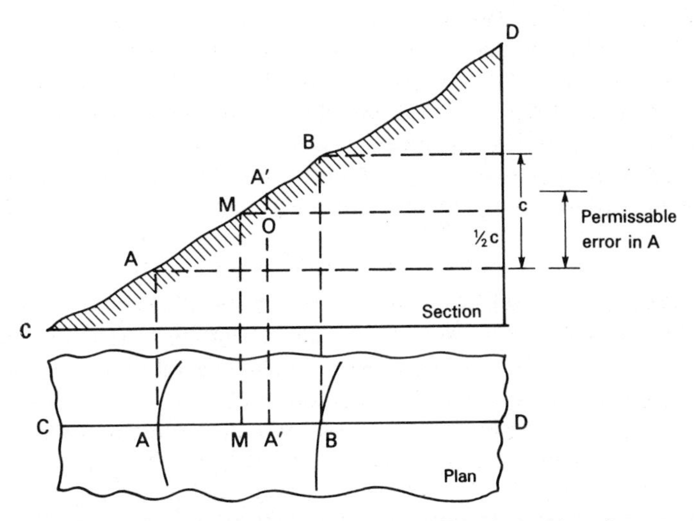

FIG. 9.6. Explanation of the allowable planimetric shift in the position of a contour in the evaluation of height accuracy.

successive contours having a vertical interval, c. To meet the specification it is necessary that A' should lie nearer A than the point AM which represents the ground point corresponding to a height $A+\frac{1}{2}c$. However, it is acceptable if this condition is stretched by allowing for a shift in the position of A' which is equivalent to the acceptable planimetric displacement. Thus if the U.S. specification states that 90% of the points tested shall lie within 0·5 mm of the correct location, this means that we are permitted to consider a position of the contour which is further from A than is the mapped position of A''. In other words, from the small triangle MOA', $MO = 0{\cdot}5$ mm and $OA' = MO\tan\alpha$. Therefore

$$\delta H = \tfrac{1}{2}c + 0{\cdot}5\tan\alpha \qquad (9.19)$$

The method of checking map accuracy by means of a sample of test points may be criticised as being an unsuitable method of evaluating the accuracy of

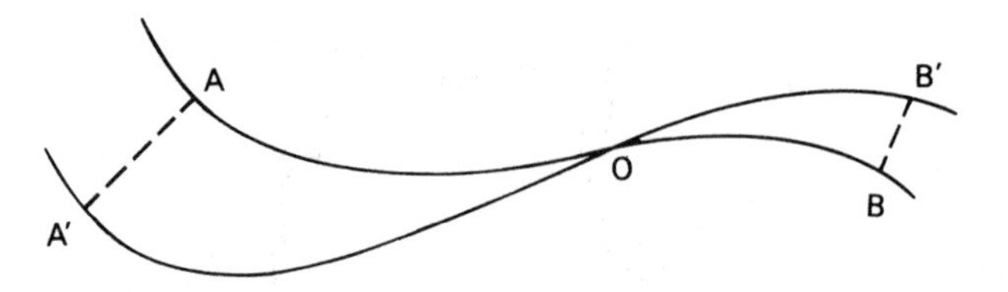

FIG. 9.7. The displacement of a linear feature.

contours. Indeed the same criticism might also be applied to the evaluation of the accuracy of other kinds of linear detail, such as the representation of roads and railways on maps.

Figure 9.7 illustrates the problem of measuring the accuracy of a linear feature by the differences between individual points as posed by Thompson (1956). The line $A'OB'$ represents the centre line of a linear symbol depicting a road or railway on a map. On the ground the correct alignment of it is through the points AOB. The question is how best to describe the lack of agreement between these two lines, especially if A and B are the only places which can be uniquely located on both road and map. Is a proper measure of the errors in the representation of this line the mean displacement at those points which can be recognised? For example, should it be defined as $(AA' + BB')/2$, or since the point O is common to both lines and has zero error, should it be $(AA' + BB' + O)/3$? Should the error be described in terms of the angle AOA', which seems to be more realistic since the mapping error evidently contains some angular misalignment as well as planimetric displacement?

During the 1950s and 1960s there were several important contributions to the study of the methods of testing contour accuracy which are germane to the problem illustrated by Fig. 9.7. The chief impetus to this work arose from the need to develop satisfactory methods of evaluating the accuracy of contours plotted by different photogrammetric methods or instruments. In particular Lindig (1956) produced a novel and comprehensive method of evaluation, and the work of Hoiz (1957) and others continued discussion of this approach. The typical problem to be solved is that illustrated in Fig. 9.8, in which two patterns of contours superimposed upon one another have the appearance of a partly unravelled strand of wool and it is necessary to describe the departures of one of these lines from the other. An important summary of the stage which had been reached by the early 1960s is contained in Blachut (1964), comprising eleven papers on the subject prepared specially for the International Society of Photogrammetry. The techniques are specifically intended for testing the accuracy of contours at the topographical mapping scales. However, there is no reason why a substantial part of the theory should not be applied to testing other linear symbols. It seems, however, that little or no work has been done on these other applications.

Two disadvantages should be stressed. The first is that some of the quantities obtained, for example in the measurement of the elements of

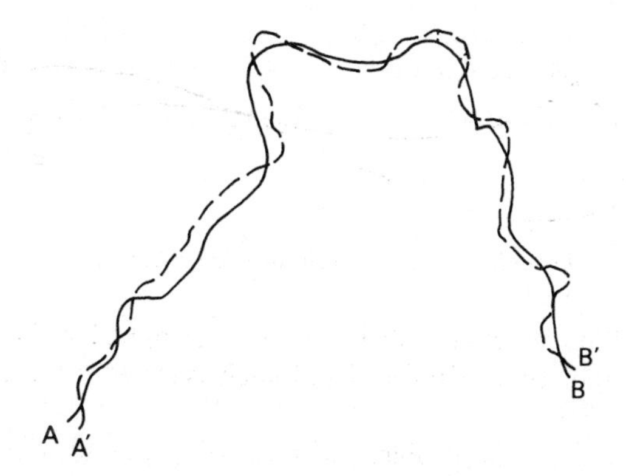

FIG. 9.8. Two contour lines *AB* and *A′B′* of the same ground surface surveyed or plotted by different methods or with different orders of accuracy.

curvature of the contours, are unfamiliar. For this reason it is difficult to understand or appreciate the significance of some of the measures obtained, especially by Lindig in his original work. The second, and more important, is that the methods which have been described require two complete sets of data, for example two complete sets of the contours plotted for the same areas. Although there is a measure of sampling involved in actually determining the various errors, the preparation of two plots of the same feature implies that the whole survey has been repeated, at least as far as photogrammetric plotting of detail is concerned. This may be a practical possibility in research investigations, for example in comparing the different results in an international mapping experiment, but it has no part to play in monitoring map accuracy as an element of production control in routine map production. There is simply not enough time to map everything twice, so that point sampling remains supreme.

THE COMPONENTS OF QUANTITATIVE ACCURACY

The positional accuracy of a point on a printed map depends in part upon the accuracy of the survey and in part upon the accuracy with which points have been located by the plotting, drawing and printing processes. The smaller the scale of the map, the less is the importance to be attached to the errors in surveying and the greater is that attached to the errors of the graphical and reproduction stages. Thus most maps of scale 1/25,000 or smaller which have been derived from modern control surveys and photogrammetric plots have errors of drawing and reproduction which are much larger than those derived from other sources.

In order to examine these propositions in detail it is desirable to consider the contribution which is made by the errors arising at each stage of the mapping process from the original control survey through to the printed map. If there are n different stages in production and each of these gives rise to a root mean square error, s_i, the combination of all of them, derived from equations (6.12)–(6.17) may be written as

$$s = \pm \sqrt{\left(\sum_{i=1}^{n} s_i^2 \right)} \tag{9.20}$$

where s is the overall or total mean square error of the map.

The pioneer work in estimating the components of quantitative accuracy in this fashion was carried out in Russia before and during World War II, notably by Volkov (1946, 1950) (working with K. A. Salishchev) with a later summary by Sukhov (1957). Glusic (1961) subsequently used much of this work and added some other relevant material to provide the only English-language account of this approach.

Although quantitative errors should be expressed numerically, the desirable tolerances needed for plotting and drawing have often been described qualitatively, using such phrases as "accurate to the limits of drawing", "plottable accuracy" or the exhortation to "plot with a precision that equals the measuring capabilities of the most critical user" (U.S. Army, 1955). Such expressions, as Crone (1953) noted, can be variously interpreted according to the outlook of the user, and are of no more guidance to the cartographer than are the "normal reasonable requirements of the specialised map user" suggested earlier. If the statement quoted above from the *Guide to the Compilation and Revision of Maps* were to be considered a serious instruction, the present book ought to have been produced by and for the U.S. Army Corps of Engineers more than 30 years ago, in order to identify what *are* the measuring capabilities of the most critical user.

Although the concept of plottable accuracy is often mentioned in the chapters concerned with drawing office practices to be found in surveying textbooks in the English language, we usually look in vain for any specific dimensions attached to it. In Britan, the nearest approach to a definitive statement is to be found in *Survey Computations* (War Office, 1926), which states that errors in plotting are generally less than 0·01 in. (0·25 mm) and those of fair drawing are generally less than 0·02 in. (0·5 mm). The first of these errors is the same as that figure quoted in the U.S. map accuracy standard as being the limit of plotting for well-defined features. The technical literature in other languages is often more forthcoming about draughting accuracy and related matters. Nevertheless there is still plenty of scope for confusion. For example, in the definition of *draughting accuracy* given in the *Multilingual Dictionary of Technical Terms in Cartography* (I.C.A., 1973), the French definition indicates that

> the average uncertainty resulting from draughting error generally amounts to 1/10 mm and is independent of the scale of the map.

whereas German language sources usually quote a root mean square error $\pm 0{\cdot}2$ mm.

The different stages of map production were summarised in Fig. 2.1, p. 14. We follow the stages listed in that diagram and comment upon the error component belonging to each. In following this procedure it is desirable to remind the reader of the revolution in surveying and mapping techniques which has occurred during the last quarter-century. Methods of fixing position by measuring distances from artificial satellites, originally developed for navigation, can now fix the positions of control points quickly and within a few metres anywhere on earth. This has eliminated much of the dependence upon the traditional methods of control surveys associated with astronomical fixing, in areas which still need to be mapped. In areas which are already well mapped the new methods are, of course, hardly required to improve the positional accuracy of the survey control, for this has already been fixed with adequate precision. Methods of digital cartography have begun to supersede the older methods of compilation and plotting, such as the use of proportional dividers, optical pantograph and process camera. Therefore some of these stages in traditional cartographic production no longer apply for new maps. However, the user who needs to measure distances or areas in a particular place does not necessarily have any choice between using maps made by the older methods and a modern edition created wholly by high technology. In short, if we have to evaluate the accuracy of the source map, we have to bear in mind such factors as its age, purpose and history of revision. This is precisely what the map analyst, examining the source material for, say, a new aeronautical chart, has to do. It is instructive to compare some of the statements made by Glusic (1961) with those by Czartoryski (1987), but, as long as we have to use source maps made in earlier times we cannot ignore the accuracy of the methods of production used quarter of a century or more ago.

Control Survey (r.m.s.e. = s_1)

Nearly all modern basic mapping is based upon correctly executed control surveys. Table 9.3 provides detail of the typical standards which should be satisfied in producing a network of points of known position and height upon which the methods of mapping the surface detail depend.

These standards were achieved by certain national surveys in Western Europe before the end of the nineteenth century. In most other parts of the world, however, the control networks were of lower standard, less extensive, incomplete or just totally lacking until much later. There are, indeed, plenty of parts of the Third World where these strictures still apply. It follows that much of what passed for control on many older maps was patchy, and in some undeveloped countries it might be restricted to a few points which had been located by means of astronomical observations with not much connecting them. The precision with which such points were fixed astronomically is of the

TABLE 9.3. *Classification and accuracy of different categories of control survey*

Category	Average distance between stations (km)	Average angular misclosure in a triangle (sec)	Mean square error in position (cm)	Mean square error in height (cm)
Primary or first-order	>100 <100	1 1	±50 ±5–10	
Second-order	15	5	±3–8	±0·03–0·1
Third-order	5	15	±2–5	±5–10
Fourth-order	1–2	20	±1–2	±1–2

order ±0″·2 to ±10″ of arc. This corresponds to ground distances of 5–500 metres in the direction of the meridian, but is less along a parallel of latitude. Modern developments in fixing position from navigation satellites have changed this picture considerably and we are in sight of obtaining world-wide facility of fixing position within ±2 metres using GPS (global positioning system).

Plotting the Control (r.m.s.e. = s_2)

Irrespective of whether the basic map is to be prepared from ground surveys, photogrammetric plots, mosaics or orthophotos, it is necessary to plot the positions of the planimetric control upon the plotting sheets which are to be used for subsequent stages of the work.

It is at this stage that the ground dimensions are reduced to the scale of a map, and it is in the plotting processes that errors in position attributable to the survey are absorbed. In modern work the mean square errors in position correspond to the values in Table 9.3, and represent only a few centimetres on the ground. Such errors are so small that, to all intents and purposes, control points may be regarded as having no positional error. This does not imply that the positions on the plotting sheet are themselves without error, for the process of locating them with respect to the grid or graticule of the map introduces some *plotting errors*.

Many of the methods of plotting depend upon the use of a *master grid*, which is a square grid plotted on draughting plastic, usually with a separation of 5·0 or 10·0 mm between the points and lines. This grid is used as the framework upon which the map grid and graticule, the neat lines and the survey control are all plotted. The various methods by which this may be done have been summarised by Maling (1973) and are illustrated in Fig. 9.9. This indicates that there are four principal ways in which the master grid may be constructed and the control points plotted, ranging from the use of a large coordinatograph through graphical construction using a steel scale and beam compass. The precision of construction and plotting depends, to a major extent, upon the number of times each point has to be relocated. All geometrical constructions

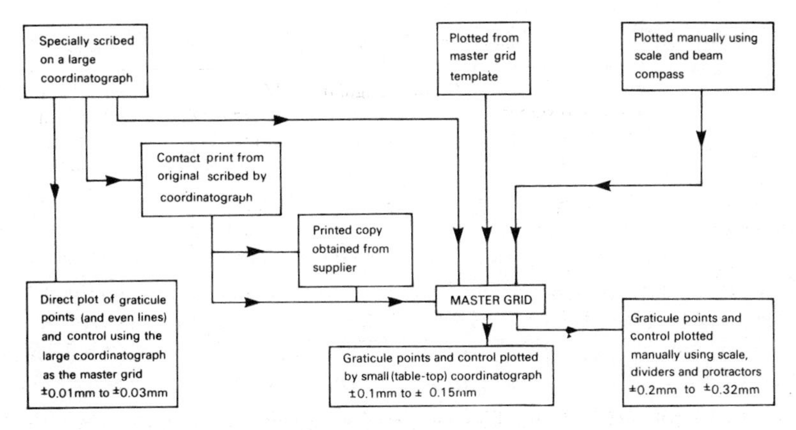

FIG. 9.9. The different methods of preparing a master grid in order to plot survey control points.

are partly cumulative and although Maling in I.C.A. (1984) has emphasised the checks and balances which are used to overcome the growth of systematic error in plotting, some construction errors are virtually unavoidable. The different methods illustrated in Fig. 9.9 increase in precision from right to left. If a large coordinatograph is used to plot the grid, graticule and control points without the need to remove the plotting sheet between any of these stages, the coordinatograph itself has served as the master grid and there are no plotting errors in the positions of the control points relative to the grid apart from those errors present in the scales and movements of the coordinatograph itself. Volkov considered this to be zero, but there is usually a small instrumental error. Glusic suggests that this is of order $\pm 0{\cdot}01$ mm for the type of coordinatograph in which the scales are operated by means of a rack-and-pinion system, but the precision falls to $\pm 0{\cdot}03$ mm for those instruments which are attached to photogrammetric plotters and operated by lead screws which drive the movements of the plotting head. The *flatbed plotters*, which are large coordinatographs designed to operate as an output peripheral to a computer, have accuracy equivalent to the best of the manual coordinatographs. For example, the accuracy of the Gerber Model 32 flatbed plotter is $\pm 0{\cdot}0225$ mm throughout its working area of $1{\cdot}2 \times 1{\cdot}5$ m with a repeatability of $\pm 0{\cdot}0125$ mm.

In all the methods of preparing the plotting sheet where, for various reasons, the work has to be done on different occasions, the mean square error increases appreciably because of the need to relocate the grid intersections on the coordinatograph. This gives rise to a mean square error in setting the plotting sheet of the order of $\pm 0{\cdot}1$ mm. Measurements of distance made by steel scale have similar precision (as we saw in Chapter 7). From the examination of the construction procedures in detail, it may be shown that the overall precision in plotting control by these less satisfactory methods will be $s_2 = \pm 0{\cdot}17$ to $\pm 0{\cdot}32$ mm. This corresponds to CMAS values of 0·26–0·49 mm.

Detail Survey (r.m.s.e. = s_3)

The errors attributable to the survey of map detail are those which may be measured on the plotting sheet after the detail has been surveyed in the field, plotted from aerial photographs and has been transferred to the plotting sheet.

Fifty years ago most detail surveys were done on the ground by chain survey tachymetry or plane table. The first two methods were appropriate for large-scale work whereas plane table surveys were usually conducted at medium or small topographical scales. Maslov and Gorokhov (1959) have investigated in some detail these sources of error in both plane table and tachymetric surveys done at 1/10,000 scale, and conclude that there is not much difference in the mean square errors of either method. The error corresponds to about $\pm 0{\cdot}32$ mm on the plotting sheet. Examples of production work quoted by Glusic are of similar precision, within the range $\pm 0{\cdot}3$ to $\pm 0{\cdot}5$ mm.

The use of aerial photography first came into use as a means of depicting and plotting planimetric detail before the 1920s, but the use of analogue plotters for this work was rare before World War II. Much of the early mapping from photography was done by graphical methods, or by the preparation of photomaps from mosaics and tracing detail from mosaics. The use of rectified prints of prepare *controlled mosaics* as used in some European countries was a preliminary to plotting the detail for large- and medium-scale maps. A typical example of the precision to be expected is provided by the Economic Map of Sweden, at 1/10,000 scale, which was produced with a photomap base derived from controlled mosaics between 1937 and 1966. The root mean square error of the photomap detail derived from these mosaics was $\pm 0{\cdot}8$ mm.

Plotting of topographical detail is now usually done in analogue photogrammetric plotters. There is abundant information concerning the precision of such instruments which is based upon both theory and practice. Thus Hallert (1960) derived theoretical expressions for the root mean square error in planimetry ($s_x = s_y = \pm 9\ \mu$m) and height ($s_z = \pm 13\ \mu$m) for a typical precision plotter such as the Wild A8 instrument. These are measurements made in the plane of the photographic image and distinguish the r.m.s.e. in the direction of the line of flight (s_x) from that in the athwartships (s_y) direction. Actual experience gained from international mapping experiments naturally indicates somewhat lower precision. For example, Blachut *et al.* (1960) quote values for the different instruments used for the Renfrew test and an average of these, $s_x = s_y = \pm 13\ \mu$m.

Obviously the effect of this upon the accuracy of the final map depends upon the scale of photography, the scale of the stereomodel within the instrument and the scale of plotting by coordinatograph or pantograph. Approximate

TABLE 9.4. *Photogrammetric mapping of detail. Root mean square errors in position and height*

Scale	Root mean square error in planimetry (m)	CMAS (m)	Root mean square error in height (m)
1/5,000	±1·0–1·5	1·5–2·3	±0·9–1·5
1/10,000	2–3	3·0–4·5	1·2–1·5
1/25,000	5–7·5	7·6–11·4	1·8–3·0
1/50,000	10–15	15·2–22·8	2·2–3·8

values based upon the use of wide-angle photography gives a mean square error of about ±0·15 to ±0·3 mm (CMAS 0·23–0·45 mm) in the plot. Table 9.4 indicates the corresponding errors in metres on the ground.

Output from orthophoto devices linked to conventional analogue plotters shows little loss in precision compared with the use of the same plotters in the graphical mode. Usually this is quoted as the root mean square errors $s_x = s_y = \pm 0{\cdot}12$–$0{\cdot}2$ mm, or a vector root mean square error ±0·2–0·3 mm. The precision of the contours which are a by-product of the same scanning process is comparable with that obtained with a topographical plotter such as the Wild B8 to draw the contours manually. Johansson (1968b) claims a standard error ±1·7 m in production mapping for the Economic Map of Sweden, produced on an orthophoto base with contours derived from the line-drop method.

Revision of planimetric detail should also be mentioned in this context. Suitable examples of the accuracy of the different methods of revision of the basic Ordnance Survey maps have been given in OS (1971, 1972), Harley (1975) and Matthews (1976). As an example of the 1/1,250 maps of urban areas, the history of which only spans the postwar period, the methods of revision have changed little. Routine accuracy tests for the sheets of Lowestoft have been published in Harley, indicating vector errors of the following magnitudes. These figures are derived from a sample of 409 points on 15 different maps.

Although these represent sheets which have been revised and republished up to six times, there is no evidence of any systematic increase in error with later editions of the map.

TABLE 9.5. *Results of routine accuracy tests of 1/1,250 scale maps of Lowestoft by the Ordnance Survey*

Parameter	Range (m)	Average (m)	Average (mm)
Root mean square error	±0·4–0·44	±0·42	+0·33
Systematic error	±0·11–0·17	±0·14	±0·11
Standard error	±0·35–0·42	±0·39	±0·31

TABLE 9.6. *Precision of the different methods for revising and overhauling the 1/2,500 scale maps by the Ordnance Survey*

Method	Standard Error not greater than		Corresponding CMAS (m)
	metres	millimetres	
Cotswold method	±2·5	±1·0	3·8
Air–ground method	±2·5	±1·0	3·8
Plotting instrument (ARG) revision	±2·5	±1·0	3·8
Plotting instrument (ARMS) resurvey	±0·8	±0·32	1·2
Graphical resurvey	±1·0	±0·8	1·5

The 1/2,500 series of maps has a much longer history, dating in part from the late 1850s. The policies and practices of revision have changed profoundly during this time. The chequered history of the series has been described by Seymour (1963) with a later summary of the attainable accuracies of the methods used during the postwar period by Matthews. The postwar "overhaul" of the series, which was completed in 1981, has comprised a number of different techniques, ranging from the "Cotswold method" of fitting existing map detail within newly plotted ground control, through a variety of different methods using aerial photography. About 75% of the 158,000 maps comprising this series have been overhauled using aerial photography in one form or another. Table 9.6 indicates the relative precision of each technique.

Compilation (r.m.s.e. = s_4)

Discussion of the methods by which the detail selected for a new map is transferred to the compilation manuscript does not lie within the scope of this book. It will suffice to comment here that the commonest method of compiling topographical maps from modern sources is to bring these to a common scale by photography and locate the pieces of film on the compilation manuscript. Assuming that there are no measurable deformations introduced by the process camera, and that there is also negligible deformation of the film upon which the source documents have been copied, the root mean square error introduced by photographic copying is

$$s'_4 = k \cdot s_3 \tag{9.21}$$

where k is the amount of reduction in scale from that of the plotted detail to that of the compilation manuscript. This is expressed as a decimal so that $k = 0{\cdot}5$ represents reduction to one-half the original scale. Less precise methods of compilation involve copying, usually by tracing, from one document and transferring this image to the compilation manuscript. This may be done over a light table, using tracing paper or matt surface plastic, transferring the image to the second document by tracing the detail a second time. Putting the root mean square error of this graphical work at $\pm 0{\cdot}08$ mm[†] we have

$$s_4 = \pm (s_3 + 0{\cdot}08^2 \cdot n)^{1/2} \tag{9.22}$$

There are four stages in map production between plotting the detail and preparing the printing plates. Consequently there are three occasions when map scale may be changed. At each or any of these stages k may take a value less than unity, such as 0·5, 0·333 or 0·25. Only reduction is considered here because it is a fundamental principle of cartography that one should never enlarge a map manuscript. This is because the source information has already been generalised so that it is impossible to recover by enlargement any of the information which was lost at an earlier stage. Moreover, any displacement of detail to improve legibility is exaggerated further by the process of enlargement. The principal reason for reducing the scale after fair drawing is to improve the overall visual appearance of the map. Ever since the process camera become available for cartography it has been standard practice to prepare drawings at two or three times the required size so that, after photographic reduction, most of the small imperfections in drawing have been removed. For example, the majority of diagrams in this book were originally drawn at twice or three times the size at which they appear. Nowadays the quality of scribed originals, combined with the uniformity of stick-on symbols, letterpress and ornament makes it possible to produce high-quality drawings which do not need to be reduced photographically. Since considerable economies in production result from copying documents by contact printing processes rather than suffering the production bottleneck of putting all the work through a single process camera, it is now less common to work at a scale larger than that of the final map.

The other factor of importance in the present context is that reduction after compilation or fair drawing reduces the planimetric errors which were created at earlier stages in the work.

Accuracy of Drawing (r.m.s.e. = s_5)

From the I.C.A. definition of cartographic accuracy already quoted, the "average uncertainty" owing to draughting is 0·1 mm. Without attempting to generate further discussion about what measure this figure is intended to represent, it is to be noted that it is more common practice to quote the root mean square error, and this is generally larger. For example, both Schmidt-Falkenberg (1962) and Imhof (1965) refer to a r.m.s.e. of $\pm 0{\cdot}2$ mm (or a CMAS = 0·3 mm). However Johansson (1968a) was unable to find any reason for

[†]This value is given by Maslov and Gorokhov (1959) to represent the root mean square error of graphical work. It is based upon data originally published in *Trudy Moskovskogo Instituta Inzhenerov Zemleistroystva*, Vyp 2, Geodezizdat, 1957, and includes the errors in plotting points and lines on a plan, construction of angles and the montage errors in fitting photographic images to plotted control.

adopting this value and, moreover, makes the obvious but important point that drawing accuracy is likely to vary for different kinds of work. There are, for example, some kinds of detail where the drawing must be of greater precision, for example scribing detail where this is intricate and dense and where it is therefore important to create a good visual impression with adequate separation between lines and symbols. In other cases, such as drawing contours of flat land, the need for such high planimetric accuracy is not critical. Johansson made a series of test measurements of production work and concluded that the drawing accuracy may vary between ±0·06 mm and ±0·18 mm. The larger of these values, which is close to the root mean square error quoted elsewhere, corresponds to the standard of drawing where the demands for accuracy are low. The values of the root mean square error correspond to CMAS values of 0·1–0·27 mm respectively.

Errors in Reproduction (r.m.s.e. = s_6)

The photomechnical stages in map production following fair drawing of the colour separation manuscripts are often ignored in the study of map accuracy. It is usually claimed that any displacements caused by the process camera are trivial; that any errors which do occur must be operating errors associated with adjusting the camera to reduce by an exact amount and setting the images of the grid, neat lines or colour registration marks against the scales of the graticule etched upon the ground-glass screen of the camera. It is reasonable to suppose that these errors are less than the width of a line, say ±0·15 mm, but may also be reduced, as in equation (9.21) by the reduction factor, k. The same standard of setting grid lines or registration marks one upon another is also likely to occur in contact printing. For contact work, however, $k = 1{\cdot}0$, so that the root mean square error cannot be reduced during execution of the work.

During the early post-war period, when the policies and methods of revising the basic Ordnance Survey maps were being formulated, experiments were made into the effects of repeated copying of materials by contact printing upon the positional accuracy of the line-work. It was done by making a positive polyvinyl alcohol copy from a glass negative of a map and then making a new glass negative from the P.V.A. image. According to Seymour (1963) the cycle was repeated ten times, corresponding to what was thought, at that time, to a century of upkeep of the series. At the completion of the original experiments, as well as in actual revision practice exemplified by the Lowestoft Experiment already mentioned, a slight thickening of some of the linework was all that could be detected. Evidently, therefore, there is very little deterioration in image quality or loss in positional accuracy which could be attributed to contact printing – provided that the media used at all stages in the work are dimensionally stable. At these stages in production it is the stability of the base materials which is of supreme importance in preserving the shape and size of a map during copying, revision and storage. We do not consider the nature of the dimensional changes here, for this subject is treated in detail elsewhere (pp. 196–207). It will suffice to state that some materials, like glass, metal and some of the plastics, change their dimensions regularly with any changes in temperature or humidity affecting the documents, whereas other materials, notably paper, do not. An essential feature of good cartographic practice is that all the materials used for the production of a multicoloured map should be dimensionally stable, even if the prime consideration is to obtain good colour registration and therefore high-quality printing.

Glusic considers that reproduction error in total give rise to r.m.s.e. $\pm 0{\cdot}1$–$\pm 0{\cdot}2$ mm, or CMAS values of 0·15–0·3 mm.

Errors in Colour Registration (s.e.s.o. = s_7)

Irrespective of the materials used in cartographic production, the manuscript map is ultimately reproduced on a series of metal printing plates and then printed on paper. Paper is dimensionally unstable, for it is greatly affected by changes in moisture content which is an inescapable consequence of printing by lithography. Erratic changes in the shape and size of paper sheets during printing may be reduced by meticulous care and attention to the details of handling paper stocks from their manufacture onwards. A description by Harris (1959) of how this was done at the Ordnance Survey more than a quarter of a century ago is still a good indication of the problems involved, and how they may be overcome.

Deformation of the image which occurs after printing is not considered here as an aspect of map accuracy, for it is studied in Chapter 10 as one of the factors which alter the map later, so that cartometric measurements may require special correction in their reduction to ground measurements. However, the errors which may result from poor colour registration during printing should be considered as representing the final element in the accuracy of the map.

In making ready a lithographic press before printing one colour plate, the position where each sheet of paper makes contact with the impression cylinder can be adjusted by means of the paper-feed controls. It is done visually with respect to individual marks plotted near the edges of every printing plate. With experience, skill and care the position of the paper can be controlled within the width of the lines forming these *registration marks*. The root mean square error of setting ought, therefore, to be of the order $\pm 0{\cdot}15$ mm as in photomechanical work, so that the image displacements between the different colours on a map ought not to exceed three times this value. In practice, however, there is occasional random displacement of a sheet or group of sheets as these pass through the press, resulting in much larger image shifts in one colour compared with the remainder of the map detail. This defect is most easily recognised in maps in which the colour registration marks are visible. However, these marks are normally placed close to the sheet edges where they

TABLE 9.6. *The effect of errors in colour registration upon linear measurements. Root mean square errors in millimetres*

Type of line measured	Length of line (mm) 10	50	100
Distance measured between points in the same colour	±0·56	±0·66	±0·93
Distance measured between points of different colours	±0·75	±0·82	±1·05

can be removed by guillotine when the maps are trimmed. Careful inspection of a map along the neat lines should indicate if there is a gross displacement of the images printed in one colour. Obviously such a map cannot be used to measure objects printed in that colour.

Glusic considers that the root mean square error $s_7 = \pm 0{\cdot}17\text{–}0{\cdot}3$ mm, or CMAS values 0·26–0·45 mm.

The influence of errors in colour registration upon cartometric measurements is also demonstrated by the following figures from Sukhov (1957), who has investigated the effect of this error upon the measurement of straight lines on maps in the scale range 1/25,000 through 1/100,000.

THE TOTAL ROOT MEAN SQUARES ERROR OF A MAP

The total root mean square error of a map may be obtained by combining the various component errors s_1–s_7 using equation (6.13). The appropriate values of these errors may be collected to form the following summary table.

Combining these standard errors we determine the following expressions for the root mean square error on each of the documents used in the production sequence. For the numerical examples given as the smallest and largest values in Table 9.7 and the "best" and "worst" values of Table 9.8, the

TABLE 9.7. *The component errors (in mm) at different stages in map production*

Stage	r.m.s.e.	Smallest value	Largest value
Control survey	s_1	±0·005 mm on a map of scale 1/1,000 or larger; otherwise negligible	
Plotting control	s_2	±0·01	±0·32
Plotting detail	s_3	±0·15	±0·30
Compilation	s_4	±0·30	±0·32
Fair drawing	s_5	±0·06	±0·18
Reproduction	s_6	±0·10	±0·20
Colour registration	s_7	±0·17	±0·30

TABLE 9.8. *The combined errors (in millimetres) at different stages in map production*

Document	Equation	Best numerical value	Worst numerical value
Plotting sheet	$s^{①} = \pm\sqrt{\{(s_1/S)^2 + s_2^2\}}$	$\pm 0{\cdot}01$	$\pm 0{\cdot}32$
Detail plots	$s^{②} = \pm\sqrt{(s^{①2} + s_3^2)}$	$\pm 0{\cdot}15$	$\pm 0{\cdot}44$
Compilation ms.	$s^{③} = \pm\sqrt{(k^2 \cdot s^{②2} + s_4^2)}$	$\pm 0{\cdot}33$	$\pm 0{\cdot}54$
Fair drawings	$s^{④} = \pm\sqrt{(k^2 s^{③2} + s_5^2)}$	$\pm 0{\cdot}33$	$\pm 0{\cdot}57$
Printing plates and single colour maps)	$s^{⑤} = \pm\sqrt{(k^2 s^{④2} + s_6^2)}$	$\pm 0{\cdot}38$	$\pm 0{\cdot}67$
Multicolour map	$s^{⑥} = \pm\sqrt{(s^{⑤2} + s_7^2)}$	$\pm 0{\cdot}42$	$\pm 0{\cdot}73$

assumption has been made that the map was produced by an up-to-date organisation. This implies that:

- the control survey is of modern origin so that s_1 may be ignored;
- the detail has been plotted at the scale $1/S$ of the finished map using geometrically rigorous, rather than approximate, photogrammetric methods;
- the entire production sequence from plotting detail through final printing has all been done at the same scale, i.e. $k = 1$ throughout;
- the map is of sufficiently large scale to ignore any effects of generalisation upon the positional accuracy.

From the value of the root mean square error for the printing plate we may obtain CMAS values of 0·57–1·02 mm as being appropriate to a map printed in one colour. These values should be compared with the following values used by the Ordnance Survey as criteria of acceptability of their large scale maps.

- All of these series are printed in one colour, excepting the 1/10,000 scale maps on which contours are printed in brown.
- Taking the final values from Table 9.8, the corresponding CMAS values for a multicolour map are 0·63–1·10 mm.

We should also note that according to the "worst" values in Table 9.8, the final root mean square error ($\pm 0{\cdot}73$ mm) is 2·3 times greater than the correspond-

TABLE 9.9. *Ordnance Survey criteria of acceptability for maps at the basic scales*

Scale and type of survey	Root mean square error: metres	Root mean square error: millimetres
1/1,250 resurvey and continuous revision	±0·4	±0·32
1/2,500 resurvey and continuous revision	±0·8	±0·32
1/2,500 overhaul and continuous revision	±2·5	±1·0
1/10,000 resurvey and continuous revision	±3·5	±0·35

ing value for the plotting sheet. During World War II, German investigation of the accuracy of Russian topographical maps at scale 1/100,000 suggested that this increase was approximately 2·5 times. This led to a method of accuracy evaluation which was later developed by the Aeronautical Charting and Information Center of the U.S. Air Force (now known as *Defence Mapping Agency Aerospace Center*, or DMACC) into a method of factor analysis. This has been described in ACIC (1962) and (1963). The method comprises the determination of the root mean square errors with which control points have been plotted on a map, either by comparison with the published coordinates of these points, or by measurement of corresponding points on the map to be tested with another of larger scale. The CMAS of the map detail is now obtained by allocating numerical factors which correspond to the age of the map, its purpose and the reliability of the mapping organisation responsible for its production. The sum of these factors is within the range 1·1 (for the best up-to-date work) through 3·5 for old and indifferent surveys. The empirically derived factor of 2·5 lies approximately midway between these extremes.

QUALITATIVE ACCURACY

A qualitative error occurs when the information provided by a symbol, or the category attributed to a locality on the map, does not agree with the facts observed on the ground. In short, the inn turns out to be the church or vice-versa. To decide whether a map has acceptable accuracy in this respect a sample of points on the map are checked against the ground and a probabilistic statement can be made about the results. This statement generally claims some minimum level of accuracy (e.g. "90 per cent of the points tested") with a high level of confidence, such as 95% in this claim. Using the form of shorthand which has arisen in the evaluation of agricultural information interpreted from Landsat imagery this example would be called a *90/95 evaluation.* If we can calculate the size of the sample, N, and the allowable number of misclassifications, X, which satisfy this statement, and if the actual number of misclassifications is equal to or less than X, then the map may be accepted as meeting the accuracy specification. The analysis is usually carried out by cross-tabulating the categories identified on the map in a contingency table, or error matrix, such as Table 9.10. In this example the observed categories are the classes of land use shown on the map and the verified categories are the actual land-use classes on the ground. Since much of the theory of qualitative accuracy testing has been developed in the field of remote sensing, especially for the study of maps produced directly from Landsat imagery, the reader will find that most of the literature on the subject refers to the evaluation of the accuracy of different methods of remote sensing. From these figures we may determine the percentage probabilities of correct classification for any category of the variable. This is provided by the entry located on the principal diagonal of the matrix, such as the total of 26 correct

TABLE 9.10. *Map accuracy test results for five land use categories*

Observed categories	Verified categories							
	Arable	Meadow	Permanent grass	Moorland	Woodland	Total	Percentage correct	Error of commission
Arable	26	1	0	0	1	28	93	7
Meadow	1	5	0	0	3	9	55	44
Permanent grass	2	0	43	1	2	48	90	10
Moorland	4	1	2	76	13	96	79	21
Woodland	0	0	2	1	29	32	91	9
Total	33	7	47	78	48	213		
Percentage correct	79	71	91	97	60			
Error of omission	21	29	9	3	40			

identifications of arable land or 29 of woodland, divided by the total number of points observed in that particular category (the column sum). For example, the arable category yields $26/33 = 79\%$ of points correctly identified; that for meadow is $5/7 = 71\%$, etc. It follows that the proportion of all land which has been correctly identified is the ratio of the sum of all the elements lying along the principal diagonal $(26 + 5 + 43 \cdots + 29) = 189$ to the total number of points tested. This is $179/213 = 0{\cdot}84$ or 84%. This figure provides an indication of the overall accuracy of the objects tested, but it is not the only information which may be obtained from the contingency table and some more sophisticated methods will be mentioned later. An important advantage of the contingency table is that it stresses the classification relationship between two different kinds of error which are inherent to any classification system.

First, there is the *error of commission*, which corresponds to the identification of the land use at a point being, for example, arable when in fact it is not arable. Secondly, and conversely, there is an *error of omission*, which occurs when the land use at a point is identified as belonging to another category when, in fact, it is arable. The error of commission therefore relates to the entries along each line of the table. For example $2/28 = 7\%$ of arable land and $4/9 = 44\%$ of meadow have been erroneously observed on the map described in Table 9.10. The error of omission relates to entries in each column of the table so that $7/33 = 21\%$ of arable and $2/7 = 29\%$ of meadow have been erroneously attributed to other classes and have been omitted from their true categories. Ginevan (1979), Aronoff (1982a, b) and Card (1982) have all emphasised the importance of this approach to evaluation of qualitative accuracy, because those methods which do not employ a contingency table tend to concentrate upon the probabilities of correct classification and implicitly ignore the errors of omission. These are particularly important to accuracy testing because they allow us to answer the question "What proportion of the total area of woodland is correctly shown as woodland on the map?" From Table 9.10 this is clearly $29/48 = 60\%$ which is also $100\% -$ error of omission.

A qualitative error in classification should not be confused with any positional errors which may also occur on the map. For example, a parcel of woodland surrounded by grassland may have been correctly interpreted and represented on the map as woodland. If, however, the map also contains quantitative errors so that the mapped position of the boundary of the woodland is incorrect, strict insistence upon positional accuracy will compound additional qualitative errors. If the map is tested in the vicinity of the erroneous boundary, some points may indicate incorrect classification because the boundary is incorrect here. Consequently an allowable displacement of the boundary is required in making the qualitative evaluation, just as a horizontal shift in the position of a contour line is used in measuring the errors in height on a topographical map. An indication of the importance of this principle is provided by the measured positional errors of some Landsat images. The images produced by bulk-processing of the tapes generated by the multispectral–spectral scanner show graticule ticks, the positions of which have been determined from the theoretical position of each picture. Therefore their accuracy depends upon the absence of any deviations in the orbital position of the spacecraft from its predetermined attitude, height and orbital direction. For example, using bulk-processed images of the Chesapeake Bay area, which much of the accuracy testing has been carried out, Colvocoresses and MacEwan (1973) identified a difference of 5 km (5 mm on the picture) between the image centre coordinates given by the orbital data and that derived from ground control identified on the image. So much for the early claims that such imagery could produce maps of equal positional accuracy to conventional mapping methods which were current at that time. It is, however, only fair to add that nowadays the accuracy of such maps is greatly improved. In the present context a human analyst would identify and compensate for such a massive systematic displacement, but it is more difficult to allow for this rather sophisticated reasoning in an automatic image analyser, which would consistently recognise the discrepancies in image detail as qualitative errors rather than as a systematic quantitative displacement.

SAMPLING STRATEGY IN QUALITATIVE TESTING

It should be stressed that the qualitative categories such as land use or land cover are nominal data; that each represents a clearly defined category of the variable and there is no quantitative relationship between or ordering of the categories. The land represented at a point on a map or a small area on the grounds either arable or it is not arable. It cannot be regarded as being "one-half arable" unless a special category of mixed land use having this specification has been included in the classification system. Similarly, because meadow is listed between arable and grassland in Table 9.10 this does not mean that it represents an intermediate category of land use between these two. Arable land does not have a higher or lower quantitative value than woodland because these happen to occur at opposite ends of the table. An important consequence is that the statistical arguments to be employed with such data are those relating to discrete data through binomial probabilities and the binomial distribution.

A contingency table cannot be properly analysed until it is known which sampling methods were used to collect the data. Because the map occupies two dimensions we must employ two-dimensional sampling strategies to test it. Table 9.11 indicates that no single sampling procedure is preferred. All of those described in Chapter 8 have been used by one or other of the investigators working in this field. However, there is increasing awareness that only the unrestricted random sample, or the unaligned stratified systematic sample offer satisfactory statistical possibilities.

TABLE 9.11. *Sampling strategies used to evaluate the qualitative accuracy of Landsat-derived maps*

	Sampling strategy				
	Unrestricted		Stratified		
Author	Random	Systematic	Random	Unaligned systematic	Combined or multistage
Hord and Brooner (1976)	√	–	–	–	–
Lins (1976)	–	√	–	–	–
Fitzpatrick (1977)	–	–	√	–	–
Van Genderen *et al.* (1978)	–	–	√	–	–
Latham (1979)*	–	√	–	–	–
Hay (1979)	√	–	–	–	√(1)
Ginevan (1979)	√(3)	–	–	–	–
Rosenfield and Melley (1980)	–	–	–	√	–
Strahler (1981)	–	–	√	–	–
Fitzpatrick-Lins (1980, 1981)	–	–	–	√	–
Aronoff (1982a)	√(3)	–	–	–	–
Aronoff (1982b)	√(3)	–	–	–	√(1)
Rosefield *et al.* (1982)	–	–	–	–	√(2)

(1) Simple unrestricted random sampling followed by stratification by class and additional random sampling within each category until the minimum sample size has been attained.
(2) Stratified unaligned systematic sampling followed by additional random sampling in the under-represented categories.
(3) Subsequent analyses are based upon acceptance sampling methods.
*Quoted by Rosenfield (1982a).

Sample Size

The desirable size of sample is related to the sampling strategy, although, as noted in Chapters 7 and 8, the method of determining sample size which is based upon equation (7.16) is only appropriate to a random sample but is often used in conjunction with other strategies. The methods is used by the U.S. Geological Survey, notably by Fitzpatrick-Lins (1980, 1981) and by Rosenfield *et al.* (1982), who write the equation

$$N = z^2 \cdot pq/E^2 \tag{9.23}$$

to find N. Here $z = 2 \approx 1{\cdot}96$ corresponding to the value for t for an infinitely large sample at the 95% probability level. The values for n given in Table 7.8 are not strictly applicable for map accuracy testing because the 90% criterion is almost universally applicable in this work so that $pq = 0{\cdot}09$ whereas Table 7.4 is based upon $pq = 0{\cdot}25$.

A somewhat different procedure for determining sample size together with an allowable number of misclassifications, X, may be developed from the theory of *acceptance sampling*, which is the branch of applied statistics concerned with the determination of whether large batches of manufactured articles are of acceptable quality. The analogy of this to the map accuracy problem is that we may consider a map to be composed of a large number of potential check points (articles) which are either correctly classified (acceptable articles) or misclassified (defective articles). In the map accuracy application, where the number of articles in a batch is virtually infinite, a sampling plan may be based upon the binomial probability density function (5.19). For this purpose we may write

$$P(c) = \frac{N!}{(N-c)! \cdot c!^{Q^c(1-Q)^{N-c}}} \tag{9.24}$$

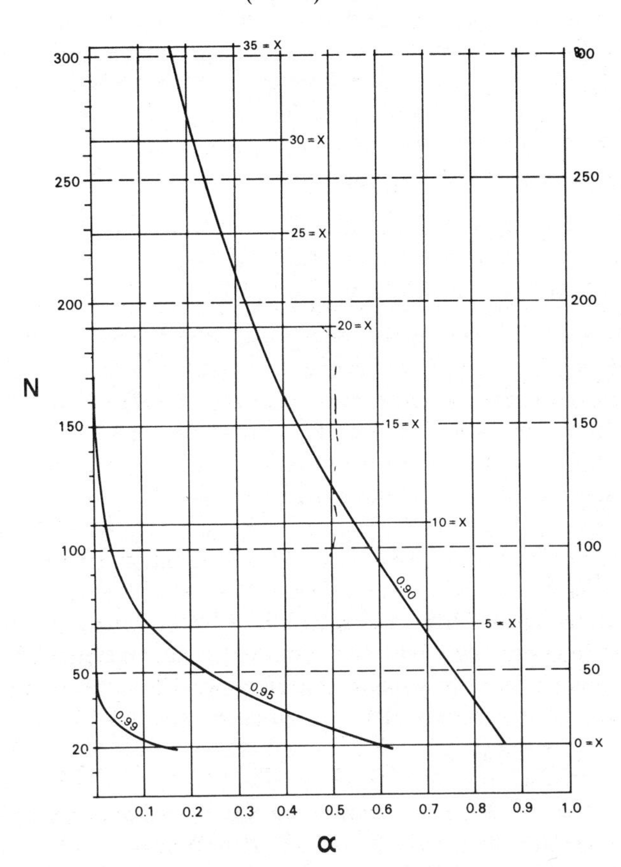

FIG. 9.10. Graph derived from Ginevan's Table 2 to demonstrate the use of acceptance sampling methods in qualitative map accuracy evaluation. This graph is based upon the assumption that the consumer's risk, $\beta = 0{\cdot}05$. The three curves are for $Q = 0{\cdot}90$, 0·95 and 0·99. The abscissa α represents the producer's risk and the ordinate, N, is the size of sample meeting the specified requirements. X is the number of allowable errors which will meet this specification.

In this equation, $P(c)$ is the probability of obtaining c correct points in the sample of size N. Q is the accuracy of the map, i.e. the proportion of correctly classified points required to meet the map accuracy specification.

We may determine two quantities using equation (9.19) which allow us to choose a suitable size of unrestricted random sample. First, we use a lower accuracy proportion, Q_L, substituting which in (9.19) allows us to determine the quantity β which represents the probability of accepting an inaccurate map, corresponding to condition (1) on page 148. In acceptance sampling this is known as the *consumer's risk*. Using similar arguments to determine a higher accuracy proportion, Q_H, we may obtain the quantity α which is the probability of rejecting an accurate map. This is condition (2) on page 148, and is the *producer's risk* of acceptance sampling. Ginevan (1979) has calculated the relationship between Q, α, β, X and N for the binomial distribution and has published extensive tables to assist the determination of sample size. Figure 9.10 has been constructed from part of these tables to show the relationship between α, X and N for different levels of probability.

In order to show how the graphs are used, let us assume that we wish to test a map according to the specification that:

1 the map accuracy standard shall apply – then $Q = 0{\cdot}90$;
2 the consumer's risk shall be $\beta = 0{\cdot}05$ (or 5%);
3 the producer's risk shall be $\alpha = 0{\cdot}3$ (or 30%).

By inspection of Ginevan's tables, or more approximately from Fig. 9.10 for these values for Q or β we find that the closest value to that required for α is 0·2998. The corresponding values are $X = 23$ and $N = 213$. In other words we require a sample of minimum size 213 points to be chosen by unrestricted random sampling. The allowable number of mistakes in classification which meets the specification given above is 23.

Further Analysis

An error matrix clearly shows the results of testing many variables, with separate results which we have seen from the elementary interpretation of Table 9.10. However, when a single measure of quality is needed, for example when comparing two different remotely sensed images of the same area both of which have been compared with a map, it is convenient to employ a single value to summarise the information contained in the whole matrix. One method is to use an analysis of variance, exemplified by the work of Rosenfield and Melley (1980), Rosenfield (1981), etc. However, in this procedure it is necessary to assume that the accuracy levels observed in each category are independent. If, as often happens in the interpretation of images, one category of the variable is consistently confused with another, this essential condition is not satisfied.

A second approach has been used by Congalton *et al.* (1983), who have used discrete multivariate analysis to generate a *normalised error matrix* corresponding to that in Table 9.10. The object is to measure the agreement between two such matrices. A normalised error matrix has line and column adjusted until its contents sum to unity. This is done through a process of iterative proportional fitting.

The third, and least-explored, line of investigation is that reported by Maxim and Harrington (1983), who have employed pseudo-Bayesian estimations in multivariate analysis for this purpose.

10

Deformations of the Medium

The impossibility of describing the surface of the *sphere* on a *plain*, has induced the *ingenious* to many projections of the sphere *in plano*, scientifically though *apparently* true approximations of the truth have been divised, which describe small portions of the sphere very accurately on a *plane*, but every attempt hitherto to describe the surface of the *whole globe* has greatly failed in one or more of the objects essentially requisite in geography.

(Alexander Dalrymple, *An Historical Collection of the Several Voyages and Discoveries in the South Pacific Ocean, 1770.*)

In Chapter 2 we saw that deformation of the mapped image may occur and that there are two major reasons for this. The first is that the image is inevitably deformed by the process of making a plane map represent the curved surface of the earth. The second cause of deformation is regular or irregular alteration in the dimensions of the printed map arising from variations in temperature and humidity during manufacture, use and storage. Both affect the results of measurement by changing the scale of the map locally. Since they are unavoidable, it is either necessary to apply some form of correction when the measurements are reduced to the ground dimensions or they may be ignored. The choice of action depends upon whether they are large enough to affect the results, and the usual variables; such as the size of the parcel, the length of the line, the accuracy of the method of measurement and the purpose of the measurement influence the choice. As always, the last is by far the most important consideration. In this chapter we examine the magnitude of the deformations which may occur; in Chapter 11 we study some methods of applying corrections where these are considered to be desirable.

REPRESENTATION OF THE CURVED SURFACE OF THE SPHERE AS A MAP

It was hinted in Chapter 2 that the simple definition of map scale is not strictly correct, for scale varies in all parts of the map and may do so in different directions, too. If the map scale were true at all points and in all directions, the map of the curved surface of the earth would be an exact replica of it, and would therefore be a curved surface, such as is represented at a very small scale by a globe. The resulting transformation to the plane surface, which we have already likened to stretching of the curved surface, gives rise to changes in scale which we will describe broadly as *linear distortion.* This, in turn, affects the

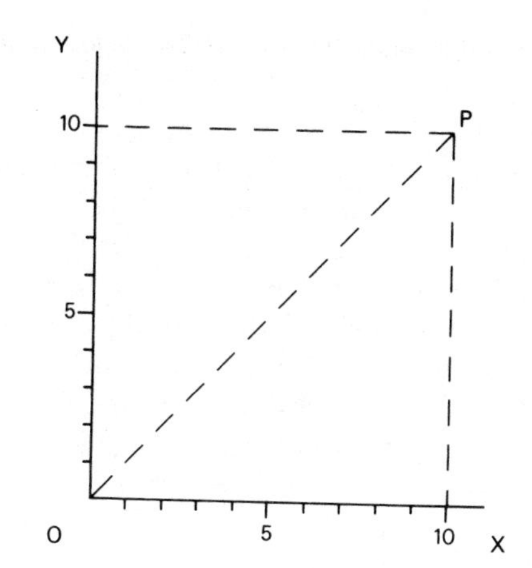

(a)

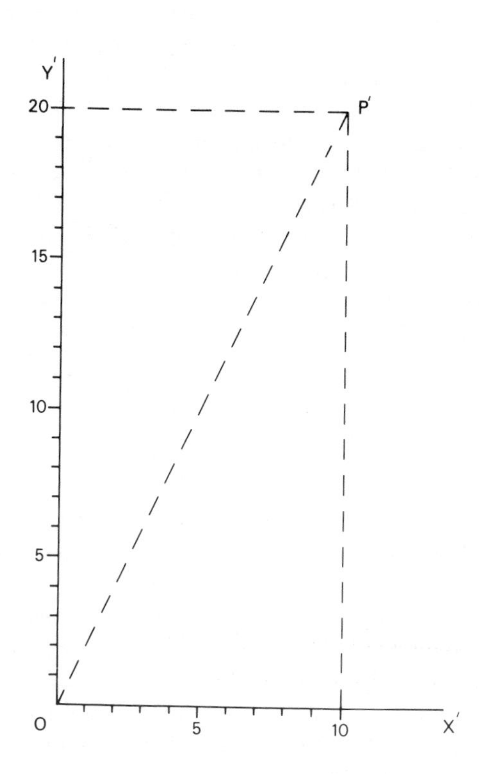

(b)

FIG. 10.1. The effect of linear distortion upon the representation of angles and areas in a square (a) which has been stretched along the y-axis into a rectangle (b).

representation of areas and angles on a map, as the following example illustrates.

Figure 10.1(a) represents a small (strictly speaking, an infinitely small) portion of the spherical surface upon which a point P is referred to another point O by means of (x, y) Cartesian coordinates. If the coordinates of P are (10, 10) units it follows that the length of the straight line $OP = \sqrt{2} \times 10 =$ 14·14 units, the angle $XOP = 45°$ and that the area of the square enclosed by the points $XOYP = 100$ square units. Figure 10.1(b) illustrates the corresponding figure which has been stretched in the direction of the ordinate by doubling the distance $O'Y'$. Thus $P' = (10,20)$ units and the length of $O'P' =$ 22·36 units. Therefore the amount of linear distortion in that direction amounts to $22{\cdot}36/14{\cdot}14 = 1{\cdot}58$ times, whereas that in the direction $O'X'$ remains unity because $O'X' = OX$ in the original square. The angle $X'O'P' =$ 60° so that there is deformation of that angle amounting to $60° - 45° = 15°$. The area of the rectangle $O'X'P'Y' = 200$ square units, which is double that of the original figure.

The consequences of stretching or the creation of linear distortion demonstrated by this simple example indicates the need to re-examine the concept of what we mean by scale and how this affects the use of maps for measurement. In the following discussion the reappraisal is descriptive, employing a minimum of mathematics. For a more complete exposition the reader is referred to the literature on map projections. Naturally pride of place is given to Maling (1973), but this book does not offer a treatment of the subject which is mathematically rigorous. The reader requiring this should refer to Richardus and Adler (1972). Some of the simpler books on map projections in the English language which are preoccupied with descriptions of how each projection is constructed, are not particularly useful for the study of the distortions and therefore the kind of decisions and choices which have to be made in cartometric work.

THE SCALE OF A GLOBE

Consider two pairs of corresponding points, A and B on the earth, which is here regarded as being perfectly spherical, and $A'B'$ upon a small globe which depicts the earth at a smaller scale.

In Figure 10.2(a) the distance between A and B on the earth is the angle $AOB = z$. In Fig. 10.2(b) the corresponding angle on the globe is $A'O'B' = z$. From the elementary geometry of the circle the arc length AB may be expressed as $R{\cdot}z$, where R is the radius of the earth and z is measured in radians. On the globe, $A'B' = r{\cdot}z$, where r is the radius of the globe. Applying the elementary definition of scale to these arc distances we find that

$$1/S = A'B'/AB = r{\cdot}z/R{\cdot}z = r/R \qquad (10.1)$$

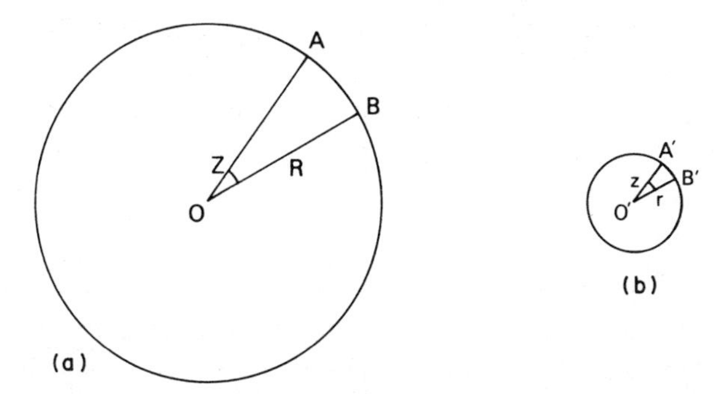

FIG. 10.2. Representation of an arc, AB, on a sphere of radius R (a) and on a smaller sphere of radius r (b).

Since the globe is an exact replica of the earth, the ratio r/R is constant for all pair of points on the surfaces and therefore the scale $1/S$ applies to *any line* measured in *any direction.*

The Principal Scale

It will be convenient to describe this relationship as the *principal scale* (μ_0) when referring to map projections. The reason for this choice of phrase is to emphasise that it is possible to preserve the simple scale relationship of (2.1) or (10.1) at certain points, along certain lines or in certain directions on the plane map. The elementary definition of scale relates to the lengths of lines which are, obviously, of finite length. However, the principal scale of a map applies to infinitesimally small parts of it. Otherwise it would not be possible to make the mental leap from these equations, which relate to finite distances, to the bald statement that the principal scale can be preserved at a point.

Since the principal scale relates to the comparison of two curved surfaces in which there is no deformation, the points or lines where this condition is realised on a map are known as *point(s) of zero distortion* or *line(s) of zero distortion.* A line of zero distortion represents an infinitely narrow zone upon the map, and a point of zero distortion represents a circle of infinitely small radius within which the principal scale is preserved. Geometrically those points or lines frequently coincide with those parts of the map which may be regarded as having been fitted to the spherical surface of a globe by wrapping the map around it, for example in the form of a cylinder or cone. However, this analogy only holds good for certain map projections. Figure 10.3 illustrates some examples of the location of the points or lines of zero distortion on some map projections in common use. The qualities which Maling (1973) regards as being the *fundamental properties* of a map projection relate to the nature and location of these points or lines with respect to the world or hemispheric outline of the map as illustrated here. Whenever we use the representative fraction to describe the scale of a map, we are referring to the principal scale,

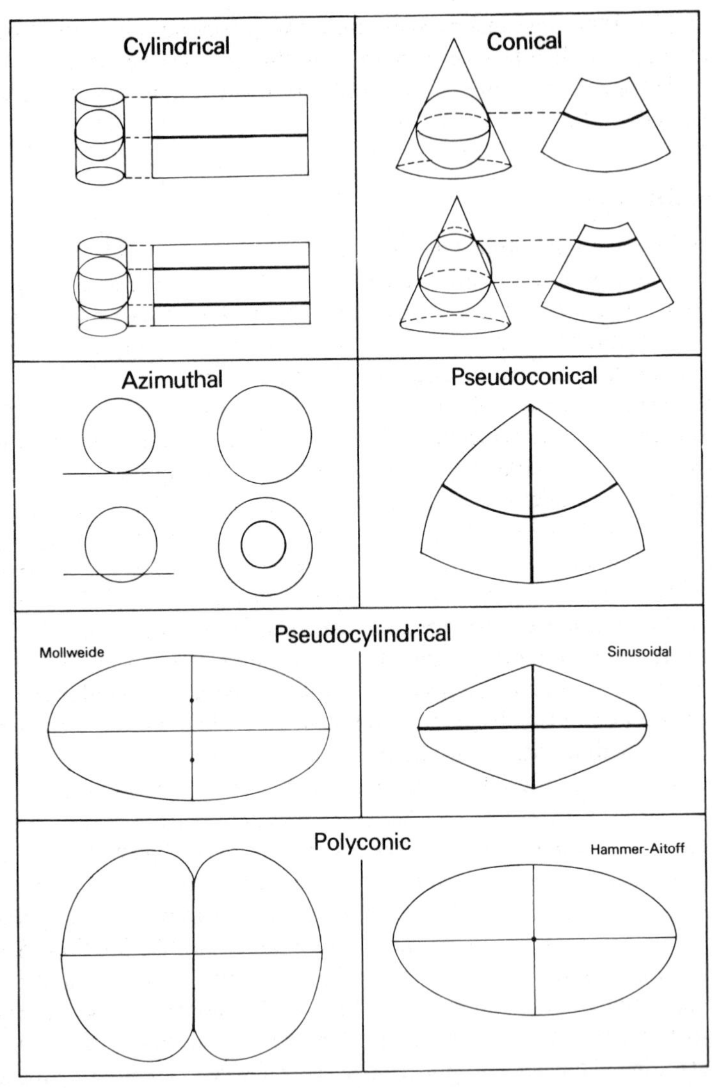

FIG. 10.3. The six principal groups of map projections in practical use, illustrating the location and nature of the lines of zero distortion and the outline of the map of a sphere or hemisphere according to results.

and it is this quantity which is used to convert measurements into their corresponding ground units. If this conversion is applied without any modification, as is customary in large-scale work, it is implied that the user has assumed that linear distortion in the map may be disregarded.

Thus far we have thought of scale as being a fraction, for example 1/5,000,000. However, both the representative fraction and the corresponding decimal 0·0000002 are inconvenient for making comparison with other scales which may be identified on a map. For this reason the principal scale is normally expressed by unity, or $\mu_0 = 1{\cdot}0000$, and all other scales are expressed as decimals of this.

Particular Scales

Beyond the points and lines of zero distortion the map scale varies continuously and takes values called *particular scales*, denoted by μ. There may be a variety of particular scales in different directions from a point on a map. However, the particular scales also change from place to place. It is possible to calculate these, together with the other distortion characteristics, from the general theory of map projections, provided that we have adequate knowledge about the projection in use. This is because it is always possible to determine two particular scales at any point; that along the meridian, h, and that along the parallel of latitude through the point, k. They are determined by comparison of the infinitesimally small arc elements on the map derived from the projection equations which are expressed in either plane Cartesian (x,y) or plane polar (r,θ) coordinates for arguments of latitude (φ) and longitude (λ) on the earth. In practice, however, it is not usually necessary to calculate the particular scales, for this has already been done for the projections which are likely to be encountered in map work. They have been tabulated in the form of Tables 10.1–10.4, and have been published in most of the encyclopaedic textbooks on the subject, such as Reignier (1957) and Wagner (1949). Snyder (1982, 1987) is also good in this respect, but only treats with those map projections used by the U.S. Geological Survey.

Since the principal scale is taken as unity, and the particular scales are expressed as decimals, the convention saves a great deal of clumsy arithmetic and, moreover, it can be applied to any maps based upon the same projection irrespective of their scales as indicated by the representative fractions. Suppose that we have a map of scale 1/5,0000,000 ($= \mu_0$), but measurement of part of it indicates that here the scale is 1/4,830,000. There is a relationship $5{,}000{,}000/4{,}830{,}000 = 1{\cdot}0352$, which is the value of the particular scale in that part of the map. Likewise we may have a map of principal scale 1/1,000,000 but measurement along a parallel indicates a scale of 1/965,997 in that part of the map. Then $\mu = 1{,}0000{,}000/965{,}997 = 1{\cdot}0352$ as before. Obviously there is an advantage it we can avoid having to calculate the individual representative fractions and simply work with the two decimal values 1·000 and 1·0352 respectively.

From the relationship between μ_0 and μ, it follows that the *scale error*, in a particular part of a map or in a particular direction, may be expressed as $\mu - 1$. In the example given above, $\mu - 1 = +0{\cdot}0352$ or $+3{\cdot}5\%$. If however, the particular scale was given as 0·9658, the scale error would be $-0{\cdot}0352$ or $-3{\cdot}5\%$.

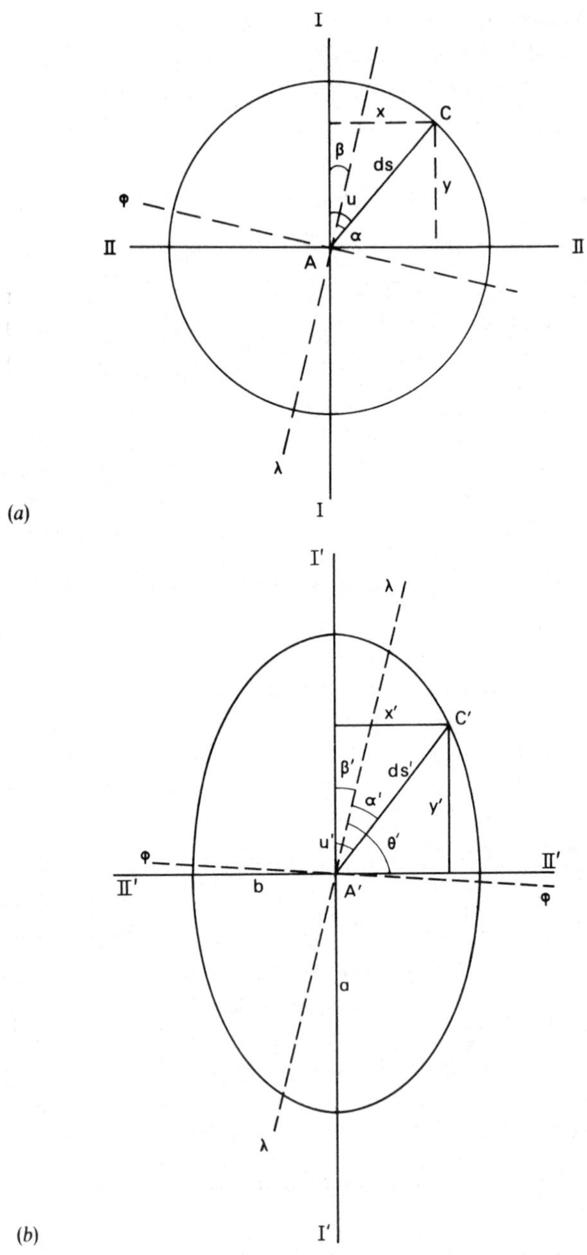

FIG. 10.4. The ellipse of distortion, or Tissot's indicatrix. (a) An infinitely small circle, of radius *ds*, centred at the point $A = (\varphi, \lambda)$ on the sphere. (b) The representation of the circle as an ellipse at the corresponding point A' on the map. The line *ds* has been stretched to *ds′* and the bearing of it, measured from the principal direction I' as the angle ω' is different to the angle μ measured from the corresponding direction I in (a). The broken lines represent the graticule lines, φ and λ intersecting at A and A'.

The Ellipse of Distortion

An essential stage in the study of the theory of deformation is the statement and proof of *Tissot's theorem*, which demonstrates that in the transformation from one surface to another there are two orthogonal directions on the first surface which remain perpendicular on the second. These are known as *principal directions*. Moreover, a circle having infinitely small radius upon the first surface is transformed into an infinitely small ellipse on the second. The resulting figure is known as *Tissot's indicatrix* or the *ellipse of distortion*, and is illustrated in Fig. 10.4. It can be shown that the principal directions are those in which the particular scales are the maximum and minimum values for that point. These scales are denoted a and b respectively and, because they define the extremes, they also correspond to the semi-axes of the ellipse of distortion.

Two further measures of deformation follow from the identification and description of the ellipse of distortion.

Area Scale

The area scale at a point is that relating unit area at that point to unit area on the curved surface of the earth. It can be shown that the area scale may be expressed as

$$p = a \cdot b \tag{10.2}$$

and in the special case where the principal directions coincide with the parallels and meridians on the map,[†] this may be written

$$p = h \cdot k \tag{10.3}$$

In the general case, however, the principal directions do not coincide with the graticule and the angle formed between the parallel and meridian at the point is some other value, θ. Then the area scale may be determined from

$$p = h \cdot k \cdot \sin \theta \tag{10.4}$$

The *area distortion* or exaggeration of area at a point is clearly $p - 1$. This may be expressed as a percentage, like linear scale error.

ANGULAR DEFORMATION

It can be shown that the maximum angular deformation which occurs at a point may be derived from the equation

$$\sin \omega/2 = (a - b)/(a + b) \tag{10.5}$$

[†]This occurs in the *normal aspect* of many projections, including all those of the cylindrical conical and azimuthal classes, but not in their transverse or oblique aspects. Nor can this happen in other classes of map projections in which the graticule intersections are not perpendicular on the map.

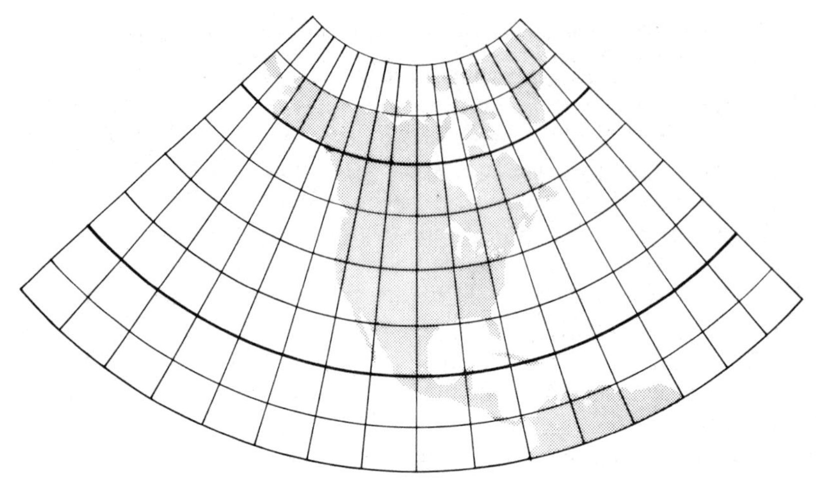

FIG. 10.5. North America on Alber's Projection having standard parallels in latitudes 20° N and 60° N.

For the special case where

$$a = b = h = k \tag{10.6}$$

We find that $\omega = 0°$ and the ellipse of distortion is a circle. This occurs at the points and along the lines of zero distortion, where $a = b = 1{\cdot}0000$. It also occurs at all points in a conformal projection, where $a = b$ throughout so that all angles are correctly represented.

A typical example of how the data relate to the particular scales and distortion characteristics of a particular map projection is given in Table 9.1. A more complete version of this table will be found in Snyder (1982, 1987). This version of Albers' projection has been used for maps of the coterminous U.S.A. at a scale of 1/2,500,000 by the U.S. Geological Survey and is illustrated in Fig. 10.5.

The following features of Table 10.1 should be noted:

1 There are two lines of zero distortion, corresponding to the two standard parallels in latitudes 29° 30′ north and 45° 30′ north. The table shows that along these parallels both of the particular scales, h and k are equal to unity so that there is no angular deformation. There are the only places on the map where this condition is satisfied.

2 The projection is described as being *equal-area*. This is an important special property of some map projections in which the area scale $p = 1{\cdot}0000$ throughout. It follows from (10·7) that in order to achieve this condition the particular scales in the principal directions must be reciprocals of one another. Thus we see that

$$h = 1/k \quad \text{or} \quad k = 1/h \tag{10.7}$$

TABLE 10.1. *Albers' conical equal-area projection; particular scales and distortion characteristics for the version suitable for a map of the coterminous U.S.A., having standard parallels in latitudes 29° 30°′ N and 45° 30′ N*

Latitude	Particular scales		Scale errors (%)		Area scale	Maximum angular deformation
φ	Meridian (h)	Parallel (k)	$\|1-h\|$	$\|1-k\|$	(p)	($\omega°$)
52	0·9721	1·0287	−2·8	+2·9	1·0000	1° 37′
50	0·9830	1·0173	−1·7	+1·7	1·0000	0° 59′
48	0·9917	1·0083	−0·8	+0·8	1·0000	0° 29′
45° 30′	1·0000	1·0000	0·0	0·0	1·0000	0°
42	1·0071	0·9929	+0·7	−0·7	1·0000	0° 24′
40	1·0091	0·9909	+0·9	−0·9	1·0000	0° 31′
38	1·0098	0·9903	+0·9	−1·0	1·0000	0° 34′
36	1·0093	0·9908	+0·9	−0·9	1·0000	0° 31′
34	1·0076	0·9925	+0·7	−0·7	1·0000	0° 25′
32	1·0048	0·9952	+0·4	−0·4	1·0000	0° 16′
29° 30′	1·0000	1·0000	0·0	0·0	1·0000	0°
26	0·9909	1·0091	−0·9	+0·9	1·0000	0° 31′
24	0·9846	1·0156	−1·5	+1·5	1·0000	0° 53′
22	0·9776	1·0229	−2·2	+2·3	1·0000	1° 18′

because this is a projection in which equation (10.7) is satisfied. In the general case,

$$a = 1/b \quad \text{or} \quad b = 1/a \tag{10.8}$$

3 Although the maximum and minimum particular scales coincide with the graticule, as indicated by the footnote on page 185, in this example the significance of them changes at the lines of zero distortion. Thus to the south of latitude 29° 30′ N and to the north of 45° 30′ N the maximum particular scale corresponds to k along the parallel, but between the standard parallels the maximum particular scale is directed along the meridians. This is a peculiarity of all those projections having two lines of zero distortion which are parallel to one another.

4 The geographical pole is a point on the earth but on Albers' projection it is represented by a circular arc. Consequently the relationship outlined above breaks down, for this is a point which has been mapped as a line. The geographical pole on this projection must be treated as a *singular point* to which the distortion theory cannot be applied. Many map projections have such points at their extremities. Obviously such parts of the map should be avoided in cartometric work.

5 Since this projection was designed for use between the parallels 22° and 52°, the scale errors within it do not exceed ±2·9%. Indeed between the parallels 25° and 50° it does not exceed ±1·5%. For many practical purposes, therefore, the whole of the centre of the map is suitable for making linear measurements, for the linear scale error is generally less than 1%. However, the question whether it may be desirable to apply small

corrections for the scale errors which are present depends upon the purpose for which the measurements are required.

Table 10.1 indicates these listed features at a glance. In order to assess the intervening scales between the tabulated values, or to study the more complicated two-dimensional distribution of the particular scales where these change with both latitude and longitude, it is useful to employ graphs or maps upon which the tabulated data have been plotted. There are a variety of ways in which this may be done, many of which have been described by Maling (1973).

THE SPECIAL PROPERTIES OF MAP PROJECTIONS

Although an infinity of different map projections could be described, probably less than 100 of them have ever been used for maps, and the majority of modern maps are based on fewer than 20 different projections. Most of these are preferred because they possess a specific mathematical property. There are three such *special properties* which are particularly important to cartometry, known as *equivalence*, *conformality* and *equidistance*. There are a number of additional qualities which also have some practical value, which are listed by Maling (1968) and which may occur together with one of those already mentioned. An important feature of these three is that they are mutually exclusive. Thus it is impossible to have a projection which is at the same time equal-area and conformal, one which is both conformal and equidistant, or an equidistant projection which is also equal-area.

Equal-area Projections

As indicated by the example of Albers' projection given in Table 10.1, the property of equivalence is that of maintaining constant area scale, $p = 1{\cdot}0000$, throughout the map. It can be shown that this special property applies to parcels of any size, not only to infinitesimally small ellipses representing Tissot's indicatrix at different points on the map. As a result, the areas of parcels of *all sizes* and *anywhere* on the map are represented without distortion.

Conformal Projections

The special property of conformality is that the particular scales are equal in all directions at each point. If equation (10.5) is satisfied then $\omega = 0°$ everywhere, and angles or bearings may be measured anywhere on the map or chart without errors arising from the nature of the projection. It should be remarked that equation (10.5) is not equivalent to the statement that $a = b = 1{\cdot}0000$ throughout the map, for that is the criterion of perfect representation which cannot be satisfied on any plane map. The particular scales therefore increase towards the edges of the map or chart from the point or the lines of zero distortion, as demonstrated, for example, in Tables 10.1–10.4. It follows, moreover, that the area scale increases according to the square of the particular scale.

Equidistant Projections

Equidistance is the condition that the principal scale is preserved throughout the map in one direction only, usually that perpendicular to the line of zero distortion or radially outwards from a point of zero distortion. This means that in the normal aspect the principal scale is preserved along the meridians, or $h = 1{\cdot}0000$ throughout the map. Since k cannot also equal unity, or be the reciprocal of h, it follows that an equidistant projection cannot also be equal-area or conformal. Consequently there are always variations both in the area scale and in maximum angular deformation on an equidistant map. However, the increase in the area scale is much less than that occurring in the conformal member of the same class of projections. For example, the area scale for the cylindrical equidistant (or plate carrée) projection is

$$p = k \tag{10.9}$$

whereas for the conformal cylindrical (or Mercator's) projection it is

$$p = k^2 \tag{10.10}$$

Similar arguments show that the maximum angular deformation of the plate carée is less than that of the cylindrical equal-area projection.

THE CHOICE OF THE MOST SUITABLE PROJECTION FOR CARTOMETRIC WORK

Ideally, of course, the most suitable map projection for a particular cartometric purpose is that allowing measurements to be made without any deformation at all. Thus an equal-area projection ought to be used to measure areas and a conformal projection to measure angles. However, there is no similar principle which can be applied to linear measurements. Equidistant projections only satisfy their special property in one direction, whereas an irregular line like a river or a coastline has many directions. The nearest we can approach to the unattainable constant-scale map is to use one which is based upon a projection in which the linear distortions are small. There is a category of *minimum-error* projections in which the sums of the squares of the scale errors, $(1 - a)^2$ and $(1 - b)^2$, integrated over the total area of the map, are made a minimum value. Several such projections have been described and used, for example, Airy's projection, described by its inventor as being by the "balance of errors", is the minimum-error member of the class of azimuthal projections.

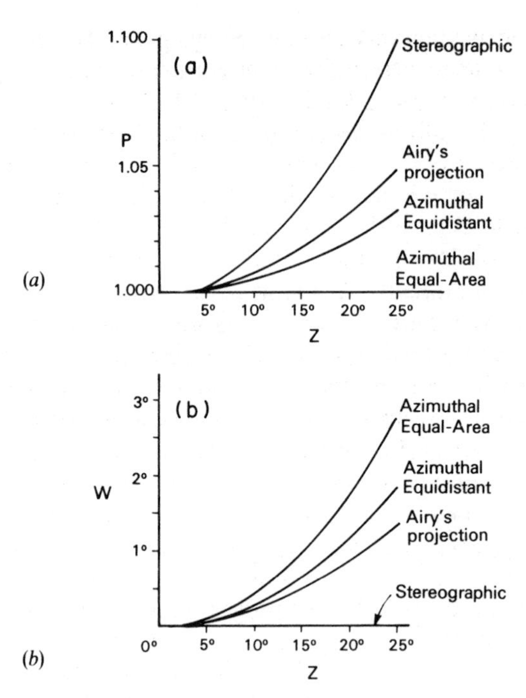

FIG. 10.6. Graphic representation of the distortion characteristics of certain azimuthal projections. (a) Plot of area scale (p) against angular distance (z) measured from the centre of the map, which is a point of zero distortion. Note that the azimuthal equal-area projection which, by definition, has $p = 1$ throughout, is represented by a line coincident with the abscissa of the graph. (b) Plot of maximum angular deformation ($\omega°$) against angular distance (z). Note that the stereographic projection, being the conformal azimuthal projection, has $\omega = 0°$ throughout. Therefore its line coincides with the abscissa of the graph.

However, the author has argued, in Maling (1973), that this projection approximates so closely to the azimuthal equidistant projection for maps covering less than a hemisphere that there is no need to use the more sophisticated minimum-error projection for linear measurement, and that it is sufficient to choose a map which has been compiled on an equidistant projection. Figure 10.6 illustrates this point.

A rather different argument applies where linear measurements are to be corrected for the local variations in particular scales by multiplying the measured distance by the reciprocal of the particular scale. In this case there is a strong case for using a conformal projection because the correction is the same in every direction from each point where it is applied. The converse to the rules which have been stated are injunctions that conformal projections should never be used for measurement of area and, secondly, that angles should not be measured on equal-area maps. However, the first of these rules creates difficulties because nearly all topographical maps and all navigation charts are based upon conformal projections. If we were to adhere strictly to this rule it would be necessary to discard many maps which would otherwise be suitable, for it is most unusual to find maps which are of similar scale and reliability, which are equally up-to-date and differ from one another only in the projection used. Because of the expense involved in compiling, fair drawing and reproducing a map, it is most unlikely that two such similar maps would both be published. Generally, therefore, the choice lies between using a map in which the projection is not the most suitable for measurement, or a map on a suitable projection, but which is old, too small in scale, or which does not show the information to be measured. It is usually wiser to select a map which is more satisfactory in the other respects, such as being more complete or up-to-date. Consequently it is possible that cartometric work may have to be done on maps which are based upon unsuitable projections. We must therefore examine the magnitude of the errors which are likely to arise from pursuing such practices.

AREA MEASUREMENT ON TOPOGRAPHICAL MAPS

It has already been stated that at the larger map scales the influence of the map projection is very small, simply because the map represents a very small portion of the earth's surface and the variations in particular scale are often too small to measure. However, it is necessary to demonstrate this for specific examples.

The majority of topographical maps are now based upon the conformal *Transverse Mercator projection*, sometimes known by the alternative name of *Gauss–Krüger projection*. Various national versions of this projection exist, but the best-known and most widely used version is the U.T.M., or *Universal Transverse Mercator*, which has almost world-wide application. In this system the earth is considered to have been divided into a series of meridional zones, or *gores*, illustrated in Fig. 10.7, each of which measures 6° in longitude and which extend from latitude 80° south to latitude 80° north. Each gore is divided by a *central meridian* so that the projection extends through 3° of longitude on either side of it. If there were no further modifications, the central meridian would be a line of zero distortion, but in most versions of the Transverse Mercator, including the U.T.M., a *scale factor* equal to 0.9996 is applied to it. This means, in effect, that a particular scale of 0·9996 has been assigned to the central meridian and consequently two lines of zero distortion have been created. These lie 180 km from the central meridian on either side of it. The effect of this upon the particular scales is shown in Table 10.2. We see that the exaggeration of areas does not exceed 0·3% even at the edges of each gore. Consequently we may usually disregard any correction to area measurements for most practical purposes. However, it cannot be ignored altogether. As an example of the need to take the area scale into consideration, we quote the example of Ordnance Survey practice used in the reduction of the

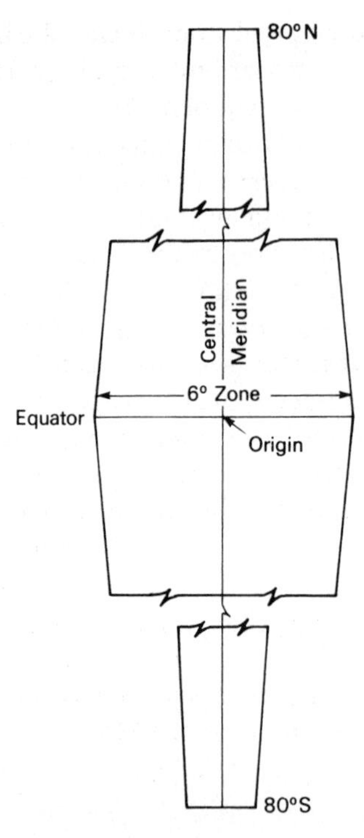

FIG. 10.7. Diagrammatic representation of a gore forming one zone of the Universal Transverse Mercator projection.

area measurements made on the 1/2,500 scale maps of Britain. Although this is made for the Transverse Mercator projection of Great Britain, which is not the same at the U.T.M., the same scale factor has been employed on the central meridians of each. The standard 1/2,500 scale map represents 2 km^2 on the national grid. If, therefore, a sheet were centred symmetrically over a line of zero distortion, the theoretical value of the area covered by it is 200·0 ha or

TABLE 10.2. *Universal Transverse Mercator projection; particular scales and distortion characteristics for a zone of width 6° and scale factor of 0·9996 on the central meridian*

U.T.M. easting in metres	Distance from central meridian (km)	Particular scales ($h = k$)	scale error (%)	Area scale (p)
500,000	0	0·9996	−0·04	0·99920
400,000 and 600,000	100	0·99972	−0·03	0·99944
320,000 and 680,000	180	1·00000	0·0	1·00000
300,000 and 700,000	200	1·00009	+0·01	1·00018
200,000 and 800,000	300	1·00072	+0·07	1·00144
100,000 and 900,000	400	1·00159	+0·16	1·00318

494·21 acres. This is obtained simply by converting 2 km^2 into hectares or acres. Elsewhere the maps, though apparently of identical format, represent different amounts of ground area because of the variations in area scale. Thus on the central meridian, the theoretical area contained within the map is $200{\cdot}0/0{\cdot}9996^2 = 200{\cdot}16$ ha or $494{\cdot}21/0{\cdot}9996^2 = 494{\cdot}602$ ha. At the other extreme, corresponding to the national grid eastings 130 km and 670 km, which are affectively the mainland extremities of England and Scotland, the particular scale is 1·0005, so that the area scale is 1·0010. Here the area of the 1/2,500 scale map is 100 ha or 493·716 acres. It will be seen that the influence of the projection is to create a difference of nearly 1 acre in the area contained within the standard 1/2,500 map, or a relative error of nearly 1/500 or 0·2%. The object of calculating these data has been described by McKay (1966) in his introduction to the methods used by the Ordnance Survey with the introduction of the Stanley–Cintel automatic reading planimeter (p. 357) and computer processing of all the measurements. The corrected area is determined for each 1/2,500 map to ensure that, when the total area of all the parcels measured has been found, no systematic error has entered the results.

AREA MEASUREMENT ON AERONAUTICAL CHARTS

The projection which is most commonly used for aeronautical charts is the *Lambert conformal conical projection*, with two standard parallels. There is no single table of particular scales and distortion characteristics which can serve for all of them, because the actual values depend upon both the north–south extent of the chart and the positions of the standard parallels with respect to the sheet edges. We therefore give three examples of the use of this projection.

International Map of the World at 1/1,000,000

We introduce the I.M.W. here as an aeronautical chart because of its long history of use in this fashion. Originally the maps of this series were based upon a modification of the polyconic projection, this being a good compromise projection which does not satisfy any of the three special properties. For want of any other cover at this scale it was used as an aeronautical chart in parts of Africa, Asia and South America up to and during World War II, but the fact that it was not conformal meant that it could not be used for accurate D.R. navigation because courses and bearings could not be reliably plotted or measured. Moreover, some countries, like Canada, which was already committed to the production of aeronautical charts of similar scale (1/1,013,760), resolutely refused to produce the corresponding I.M.W. sheets because of the additional cost which would be involved. Following the U.N. Technical Conference on the *International Map of the World* held in 1962, it became acceptable to use the Lambert conformal conical projection for this

TABLE 10.3. *International Map of the World at 1/1,000,000. Particular scales and distortion characteristics for the version of the Lambert conformal conical projection for zone N (between latitudes 52° and 56°)*

Latitude φ	Particular scales $h=k$	Scale error (%)	Area scale p	Maximum angular deformation $\omega°$
56°	1·00034	+0·03	1·00069	0°
55° 20′	1·00000	0·0	1·00000	0°
55°	0·99988	−0·01	0·99976	0°
54°	0·99973	−0·03	0·99946	0°
53°	0·99988	−0·01	0·99977	0°
52° 40′	1·00000	0·0	1·00000	0°
52°	1·00033	+0·03	1·00667	0°

series as the alternative to the polyconic projection. It is now accepted and used regularly as an aeronautical chart.

The neat lines of the standard I.M.W. sheets measure 4° in latitude and 6° in longitude. The standard parallels are located at 1/6 and 5/6 of the north–south extent of the map. Thus for the maps lying in zone NN, which lies between latitudes 52° and 56° N, the standard parallels are located in latitudes 52° 40′ N and 55° 20′ N. The resulting particular scales are given in Table 10.3. The largest values for the particular scales occur on the northern and southern edges of the map; the smallest on the middle parallel which, in this zone, is 54° N. The values for area scale indicate that the exaggeration of area is only marginally greater than that occurring on the extreme edges of a U.T.M. zone. Consequently there is no need to make any corrections for variation in area scale.

R.A.F. Constant Scale Plotting Sheet, Series PL2(N)

This series of aeronautical plotting charts at 1/2,000,000 scale provides cover of Europe, Africa and southern Asia, together with most of the North Atlantic and Indian Oceans. Most of the sheets extend through 14° of latitude and have standard parallels which are 9° 20′ apart. As an example of the particular scales we have selected in Table 10.4 those for a chart in the zone which extends from latitude 45° N to 59° N. As before, the particular scales are greatest at the extremities of the chart, but the scale error is still much less than 1%. Indeed this is why the chart series are described as being of "constant scale". To all intents and purposes, therefore, it is unnecessary to apply any corrections to, or to use any special techniques in making, linear measurements. This confirms the utility of the practice in air navigation of measuring distance on a Lambert chart with the aid of a plastic scale whereas, on Mercator's projection, as described in Chapter 22, dividers are essential.

TABLE 10.4. *R.A.F. constant scale plotting chart at 1/2,000,000. Particular scales and distortion characteristics for the version of the Lambert conformal conical Projection used for the zone extending from latitude 45° N to 59° N with standard parallels in latitudes 47° 20′ N and 56° 40′ N*

Latitude φ	Particular scales $h=k$	Scaler error (%)	Area scale p	Maximum angular deformation $\omega°$
59°	1·00442	+0·44	1·00887	0°
58°	1·00229	+0·23	1·00458	0°
56° 40′	1·00000	0·0	1·00000	0°
54°	0·99725	−0·27	0·99450	0°
52°	0·99668	−0·33	0·99337	0·
50°	0·99725	−0·27	0·99467	0°
47° 20′	1·00000	0·0	1·00000	0°
46°	1·00209	+0·21	1·00000	0°
45°	1·00398	+0·40	1·00797	0°

Exaggeration of area is similarly small, and therefore the series could be used for area measurement, probably without any need to make special corrections.

U.S. Geological Survey Maps of the U.S.A. at 1/2,500,000 and Smaller

Some of the maps of the U.S.A. at scales 1/2,500,000 or smaller published by the U.S.G.S. are based upon the Lambert conformal conical projection with standard parallels located in latitudes 33° N and 45° N, a separation of 17° in latitude between them. For a map of the coterminous U.S.A. a latitude range of 24° is needed to cover the country. Table 10.5 is a shortened version of that in Snyder.

Because of the greater north–south extent of the map compared with the two other examples studied, area exaggeration has increased to more than 3% at the edges, and is even greater than 1% in the middle of the map. This indicates that we have now more or less reached the limiting size of map on which the area scale may be ignored, and that it would be necessary to consider whether a map on a different projection would be more suitable as the base for area measurements. In this particular case the obvious solution is to use one of the maps produced by the U.S.G.S. which are based upon Albers' projection, but this happens to be one of the very few exceptions to the rule that maps of similar quality are seldom prepared on different projections.

Measurements on Mercator's Projection

Strictly speaking we should now consider the equivalent effects upon measurements made Mercator's projection, for this is the base for all nautical charts and therefore of great importance to any consideration of oceano-

TABLE 10.5. *Version of the Lambert conformal conical projection with standard parallels at 33°N and 45°N used by the U.S. Geological Survey for small scale maps of the U.S.A. Particular scales and distortion characteristics.*

Latitude φ	Particular scales $h = k$	Scale error (%)	Area scale p	Maximum angular deformation $\omega°$
52	1·02222	+2·22	1·04494	0°
48	1·00725	+0·73	1·01456	0°
45	1·00000	0·0	1·00000	0°
44	0·99828	−0·17	0·99656	0°
40	0·99464	−0·54	0·98932	0°
36	0·99594	−0·41	0·99190	0°
33	1·00000	0·0	1·00000	0°
32	1·00193	+0·19	1·00386	0°
28	1·01252	+1·25	1·02520	0°
24	1·02774	+2·77	1·05625	0°
22	1·03712	+3·71	1·07563	0°

graphic cartometry. Despite its undoubted value as the projection upon which graphical D.R. navigation may be most easily plotted, the rapid increase in the particular scales make this a rather difficult projection to use, and requires either special techniques or special corrections to obtain acceptable results. Because this projection is so important, we devote the whole of Chapter 22 to the study of the special cartometric problems arising from its use.

DEFORMATION OF THE MAP BASE

The Coefficient of Linear Expansion

All substances expand with increasing temperature, and contract with decreasing temperature. The length, l_t at the temperature t_t of a solid which was originally l_0 at temperature t_0 may be expressed as

$$l_t = l_0(1 + \alpha \cdot \delta t) \tag{10.11}$$

where $\delta t = t_t - t_0$, within the comparatively small range of atmospheric temperature variations and where no change in physical state occurs. The constant α is known as the *coefficient of linear expansion*. This linear equation of elementary physics implies three further conditions; first, that expansion is uniform in all directions; secondly, that it is instantaneous; and third, that it is reversible, so that when the temperature of the substance returns to the initial temperature, t_0, the length of it immediately returns to l_0.

In cartography this physical process is considered to be of critical importance because it controls the dimensional stability of the media chosen as the bases for plotting sheets and fair drawings, as well as additional documents such as masks and combinations of documents. Some media are relatively stable. For example, the coefficient of linear expansion for plate glass is $\alpha = 8 \times 10^{-6}/1°C$. Metals also have comparatively small values for α. For example

Copper, $\alpha = 16.7 \times 10^{-6}/1°C$.
Zinc, $\alpha = 25{\cdot}8 - 26{\cdot}3 \times 10^{-6}/1°C$.
Aluminium, $\alpha = 25{\cdot}3 \times 10^{-6}/1°C$.

Glass and metals are unaffected by any change in the humidity of the environment, or even by immersion in water and photographic solutions. It was for these reasons that metal and glass were preferred as the bases for cartographic work before the so-called dimensionally stable plastics became available in the 1950s.

Most organic substances undergo dimensional change as they absorb or lose water from the atmosphere. Therefore it is necessary to create a second equation like (10.11) which equates variation in length, $l_h - l_{h'}$, of a solid which undergoes a change in relative humidity, $\delta h = h - h'$, viz.

$$l_h = l_{h'}(1 + \beta \cdot \delta h) \tag{10.12}$$

Here β is the *coefficient of hygrometric expansion.*

Values of β have been determined for the various plastics and emulsion bases used in cartography and the graphic arts. For example, $\beta = 0{\cdot}8 - 1{\cdot}6 \times 10^{-6}/1\%$ for the dimensionally stable vinyl plastics, such as Astrafoil and Astralon. It is slightly less ($\beta \simeq 0{\cdot}6 - 1{\cdot}0 \times 10^{-6}/1\%$) for the polyester plastics, such as Mylar or Melinex.

Two values for β have been quoted for each example of plastics, for there is a difference in the behaviour of each sheet which results from *calendering*, or rolling of vinyl plastics and the extrusion of polyester plastics during manufacture. Therefore it is necessary to distinguish between the coefficient of hygroscopic expansion, as well as that for thermal expansion, in the longitudinal direction (of rolling) from that in the transverse direction.

It was stressed in Chapter 9 that preoccupation with dimensional stability is the result of the need to achieve and maintain precise colour registration, rather than any particular desire to produce accurate maps. This is because a map is printed on paper, which is much less stable than the other materials mentioned. It changes its shape and size as the environment changes, for paper is affected by both temperature and humidity in printing, storage and use. Paper reacts in a most irregular fashion compared with the other materials, and these changes may be large. Moreover, neither the uniformity of expansion, nor the reversible nature of equations (10.11) and (10.12) apply to these changes. In the days before suitable plastics were available, it was necessary to mount a sheet of hand-made drawing paper upon a metal plate (usually copper or zinc) to control the shape and size of the fair drawing.

The Composition and Characteristics of Paper

The basic ingredients of paper are fibre, mainly composed of cellulose. The principal sources of the fibre content include rags, esparto grass, wood, straw, reclaimed (repulped) material and plastics. These materials are used in various combinations and proportions to produce paper with specific characteristics. The paper most commonly used for printing maps by lithography is known as *high wet strength* paper, which is a combination of rags, grasses and wood. Special synthetic fibre papers which are intended to provide maps of immense strength and durability are made from pulp containing a large proportion of plastics fibres. The fibres within paper vary both in length and in their cellulose content. The longer the fibres the stronger the paper; the higher the cellulose content, the less treatment required in the manufacturing processes and the fewer the additional additives needed. During manufacture the cellulose fibres are cleaned and *beaten* to produce the felted material known as *stuff.* For the so-called *machine-made* papers this is passed through a strainer into a box from which it is fed to an endless copper mesh or web. The web oscillates, the water drains through it and the fibres are shaken so that the long fibres lie in the direction of the machine "like sticks in a flowing stream" as Winterbotham (1934) described the process. The direction of the long fibres is called the *machine direction* or *grain direction* of the paper. Finishing of the paper includes the calendering process, which effectively reinforces this preferred direction of the fibres and introduces a characteristic tendency for the paper to curl in this direction, too. This provides a convenient method of identifying the grain direction of a sheet of paper. In the production of hand-made paper, a skilled hand shakes the tray in the felting process to ensure that the fibres take up no particular orientation.

It is the effect of humidity which exerts the greatest influence upon the paper dimensions, for there is a functional relationship between the relative humidity of the environment and the moisture content of the paper. This is illustrated by Fig. 10.7, which has been prepared for typical room temperatures of 18–20°C. For a relative humidity of 50–55% the paper contains 6–7% of its total weight of water, which is held both mechanically and chemically in its fibres. As they become moist the fibres swell more across the grain than along it, the paper stretches less in the machine direction than across the grain. The differential expansion of machine-made papers is illustrated by Fig. 10.7(b). As paper dries, so it shrinks. But the rates of expansion and shrinking are not the same, for the rate of shrinking lags behind that of expansion. This means that, after a cyclical change in humidity, the paper does not readily return to its original dimensions.

Since there is no preferred direction for the fibres in hand-made paper, it changes its size more or less uniformly in all directions. Hand-made paper was formerly used for printed maps and charts when printing was done directly from copper plates. It was still used as the base for some fair drawings until the postwar development of dimensionally stable plastics provided the draughtsman with these new materials. It is therefore still quite common to find original drawings, for example of estate plans, which were produced on hand-made paper bearing Watmans' celebrated watermark. Today this paper is no

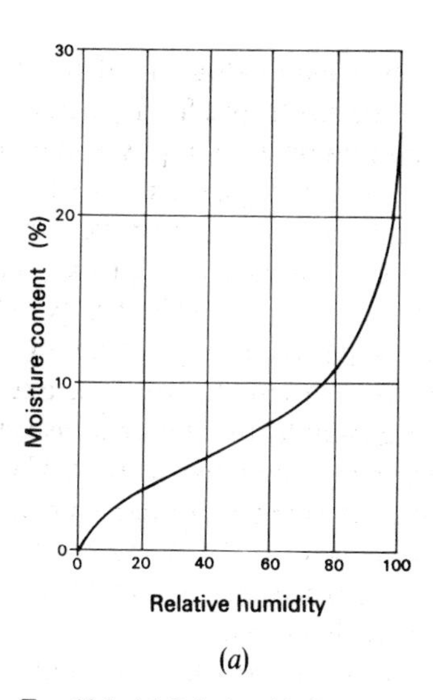

(*a*)

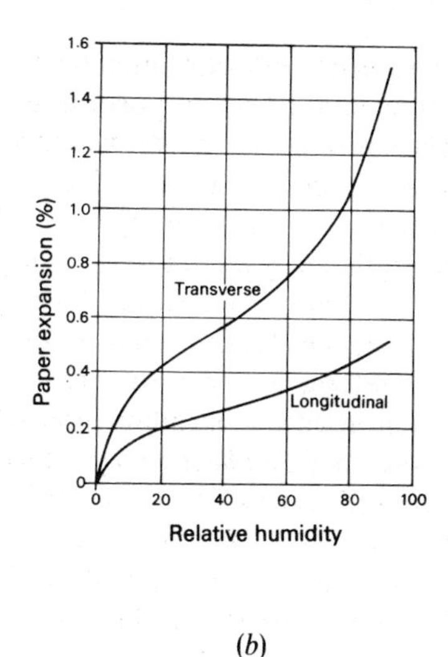

(*b*)

FIG. 10.8. (a) Relationship between relative humidity and moisture content of paper used for printing maps. (b) The differential expansion of machine-made paper.

longer used in map making; indeed large format sheets are no longer made. As a draughting medium it has been entirely replaced by plastics.

In lithographic printing the paper is fed into the press in such a way that the machine direction lies along the press cylinders, so that the paper passes through the printing press at right angles to the direction in which it was manufactured. This facilitates the adjustments which have to be made for colour registration. If a sheet is to be permanently folded, for example the folded plate of an atlas map, the grain direction should be parallel to the binding spine and therefore to the fold.

The kind of paper most commonly used for printing multicolour topographical maps is high wet strength paper of weight 92 g.s.m. (grams per square metre) or thereabouts. For some uses, particularly by the military and in some sports such as mountaineering, even this weight of paper is somewhat inadequate. There are distinct advantages in using maps which are not going to be torn when used in wind and rain. Paper which has suitable properties for such uses has plastics fibres in its content. This is exemplified by the synthetic paper called *Syntosil*, which was first produced in Switzerland during the 1950s. This Ordnance Survey refer to a similar material as *tear- and water-resistant material*. It has the advantage from the cartometric point of view that its plastic content swells less with increasing humidity so that distortions are reduced. It has the disadvantage of being much more expensive than ordinary paper. The availability of maps printed on such materials varies. The Ordnance Survey has only printed a few special maps, notably some of the 1/25,000 *Outdoor Leisure Maps* of the English Lake District and similar areas, where their relative indestructibility is a boon to the walker and climber, but it is understood that the production of maps on this material has now ceased. In Switzerland, Eidgenössische Landestopographie have produced virtually all their topographical maps in two versions; one printed on ordinary paper and the other on Syntosil.

The Dimensional Stability of Maps

The main source of information relating to the dimensional stability of printed maps is from measurements of known distances within the grid, along a scale line, or between the corners or neat lines of a map. From the knowledge of what these distances ought to be, it is easy to determine the magnitude of the errors attributable to paper deformation within the precision that can be obtained from particular measurement techniques. McCaw (1936a, b) demonstrated from first principles how such deformation affects measurement of areas and bearings on maps.

Table 10.6 is taken from Kishimoto (1968), who has carried out a number of tests on stability for different kinds of maps. This example relates to a topographical map of scale 1/100,000 published by Eidgenössische Landestopographie. The amount of deformation is usually stated as a coefficient derived from the formula

$$q = (l_0 - l)/l_0 \quad (10.13)$$

where l_0 is the correct value for the length of a line and l is the measured length of it. Thus, for the northern neat line of the map measured in the first trial, we have

$$q = (700 - 698{\cdot}6)/700$$
$$= -0{\cdot}002 \quad \text{or} \quad -1/500 \quad \text{or} \quad -0{\cdot}2\%$$

It is generally accepted that the following values, which were published by Reignier (1957) on the basis of measurements made at the Institut Géographique National, are representative of experience elsewhere.

TABLE 10.6. *Results of measurements of the side lengths of 1/100,000 map carried out by Kishimoto. Theoretical side lengths are: North, 700·0 mm; South, 700·0 mm; East, 480·0 mm; West, 480·0 mm*

Trial	Measurements				Linear deformation				R.H.	Temp.
	North	South	East	West	North	South	East	West	(%)	(°C)
1	698·6	698·8	478·8	478·8	−1·40mm −0·20%	−1·20mm −0·17%	−1·20mm −0·25%	−1·20mm −0·25%	50	20
2	699·2	699·3	479·4	479·5	−0·80mm −0·11%	−0·65mm −0·09%	−0·60mm −0·13%	−0·50mm −0·10%	70	18·5

- At the time of printing, the paper is warm and moist. Consequently it is stretched up to 1.5% in the longitudinal direction and 2·5% in the transverse direction.
- When the paper subsequently dries, it shrinks by 0·25–0·5% in the longitudinal direction and up to 0·75% in the transverse direction.
- Consequently the expansion resulting from these changes may reach a maximum of 1·25% in the longitudinal direction and 2·5% in the transverse direction.

Obviously these figures can only indicate the range of possibilities, because the actual amount of change recorded in a particular sheet of paper depends upon its composition, the number of printing runs, the rate of drying after printing and conditions of storage. Factors such as keeping a map rolled, rather than storing it flat in a drawer, also affect the final dimensions.

The following examples of experimental work indicate the range of variability which may be experienced. Each of the writers has investigated somewhat different environmental conditions. Curiously none of them have attempted to answer more than one question, but between them we have a fairly comprehensive picture of how measurements are affected

1 When the same map is subjected to a range of different temperatures and humidities.
2 When several similar maps are tested in the same way.
3 When a folded map is compared with a flat map.
4 How deformation varies in detail within the body of the map.

Sheppard's Measurements of Aeronautical Charts

Sheppard (1953) made a series of simple measurements to test the stability of a typical aeronautical chart under different conditions of temperature and humidity. He tested a copy of the 1/500,000 aeronautical chart series G.S.G.S. 4072, using the sheet for southern England upon which he measured the length of the Greenwich meridian between latitudes 50° 40′ N and 52° 40′N. This meridional distance corresponds to 120 nautical miles if we assume the earth to be truly spherical, or 120·08 nautical miles on the Airy spheroid. On the chart this is a straight line of approximate length 444 mm. Sheppard used the graduated perspex scale intended for use with 1/500,000 scale charts which was standard issue to R.A.F. navigators at that time. Figure 10.9 summarises the 17 sets of measurements made by him. The air temperature observed at the time of measurement is recorded in degrees Fahrenheit. It can be seen that there is a good linear relationship between the measured distance and relative humidity (correlation coefficient, $r = 0{\cdot}979$) but that between measured distance and temperature is weak ($r = -0{\cdot}21$). The mean of all the measurements is 119·75 nautical miles, corresponding to an error of −0·25 nms, which is about 0·9 mm on the chart. This represents a percentage error of 0·75%, which is larger than most of the other experiments described here but is probably explained by Sheppard's use of a perspex scale to make the measurements. Such scales were considered to be somewhat unreliable.

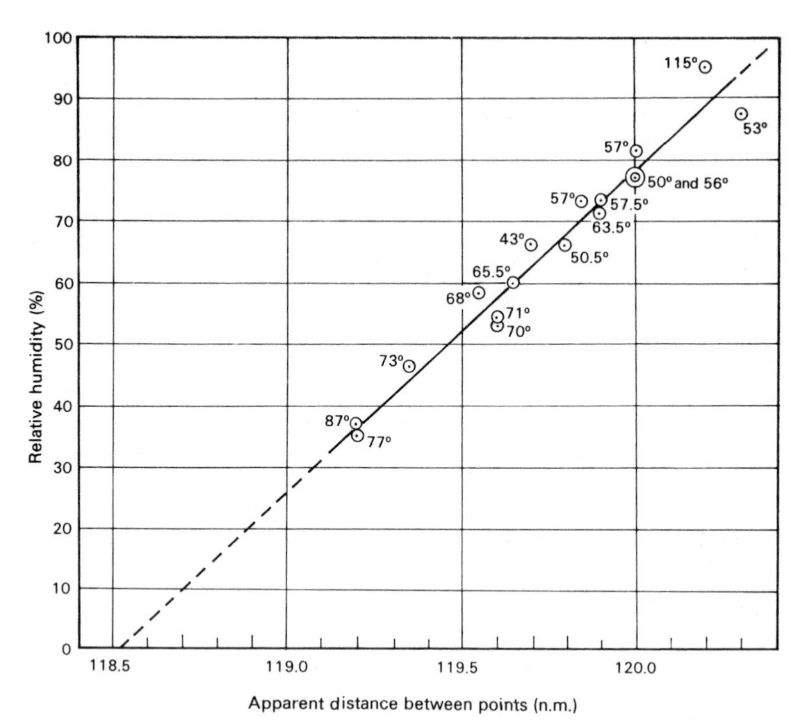

FIG. 10.9. The length of part of the Greenwich meridian as measured by Sheppard on an aeronautical chart at different temperatures and humidities. Temperatures are expressed in degrees Fahrenheit.

Libaut's Measurements of the Scale Lines on French Topographical Maps

Libaut (1961) tested the stability of three different kinds of topographical ma published in France by measuring the length of the graphic scale reproduce on each. The measurements were made by steel scale.

(a) CARTE DE FRANCE AU 1/80,000 (CARTE DE L'ETAT-MAJOR)

This was the basic mapping of France from the mid-nineteenth century unt 1940. The original Etat-Major maps were engraved on copper and printed in single impression in black by copper-plate methods. This is a drier proce than lithography, though some printers were accustomed to sprinkle a litt water on each sheet of paper before closing the press. Libaut drew a rando sample of 47%, this being 125 maps from the total of 267 forming the serie Since each numbered map comprises four quarter-sheets, and since each ha its own graphic scale, Libaut's sample comprised 500 separate measurement The scale represents a distance of 10 km on the map and ought, therefore, t measure 125·0 mm. Libaut obtained a mean length of 124·72 ± 0·015 mm. Th represents an average error of −0·28 mm or 0·22%, and a range of 1·5% between the extreme measurements 123·9 mm and 125·8 mm (Libaut did reject one measurement of 122 mm, and replaced the sheet on which it occurred by another). The frequency distribution of the measurements is symmetrical about the mean but the amount of kurtosis is greater than should occur in the normal distribution.

(b) 1/50,000 SCALE MAPS DERIVED FROM THE ETAT-MAJOR SERIES

These maps were produced to provide cover for parts of France which had not been published on the Type 1922 maps until that series of 1/50,000 maps had been completed. The maps had been derived from the Etat-Major series by photographic enlargement and were printed by lithography. A sample of 150 measurements were made, giving the mean length 120·06 ± 0·02 mm for the length of the scale line, whereas this should have been 120·0 mm. Therefore Libaut's measurements show a mean error of + 0·06 mm, or + 0·05%. There is a range of 1·3% between the two extremes of 119·6 mm and 121·2 mm. The frequency distribution of the measurements is positively skewed.

(c) CARTE DE FRANCE AU 1/100,000, TYPE 1954

Publication of this series had only just commenced at the time when Libaut was making these measurements. He therefore used a sample comprising each of the 35 sheets already published. The theoretical length of the 10 km scale line is 100·0 mm. Libaut obtained a mean length 100·01 ± 0·021 mm, representing an average error of + 0·006 mm or 0·01% and a range of 0·7% between the extreme values of 99·80 mm and 100·50 mm. This distribution is also positively skewed.

Kishimoto's Measurements of Swiss Maps

Kishimoto (1968) tested 42 different maps by measuring the distances between the corners of each sheet using a small coordinatograph. The results are given in Table 10.7.

Eight folded maps are included in Table 10.7. It would be reasonable to suppose that folding of the paper reduces the accuracy of measurement, so that it is unwise to use folded maps for cartometric work. Often, however it is difficult to avoid using maps which have been folded; for example, in making measurements from manuscript archives where there is no alternative to working with folded documents. Moreover, atlas maps and nautical charts have usually been folded in one direction. Many small- or medium-scale maps are only available from retailers in folded versions. It is sometimes possible to order special "flat and unmounted" copies of maps from the publishers, but the occasional user who is dependent upon retail purchases may be obliged to

TABLE 10.7. *Kishimoto's measurements of the neat lines of Swiss maps*

Title	Scale	No. of sheets tested	Average linear deformation (%)	
			Longitudinal	Transverse
Cadastral General Plan of Zürich	1/2,500	6	−0·67	−0·39
	1/5,000	1	−0·06	+0·13
	1/10,000	1	−0·86	−0·74
Cadastral General Plan of Canton Berne	1/10,000	3	−0·08	−0·09
Fortification Map	1/10,000	4	−0·06	−0·28
Map of Canton Zürich by Wild	1/25,000	4	−1·82	−0·70
Siegfried Atlas	1/25,000	4 (pre-1881)	−0·90	−0·54
		1 (1914)†	−0·07	−0·12
National Map of Switzerland	1/25,000	5	−0·11	−0·09
		2†	−0·28	−0·31
	1/50,000	1	−0·04	−0·10
		4†	−0·26	−0·17
	1/100,000	5	−0·12	−0·12
		1†	−0·22	−0·10

†Indicates folded maps

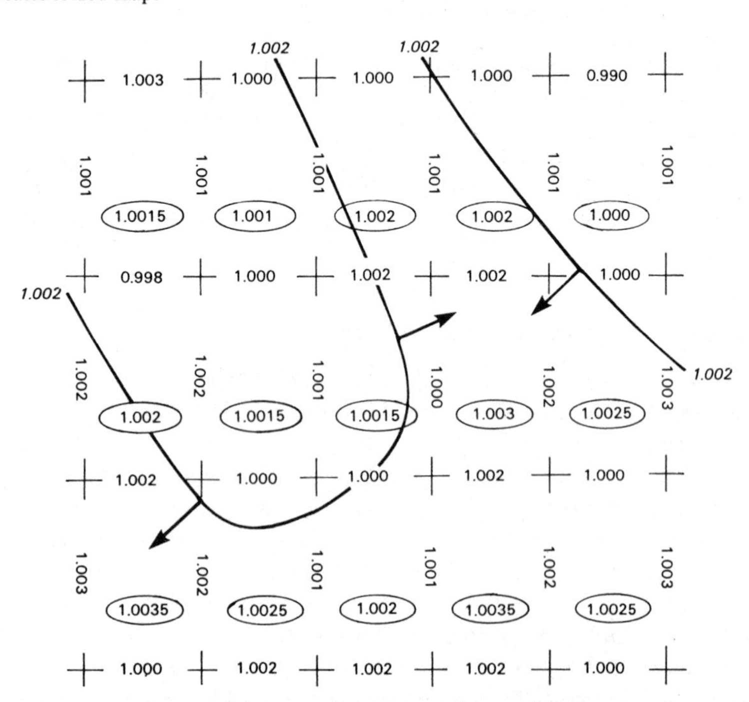

FIG. 10.10. Volkov's method of investigating local deformations in paper. Construction of the isograms of equal area distortion from measurements made on a rectangular grid.

make measurements upon maps which have been folded. These may comprise six or more separate folds.

Although it is desirable to consider the loss in accuracy which results from measuring distances or areas across folds, Kishimoto's small sample is the only quantitative information on the subject which is known to the author. Table 10.7 indicates that, excepting the anomalous example of an old Siegfried Atlas map, the measurements of the neat lines of Swiss topographical maps indicates that folding has increased the linear errors by about 0·1% more than that experienced with flat maps. In this context it seems that folded maps printed on Syntosil may be more suitable for cartometric work, because it is relatively easy to smooth away any existing folds when such a map is placed on a table or drawing board. However, the author has no quantitative information to substantiate this opinion. There are still useful investigations to be made into the effects of rolling and folding paper and synthetic base maps upon cartometric measurements made from them.

Volkov's Investigation of Local Deformation

The tests described so far have all been made on certain parts (usually the edges) of printed maps, and do not indicate the presence of local deformation

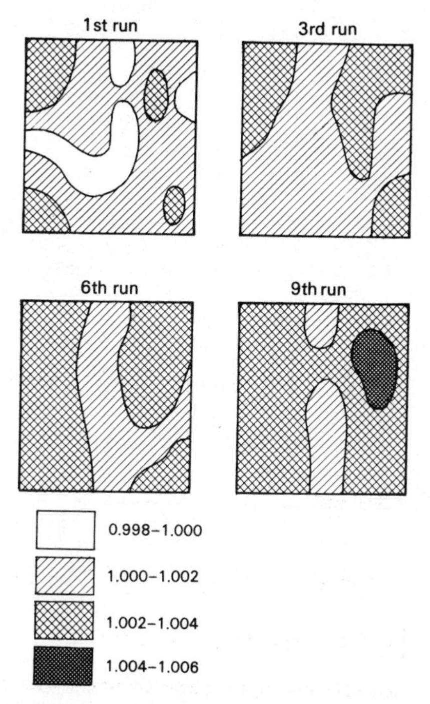

FIG. 10.11. Volkov's method of investigating local deformation in paper. Isograms of measurements made on the same sheet of paper after different numbers of impressions.

within the body of the map. Volkov (1950) investigated this in the following manner. He used a sheet of Watmans paper mounted upon an aluminium plate. A grid was plotted on this base, the intersections serving as the corner points of 180 rectangles each with dimensions 28 × 45 mm, as illustrated in Fig. 10.10. A lithographic plate was made from this original, and the distances between the crosses along the rows and columns of the grid were carefully measured on this by steel scale. A series of impressions were now made on paper of different weights, simulating multicolour printing by running the paper through the press nine times. Four different colour impressions were made, but, in order to save cluttering the paper with too many images alternate runs were made through a clean press with damped plate and rollers. The printed grid was measured on the paper after each pass through the press. If the distance between two crosses was 28·2 mm on the printing plate, bu

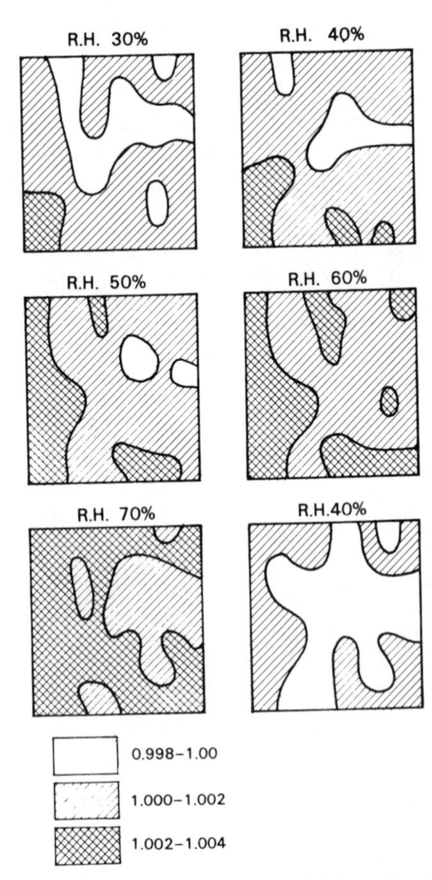

FIG. 10.12. Volkov's method of investigating local deformation in paper. Isograms for the same sheet of paper, the relative humidity of which was raised from 30% to 70% and then allowed to return to 40%.

28·3 mm on the paper, the deformation was 28·3/28·2 = 1·0035. From comparison of such linear measurements an average linear distortion was obtained for the sides of each rectangle. Thus in the top left rectangle of Fig. 10.11, the two average values are (1·001 + 1·001)/2 = 1·001 and (1·003 + 0·998)/2 = 1·0005. It follows that the distortion in the area of that rectangle is 1·001 × 1·0005 = 1·0015, which value is assumed to refer to the geometrical centre of the rectangle. Having determined the mean area deformation of each rectangle, isograms for equal amounts of deformation were interpolated and plotted for the whole grid. The patterns for the isograms following different numbers of press runs is illustrated in Fig. 10.11. Similar measurements were made on one sheet for different conditions of relative humidity, which was increased in steps of 10% from 30% through 70%. The measurements were repeated when the paper had returned to 40% relative humidity. The results are illustrated in Fig. 10.12. They show that although the magnitude of the changes did not generally exceed 0·6%, the actual pattern of the isograms may be quite different and, as demonstrated by the two sets of measurements at 40% relative humidity in Fig. 10.12, change significantly as the paper changes its dimensions. The experiments show that local distortion does occur, and that it may be as much as 0·4% or 1/250. Since this is greater than the reading resolution of the vernier of a polar planimeter, it is too large to be ignored. It follows that *measurement control* is required to reduce the local deformations of the paper to acceptable limits. Moreover, this control can do much to reduce the distortions introduced by the map projection if the scale errors are fairly small and do not change rapidly from place to place on the map.

11

Correction of Deformation by Measurement Control

The method I took for the doing it was, by weighing in nice Scales that part of the *Sheet-Map* *England*, copied from Mr Adam's (which I esteem the best) that represents the Land, an comparing the weight of the whole with that of a Circle taken out of the middle of the Map whos diameter was 138[2] Miles or two Degrees, which was the greatest that the Kingdom can affor being so much between the Seavern Sea in the West, and the Inlet by Lyn-Regis, in the East..

...I found that the Land of the whole Map, together with the Isles of Wight, Anglesey and Ma weighed just four times as much as the said Circles; and consequently that the Acres in the whol Kingdom, which are by computation 9,645,000 whence the whole Kingdom must be 38,660,00 Arces, and this I believe to be no wide conjecture:.... When my hand was in, I thought it migh tend to the same end, or otherwise be serviceable to you, to give the acres of each County England, which I have derived from the same method of Weighing, having cut a six sheet Map pieces for that purpose, in which each 40,000 acres weighed about a Grain. In this I took care avoid two inconveniences; the one, that the Map consisting of several Sheets of Paper, they wer found to be of different thicknesses or compactness, so as to make a sensible difference, whic obliged me to examine the proportion between the Weight and Acre in each sheet: The other wa that the moisture of the Air imbibed by the Paper, did very notably increase its weight; whic made me very well dry the pieces before I weighed them, that so I might be assured there was n Error upon that account; and in so doing, I found that in a very few Minutes of time, their weigh would sensibly increase by their reimbibing the humidity out of the air. This Method I concei exact enough for the uses you design, and that I have not much erred will appear by the content this tryal with the former: The Acres of each County *are* as followeth. Which you have in th General Sheet. In all 39,938,500 Acres....

I am your very humble Servant,

Edmund Halley.

(John Houghton, *A Collection for Improvement of Husbandry and Trad*
Num. 25, *January* 20, $169^{2/3}$

Measurement control is important in cartometric practice because it correct for most of the measurement errors arising from paper distortion and, to certain extent, it also reduces those errors which arise from some of the dis placements in position caused by the map projection. Control depends upo knowledge of the dimensions of certain features printed on a metric map, suc as the graphic scales, the spacing of the grid or graticule lines and the know dimensions of the neat lines of the map or chart. By relating measurements t this control, much of the paper deformation is automatically corrected. This i because both detail and control behave in the same way as the map is expose to the fluctuations in temperature and humidity which cause the paper t change shape and size. Even if we recognise that there may also be loca deformation, as described in Chapter 10, this principle remains true within small portion of the printed map, say in a square with sides 10 × 10 cm, where both graticule and map detail are stretched or shrunk by the same amount. Consequently we may measure deformation of the detail by measuring deformation of the control. Although projection distortions, as well as image displacement occurring on aerial photographs, may also be removed by similar methods, it is necessary to determine how rapidly the changes in particular scales on the map affect measurements, and from this to select a limiting size for the unit quadrangle, or the width of a zone on the map which may be treated in a particular way. Since some of the displacements of detail resulting from these causes may be rather large, and since they are invariably systematic within the confines of a typical small-scale map or chart, it may be necessary to make the unit quadrangle so small, or the zone so narrow, that the method becomes unworkable in its simplest form. The difficulties which may arise when it is necessary to measure area on small-scale maps or charts based upon Mercator's projection are typical of the problems which arise in this respect.

The simplest example of the use of the control contained upon the map is to refer all distance measurement of the graphic scale printed on the map, rather than use a steel or plastic scale, subsequently converting the results into ground measurements by calculation. In the description of the card method of distance measurement, the British Army manual on map reading (War Office, 1957; Ministry of Defence, 1973) recommends that the marked card should be laid against the graphic linear scale printed on the map so that any stretching or shrinking which has affected the whole map may be taken into consideration. Compare this procedure with measurement of the length of a line using a separate scale, such as the experiment by Sheppard (1953) which was described in Chapter 10. In order to measure paper distortion on charts under different environmental conditions, he measured a line of known length using a separate scale. He therefore assumed that the plastic scale would remain unaltered by those changes in temperature and humidity which were affecting the map. If he had measured the portion of the Greenwich meridian by beam compass transferring each setting to one of the meridional scales on the chart, he would (ignoring other sources of error and assuming no significant alteration in the length of the beam compass with changes of temperature or humidity) have obtained a constant value of 120·0 mm, because the printed scale used for comparison would have stretched or shrunk by the same amount as the rest of the chart. The assumption that the beam compass remains invariable throughout the experiment is reasonable in this context because the coefficients of thermal linear expansion of its components are small compared with that of paper, and the instrument is little affected by changes in humidity.

It follows that the correction procedures to be studied in this chapter apply particularly to the kinds of cartometric measurement which we make by comparing dividers with a steel scale, or using an instrument with its own scales, such as a chartwheel or planimeter. If the results of the measurements are reduced to their ground equivalents by making additional measurements on the same map, then an appreciable amount of the potential error may be eliminated.

The graphic scale is usually located near an edge or corner of the map. Therefore it may be situated where dimensional changes are exceptional. On some navigation charts, additional scales are provided in the form of closely subdivided parts of the graticule, within the body of the chart. These are more useful gauges for paper distortion because they are located well away from the edges of the paper.

As we have seen in earlier chapters, a grid appears on the majority of modern land maps of scale 1/500,000 and larger. Because it comprises a uniform network of squares having precisely defined linear dimensions, it is simple to calculate the lengths of the sides, or the area enclosed within a group of contiguous grid squares, irrespective of whether this information is required in units of ground measurement or expressed in millimetres or square centimetres on the map. Moreover a large-scale or medium-scale map with a grid commonly has neat lines coinciding with this grid. For example, all Ordnance Survey maps of scale 1/625,000 and larger have neat lines formed by the national grid. Knowing that the neat lines of such maps are so accurately defined, and knowing the detailed properties of the projection in use, such as the parameters of the U.T.M. or the Ordnance Survey Transverse Mercator projection, it is possible to determine the ground area contained with a map to a high order of accuracy. This practice was briefly described in Chapter 10 with reference to the way in which the Ordnance Survey reduce areas after all the parcels on a particular map have been measured.

A graticule comprises lines representing equal values of latitude and longitude, both of which are expressed in angular measures, the conversion of which into linear measurements complicates the calculations. Because the meridians converge towards the poles the quadrangle formed by a pair of meridians and a pair of parallels is neither square nor rectangular, so that its area is less easy to calculate. The curved surface area of a spherical or spheroidal quadrangle may be determined by various formulae which are derived later in this chapter. As in most terrestrial studies of this nature, there is choice of the reference figure to be used as the basis for calculation. The solution which is based upon the assumption that the earth is a perfect sphere is simpler to understand and calculate, but it is less accurate than the solution based upon the assumption that the earth is a spheroid. Formerly it was usual to use special tables which had been calculated for the figure of the earth employed in a particular part of the world to carry out various geodetic computations. Nowadays, of course, the calculation of arc distances and areas contained within spherical or spheroidal quadrangles can be easily done with a scientific pocket calculator. We describe the various equations later. Therefore tables for the areas of quadrangles have become as obsolete as are the huge volumes of seven- or eight-figure natural and logarithmic trigonometric functions which were once an essential requirement for all survey and cartographic computations.

Occasionally the true dimensions of the neat lines are recorded on the map or chart. An important example of this practice is the indication on nautical charts of the precise measurements of the original plotting sheet. Since these are based upon the normal aspect of Mercator's projection, the neat lines form a rectangle and only two measurements are needed to describe the original dimensions of the chart. For example, the entry

(38·97 × 24·90)

located in the margin of the chart close to its south-east corner indicates the measurements of the original drawing in inches, or, since the adoption of the metric system for British charts, the equivalent in millimetres. American charts have a more explicit statement printed just inside the chart border near the south-east corner, such as

(Inner neatline 108·25 cm N.S. × 82·90 cm E.W.)

The procedure is also used for large-scale charts which are supposed to be based upon the gnomonic projection.[†] Although the parallels and meridians are still represented by straight lines, because of the convergence of the meridians and the different lengths of the northern and southern edges of the chart, three figures are needed to describe the dimensions of the neat lines. These are written in the form

$$\left(\begin{matrix}972{\cdot}5\\980{\cdot}2\end{matrix} \times 635{\cdot}0\,\text{mm}\right)$$

Some maps have arbitrary neat lines which are unrelated to the grid or graticule. If it is known that the rectangle containing the map has precise dimensions, these may be used as control measurements. For example, the popular edition of the One-inch to One-mile Ordnance Survey map measures 27·0 × 18·0 in and is also subdivided into squares (which are not to be confused with grid squares) measuring 2·0 × 2·0 in or 2·0 × 1·0 in along one edge. These units may be employed as control in the use of these maps, provided that it is appreciated that they do not bear the same precise relationship to measurements on the ground as do the components of the national grid. Thus such a square may be measured on a map and its side lengths compared with the nominal 1·0 or 2·0 inches, purely to correct for paper deformation within that square, but it is not wise to assume that, for example, at 1/63,360 scale a square of area 4·0 square inches corresponds exactly to 4·0 square miles without taking into consideration the special properties of the map projection. Before the introduction of the

[†]But see Maling (1973) on this subject.

national grid as the geometrical basis of all Ordnance Survey work, it was necessary for revisers to construct their own grids for local use. Thus the *OS instructions to Field Revisers*, 1/2500 scale, 1932, contained the statement:

> As a guide to the amount of contraction or expansion that has occurred on a plan to the date of preparation of the plate, small red ticks are printed on the red traces dividing the north and south edges of plans into three and the east and west edges into two equal divisions of 40 chains approximation, with two smaller internal crosses. Consequently these marks appear at each internal trace corner, and give the reviser a forty chain ratio as a general guide as to the amount of plus or minus ground measurement to be expected in supplying new work. The variation of paper is not even over the whole of its surface, however, and therefore the amount of paper error to be equated in a particular area must be arrived at by local ground measurements and intersections between good points of old detail on the trace.

The majority of modern maps with arbitrary neat lines and no indication of any grid or graticule belong to the category of non-metric maps which we have already suggested should not be used for cartometric work. In the event of having to use such maps, the only form of control which is likely to be of any use is the known area of some mapped feature which may be obtained from some other source. For example, if the map shows the boundary of a country, a county or a parish, and we already know the area of the unit from other sources, we may use this as a control for additional measurements of areas of other distributions. At best, however, this is a poor substitute for the use of the grid or graticule as the control.

CORRECTION TO LINEAR MEASUREMENTS

The magnitude of the deformation may be characterised by the relative error, q, as in (10.13)

$$q = (l_0 - l)/l_0 \tag{10.13}$$

where l_0 is the theoretical length of the line on the map or plan, for example the length of a portion of the grid representing the side of one or more grid squares and l is the length of that line measured on the map. For example, if

$$l_0 = 4000\text{ m} \quad \text{and} \quad l = 3980\text{ m},$$

$$\begin{aligned} q &= (4000 - 3980)/4000 \\ &= +0{\cdot}005 \\ &= 1/200 \end{aligned}$$

We may write (10.13) in the form

$$l_0 = l/(1 - q) \tag{11.1}$$

Multiplying the numerator and denominator of the right-hand side by $(1 + q)$ and ignoring the small terms in q^2, we obtain

$$l_0 = l + l \cdot q \tag{11.2}$$

where $l \cdot q$ is the correction to be applied to the line l. For example, if $l = 323{\cdot}0$ m and $q = +1/200$,

$$l_0 = 323{\cdot}0 + 323/200 = 324{\cdot}6\ \text{m}$$

The Effect of Paper Distortion upon Area Measurement

Suppose that we measure the area of a triangular parcel on the map, using the linear measurements base l and height h. Then from the simple and well-known equation of Euclidean geometry, the area of the parcel is

$$A = (l \cdot h)/2 \tag{11.3}$$

If these distances have been measured on a distorted map, the true area of the parcel should be

$$A_0 = (l_0 \cdot h_0)/2 \tag{11.4}$$

From (11.2) we may write

$$A_0 = 1/2(l + l \cdot q)(h + h \cdot q)$$
$$= (l \cdot h)/2 \times (1 + q)^2$$

Substituting (11.3) and ignoring the small terms in q^2, we obtain

$$A_0 = A + 2A \cdot q \tag{11.5}$$

It can be shown that this expression is true for any shape of parcel.

Let us now consider the case where the deformation has been measured in the two directions corresponding to the map grid. We shall call these q_x and q_y. Then from equation (10.2) it follows that the displacement of the grid line corrected for paper deformation will be

$$\Delta x_0 = \Delta x/(1 - q_x) \tag{11.6}$$

and

$$\Delta y_0 = \Delta y/(1 - q_y) \tag{11.7}$$

Multiplying the numerators and denominators of the right-hand parts of these equations by $(1 - q_x)$ and $(1 - q_y)$ respectively, and again neglecting the terms in q_x^2 and q_y^2, we obtain

$$\Delta x_0 = \Delta x(1 + q_x) \tag{11.8}$$

and

$$\Delta y_0 = \Delta y(1 + q_y) \tag{11.9}$$

Now

$$\ell_0^2 = \Delta x_0^2 + \Delta y_0^2$$
$$= \Delta x^2(1 + q_x)^2 + \Delta y^2(1 + q_y)^2 \tag{11.10}$$

Once more neglecting the higher terms, we obtain

$$l_0^2 = l^2 + 2(\Delta x^2 \cdot q_x + \Delta y^2 \cdot q_y) \tag{11.11}$$

The second term in equation (11.11) is the correction to be applied to the line l measured on the map.

If $q_x = q_y$ or the difference between them is small, then we may use the mean value of q. From (11.12) we find that

$$l_0^2 = (\Delta x^2 + \Delta y^2)(1 + q)^2$$
$$= l(1 + q) \quad \text{as in (11.12)}$$

We now return to the consideration of the triangular parcel of area A. In order to apply the correction for paper deformation, we may determine the corrected lengths of the sides l and h. Then from equation (11.11), these values will be

$$l_0^2 = l^2 + 2(\Delta x_l^2 \cdot q_x + \Delta y_l^2 \cdot q_y)$$
$$h_0^2 = h^2 + 2(\Delta x_h^2 \cdot q_x + \Delta y_h^2 \cdot q_y)$$

We obtain the corrected values for the parcel area, by squaring equation (11.4)

$$\begin{aligned} A_0^2 &= (l_0^2 \cdot h_0^2)/4 \\ &= l^2 \cdot h^2/4 + l^2/2(\Delta x_h^2 \cdot q_x + \Delta y_h^2 \cdot q_y) + h^2/2(\Delta x_l^2 + \Delta y_l^2 \cdot q_y) \\ &\quad + 4(\Delta x_l^2 \cdot q_x + \Delta y_l^2 \cdot q_y) \cdot (\Delta x_h^2 \cdot q_x + \Delta_l^2 \cdot q_y) \end{aligned} \tag{11.12}$$

We may disregard the last terms in this equation because after multiplication this is very small compared with the other quantities q_x^2, q_y^2 and q_x, q_y. Now

$$\left.\begin{aligned} \Delta x_l &= l \cdot \cos \alpha_l \\ \Delta y_l &= l \cdot \sin \alpha_l \\ \Delta x_h &= h \cdot \cos \alpha_h \\ \Delta y_h &= h \cdot \sin \alpha_h \end{aligned}\right\} \tag{11.13}$$

where α_l and α_h are the directions of l and h. Then

$$\begin{aligned} A_0^2 &= l^2 \cdot h^2/4 + \tfrac{1}{2} l^2 (h^2 \cdot \cos^2 \alpha_h \cdot q_x + h^2 \cdot \sin^2 \alpha_h \cdot q_y) \\ &\quad + \tfrac{1}{2} h^2 (l^2 \cdot \cos^2 \alpha_l \cdot q_x + l^2 \cdot \sin^2 \alpha_h \cdot q_y) \end{aligned} \tag{11.14}$$

Since the directions of the base and height of a triangle are perpendicular to one another, it follows that

$$\cos^2 \alpha_l = \sin^2 \alpha_h$$

and

$$\cos^2 \alpha_h = \sin^2 \alpha_l$$

so that we obtain

$$A_0^2 = A^2 + 2A^2(q_x + q_y) \tag{11.15}$$

Because the second term of this expression is very much smaller than the first, we may simplify it further and write

$$A_0 = A + A(q_x + q_y) \tag{11.16}$$

and clearly the term $A(q_x + q_y)$ is the correction in area resulting from paper deformation. Equation (11.16) is true for a parcel of any shape and, as we have already seen, it does not depend upon the orientation of the parcel on the map. If the coefficients are equal, or

$$q_x = q_y = q,$$

or the calculated mean coefficient

$$q = 1/2(q_x + q_y)$$

then

$$A_0 = A + 2A \cdot q$$

which is the same as (11.5). The following example shows the calculations to be made:

$$A = 52 \cdot 15 \text{ ha}, \quad q_x = +1/200, \quad q_y = -1/100$$

Then

$$A_0 = 52 \cdot 15 + 52 \cdot 15(1/200 - 1/100) = 51 \cdot 89 \text{ ha}$$

or

$$A_0 = 52 \cdot 15 - 104 \cdot 3/400 = 51 \cdot 89 \text{ ha}$$

In the second example, $q_x = q_y$

$$A = 38 \cdot 40 \text{ ha}, \quad q_x = q_y = q = +1/200$$

Then

$$A_0 = 38 \cdot 40 + 76 \cdot 8/200 = 38 \cdot 78 \text{ ha}.$$

In the method of determining area by calculating the areas of simple geometrical figures, described in Chapter 16, the corrections should not be made separately to each component figure, but should be applied to the whole parcel using equation (11.16). According to this expression we obtain a correction for paper deformation, the value of the area obtained from the method of squares or by planimeter with estimation of the divisions obtained from tracing round a group of grid squares on a map which is not distorted. If the planimeter constant has been determined by the calibration procedure using a jig (described in Chapter 17) or by tracing the unit quadrangle on an undeformed map, and is then used for measurement of a parcel on a deformed map, we should use of the equation

$$A_0 = A + 2pq$$

However, a better method would be to determine the value of the vernier unit by measuring a parcel of known area in the following fashion.

DETERMINATION OF AREA BY THE SAVICH METHOD

The *Savich method* is the term used in Russia to describe such methods of controlling area measurements. The ideal control should comprise one calculation and three independent measurements as follows:

(a) First the area of a group of grid squares or a graticule quadrangle containing most of the parcel, is determined by calculation. The result is the standard, A_0, against which the actual measurements may be compared.

(b) This unit figure is measured using the same technique, instrument and map which is to be used later to measure the component marginal parts of the parcel. We call this area A. It follows that $A_0 - A$ is the error in the measurement of the area of the quadrangle. This error is composed partly of the technical errors in measurement, some arising from paper deformation and some being the influence of the map projection.

(c) The required parcel lying within that grid or graticule quadrangle is now measured separately. This is the area a. To many map users this may seem sufficient, for it is this area which is needed and if $A \approx A_0$, evidently the amount of deformation is small.

(d) In order to maintain a check upon the various errors of measurement, that part of each control figure which does not lie within the parcel, b, is measured separately. There ought to be the simple relationship that

$$A = a + b \tag{11.17}$$

and if we accept this as being true, we also assume that

$$b = A - a \tag{11.18}$$

If, however, we make this assumption in practical work we have no knowledge whether the measurements of a or b are accurate. In short, we have no protection against making any gross error of measurement, whereas if a and b

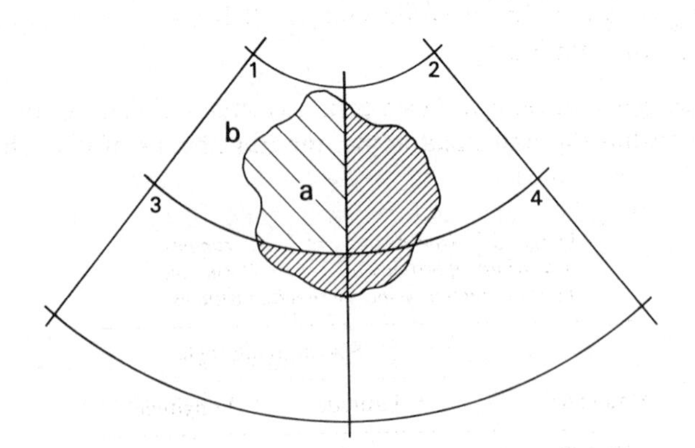

FIG. 11.1 Representation of the parcel to be measured (shaded) within the graticule of the map to be used for control purposes. Consider the quadrangle (1) which has theoretical area A_0 and measured area A. The portion of the parcel which is needed from this quadrangle is a; the part of the quadrangle lying outwith the parcel, b, should be measured separately. The other quadrangles, (2), (3) and (4) also contain portions of the required parcel and should be treated similarly.

have both been measured independently, we are immediately aware of any errors.

THE FORMAT AND SIZE OF THE UNIT QUADRANGLE

A major factor in controlling area measurements on small-scale maps is the determination of the surface area for each quadrangle wholly contained within the parcel, together with any containing marginal elements which have to be measured in parts as well. Published tables of area are, of course, restricted to listing a few sizes of unit quadrangle, and these do not necessarily correspond to size of the unit figure which is controlled by the spacing of the parallels and meridians on the map. For example those compiled by Amiran and Schick (1964) are limited to 10° × 10° and 1° × 1° figures, together with a graph to extract the areas of 5′ × 5′ quadrangles. It follows that the user who is committed to using tables is obliged to make use of one or other of these standard sizes as the unit quadrangle. If, on the other hand, the area enclosed by any format of quadrangle is determined by calculation, it is possible to choose a control figure of more suitable format. In making the choice the following considerations are important:

- The dimensions of the grid or the graticule quadrangle should be small enough to neglect the effects of local paper deformation.
- The size of the grid or the graticule quadrangle should be small enough to allow the changes in particular scale of the projection to be taken into consideration as a single correction.
- It is desirable that the quadrangles should have uniform dimensions, corresponding to the parallels and meridians shown on the map.
- Any need to plot and draw additional parallels and meridians should be reduced to the minimum.

It has been suggested that if we wish to neglect the influence of local paper deformation within the unit quadrangle, the dimensions of it ought not to

TABLE 11.1. *Dimensions of quadrangles measuring approximately 10 × 10 cm on maps of different scales in middle latitudes*

	Size of quadrangle	
Map scale	Latitude	Longitude
1/100,000	5′	7′·5
1/200,000	10′	15′
1/500,000	30′	45′
1/1,000,000	1°	1° 30′
1/1,500,000	1° 30′	2° 15′
1/2,500,000	2° 30′	3° 20′

exceed about 10 × 10 cm. These correspond to the subdivisions of the graticule given in Table 11.1.

The combination of these dimensions with the changes in area scale on different map projections, indicates that for maps of scale 1/1,000,000 and larger compiled on the projections used for topographic mapping, such as the U.T.M., paper deformation plays the decisive role in the choice of the width of the latitude zone and for the dimensions of the unit quadrangle. At scales smaller than 1/1,500,000 the property of the map projection becomes increasingly important.

In the use of control on small-scale maps and charts, especially those in which the particular scales or the area scale change rapidly with latitude, as on the normal aspect of Mercator's projection, it is necessary to determine the dimensions of the latitude zone within which the range of scale is no larger than the technical errors of area measurement. The procedures for area measurement on Mercator's projection are described in detail in Chapter 22. Here it is sufficient to note that because the particular scales of most normal aspect map projections vary with latitude, it is often easier to work in *zones* rather than quadrangles.

CONTROL OF AREA MEASUREMENTS BY MEANS OF ZONES

The use of zones comprises subdivision of the parcel into a series of belts, the widths of which are usually determined by the special properties of the map projection and the rate of change of area scale with latitude. The procedure is similar to that for more symmetrical quadrangles; indeed the zone is nothing more than a quadrangle which is much elongated in the longitudinal direction. In a sense, too, the method is similar to the use of measurement by means of strips, described in Chapter 20, but with the differences that first, the orientation of each zonal strip is determined by the nature of the deformations. Because this method is usually chosen to overcome the variations in area scale of the projection and because, on most useful map projections, these vary with latitude, the zones are usually defined by two parallels of latitude. Secondly the width of the strip may vary from one zone to the next simply because this is controlled by the way in which the area scale changes.

Since the technical errors of measurement occur only in those parts of the parcel which have actually been measured, obviously the smaller the proportion of the total area which has to be measured, the smaller is their influence upon the result. It follows that we may increase the accuracy if the unit size of strip used to subdivide the parcel is made smaller. But this may involve plotting and drawing additional parallels on the map or chart. This work is also prone to error. Ignoring the possibility of any errors in the original graticule, the mean square error in position of a point plotted to provide an additional parallel of latitude is, from Chapter 9, about $\pm 0{\cdot}13$ mm. We may demonstrate the effect of such an error upon a quadrangle of area 1,200 km^2, in

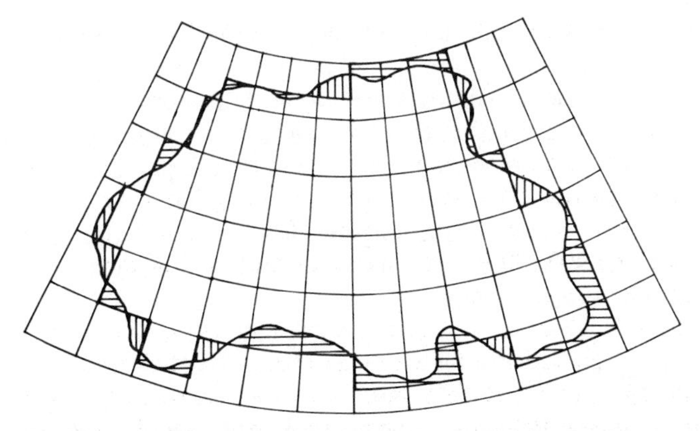

FIG. 11.2. Control of area measurement of a large country by means of zones. Each latitude zone is bounded by two meridians which have been selected in a similar fashion to the give-and-take lines used to determine the lengths of strips used to measure parcels on large-scale maps. Strictly speaking no measurement is required, for the area contained within each zone between the two bounding meridians can be calculated, and if the choice of these is satisfactory, the residual portions within and without the ends of each zone are equal. (Source: Volkov, 1950b).

TABLE 11.2. *Errors in the determination of area because of inaccurate construction of the graticule on the map*

Scale	Relative error in area of quadrangle	Error in km^2 in the quadrangle
1/100,000	1/1,400	0·21
1/200,000	1/1,400	0·85
1/500,000	1/570	2·1
1/1,000,000	1/280	4·3
1/1,500,000	1/190	6·3

the approximate latitude 55°. If the additional parallel defining a zone also comprises one of the boundaries of the parcel to be measured, its area may be distorted by as much as the amounts tabulated.

We see that at the smaller scales the amount of distortion is greater than the relative error which we expect in using a polar planimeter. It is therefore evident that it is most undesirable to construct *both* parallels and meridians for the same quadrangle. However, we usually have to disobey this rule if we wish to construct suitable zones.

CALCULATION OF THE AREA OF A QUADRANGLE

On large- and medium-scale maps it is, of course, usual to employ a block of grid squares as the unit area for both calibration and calculation and, as we have seen, we may often determine the area of such a figure by mental arithmetic. By contrast, the standard against which measurements made on maps or charts showing a graticule only must be compared is the calculated area of a quadrangle formed by a pair of meridians and a pair of parallels. Because of the convergence of the meridians it is not possible to regard this figure as either square or rectangular. The following pages explain in some detail how the area of such a quadrangle may be calculated. We must therefore consider how to calculate such figures for both the spherical and the spheroidal reference surfaces. Naturally, we find the spherical solution to be mathematically the more easy.

Area of a Spherical Quadrangle

Many textbooks on elementary calculus derive the equation for the surface areas of any solid of revolution as an example of integration.

Here we derive the expression for the area of a zone lying between two parallel planes, which correspond on the earth to two parallels of latitude. In Fig. 11.3 the sphere is formed by the rotation of the circle $x^2 + y^2 = R^2$ about the y-axis.

If θ be the angle AOP, the coordinates of P may be expressed as $(R \cdot \sin \theta$, $R \cdot \cos \theta)$ and the length of the arc $AP = ds = R \cdot \theta$.

The whole surface area of the zone contained between the planes HH' and KK' is twice the surface area which is generated if B is rotated to B' about y

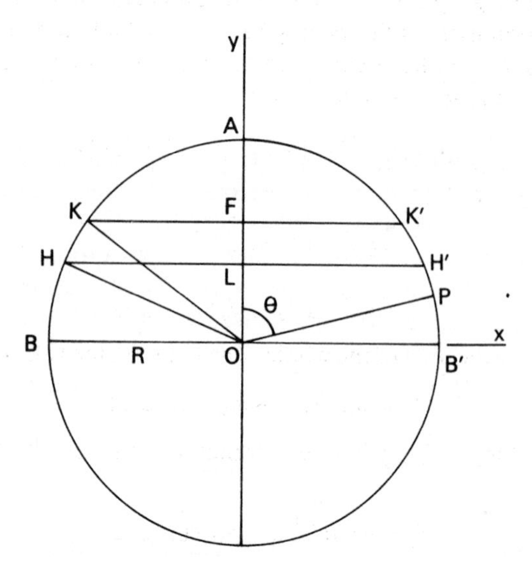

FIG. 11.3. The determination of the area of a zone of a spherical surface.

From the general expression for the surface area of any solid of revolution,

$$A = 2\pi y[1 + (\mathrm{d}y/\mathrm{d}x)^2]^{1/2}\cdot \mathrm{d}x \tag{11.19}$$

and in the special case of the sphere this may be written

$$A = 2\pi x(\mathrm{d}s/\mathrm{d}\theta)\mathrm{d}\theta \tag{11.20}$$

If the planes HH' and KK' correspond to those of two parallels of latitude, φ_1 and φ_2, we may write the corresponding colatitudes as $\chi_1 = HOA$ and $\chi_2 = KOA$. Then the total area of the zone contained between these planes is

$$A = 2\pi R^2 \int_{\chi_1}^{\chi_2} \sin\chi\cdot \mathrm{d}\chi \tag{11.21}$$

$$= 2\pi R^2(\cos\chi_1 - \cos\chi_2)$$

$$= 2\pi R(R\cdot\cos\chi_1 - R\cdot\cos\chi_2) \tag{11.22}$$

From Fig. 11.3,

$$R\cdot\cos\chi_1 = OL$$

and

$$R\cdot\cos\chi_2 = OF$$

Therefore

$$A = 2\pi R(OL\text{–}OF)$$

or

$$A = 2\pi Rh \tag{11.23}$$

where h is the distance $OL\text{–}OF$. Geometrically this is the height of the cone whose frustum approximates to the two arcs HK and $H'K'$. Equation (11.22) is clumsy to use with geographical coordinates and better solutions are possible. In (11.22) the term in parentheses may be converted into functions of latitude as $(\sin\varphi_1 - \sin\varphi_2)$ and from the standard trigonometric expression relating the difference in the sines of two angles,

$$(\sin\varphi_1 - \sin\varphi_2) = 2[\cos(\varphi_1 + \varphi_2)/2]\cdot\sin[(\varphi_1 - \varphi_2)/2]$$

so that, putting

$$\varphi_M = (\varphi_1 + \varphi_2)/2$$

and

$$\delta\varphi/2 = (\varphi_2 - \varphi_1)/2$$

the expression for the area of the whole zone between the two parallels is now

$$A' = 2\pi R^2(2\cdot\cos\varphi_M\cdot\sin\delta\varphi/2) \tag{11.24}$$

or, for a part of this zone, which is a quadrangle or zone of width $\delta\lambda$ radians, the area is

$$A'_q = \delta\lambda\cdot R^2(2\cos\varphi_M\cdot\sin\delta\varphi/2) \tag{11.25}$$

It remains to allocate a suitable value for the radius of the sphere, R.

THE CHOICE OF SPHERICAL RADIUS

This problem is not confined to the present study, for it also occurs in oth aspects of the study of geodesy and map projections, whenever it is considere appropriate to simplify the algebra or the calculations by making th assumption that the earth is truly spherical. However, there is an almost casu approach in many works to the choice of the numerical value of R, the radius the sphere, which is required to convert from results in angular measures int linear units. It follows that unless the adopted value for R is clearly stated, it easy to introduce a systematic error in the calculations. Indeed, I go so far as assert that a great deal of effort has often been put into the derivation of som rather complicated equations, but much of the resulting mathematic sophistication has been lost by introducing a wholly unsuitable numeric value for R. We must therefore consider the possibilities which are availab to us.

We saw, in Chapter 2, that the closest approximation to the earth's shap and size which is commonly used in cartography and surveying is that th earth approximates to an ellipsoid of rotation, or spheroid. One of th mathematical properties of the ellipsoid of rotation is that every point on i surface has two radii of curvature (which we have denoted as ρ and ν), an these radii vary with latitude. At the next, lower, level of approximation w may, as here, assume that the earth is a true sphere. The correspondin spherical body is to be regarded as being tangential to the spheroid at particular point or along a particular line, and is called an *auxiliary spher* There are a number of different ways of defining it.

1 The simplest choice for the radius of an auxiliary sphere is to use *one of th semiaxes of the ellipsoid*, or some combination of both of them to provide single value. All of the following have been used:

- The equatorial radius of the spheroid. This corresponds to the use c the major semiaxis, a, so that the sphere is tangential to the spheroid a the equator.
- The polar, or minor semiaxis, b.
- A combination of both a and b, this being either the arithmetic mear $(a+b)/2$ or the geometric mean, $\sqrt{(ab)}$.
- In a triaxial ellipsoid the equator is also an ellipse, requiring definitio by two axes. If, therefore, the figure of the earth were to be regarded as triaxial ellipsoid it would be necessary to take three axes int consideration, whereas in an ellipsoid of rotation the equator is a circl so that, as we have seen, the body is completely defined by only tw axes. In order to make approximation of the triaxial body it i therefore usual to take twice the value of the major semiaxis a, to give redius which is either the arithmetic mean $(2a+b)/3$ or geometri mean $(2ab)^{1/3}$.

2 We may use *the radii of curvature of the spheroid for some reference latitude.* For example,
   - The value of ρ or ν may be taken for the latitude 45°, this being chosen because this latitude lies midway between the equator and the poles, and corresponds to an auxiliary sphere which intersects the spheroid in this latitude.
   - More commonly, the values for ρ or ν are taken for the mean latitude of the quadrangle. Then we calculate the radii of curvature for this latitude and use ρ_M or ν_M.
   - Again we may use either the arithmetic or geometrical means $(\rho + \nu)/2$ or $\sqrt{(\rho\nu)}$ from them. The last of these is known as the "Gaussian curvature".

3 The radius may be determined for *a sphere having the same volume* as the chosen figure of the earth. This radius may be determined from the expression

$$R = a(1 - f/3 - f^2/9) \tag{11.26}$$

In this equation, f is the flattening of the spheroid, as defined in (2.2).

4 The radius may be determined for *a sphere having the same surface area* as that of the chosen figure of the earth. This is also known as the *authalic sphere* and

$$R^2 = (a^2/2\pi)\cdot\{1 + (1 - e^2/2e)\cdot\ln[(1 + e)/(1 - e)]\} \tag{11.27}$$

5 The *rectifying sphere* has meridional length equal to that of the spheroid. Wray and Weiss (1982) derive it from the expression

$$R = 8a^2b^2/(a + b)^2(1 + 9n^2/4 + 225n^4/64 + \cdots) \tag{11.28}$$

where

$$n = (a - b/a + b) \tag{11.29}$$

6 Frolov (1966) makes use of the expression

$$R^2 = ab[1 + e^2(2\sin^2\varphi - \tfrac{1}{2})] \tag{11.30}$$

Since this equation contains a term in latitude, it varies with φ. For the International Spheroid, the range in from 6,356,930 metres at the equator to 6,399,666 metres at the poles.

The variability of R resulting from so many different definitions gives rise to markedly different values for the spherical distance. The methods outlined above have been used to determine values of R which are based upon the International Spheroid (1924). The results are listed in Table 11.3. The range of the calculated values for the areas of 1° × 1° quadrangles is of the order of 0·75–1% of their areas. This is much too large to be ignored.

There are two possible solutions. The first is *to specify precisely which value for the radius of the sphere has been adopted.* In earlier work, such as Maling

TABLE 11.3. *The values of spherical radius, R, determined from the International Spheroid by different methods*

Definition of spherical radius	R (metres)	Areas (km²) of 1° × 1° quadrangles with mean latitudes		
		15°	45°	60°
Areas of the 1° × 1° quadrangle determined for the International Spheroid		11,900·8	8,762·9	6,217·3
Major semiaxis, a	6,378,388	11,969·2	8,763·3	6,196·1
Arithmetic mean of a and b	6,367,650	11,929·7	8,734·4	6,175·7
Geometric mean of a and b	6,367,641	11,928·9	8,733·8	6,175·3
Arithmetic mean of three axes $(2a+b)/3$	6,371,229	11,942·3	8,743·6	6,182·2
Geometric mean of the three axes $(2a\cdot b)^{1/3}$	6,371,221	11,942·3	8,743·6	6,182·2
g.m. for $\varphi = 15°$	6,359,778	11,900·8	–	–
ρ for $\varphi = 45°$	6,367,586	11,928·7	8,733·6	6,175·2
ν for $\varphi = 45°$	6,389,135	12,009·5	8,792·9	6,217·0
g.m. $\varphi = 45°$	6,378,351	11,969·0	8,763·2	6,196·1
ρ for $\varphi_m = 60°$	6,383,727	11,989·2	8,778·0	6,206·5
ν for $\varphi_m = 60°$	6,394,529	12,029·8	8,807·7	6,227·5
g.m. $\varphi_m = 60°$	6,389,126	12,009·5	8,792·8	6,217·0
Sphere of equal volume	6,371,221	11,942·6	8,743·6	6,182·2
Authalic sphere	6,371,228	11,942·3	8,743·6	6,182·2
Rectifying sphere	6,367,655	11,928·9	8,733·8	6,175·3
Frolov's equation				
For $\varphi_m = 15°$	6,359,802	11,899·5	–	–
For $\varphi_m = 45°$	6,378,334	–	8,763·1	–
For $\varphi_m = 60°$	6,389,009	–	–	6,175·3
Range	26,940 metres	101·1 km²	64·1 km²	52·3 km²

(1973), the author has recommended the use of either the sphere of equal volume or the authalic sphere, which give similar values for R. These are adequate for constructing map projections to be used for small-scale maps, where the discrepancies are wholly absorbed within the graphical resolution of plotting. For other jobs, particularly those associated with the determination of offshore boundaries as specified by the Law of the Sea Conference, such uncertainty is simply not accurate enough. Comparison of these results with those given in Tables 11.4 and 11.5 shows that none of the determinations for the areas of unit quadrangles correspond exactly with those derived for the same latitudes on the spheroid. In fact the best results are to be obtained from equation (11.30), which needs to be determined for each mean latitude, together with the values for R determined as the geometric mean of the two spheroidal radii, again for the mean latitude of the quadrangle. It follows that there is much to be said for the second procedure; *to determine the areas of unit quadrangles using the spheroidal assumption and a known figure of the earth.*

THE AREA OF A SPHEROIDAL QUADRANGLE

For an infinitely small quadrangle in latitude φ, we have the arc element on the meridian of length $\rho \cdot d\varphi$ and along the parallel of length $\nu \cdot \cos\varphi \cdot d\lambda$. Hence the area of the infinitely small rectangle is

$$dA = \rho \cdot \nu \cdot \cos\varphi \cdot d\lambda \cdot d\varphi \tag{11.31}$$

or, substituting for ρ and ν from equations (2.4) and (2.5)

$$A = \lambda ab \int_0^{\varphi} \cos\varphi \cdot d\varphi/(1 - e^2 \cdot \sin^2\varphi)^2 \tag{11.32}$$

The solution of this integral involves some fairly complicated algebra which is not given here. In the end we obtain the following expression for the area of a spheroidal quadrangle:

$$\begin{aligned} A = \lambda \cdot ab[&\sin\varphi - 1/3n(2+n) \cdot \sin 3\varphi + 1/5n^2(3+2n)\sin 5\varphi \\ &- 1/7n^3(4+3n)\sin 7\varphi + \cdots] \end{aligned} \tag{11.33}$$

The term n has the value already determined in (11.29). An alternative solution by McCaw (1942) leads to an expression for the area of a zone extending from the equator to the parallel φ as

$$\begin{aligned} A = \pi a^2(1-e^2)\{&\sin\varphi/(1 - e^2 \cdot \sin^2\varphi) \\ &+ (1/2e)\ln[(1 + e \cdot \sin\varphi)/(1 - e \cdot \sin\varphi)]\} \end{aligned} \tag{11.34}$$

We may derive the area of the zone between two parallels of latitude from the difference between two such expressions. This is best expressed by expanding e in series to obtain from (11.33)

$$\begin{aligned} A_q = \delta\lambda \cdot a^2(1-e^2)\{&2A \cdot \sin\delta\varphi/2 \cdot \cos\varphi_M - B \cdot \sin 3\delta\varphi/2 \cdot \cos 3\varphi_M \\ &+ C \cdot \sin 5\delta\varphi/2 \cdot \cos 5\varphi_M - D \cdot \sin 7\delta\varphi/2 \cdot \cos 7\varphi_M + \cdots\} \end{aligned} \tag{11.35}$$

where φ_M is the mean latitude of the quadrangle or zone.

The coefficients A–D are determined from the following expansions in e:

$$A = 1 + \tfrac{1}{2}e^2 + 3/8e^4 + 5/16e^6 + \cdots \tag{11.36}$$

$$B = 1/6e^2 + 3/16e^4 + 3/16e^6 + \cdots \tag{11.37}$$

$$C = 3/80e^4 + 1/16e^6 + \cdots \tag{11.38}$$

$$D = 1/112e^6 + \cdots \tag{11.39}$$

Since $e^2 \approx 0{\cdot}0067\ldots$, higher powers in e are very small values. Thus by neglecting all terms higher than e^2, equation (11.35) may be rewritten in the form

$$A_q = \delta\lambda\, a \cdot b(2\cos\varphi_M \cdot \sin\delta\varphi/2 - e^2/6\cos 3\varphi_M \cdot \sin 3\delta\varphi/2) \tag{11.40}$$

which is the equation needed to calculate the area of a quadrangle of width $\delta\lambda$ radians. It may be seen that it differs from the spherical solution (11.25) only by using the product of the semiaxes $a \cdot b$ in place of R^2 together with the second term in the parentheses. Frolov (1966) has argued that this expression is accurate enough for most cartographic applications.

Another simplified version, known as *Carpenter's rule*, is described briefly in Gannett (1881), which document was reproduced in full by Proudfoot (1946). Carpenter's formula was intended to determine the area of 1 square degree from

$$A = \pi/90(\lambda_1 - \lambda_2)\nu\rho m \cdot \sin\tfrac{1}{2}(\varphi^2 - \varphi_1)\cos\tfrac{1}{2}(\varphi_2 - \varphi_1) \qquad (11.41)$$

where $m = 1 \cdot 004285$ and the difference in longitude is expressed in degrees.

A remarkably simple method of computing the areas of unit quadrangles on the spheroidal surface has been devised by Balandin (1985), which makes it possible to determine the areas of unit quadrangles on a pocket calculator without having to book any intermediate results, such as the values of $A \ldots D$ in equations (11.36)–(11.39). He has shown that the term $e^2 \cdot \sin^2\varphi$ never exceeds 0·007, and that one may write without much loss of accuracy that

$$\cos\varphi \cdot d\varphi/(1 - e^2 \cdot \sin^2\varphi)^{1/2} = (1/e\rho) \cdot \arcsin(e \cdot \sin\varphi) \qquad (11.42)$$

where ρ is the conversion from radians into degrees.

He then derives a constant K_0 which for the Krasovsky figure of the earth $= 0 \cdot 163133 \approx 2e$. Therefore:

$$\cos\varphi \cdot d\varphi/(1 - e^2 \cdot \sin^2\varphi)^{1/2} = (1/K_0\rho) \cdot \arcsin(K_0 \cdot \sin\varphi) \qquad (11.43)$$

and it follows that

$$A = C \cdot \delta\lambda |\arcsin(K_0 \cdot \sin\varphi)|_{\varphi_1}^{\varphi_2} \qquad (11.44)$$

Here

$$C = b/K_0 \cdot \rho^2 = 75{,}456 \cdot 835$$

The procedure to find A is to solve (11.44) for each of the bounding latitudes φ_1 and φ_2. Then the area of the quadrangle is the difference between the two results. For example in order to determine the area of the 1° quadrangle between latitudes 30° and 31°, we find that the two values for arc sin $(K_0 \cdot \sin\varphi)$ are

363,676·03
and 353,033·40

10,642·63 square kilometres.

This corresponds well with the value of 10,642·24 obtained by calculation of all the terms in equation (11.35); moreover it can be done in a few moments on any scientific pocket calculator. The results of the different methods of determining the spheroidal quadrangle may be compared in Table 11.4.

TABLE 11.4. *The areas (in km²) of 1° × 1° quadrangles computed for different latitudes on the International Spheroid using some of the methods described above*

Method	Mean latitude of quadrangle, φ_M		
	15°	45°	60°
Full expression (11.33)	11,900·8	8,762·9	6,217·3
McCaw's full equation (11.34)	11,900·6	8,762·9	6,217·3
Frolov's simplification (11.40)	11,900·9	8,762·8	6,217·1
Balandin's method (11.44)	11,900·5	8,762·5	6,217·0

TABLE 11.5. *The sizes of 1° × 1° quadrangles (in km²) computed for different figures of the earth*

Figure	Latitude		
	15°	45°	60°
Clarke, 1,858	11,914·7	8,747·8	6,196·3
Clarke, 1,866	11,914·5	8,747·6	6,196·1
Clarke, 1,880	11,914·3	8,748·7	6,196·2
International	11,915·5	8,748·2	6,196·4
Krasovsky	11,915·2	8,747·9	6,196·2
WGS 66	11,914·9	8,747·6	6,196·0
WGS 72	11,914·8	8,747·6	6,195·9
GRS 80	11,914·8	8,747·6	6,196·0
NWL-9D	11,914·9	8,747·6	6,196·0

In all the calculations made for the spheroid, the value for e^2 as well as a and b depend upon the figure of the earth which has been adopted for use in a particular country or continent. We have already tabulated those figures in commonest practical use in Chapter 2, p. 19. Table 11.5 demonstrates the results to be obtained for different figures of the earth when different values of a, b and e are used to determine the 1° × 1° quadrangles already considered in our study of the spherical assumption.

Tables 11.4 and 11.5 show how consistent are the results of calculating the areas of 1° × 1° quadrangles on the spheroidal surface even using different formulae and figures of the earth. Compared with the irregularity of the results obtained for the spherical assumption, which derives from uncertainty about which is the most suitable value of R to use for a particular job, this consistency provides a strong argument for seeking the spheroidal solution in preference. Since it is now so easy to determine the area of the spheroidal quadrangle from Balandin's formula there seems to be no reason to employ the spherical approximation. It is treated here in detail to indicate the pitfalls which may arise in its use.

12

Definition of the Feature to be Measured

The two longest rivers in the World are the Amazon (Amazonas), flowing into the South Atlantic, and the Nile (Bahr-el-Nil) flowing into the Mediterranean. Which is the longer is a matter of definition rather than measurement.

(*The Guinness Book of Records*)

The most difficult step to take is often the first. In cartometry it is the task of defining the object to be measured – identification of the terminal points of a line, or the exact perimeter of a parcel of land. These problems seem simple enough at the large scale of an estate survey, where the neatly defined fields are bounded by straight wire fences, but in the measurement of natural boundaries, such as a coastline or river bank, it is less easy to be certain where it should be drawn. For example, the sea may penetrate many miles inland at high water in an estuary having a large tidal range. What represents the downstream limit of the river flowing into this estuary, and where is the upstream limit of the sea? Usually a set of rules have to be followed to treat consistently with uncertainties of this sort. This chapter considers some examples of the rules which have been made and the difficulties which arise in attempting to define a coastline.

We have already seen in Chapter 5 that measurements of the length of a feature may differ appreciably, notwithstanding the fact that they appear to have been made with equal care and apparently for similar purposes. There appeared to be several reasons for these discrepancies, such as the variation in the scales of the maps used, the setting of dividers and how the feature had been defined. Definition was alone responsible for some large discrepancies in the measured lengths of the coastline of the U.S.A. recorded in Table 5.7, p. 73. Some of the evident reasons for these discrepancies were summarised on pp. 73–74.

There are two kinds of discrepancy which need to be considered:

- Those arising from the way in which the data have been assembled and presented.
- Those arising from different, but not necessarily incorrect, as of identifying the position of a coastline in an estuary or the course of a river through a lake.

The first may be described as *errors in presentation*; the second as discrepancies arising from *definition*. The first group may arise from faulty selection of data and not from any defects in the measurement procedures. Indeed, they commonly occur in the use of area data which have been selected from another source and no actual measurements have been made. They often appear to be discrepancies resulting from the way in which the feature has been defined, but the author maintains that the errors in presentation differ from typical ambiguities in definition and are, indeed, gross qualitative errors.

ERRORS IN PRESENTATION

In the following example we attempt to explain why reference books and atlases seldom agree about the area of a country, even those for which good source maps exist and reliable measurements have been made. We have therefore chosen the example of the area of the United Kingdom because these data ought to be both consistent and reliable. The measurements originate as those made by the Ordnance Survey using the same methods to measure the area of every parcel of land shown on the basic scale maps of the country. The areas of the individual parcels are combined to determine (and check) the total area contained within each map sheet and, ultimately the areas of every country or region and those for the component countries are derived from them.

Table 12.1 gives details of the areas of the component countries in two forms; the area for each country which includes that of any inland water and that for dry land only. These figures are from the 1974 edition of *Encyclopaedia Britannica.* Table 12.2 lists some examples of the area of Great Britain taken from apparently reliable reference works published between 1945 and 1985.

The area of Great Britain has not changed significantly for any natural reasons, nor have there been any changes in political boundaries of the major components during the period under discussion. Therefore the differences between Tables 12.1 and 12.2 arise either from differences in opinion about how the place "Great Britain" should be defined, or from lack of care in selecting the data. Where it has been necessary to convert the values to the metric system, this has been done by the author using the same conversion factor, 1 square mile = 2·58,998 square kilometres throughout. Consequently there is

TABLE 12.1. *The area of the United Kingdom (in km²)*

	England	Wales	Scotland	N. Ireland	Isle of Man	Channel Islands	Total
Area excluding inland water	129,637	20,640	77,179	13,484	572	194	241,705
Area including inland water	130,362	20,764	78,772	14,129	572	194	244,785

TABLE 12.2. *The area of Great Britain (in km²) according to different authorities, 1945–1985*

Source	Area	Remarks
The Statesman's Yearbook, 1946	244,181	Area of N. Ireland excludes inland waters; other areas include inland water.
Chambers's Encyclopædia, 1950	230,614	Excludes N. Ireland.
Whitaker's Almanac, 1951	244,201	Area of N. Ireland excludes inland water; other areas include inland water.
Oxford Atlas, 1951	244,181	See first entry above.
Columbia Lippincott Gazetteer and Geographical Dictionary, 1952	299,848	Areas for England, Wales and Scotland only; all areas including inland water.
Chambers's World Gazetteer and Geographical Dictionary, 1954	243,421	Excluding Isle of Man and Channel Islands; area of N. Ireland excludes inland water; all others include inland water.
Advanced Atlas of Modern Geography, Bartholomew, 1956	241,772	Excludes inland water.
Encyclopædia Britannica, 1959	243,187	Area of Scotland excludes inland water; all others include inland water.
Stanford's Whitehall Atlas, 1960	244,013	Includes inland water; excludes the Isle of Man and the Channel Islands.
Readers' Digest Great World Atlas, 1962	243,995	Ditto.
The Times Atlas, 1967	244,782	Total in square miles rounded to nearest 100 square miles; all areas include inland water.
Pergamon World Atlas, 1968	244,780	Preceding value rounded to the nearest 10 km².
The Statesman's Yearbook World Gazetteer, 1975	229,868	Areas of England, Wales and Scotland only; including inland water.
Whitaker's Almanac, 1980	241,676	Excluding inland water.
	244,759	Including inland water.
The Statesman's Yearbook, 1980–81	230,609	Excludes N. Ireland but includes Isle of Man and Channel Islands.
The Great Geographical Atlas, Mitchell Beazley, 1982	244,102	Excluding the Isle of Man and Channel Islands; includes inland water.
The Europa Yearbook, 1985	244,103	Ditto.

no reason to suppose that there are any variations in Table 12.2 which result from rounding errors.

The comments on the right-hand side of Table 12.2 indicate the most likely reasons why each of the tabulated values vary in some way from those given in Table 12.1. The majority of these have occurred for two reasons; first, in the use of inconsistent data by combining some areas which include inland water with those which exclude it. This is little short of carelessness. Secondly, there are evidently differences of opinion about which of the three component countries, Northern Ireland, the Isle of Man and the Channel Islands ought to be included. These are, of course, matters of definition.

The majority of the differences recorded in Table 12.2 are small. For example, the area of Great Britain excluding the Isle of Man and the Channel Islands in only 0·3% smaller than that including them. It is only when Northern Ireland is excluded from the total that the area is reduced by as much as 6%. The inland water component is only 1·3% of the total area given in Table 12.1.

Most of the reference works listed here quote the areas of Great Britain to the nearest square mile or square kilometre, thereby giving a spurious impression of accuracy. The quality of the measurements themselves is not in doubt, but errors have arisen from using them incorrectly. If the areas are really needed to the nearest square kilometre, much greater care ought to have been taken to select data which are consistent with the intended definition of the country, but it seems better to quote areas rounded to the nearest 100 square kilometres or 100 square miles, because this convention indicates that there may be a small measure of approximation in the results.

The decision to include Northern Ireland or exclude the Isle of Man is a matter of definition and the intended purpose for using the figures. From the geographical point of view it might be argued that the Isle of Man is part of Britain, but that the Channel Islands are part of France. The choice of including the area of inland water also depends upon how we wish to describe the country. Regarded as a geographical object, the country comprises both dry land and inland water. Therefore the areas given on the second line of Table 12.1 would provide the appropriate description. Regarded as an administrative unit, the area of inland water may be meaningless, for if the law admits that nobody owns it, then rates and taxes cannot be levied upon it, and fiscally the area does not exist. Agricultural and demographic statistics related to area may be misleading if they are calculated from areas including a substantial fraction of inland water. As a political unit, however, the area of a country may include not only the inland waters but also parts of the marginal seas enclosed by the baselines used to determine the limits of sovereignty under international law. From the pragmatic and commonsense point of view, as well as any geographical definition, these are parts of the sea. The areas recorded in Tables 12.1 and 12.2 do not contain any example of this kind of definition. However, we refer later to the inclusion of sea areas within the boundaries of the United States as these were defined at the time of the Sixteenth Census of Population held in 1940. Because of this, the country had to be described by three sets of measurements; *land, inland water* and *water areas other than inland water.*

If we had found in Table 12.2 that the recorded areas consistently fell into one or other of the categories, it could be argued that the variations in recorded area arose from differences of definition. However, we have seen that many of the figures in Table 12.2 are inconsistent, so that the discrepancie must be dismissed as being gross qualitative errors, or as errors i presentation.

THE DEFINITION OF A COASTLINE

It would be a daunting task to attempt to summarise all the different ways i which every feature on a map might be defined for cartometric purposes. Ther would be as many different definitions to consider as there are different kind of map use, and such an encyclopædic approach would not be appropriate. A in Chapter 5 we concentrate attention upon the definition of the coastline There is an important practical reason for treating with the definition of th coastline in detail. In order to delimit maritime boundaries under interna- tional law, it is necessary to define and locate coastal baselines precisely. Sinc many of the proposed definitions and methods of location of maritim boundaries are cartometric this represents one of the major applications of th subject in years to come. A later chapter is devoted to the study of some of th problems and their practical solution. First, however, we need to investigat how this has been done for other, usually geographical, purposes. We want t know why particular kinds of definition have been employed. Where specifi rules are known to have been applied, we should investigate whether thes have been employed consistently.

Coastline or Shoreline?

Definition of the coastline seems to be straightforward enough. For example Shalowitz (1962) describes it as *the line of contact between land and sea.* Using modern jargon, the definition might read as *the interface between land and sea,* or even the *configuration made by the meeting of land and sea* used by Ellis (1978).

Shalowitz considered that the words coastline and shoreline are syn- onymous. Many geomorphologists, for example Bird (1984), use the word shoreline strictly to mean the water's edge, migrating to and fro within the intertidal zone and the word *coast* represents a zone of variable width, including the shore and extending to the landward limit of penetration of marine influence.

The majority of the seas of the world are influenced by tides, which are translated in shallow water into a horizontal movement of the shoreline. Consequently the element of time enters the definition, because it is necessary to know whether the coastline is to be referred to a particular state of the tide. Is it necessary, or desirable, to regard the coastline as a particular configuration of the water's edge, frozen at some instant in time? This is how it will appear on an aerial photograph or a satellite image. Alternatively do we wish to use a posi- tion of the shoreline which represents the average tide or one of its extremes?

The choice generally lies between using either the high-water line or the low-water line (to be defined more precisely on pp. 238–246) because one or both of them appears on the map or chart to be used for measurement. Both of these lines are significant to certain map users. Those who are concerned with distributions on land are more likely to use the high-water line as the preferred boundary. The expert on international law, who is concerned with the demarcation of limits of sovereignty, is more interested in the low-water line, for it is from this boundary that baselines are defined and distances are measured to locate the different maritime zones.

Estuaries, Bays and Inlets

In the literature relating to coastline measurement the most common problem is how estuaries and other indentations of the coastline should be treated.

There are two schools of thought about how best to solve this and to define such features consistently. These may be described as *geometrical* and *geographical*. A geometrical rule specifies some limiting distance, for example, $\frac{1}{4}$ statute mile, 1 kilometre, 10 nautical miles or 24 nautical miles, which is used as a measure to truncate an estuary, inlet or creek at the place where it first becomes narrower than the specified limit. The chosen limit depends upon the nature of the coastline, the scale of working and the purpose of the measurement.

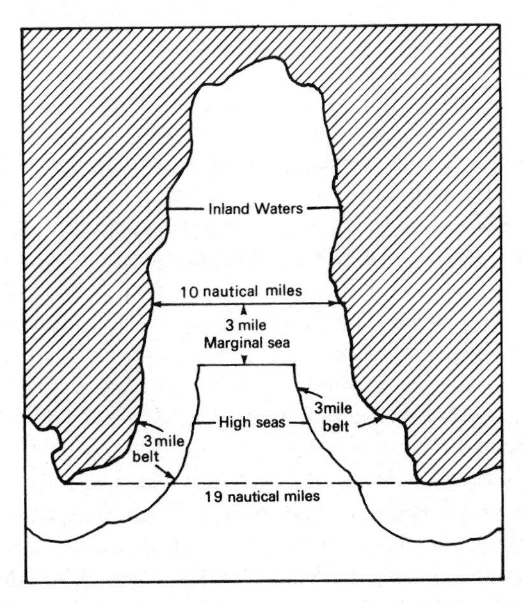

FIG. 12.1. The application of the 10-mile rule to a bay as this has been used with marginal or territorial seas of width 3 nautical miles. (Source: Shalowitz, 1962).

The geographical approach applies limits which are based upon the difference between land and sea; fresh water and salt water. A case may be made for taking the mouth of a river at the limit of tides, as exemplified by the definition of the tidal shoreline of the U.S.A. illustrated in Fig. 5.2. Thus all parts of a tidal estuary are regarded as being part of the sea, irrespective of how narrow it may be. Rigid adherence to this frequently leads to conclusions which are patently absurd because the limit is placed many tens or hundreds of kilometres from the open sea, far from where a commonsense view would regard the coastline to be situated. In Chapter 5 we cited the example of the tidal limit of the Amazon which is more than 1,000 km from the open ocean.

It is logical to argue that, because river water is nominally fresh, the boundary corresponding to the coastline ought to occur where fresh water becomes salt. However, this is usually represented by a zone of brackish water rather than a sharp demarcation; moreover, the zone migrates up and down the estuary with the rise and fall of the tide. Furthermore some of the major rivers of the world pour such huge masses of fresh water into the sea that the surface water is still comparatively fresh many miles offshore from the mouth of a river the size of the Congo. The opposite occurs in the Arctic where the winter freeze-up so reduces the flow of river water that in winter the sea penetrates far inland. For example, the Colville River in Alaska exhibits this seasonal change in character for about 50 km upstream of the summer limit for salt water.

SOME EXAMPLES OF COASTLINE DEFINITION

The Mainland Coast of Scotland

There are two modern measurements of the length of the coastline of mainland Scotland which are sufficiently well documented to provide an instructive example of how definition may give rise to different results. Both were made on Ordnance Survey maps at 1/63,360, so that map scale is not a variable in the discussion.

The first measurement was made by the Scottish Development Department in 1973. This result indicated that the mainland coast of Scotland is 3,840 km in length. Incidentally, it is interesting to record that Karo (1956) gives the length of the coastline of the mainland of Scotland as being only 716 km. Later, Baugh and Boreham (1977), of Aberdeen University, also measured the coast at this scale. They consider their best estimate of length, derived from various methods of reducing the data, to be 5,340 ± 20 km. Consequently it is 28% longer than the S.D.D. measurement. The shortest length obtained by them was 5,285 km. Admittedly the methods of measurement were not the same, but we know how the coast was defined by each team. According to Copland (1977) of the Scottish Development Department:

In our case, the measurement was made in the context of national planning for North Sea oil and gas developments, with an implicit need for unrestricted acres to the sea. We therefore excluded narrow inlets and estuaries less than one-quarter of a mile wide, and this is obviously the main cause for the apparent discrepancy.

This geometrical definition of the coastal limits is quite different to that of Baugh and Boreham, who defined the coastline as extending as far inland as either the tidal limit or to the lowest bridging point of a river, whichever was situated closer to the open sea. However, they promptly broke these rules in their treatment of the estuaries of the Forth, Clyde and Tay; at Ballachulish and Connell on the west coast; at Kyle of Tongue in the north and Cromarty Firth on the east coast. In short, they were inconsistent in their treatment of these difficult examples.

A similar set of rules were used by the Countryside Commission (1968) for the measurement of the length of the coastline of England and Wales, also at the one-inch scale and who had similar difficulties in deciding how to deal with estuaries. Their solution reads

the coastline was measured afresh, by the Commission, along the High Water Mark shown on the 1 inch to 1 mile Ordnance Survey Maps where this had a direct frontage to the sea and included inlets only where they could properly be regarded as 'arms of the sea', for example below the lowest ferry point, or natural ferry points if no actual ferry exists.

It would be interesting to know the rules to identify a non-existent natural ferry point.

The Coastline of Australia

The *1 kilometre rule* was used by Galloway and Bahr (1979) for their measurement of the length of the coastline of Australia. They describe their rules for defining coastal features as follows

we generally followed the coastline as marked on the 1:250,000 maps but estuaries and other inlets were arbitrarily cut off at the most seaward point where their mapped width was less than 1 kilometre. We regarded mangroves as part of the land with the coastline following their seaward edge fringes: inter-mangrove channels were treated as estuaries. Straits less than 1 km wide were regarded as inshore waters lying within the coastline and thus a few islands were incorporated with the mainland. All coral reefs were excluded.

The 1 km cut-off width for estuaries and straits was adopted because numerous islets and sheltered inshore waters became relatively open sea at about this width.... Furthermore the 1 km rule obviates the difficulty of how to treat lagoons sporadically open to the sea since even the widest opening of their mouths is less than 1 km.

Galloway and Bahr repeated their measurements on maps of scale 1/2,500,000. At this scale 1 km on the ground corresponds to 0·4 mm on the map. Therefore the limiting width used by them at 1/250,000 is hardly wider than the line gauge used to depict the coastline at the smaller scale. Consequently the narrow inlets which were cut off at the larger scale had been automatically removed by generalisation in drawing the smaller-scale map. Clearly it is unrealistic to apply a narrow limiting width if the measurements are to be made on very small scale maps.

The Coastline of the United States

The third example of the application of a geometrical rule relates to the use of the *10-mile rule*. This was originally proposed to define bays and inlets which might be considered to be part of the internal waters of a state, thereby reducing the likelihood of international disputes over fishing rights. At that time the limit to territorial waters was still 3 nautical miles. It follows that by considering any bay or inlet measuring less than 10 nautical miles from one headland to the other to be wholly within the territorial waters of a State left no possibility for a narrow strip of the high seas to enter the bay, such as might arise if the territorial waters were defined as lying 3 miles offshore from each side of the inlet.

By the time of the Sixteenth Census of Population of the U.S.A. in 1940 it had become desirable to remeasure the area of the country. This work was carried out by the U.S. Bureau of the Census, and a detailed report was subsequently published in Proudfoot (1948). He has written:

> The knottiest problem was that of setting outer limits for the United States. There were insistent demands for the inclusion of the Great Lakes in the area of adjoining States. There were questions relative to the inclusion or exclusion of Long Island Sound, Delaware and Chesapeake Bays, the Straits of Juan de Fuca and Georgia, and of the coastal waters to a 3-mile or some other limit. Since the area of several of the large coastal embayments and of the Great Lakes to the Canadian boundary were legally accredited to the adjoining States, it seemed that these waters required remeasurement. However a definite rule for the manner of their inclusion or exclusion was lacking. Gannett in 1881 included Chesapeake Bay but excluded Delaware Bay, Long Island Sound, the Great Lakes and the Straits of Juan de Fuca and Georgia. At later dates Delaware Bay and the Great Lakes were included as a footnote to the land and inland water area of the States, whereas the other areas contine to be excluded.

Proudfoot therefore adopted the following geometrical rules for delimiting the United States:

- Where the coast was regular it was to be followed directly unless there were offshore islands within 10 nautical miles.
- Where there were bays having headlands of less than 10 and more than 1 nautical mile in width, a straight line connecting the headlands was used to define the limits.
- The *semicircular rule* was used to define bays. Thus a bay forming a shallow indentation in the coast was only treated as described above if the area contained within the line joining the headlands was greater than the area of the semicircle formed by that line as diameter. This rule is illustrated in Fig. 12.2.
- Offshore islands also required special treatment. Thus two or more islands less than 10 but more than 1 nautical mile from the shore were connected by straight lines, and other straight lines were drawn to the shore from the nearest point on each island.

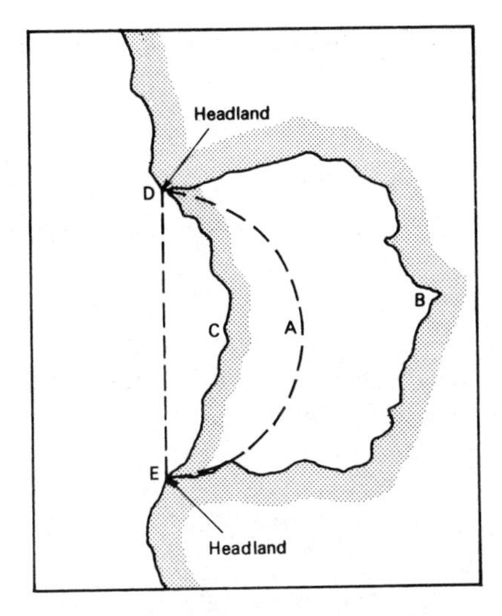

FIG. 12.2. The semicircular rule applied to two bays, DBE and DCE. DAE is a semicircle with diameter DE. If the area of the bay is greater than that of the semicircle, as in DBE, the feature is regarded as being a true bay. If it is less, as in DCE, the feature is considered to be a minor indentation in the coast.

- In the case of the Great Lakes, the same four rules were used but the limiting distance was 1 nautical mile rather than 10 nautical miles.

These rules formed a major part of the United States proposals which had been formulated by S. W. Boggs as the American case presented to the 1930 Conference on the Codification of International Law. They have remained the bases for many of the proposals which were adopted in the Geneva Conference of 1958 following the first of the United Nations Conferences on the Law of the Sea and, with modifications, have been employed in the Draft Convention of 1982 following the work of UNCLOS III. The way in which these rules were applied by Proudfoot to part of the Atlantic coast of the U.S.A. is illustrated in Fig. 12.3.

The consequence of the method of definition was to include appreciable area of "external" water within the confines of the U.S.A. The results of the measurements therefore were listed under the following headings:

Land:	7,710,500 km^2
Inland water:	117,200 km^2
Water other than inland water:	192,600 km^2

Inland water was defined to include:

Permanent water surface, such as lakes, reservoirs and ponds having 40 acres or more in area; streams, sloughs, estuaries and canals one-eighth of a statute mile or more in width; deeply indented embayments and sounds, and the coastal waters behind or sheltered by headlands separated by less than one nautical mile of water; and islands having less than 40 acres of area.

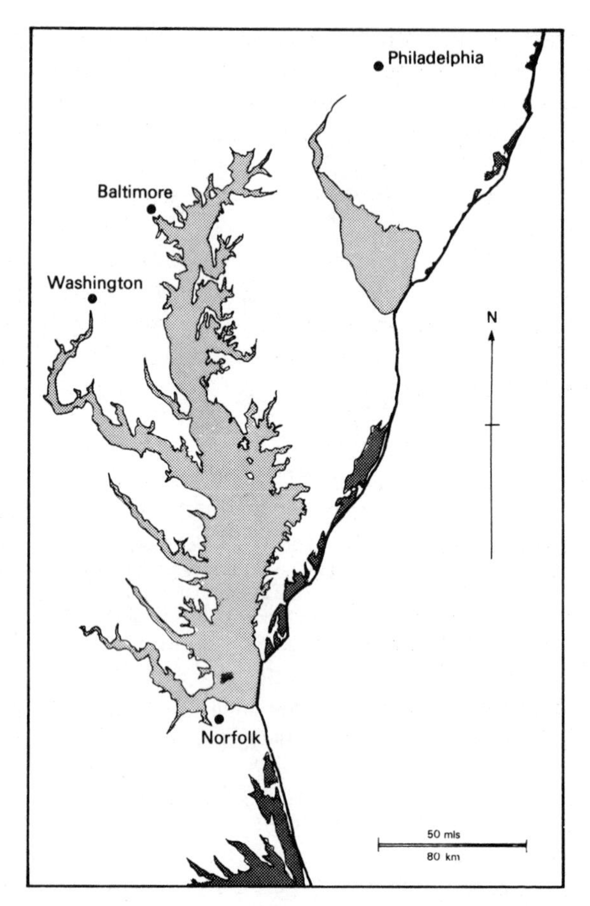

FIG. 12.3. Part of the Atlantic coast of the U.S.A. showing the coastline adopted by Proudfoot for the measurement of the area of the country for the Sixteenth Census of Population. The shaded areas represent parts of the sea included as parts of the country.

The third category represents the addition of 2·4% to what would customarily be regarded as the area of the country. More than 80% of this additional category comprises the waters of the Great Lakes extending to the Canadian boundary through them.

THE ROLE OF THE TIDELINE IN THE CARTOMETRIC DEFINITION OF THE COAST

There is curiously little reference in Proudfoot to tides and how these might affect definition of the coastline. Perhaps this is because the measurements were made at a comparatively small scale (1/500,000) using aeronautical charts. In much cartometric work, however, the effect of tidal changes and definition of the correct tideline is considered to be important.

On a steep cliff, lapped by deep water, the intertidal zone is a vertical feature recognisable by the changes in flora and fauna between the top and bottom of it. In such a place there is hardly any planimetric difference in the position of the shoreline at different states of the tide. Consequently the position of the coast at all states of the tide may be located and plotted on a map or chart with an accuracy similar to that of other well-defined natural features. On a gently shelving beach or mud flat, the intertidal zone is almost an horizontal feature through which the tideline migrates several times per day. A large tidal range combined with a gentle gradient gives rise to an extremely broad intertidal zone. For example, more than 20 km width of mud and sand is exposed at low spring tides at the head of the Baie du Mont St Michel in northern France.

One problem of coastal definition is to decide which tidal limit should be regarded as the coastline; whether the intertidal zone ought to be included or excluded from what is to be described as "land". Obviously this choice affects the results of area measurement. For example, it was maintained in France at the beginning of this century that the intertidal zone should be measured as part of the land. The effect was to increase the area of metropolitan France by 2,511 km^2. This is only about 0·5% of the total area of the country, but locally the differences are more dramatic. For example, the area of the Ile de Ré, which measures 157 km^2 at high tide is increased by about 70% to 267 km^2 at low water.

Tidal Limits

The terms used to describe tidal limits and ranges are shown in Table 12.3.

The two extremes, *highest observed high water (H.O.H.W.)* and *lowest observed low water (L.O.L.W.)* represent exceptionally high and low tides associated with abnormal meteorological conditions, such as the storm surges affecting the southern part of the North Sea on 31 January and 1 February 1953. Excluding such unpredictable meteorological components, the highest and lowest tides which may be experienced during a single 19-year lunar cycle are known as the *highest astronomical tide (H.A.T.)* and the *lowest astronomical tide (L.A.T.)* respectively.

It is a cartographic convention that both land maps and nautical charts represent the coastline by means of a line corresponding to high water. A feature of all nautical charts is that they always show depths which have been reduced to the level of low water. The reason for this is that the nevigator can see where it is safe to take a ship at any state of the tide simply by looking at the soundings shown on the chart and noting whether the depth of water recorded is greater than the draught of the vessel. The level to which the soundings have been reduced is known as *chart datum*.

TABLE 12.3. *Tidal limits and ranges*

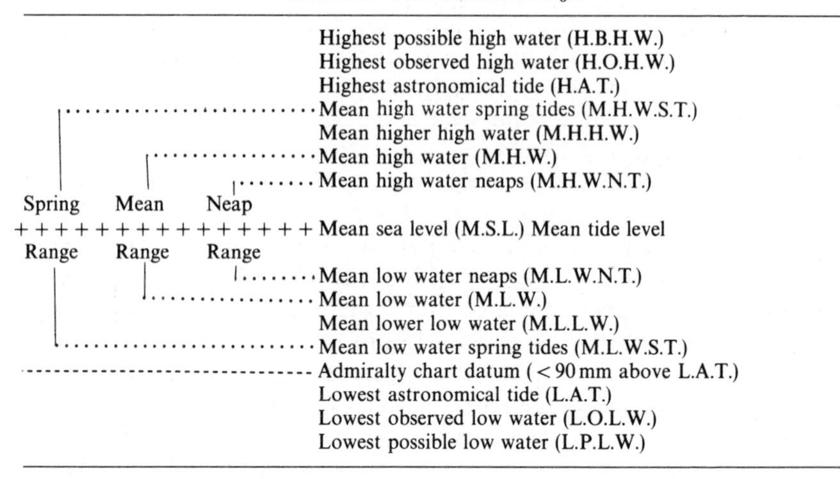

The International Hydrographic Conference of 1919 decided that chart datum should be a plane so low that the tide does not frequently fall below it. Although this conference also attempted to define "international low water" to serve as chart datum everywhere, this proved to be impossible because of the variations in the character of tides in different parts of the world. Each of the following levels have been used as chart datum:

- *Mean sea level* has been adopted for use in those seas where the tidal range is negligible, such as the eastern Mediterranean, Black Sea and the Baltic.
- *Mean low water* is used as the tidal datum on the east coast of the United States of America.
- *Mean lower low water* represents a lower datum than M.L.W. and is used primarily along the west coast of North America, where the tides are diurnal rather than semi-diurnal and one tide per day is appreciably lower than the other.
- *Mean low water springs* is the lowest level which is easily determined by observation because it occurs relatively frequently. It has therefore been adopted as chart datum more commonly than any of the other tidal limits. However, this is now being replaced by the lowest astronomical tide (L.A.T.) or a datum plane very close to this level.
- *Lowest observed low water* and *lowest possible low water* are used as chart datum in some countries, notably by France and former French colonies.

Although the last of these definitions appears to agree more logically with the international definition, the identification of it depends upon freak weather conditions, some volcanic or seismic cataclysm. Therefore it is not a level which is likely to be attained regularly or consistently. The preferred level for chart datum introduced for Admiralty Charts in 1964 is L.A.T. This is the lowest level which may be predicted within a single Metonic cycle of 18·6 years. There is no simple formula or rule to calculate it from observations or tidal constituents. It can be obtained only by studying tidal predictions covering the whole Metonic cycle. Since L.A.T. is so difficult to obtain, chart datum has been more broadly defined as *the level below which no predicted tide shall fall by more than 0·3 ft (9 cm).*

Since the tidal range varies from zero to more than 20 metres from place to place, it follows that chart datum is not a constant height, the height of chart datum varies continuously along the coast so that every sounding ought to be reduced by an unique amount. Generally, however, shallow water depths are only recorded to the nearest decimetre, so that it is more convenient to select a local datum and reduce all soundings in a given area to this level. Consequently the stepped line in Fig. 12.4b represents chart datum changing abruptly by small amounts every few kilometres along the coast.

Coastlines Referred to Tides

The coastline which is shown on most topographical maps and nautical charts by means of a continuous line is the high-water line. The *coastlining* methods used in hydrographic surveying are intended to locate this by conventional field survey or photogrammetric methods. It should be stressed that for marine surveys and nautical chart production it is not usually required to determine the line precisely at M.H.W., M.H.W.S.T. or H.A.T., but to determine an approximate boundary between M.H.W. and H.A.T. land surveyors will hold different views about the validity of this practice, because accurate location of property and administrative boundaries may depend upon determination of a particular high-water line.

The lower limit of the intertidal zone is usually portrayed on maps and charts by means of a broken line or by the edge of the stippled area depicting sand and mud with rock drawing to indicate the extent of any wave-cut platform which is exposed at low water. On a nautical chart the low-water line is that depicting the planimetric position of chart datum, for this is the level to which the soundings have been reduced and therefore represents zero depth at the appropriate state of the tide. The delineation of this line is therefore an "office compilation", being a boundary interpolated between the soundings and drawn at the time when these are reduced to chart datum and plotted. On many hydrographic charts it is not a line which has been identified and surveyed in the field. However, it is necessary to qualify this statement with reference to the work of the U.S. Coast and Geodetic Survey (now the Coastal Mapping Devision of the Office of Marine Surveys and Maps in the National Oceanographic and Atmospheric Administration) who use both field surveys and photogrammetric methods to determine the low-water line. This is because in some States and for some purposes the low-water line represents the riparian boundary to private property, and the correct demarcation of this line is important to landowners. The surveys are carried out in the field by levelling from bench marks on land to the appropriate height below datum and then locating this line using the same methods employed in engineering surveys to locate a single contour, as described by Phillips (1971).

The accuracy of the delineation of the low-water line by interpolation depends to no small measure upon the density of soundings which have been made in the area, the reliability of the tidal information used to reduce them to chart datum and the gradient of the beach. The chart user who is not experienced in methods of marine cartography should not judge the reliability of the delineation of the low-water line solely upon the density of soundings shown on the published chart. It is standard practice in chart compilation to make a careful selection of only a few soundings which are representative of the configuration of the sea bed. Standard works on hydrographic surveying, e.g. Admiralty (1948, 1964), illustrate how extremely dense networks of depth information derived from many traverses with echo-sounder may be reduced to only one or two numbers on the published chart. Moreover the metric charts introduced by the Hydrographic Department from 1967 onwards show even fewer soundings than did earlier charts of similar scale. Modern charts which have been surveyed or revised recently are usually based upon a dense network of soundings, and therefore the low-water line has been located accurately by this process. There are, however, many places for which only much older charts are available, which were surveyed by hand-held lead-lines. The individual soundings were more widely spaced so that there is some uncertainty about the position of a line interpolated between them. Haslam (1977) has stressed the fact that a great deal of hydrographic work done in recent decades has been needed to take into account the increased draught of modern ships, and has therefore been concentrated in sea areas once thought to be deep enough to offer no obstacle to any vessel afloat, where the depth of the sea is more than 20 metres. This work has been carried out at the expense of modern surveys close inshore. It is precisely because the older inshore surveys have not yet been superseded that it is necessary to emphasise the variation in the quality of delineation of the low-tide line.

The uncertainty which surrounds the interpolation of the low-water line between soundings depends to a considerable extent upon the gradient of the sea bed. Obviously there is less uncertainty about its position where the gradient of the beach is steep than where the offshore gradient is slight. Phillips (1971) has described the example of part of the foreshore of Louisiana which is more than 1 km wide although the tidal range is only 0·4 m, but this is exceptionally gentle. Usually gentle gradients giving rise to broad mud flats are associated with large tidal ranges as, for example, in the upper part of the Bristol Channel and the lower Severn estuary. Moreover the steplike changes in chart datum shown in Fig. 12.4(b) should, in theory, be reflected by

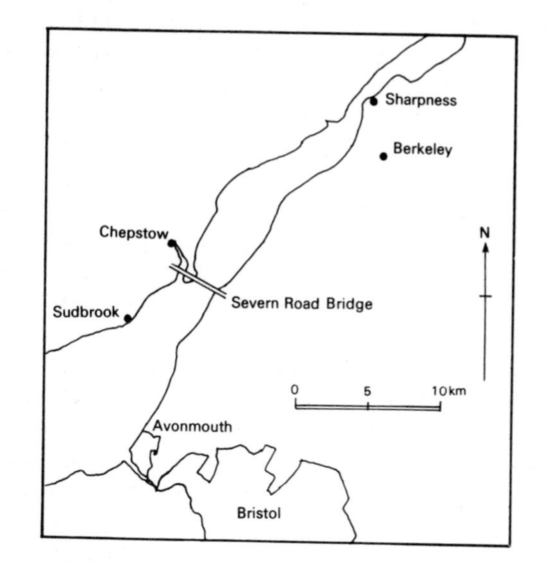

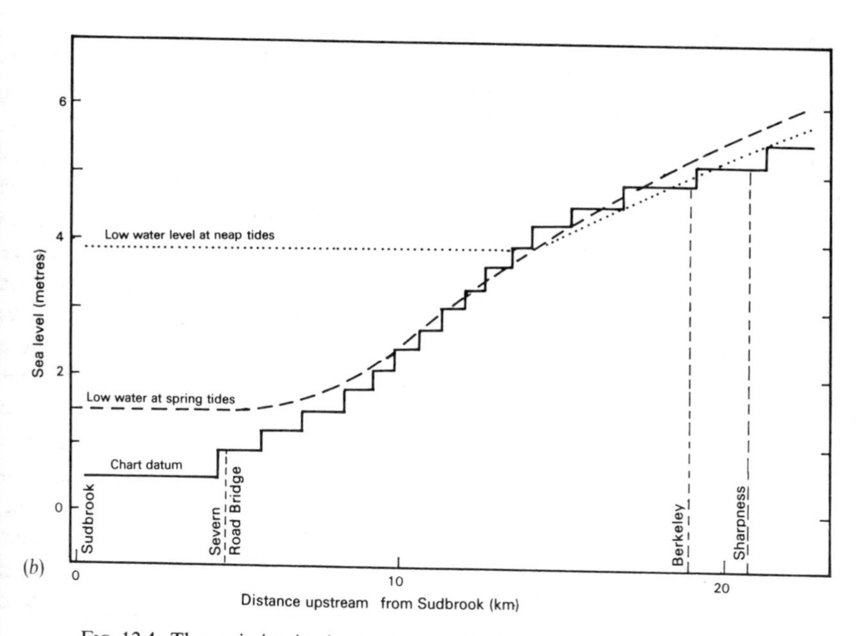

FIG. 12.4. The variation in chart datum in the upper part of the Bristol Channel (or lower Severn estuary). Figure 12.4(a) indicates the location of places indicated in Fig. 12.6(b). (Source: LeBlond, 1981).

corresponding planimetric displacements in the position of the low-water line. This is generally too small to be recognised at the scale of the typical nautical chart (1/100,000–1/300,000), but in view of the high accuracy of measurement which may be needed to identify and locate maritime boundaries in or near productive oil or gas fields (described in Chapter 23, pp. 525–551 the planimetric effects of small changes in chart datum should not be dismissed as having no practical value for the future.

Thus far we have been concerned with the representation of the low-water line as this is depicted on nautical charts. Land maps treat offshore detail in different ways; indeed on some maps it is ignored altogether. However, the emphasis to be placed upon the identification of the low-water line depends upon the legal and administrative character of the foreshore of a particular country. Thus the Ordnance Survey have the statutory requirement to map the foreshore as far out to sea as the low-water line where certain national or local responsibilities end. Moreover, they have to meet two different legal requirements of definition; that for England and Wales differs from that for Scotland. In England and Wales the limits to the foreshore are mean high water and mean low water. The reason for this stems from a judgement made by the Lord Chancellor in 1854. He maintained that the foreshore was bounded by the tidelines of medium or average tides, notwithstanding the fact that the sea extends some distance further inland than M.H.W. at every tide which is greater than the average. Under Scottish law the limits of the foreshore are mean high water springs and mean low water springs, corresponding more closely to nautical practice. It follows that in Scotland the low-water line shown on Ordnance Survey maps ought to correspond to chart datum on older charts. Neither the Ordnance Survey nor the Hydrographic Department appear to have the responsibility of surveying the zone lying between M.L.W. and L.A.T. in England and Wales. The Ordnance Survey Review Committee (the Serpell Committee) identified this "grey area" and argued for the provision of a special mapping unit to undertake coastal surveys, but seven years after publication of their report there is still no sign of this recommendation being implemented. As things stand, therefore, no government department has any responsibility for accurate mapping of that part of the foreshore, *although this is precisely where any baselines required by international law must be located.*

In the U.S.A. there are differences between the State and Federal laws relating to property boundaries facing tidal waters. Consequently the work of the U.S.C.&G.S., or its modern successor in the C.M.D., has legal and fiscal importance which is additional to nautical charting requirements. In its own way the American experience in these matters is no less complicated than those described for Britain. It has already been mentioned that the tidal regimes of the Atlantic and Pacific coasts of North America are different. Consequently a variety of local practices and customs have become enshrined in State laws which require the determination of different tideline boundaries. The complex-

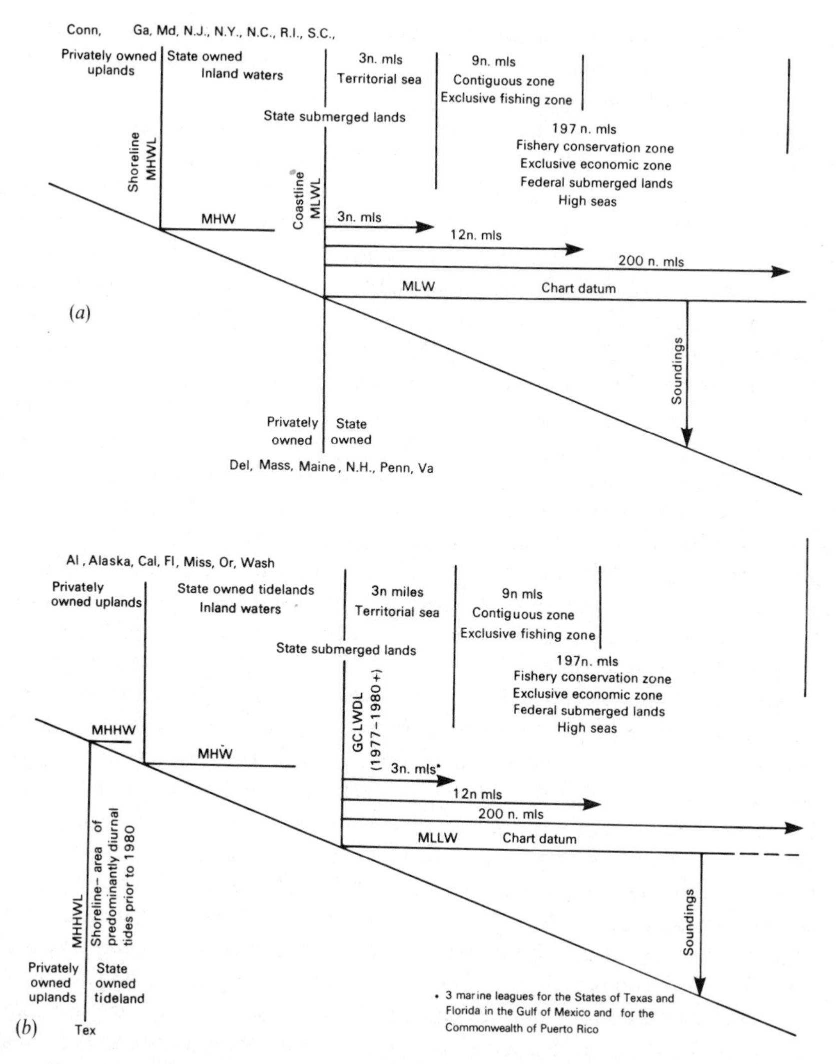

FIG. 12.5. The relationship between chart datum and different levels of the tide to land and maritime boundaries of the U.S.A.: (a) along the Atlantic coast; (b) on the Gulf and Pacific coasts. (After Hicks, 1986).

ity of these has been summarised by Hicks (1986). Undoubtedly they are best shown diagrammatically, as in Fig. 12.5.

The need to establish title to the "submerged lands" at the time when exploration for oil began along the shallow coastal waters of the Gulf of Mexico in the late 1940s encouraged the use of specialised photogrammetric mappings methods. The U.S. Coast and Geodetic Survey made a special low-water survey of the coastline of Louisiana employing infra-red photography.

This work has been described by Shalowitz (1962). At about the same time, the need to map the low-water line of Britain led to a similar photogrammetric solution being adopted by the Ordnance Survey.

The use of an infra-red-sensitive photographic emulsion is critical to the method. Light is strongly reflected from a water surface in the visible and ultraviolet parts of the spectrum, but most of the incoming energy at the infra-red wavelengths is absorbed by water. Consequently a sheet of water which is only a few centimetres deep reproduces on monochrome infra-red film as a featureless dark grey image which is strongly contrasted to the reflected images of the land surface. This property has proved to be of great value in separating land from water in several special applications because, for example, it is easy to distinguish muddy water from wet mud in mangrove swamps and similar localities. Infra-red photography taken at low tide has been used by the Ordnance Survey and by the U.S.C.&G.S./C.M.D. ever since these early experiments. In Britain the photographs are taken at scales of 1/7,500 or 1/28,000 for plotting at the basic scales of 1/2,500 and 1/10,000 respectively. American practice is to take photography at 1/20,000 or 1/40,000 for mapping at 1/20,000 scale. The main difficulty about acquiring the necessary photographs is that little time is available for photography on any particular tide. If the photographs are taken much before or after slack water when the tide is turning, the image of the water's edge does not represent the conditions obtaining at the required tidal limit. Therefore accurate timing of the sortie is a special characteristic of the work. Moreover, it is not practicable to photograph more than a few kilometres of coastline from one aircraft on any given occasion. It is therefore difficult to obtain photographs of the whole shoreline for precisely the same datum level. Phillips (1971) suggests acquisition of two sets of photography taken on occasions when the waterline is slightly higher and slightly lower than M.L.W. The required mean tideline may be interpolated between two mapped versions of the shoreline derived from the two sets. However, this practice can only be applied for mapping the mean tideline and it is of little use when the intension is to locate M.L.W.S. or L.A.T. Further limitations of the photogrammetric solution are that at least one-half of the suitable tides occur at night or at times of the day when the sun's altitude is too low for satisfactory photography. Combined with the usual delays resulting from unsuitable weather and unserviceable aircraft, it is not surprising that aquisition of this kind of photography may be extremely slow.

13

Measurements on Aerial Photographs

He set aside his stereoscope and reached for his measuring magnifier – a small precision instrument something like a jeweller's glass (it was actually intended for counting the threads of textiles to check the regularity of weave, which gives an idea of its magnification and accuracy). He placed it on one of the prints, and under the crystal-clear lens, the metal scale calibrated in millimetres rested on the image of something that looked like a slim grey splinter, with a darker knob jutting up half way along it. That "splinter" was in fact a newly launched 500-ton submarine which was being fitted out and would soon be fighting in the waters of the Atlantic....

"See what you make of her, Bunny," he said.

Assistant Section Officer "Bunny" Grierson, his indispensable assistant and partner was responsible for checking every measurement and conversion, so in this way no interpretation was made without a double check.

"215 feet," said Bunny in a few moments.

(Constance Babington Smith, *Evidence in Camera*, 1958)

In most textbooks on photogrammetry, written since the 1950s, the emphasis has rightly been upon the three-dimensional coordinate geometry of the stereoscopic model and the significance of stereoscopic parallaxes as being measures of model deformation and ground height. This approach is essential for an understanding of the rigorous methods of photogrammetry and mapping by analogue or analytical plotter. However, it has no place in this chapter because we are only concerned with the use of aerial photographs as substitutes for maps. Since we must attempt to answer the question "Do the principles and methods of cartometry work as well on photographs as they do on maps?", we must concentrate upon the geometry of image formation on the single aerial photograph and discover how this differs from a map.

Cartometric measurements made on a contact print resemble the graphical methods of plotting map detail which were used for mapping from aerial photographs up to the late 1950s. Therefore it is the older works on photogrammetry, such as the books by Hotine (1931), Hart (1940) and Trorey (1950) which treat with the geometry of the single photograph and provide a more useful introduction to the present chapter than do any of the modern books.

In Chapter 2 we stressed that survey-quality photography is desirable for cartometric work. Nevertheless we shall discover that there is some conflict between the needs for survey photography and the requirements for cartometric measurement, for these are not the same. For example, the types of lens which are suitable for cartometric purposes are not the same as those most commonly used for air survey. We shall see that the longer is the focal length of the camera lens, the smaller are the cartometric errors resulting from camera tilt, surface relief and the positions of features on the photographs. On the other hand there are cogent reasons for using shorter focal length lens cameras for survey photography. Sometimes the cartometric user may be able to specify some of the preferred requirements for a particular job, such as the scale of photography, the amount of overlap, and the orientation of the strips, but, as we shall see, there are constraints upon the range of these possibilities. Generally, however, it is too expensive to commission special photography. Then the interpretation and associated cartometric work has to be done using photographs which were originally taken for mapping.

Nearly all survey cameras are designed to take photographs of size 9 × 9 in. (228 × 228 mm) or 180 × 180 mm (7·1 × 7·1 in.). The larger format has evolved as the preferred size in the English-speaking countries so that it is most commonly used in the British Commonwealth and in the U.S.A., whereas the smaller format is preferred elsewhere. Throughout this book we refer to the 228 × 228 mm format unless otherwise stated. Three different types of camera lens are used, as shown in Table 13.1. Because of the importance of camera calibration, these lenses are not interchangeable within the same camera body.

The great majority of survey photographs taken in the last 40 or 50 years have been taken through wide-angle lenses. Indeed, some photogrammetric plotters can only be used with wide-angle photography, thereby accentuating dependence upon this single variety. Normal-angle photography is now little used for mapping. There are certain economic as well as technical advantages in using super-wide angle photography for mapping. Consequently this has gained much in popularity since suitable cameras became available in the 1950s. However, its suitability for cartometric work is even less than that of wide-angle photography.

Most wide-angle photography is taken within the scale range 1/4,000 through 1/60,000. The limit to using even larger scales is created by image movement resulting from the forward motion of the aircraft and the need therefore to fly above a critical minimum height so that blurring of images is less than the resolution of the photograph. This critical height depends upon both shutter speed and the ground speed of the aircraft, but usually it is about

TABLE 13.1. *Focal lengths of lenses used in survey cameras.*

	Format	
Lens type	228 × 228 mm	180 × 180 mm
Normal angle	304·4 mm (12 in.)	210 mm
Wide angle	152·4 mm (6 in.)	115 mm
Super-wide angle	88 mm (3·5 in.)	73 mm

600 metres (2,000 feet). The smaller scale limit is created by the economical ceiling for aircraft which are suitable for survey photography. For most civil flying it may be set at about 9,000 metres (or 30,000 feet) but obviously this height may be greatly exceeded if suitable military aircraft are available. Because the quoted scale range lies within that of large-scale to medium-scale maps, and the comparatively small format of the aerial photograph compared with that of a map, it is evident that earth curvature does not represent a significant cause of image deformation upon a single photograph to be measured by graphical methods. In other words, we may ignore the projection. This is not to say that the effects of earth curvature are not present. As on large-scale and medium-scale maps they are simply too small to detect by the methods of measurement at our disposal and are, in any case, far smaller than the image displacements which are caused by camera tilt and surface relief. In the geometrically rigorous methods of photogrammetry it is sometimes desirable to make corrections for earth curvature. Consequently some of the later designs of analogue plotter contain devices which simulate and correct it. Numerical corrections may be applied through the subroutines normally available in an analytical plotter. The reason for the difference in approach is that, for rigorous photogrammetric methods, the stereoscopic model may be formed in the instrument at a scale which is substantially larger than that of the original photography. Consequently image displacement owing to earth curvature becomes a measurable quantity. Conventional photographs taken from space vehicles are, of course, of much smaller scale because flying height has been increased from less than 10 km to several hundred kilometres. Obviously earth curvature must be taken into account in the use of these.

It was stated in Chapter 2 that for continuous stereoscopic cover with vertical aerial photography we specify both fore-and-aft and lateral overlap between adjacent photographs. Figure 13.1 illustrates the combination of 60% fore-and-aft overlap with 20% lateral overlap. It may be seen that 24% ground has been photographed six times, a further 16% occurs on four photographs, 36% of the ground has been photographed three times and only 24% has been photographed twice. Those parts which occur on three successive photographs in a strip (representing three separate bands of approximate width 4·5 mm each) are known as the *common overlap*. The overlap of these parts with the corresponding common overlap of adjacent strips creates the *supralap* containing images which are common to six photographs. These parts of the photographs are particularly important in aerial triangulation because a point within the supralap serves to connect, or *tie*, the strips together. The conditions illustrated in Fig. 13.1 represent the ideal case but this is not always attained so exactly. Owing to the small variations in performance of the aircraft, which may wander slightly off course, change height or fly in a tilted attitude, the regularity of continuity of overlap illustrated here changes from one photograph to the next.

We need to define a suitable *working area* for photographs having specified

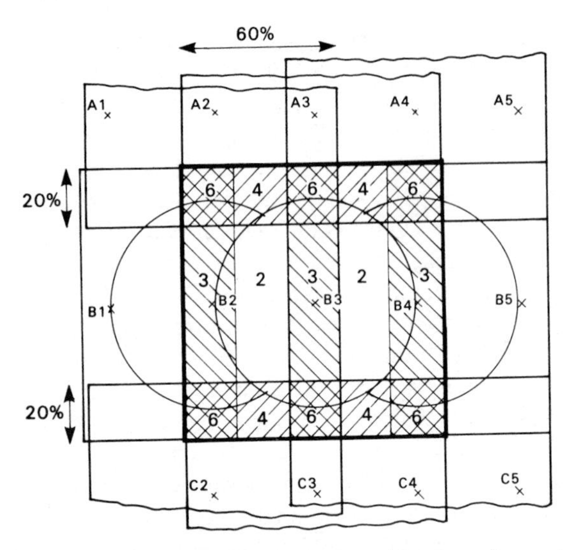

FIG. 13.1. The repetition of photographic cover resulting from the combination of fore-and-aft overlap with lateral overlap within a block of vertical aerial photographs. The principal points of 14 adjacent photographs are indicated by the crosses labelled *A*1 through *A*5, *B*1 through *B*5 and *C*2 through *C*5; *A*, *B* and *C* being three parallel strips of photography. There is 60% fore-and-aft overlap with 20% lateral overlap. The numbers 2, 3, 4 and 6 indicate the number of times different parts of the ground covered by the photograph *B*3 also appear on adjacent photographs.

amounts of overlap so that we may quickly decide which is the most suitable print to use to measure a particular feature. We shall show that the deformations in the photographic image increase radially outwards from the centre of each photograph. Consequently the nearer the measurements of a feature can be made to the middle, the smaller will be the error resulting from these causes. This is the *principle of centrality*. For aerial photography of format 228 × 228 mm, having 20% lateral overlap or thereabouts, it seems that a circle of radius 90 mm centred on the principal point represents a suitable dimension for the working area.

THE VARIABLE SCALE OF THE VERTICAL AERIAL PHOTOGRAPH

In the derivation of equation (2.2), we assumed that both of the ground points, A and B, lie upon a plane surface. It is now necessary to show how ground height also affects the scale of the photograph. We imagine the points C and D on the ground which are at the same height, situated on the edge of a vertical sea cliff, while A and B are located on the beach immediately beneath C and D respectively. On the ground, therefore, $AB = CD$. The height of the cliff is h. The corresponding image points, c and d on the photograph complete the set of similar triangles cOP, dOp, COP, DOP, so that, using the same arguments

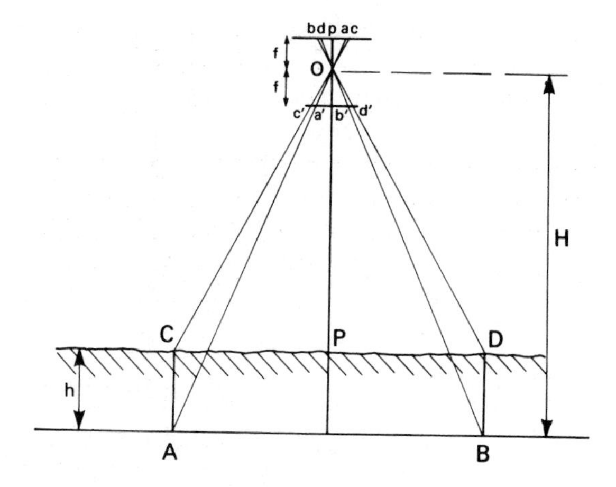

FIG. 13.2. Variation in the scale of a vertical photograph resulting from variation in ground height, h.

presented in Chapter 2,

$$cd/CD = 1/S' = f/(H-h) \tag{13.1}$$

The points c and d both lie further from p than points a and b. Since $cd > ab$, it follows that $S' < S$. Therefore we have identified that the photograph has different scales and that $1/S'$ is larger than $1/S$. The same condition could be satisfied for any other combinations of points, so that the scale of the photograph is not constant but varies within the range $f/(H-h_0)$ and $f/(H-h')$, where h_0 and h' are the lowest and highest ground within the area covered by the photograph. Equation (13.1) is therefore an algebraic expression for the description of the condition of perspective representation given on page 26.

It should be stressed that variable scale is independent of the manufacture of the camera or the quality of the flying. Construction of Fig. 13.2, in which the line pOP is perpendicular to both the planes of the photograph and those on the ground containing the pairs of points CD and AB, implies that there are no errors arising from either of these causes. We must therefore reiterate that *variable scale is a fundamental relationship of conventional photograph.* When we refer to the scale of a photograph, we either mean this as a convenient approximation without attempting to define it precisely, or, where precision is needed, we mean that scale which refers to a specific height, such as h_0, h' or some intermediate value such as h_M.

IMAGE DISPLACEMENT RESULTING FROM SURFACE RELIEF

Since, in Fig. 13.2, $pc > pa$ and $pd > pb$, the small distances $(pc - pa)$ and $(pd - pb)$ on the photograph indicate displacements of the images of the higher points of objects with respect to those which are at lower altitudes. Although this may be seen on any vertical aerial photograph showing tall buildings, factory chimneys, or even isolated trees, we can only assume that it is present in surface relief because we can see only the top of an object when this is the ground. The geometry of this displacement may be illustrated in greater detail in Fig. 13.3, where an isolated object, AA'', of height h, represents an isolated feature such as a chimney rising from flat land.

It is assumed that the photograph is truly vertical, so that the principal axis, pON, is coincident with the vertical, or plumb-line direction from the perspective centre, C. The point N on the ground is called the *plumb point* or *nadir point*, and the corresponding point, n, on the photograph is the *nadir homologue* or *photo plumb point*. It can been seen that the principal point, p, here coincides with the nadir homologue, which is the essential condition for truly vertical photography. The differences in position between n and p resulting from tilt will be discussed later.

In Fig. 13.3 the positive image of the top of the object is a'', of distance r'' from the nadir homologue and the base of the same feature is the image point a, of distance r from n. Since the triangles $AA'A''$ and Spa are similar, we have

$$f/r = h/[(r'' - r)H/F] \tag{13.2}$$

Putting $\delta r = r'' - r$, we may solve this expression for δr, which simplifies to

$$\delta r = r'' \cdot h/H \tag{13.3}$$

In other words, for given values of H and h, the displacement δr increases linearly from the centre of the photograph to its edges. It is useful to construct a graph, such as that illustrated by Fig. 13.4, from equation (13.3) for particular types of photography of the same nominal scale. It illustrates wide

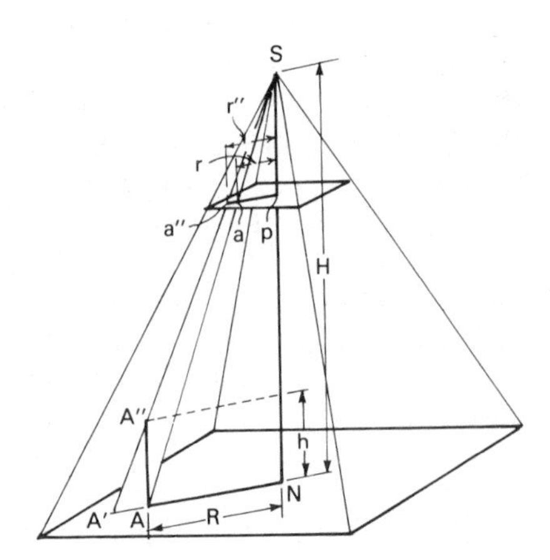

Fig. 13.3. Relief displacement on a truly vertical photograph caused by a vertical object AA'' on the ground.

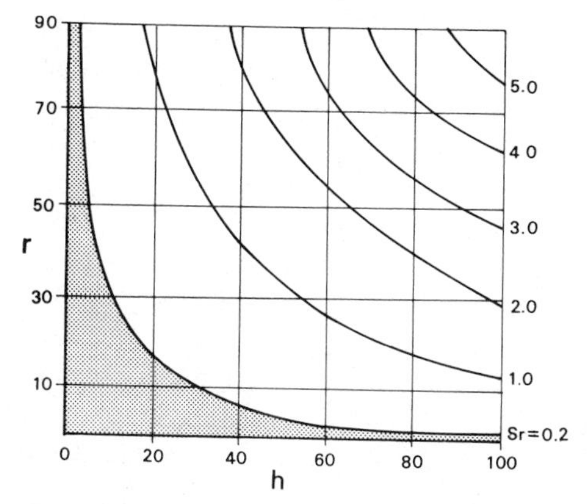

FIG. 13.4. Curves showing the amount of relief displacement, δr, in millimetres, resulting from variation in the distance r from the centre of the photograph and upon surface relief differences, h. Numerical values represent the conditions for wide angle photography at scale 1/10,000, where $H = 1524$ m.

angle photography of scale 1/10,000, so that $H = 1524$ m. Curves for the amount of displacement, δr, are drawn for every 1 mm. The shaded area indicates the part of the photograph and the amount of relief which gives rise to values of $\delta r < 0{\cdot}2$ mm, which are not going to have a material effect upon graphical measurements made on a print.

THE EFFECT OF CAMERA TILT

Superimposed upon relief displacement is deformation caused by camera tilt. We define this as the dihedral angle formed between the inclined plane representing the photograph, which we imagine to be continued until it intersects a horizontal plane representing flat ground. Thus, if the two planes are represented in section by the lines *pnC* and *PNC*, as illustrated in Fig. 13.5, the angle of tilt is $pCP = \theta$. It is easy to see from Fig. 13.5(b) that θ is also represented by the angles *Ppn*, *PON* and *pOn*. It is important to appreciate that θ is not related to any particular direction in the aircraft. It is the resultant of the three rotations about the *X*, *Y* and *Z* axes, which is the method of relating cameras tilt to the direction of flight and used in photogrammetric instruments. In Fig. 13.5 the line *Cpn* through the plane of the photograph indicates the direction of maximum tilt. The trace of this section, or the line *pn*, is called the *principal line*, and this is used as a datum from which measurements of image displacement, scale change and angular distortions may be referred.

Since the principal axis *pOP* coincides with the optic axis within the camera, this line is perpendicular to the plane of the photograph at the principal point, *p*. The downward direction from the perspective centre, *O*, has already been described as the plumb-line. It follows that *ON* is perpendicular to the ground plane at the nadir point, *N*.

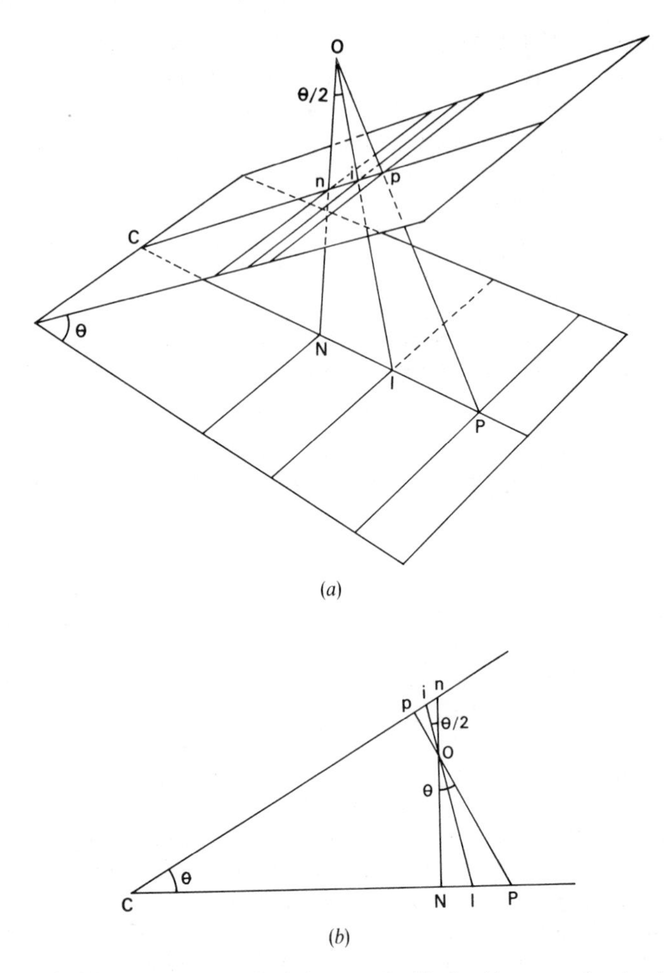

FIG. 13.5. The geometry of the tilted photograph of flat land in perspective view (*a*) and in section (*b*). Note that diagrammatically (*a*) represents the geometry of the positive image, whereas (*b*) represents the negative image. The difference between them is the position of the perspective centre, *O*, this being the camera lens, with respect to the image plane containing the points *p*, *n* and *i*.

We now consider the line *iOI* which bisects the angle *pOn* in Fig. 13.5. This identifies the point *i* on the plane of the photograph which is situated on the principal line between *p* and *n*. Since $pOn = \theta$, we have defined $pOi = \theta/2$. We call the point *i* the *isocentre* of the photograph, and *I*, which is the homologous point in the ground plane is the *isocentre homologue*. The point *i* has two important geometrical properties in the tilted photograph. First, the scale along the *isometric parallel*, which is a line perpendicular to the principal line passing through *i*, is f/H, or the scale which we would have determined for that photograph if it were vertical. Secondly it can be shown that angles measured at the isocentre of a tilted photograph of plane ground are not affected by the tilt. Moreover this is the only place on the photograph where this condition is satisfied. In other words, if we were able to measure a series of horizontal angles on the ground at the point *I*, these would be the same as the corresponding angles measured at *i* on the photograph.

We have now identified three different points on a tilted photograph, all of which lie in the same straight line.

- The *principal point* is the optical centre of the photograph, defined as the point where the optic axis intersects the focal plane. This point has been identified by camera calibration and can always be recovered by means of the fiducial marks reproduced on each photograph.
- The *nadir homologue* is the image of the point vertically beneath the perspective centre at the instant of exposure. It has the property that displacement of images owing to surface relief increases radially from this point.
- The *isocentre* lies between these two points and is equivalent to a point of zero distortion on a tilted photograph of flat land.

In order to indicate the separation of these points on a typical aerial photograph, Table 13.2 provides numerical values for wide angle photography and tilts θ to 4°. For the small tilts tabulated here the position of the isocentre may be regarded as lying midway between *p* and *n*, although this is not true for the large tilts of high oblique photography. An assumption is often made in mapping that all three points may be represented by the principal point. Table 13.2 indicates the degree of approximation which may be introduced by making this assumption even for modest amounts of tilt. However, the approximation has important practical consequences because the position of the principal point can always be located accurately with reference to the fiducial marks portrayed on every photograph. The positions of the nadir point and isocentre can be located only by tedious graphical construction, or by using sophisticated photogrammetric instruments, the

TABLE 13.2. *Distance in millimetres between the principal, point, p, the nadir homologue, n, and the isocentre, i, for near vertical wide-angle photography, where $f = 152{\cdot}4$ mm*

θ	*pn*	*pi*	*ni*
0°	0·0	0·0	0·0
1°	2·7	1·3	1·3
2°	5·4	2·7	2·7
3°	8·1	4·0	4·0
4°	10·8	5·4	5·4

availability of which largely eliminates any need to make approximate measurements on the single photograph.

PRACTICAL CONSIDERATIONS OF MEASUREMENT

There are three ways in which aerial photography may be measured in the process of making a quantitative interpretation.

- Direct or indirect measurements may be made using simple equipment upon paper prints or transparencies.
- Direct or indirect measurements, using counting, may be made using transparencies set in an image analyser.
- Direct measurements may be made with the stereoscopic model created from transparencies of a pair of overlapping photographs set in a photogrammetric plotter.

The first two methods inevitably involve some degree of approximation in the results obtained, because the single aerial photograph contains the deformations owing to camera tilt and surface relief which have been described.

The third method comprises reconstruction of the unique conditions of tilt and flying height which occurred at the time of photography, creating a three-dimensional model within the instrument which is sensibly free from those distortions affecting the single photograph. The output from a photogrammetric plotter may be in the form of readings of the (X, Y, Z) system of model coordinates within the plotter itself, or a plot of the plane (X, Y) values upon a drawing table by pantograph or coordinatograph. There is no doubt that, in terms of accuracy and repeatability, measurements made in a plotter are superior to any others made from aerial photographs. The typical precision of photogrammetric work was considered on page 162 as one element in the quantitative accuracy of a map. Similar precision would be obtained in the use of the plotter to make measurements of specific objects. If the model coordinates can be output on tape in machine-readable form at the touch of a switch, the method of collecting data is fast, too. Under these circumstances the digitised coordinates may be used to calculate distance and area just as digitised coordinates may be extracted from a map or chart. However, it was noted in Chapter 2 that it is a luxury to which the specialist user is seldom accustomed, for photogrammetric instruments are seldom available for work which is largely associated with interpretation.

The image analyser possesses the great advantage that laborious counting procedures are quickly executed. This makes it especially useful for measuring area, although in Chapter 21 we describe in some detail how the length of a coastline may be measured using the Landsat M.S.S. imagery as the source. However it is important to stress that the hardware does not correct for any of the deformations in the way an analogue photogrammetric plotter does this, so that unless geometrical corrections can be applied during computer processing of the data, the final accuracy of the work is no better than that obtained by cruder graphical means. It follows that measurements made graphically upon a paper print are the cheapest and therefore the most commonly made. It is these procedures which need to be studied in detail.

Graphical Methods of Measurement

In their study of the accuracy of measuring areas directly from aerial photographs, Smirnov and Chernyaeva (1964) list the following six factors which may affect the results.

- errors arising from camera tilt,
- errors arising from variation in surface relief,
- errors arising from inaccurate determination of the scale of photography,
- errors arising from inaccurate identification of points or in delineating a boundary,
- errors arising from the deformation of the photographic materials,
- errors arising from the incorrect use of the instruments and the methods used to make the measurements.

From this list the first three are peculiar to the use of the single aerial photograph as the medium for measurement. We treat with them in detail in the next section. The remainder of the factors are common to other kinds of cartometric measurement and have been examined elsewhere in this book. However, brief comment about the last two is appropriate here. A photograph may be available in any of the following forms:

- as a paper print,
- as a glass diapositive,
- as a film diapositive (or transparency),
- as the original negative.

Only the paper print is to be viewed by reflected light. The others must be viewed by light transmitted through them, and therefore the work has to be done over a light table. All photogrammetric plotters make use of diapositives or transparencies.

For the reasons given in Chapter 10, the dimensional stability of paper prints is not to be trusted. We have seen that this applies to maps and charts, but is especially important with reference to ordinary bromide prints because processing of them involves complete immersion in solutions, followed by washing so that the paper is thoroughly soaked after the image has been formed. Finally the prints are dried and may, if glazed, pass through an additional process equivalent to cooking them. In order to overcome the tendency of paper to change shape and size erratically during such processes, there have been attempts to produce dimensionally stable prints by using special paper. The laminated foil cards contain a layer of thin aluminium foil embedded in the paper like meat in a sandwich. This makes the print comparatively stable, but is expensive. Waterproof paper, which is intended to resist soaking during processing, is less successful but cheaper.

Glass diapositives are by far the best medium for measurement because of their high dimensional stability, but they are now becoming increasingly rare. This follows the decision made in the early 1970s by the major photographic firms to abandon all production of glass plates coated with photographic emulsion. Today, therefore, all new photography must be printed as transparencies on a base of polyester plastic. This is not as stable as glass, for, as shown in Chapter 10, it is affected slightly by both temperature and humidity.

An original negative represents an archive which ought to be preserved in a satisfactory state and only used for making copies. Consequently, the use of original negatives for actual interpretation is justified only in some military work, or in an emergency when it is essential to study the photographic cover immediately without waiting for completion of the second stage of processing. However, many modern colour films may be used directly from development. For example, Ektachrome 2448 and infra-red (false colour) films are processed directly to the positive image. Therefore there is a strong temptation to work directly on this original material rather than make copies of it.

Survey cameras manufactured in Britain have a register glass which holds the film flat during exposure. This shows a *réseau*, which is a square pattern of small crosses at intervals of 10·00 mm and having a root mean square error in position of $\pm 2\,\mu$m or smaller. The grid is centred upon the principal point and its axes are orientated to the fiducial marks along the sides. Because of its position within the camera, this extremely precise grid is reproduced on every photograph taken through it. Its purpose is to provide a framework of points which are used in analytical photogrammetry, notably in analytical aerial triangulation, to serve as a gauge against which to measure film deformation, just as we have learned to use a map grid to do this in cartometry. Therefore we may also use the réseau in quantitative interpretation to discover whether a print has been deformed by a measurable amount. For this work it is sufficient to assume that the crosses are 10·0 mm apart, whereas in aerial triangulation it is necessary to use a calibration chart which indicates the location of each réseau cross much more precisely. The value of the réseau for testing prints before measurement is a good enough reason for insisting upon survey-quality photography for cartometric work. However, corresponding measurements can be made on photographs taken through cameras without a réseau if the positions of the fiducial marks are known. These have to be established in camera calibration and the measurements are recorded upon the calibration certificate.

Apart from its use as a means of measuring film deformation, the réseau provides a regular pattern of points on each photograph. This may be used as the basis of systematic sampling and even area measurement without having to prepare a special overlay.

The majority of the instruments used for cartometric work may also be used with photographs. However, dividers are unsuitable for work with negatives or transparencies because they are likely to scratch the emulsion. Glass scales are used with photography, particularly if these are to be viewed over a light table. Generally these scales are quite short, for they are usually intended to measure objects appearing on the photographs rather than long lines stretching across their full extent. For example, the scale provided with the Casella parallax bar comprises an optically flat glass plate with scale graduations for 30 mm, about one-half of which is subdivided into units of 0·2 mm. These are even more useful when combined with some kind of magnifier such as a linen tester. Long glass scales are available from firms which specialise in making photogrammetric plotters, for these scales are used to measure the distances between the fiducial marks of a photograph as described above. This scale must be long enough to measure the diagonal distance between corner fiducials, which is about 270 mm on a 9 × 9 in. photograph. Such scales are provided with reading microscopes attached to a slide. They are rather cumbersome to use for measuring short distances. See Colwell (Ed.) (1960) for descriptions of various other measuring devices used by photographic interpreters.

Practical Determination of Photo Scale

The formulae (2.6) and (13.1) which may be used to calculate scale, require information about three quantities. The focal length of the lens, f, is determined by the manufacturer during the calibration of the assembled camera. This is known to within 1 part in 1,000 and it does not change much during the working life of the camera. The ground height may be known within 1 metre or thereabouts for flat land which has been surveyed, but h may be an extremely variable and uncertain quantity. There is no really foolproof method of determining flying height, H, accurately by independent measurement. The commonest method is to measure the height of the aircraft by aneroid altimeter. This measures atmospheric pressure, but is also sensitive to a lesser degree to other variables such as air temperature and humidity. Moreover, there may be a significant time lag between the aircraft reaching a particular height and the altimeter registering this information. Although the majority of survey cameras contain an altimeter, the reading of which is automatically photographed every time an exposure is made, the information provided by it cannot always be taken as being reliable. The more sophisticated kind of instrument which measures *differences* in atmospheric pressure between exposures of the camera is known as a *statoscope.* This provides a measure of the small changes in flying height between one photograph and the next but it does not, itself, provide a measure of H. The radar altimeter, which measures the downward distance from the aircraft to the ground is the most reliable measure of $H - h$. However, this is not always pointing directly downwards, for if the aircraft is tilted the measured distance is along an obliquely sloping line and is not the true vertical. It follows that equations such as (2.6) and (13.2) can only provide an approximate value for scale. The more reliable method of scale determination is to find $1/S$ by means of linear measurement and determine H, if this is required for other purposes, by substitution in one of these equations. Of course the value of the scale determined by measurement reflects the uncertainty which is introduced by the fact that it is a variable quantity. Thus, in Fig. 13.2, if the line *ad* is measured on the photograph, the scale determined will be larger than $1/S$ but smaller than $1/S'$. The best that we can hope to attain in measuring the distance between two points with different ground heights is to find the scale of the plane lying midway between them. In this example, if the ground heights are h_A and h_D the use of the measured line *ad* compared with the known distance *AD* provides the scale

$$1/S_M = f/[H - \tfrac{1}{2}(h_A + h_D)] \tag{13.4}$$

Differences between $1/S$, $1/S'$ and $1/S_M$ will be small if the differences in height $(h_A - h_D)$ is small compared with H.

So far we have not commented about the location of the line used to make the scale determination or the preferred length of it. Here we introduce the other concept which is fundamental to the theory and practice of making measurements on the single aerial photograph, which may be called the *principle of symmetry* to support the principle of centrality already described. Both of these principles follow logically from the investigation of relief displacement. The first, which is clearly shown by Fig. 13.3, is that displacements are small near the centre of a photograph but increase radially outwards. It follows that the best line to measure is one lying near the centre of a photograph. However, there are practical difficulties in measuring short distances. These arise from the small errors in identification of the terminal points and setting the scale. We show elsewhere (p. 278) that the error of measurement of the length of a line on a photograph or a map using a metal (or glass) scale is about $\pm 0{\cdot}2$ mm and is independent of distance. Therefore the longer the line, the greater the relative precision with which scale may be determined. It follows that there are advantages in using a line which extends throughout the working area of the photograph rather than attempting to obtain a result by measuring a very short line, even if this is located near the centre of the working area.

If we attempt to measure the length of a straight line which is about 100 mm in length and one end of it is located near the centre of the photograph, the other end must, perforce, lie beyond the edge of the working area, as illustrated in Fig. 13.6(b). In this case the amount of distortion at the point *a* is likely to be small because it lies near the centre of the photograph, but the displacement of the point *b* is likely to be large because both tilt and relief displacements are greater near the edge of the working area. On the other hand, if the line is

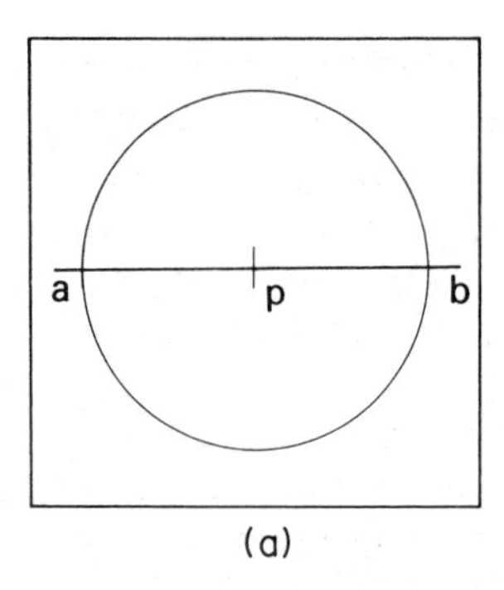

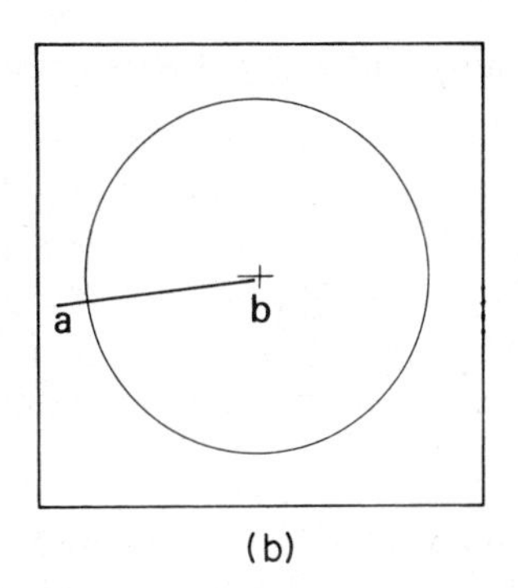

(a) (b)

FIG. 13.6. The principle of symmetry applied to linear measurements made on a single aerial photograph. In (a) the line *ab* passes through the principal point, *p*, of the photograph and is bisected by it. In (b) the line *ab* is asymmetrical with respect to the principal point.

ocated in such a way that it passes close to the centre of the photograph (the principle of centrality) and is also located so that $ab \approx bp$ (the principle of symmetry), displacements owing to tilt tend to cancel one another (if the line *ab* approximates to the principal line) or even vanish (if *ab* coincides with the isometric parallel). Again, if the ground height near the centre of the photograph lies approximately midway between h_A and h_B, the amount of relief displacement at either terminal point will be equal and opposite, referred to the plane through *p* of height h_P. The third condition is the general case where the line is neither central nor symmetrical, as illustrated in Fig. 13.6(b). We shall investigate the deformation which is likely to affect this line and will, in so doing, confirm the utility of the two concepts of centrality and symmetry.

The Accuracy of Measurement of Distance and Area on Aerial Photographs

Practically all the work which has been published on this subject appeared in the U.S.S.R. during the 20 years following the end of World War II. Because it was published in Russian, the work is little known in Western countries. It is therefore desirable to present some of the more important work in some detail. About a dozen different writers have contributed to these studies, but most of them were primarily concerned with the accuracy and the influence of camera tilt upon area measurements only. Moreover, several of the studies only investigated special cases – for example, the deformation of a square parcel having one side coincident with, or parallel to, the isometric parallel. The most comprehensive study of the accuracy of both linear and area measurements was made by Demina (1960). It is her work which is most often quoted in later Russian books on photographic interpretation, and whose analysis we follow here.

The Influence of Tilt and Relief upon Linear Measurements

We intend to study the measurement of the length of a straight line, l', on tilted photograph compared with the undistorted length of it, l, such as woul be represented upon a truly vertical photograph of a plane ground surfac taken from a height which is accurately known. We call the difference betwee the two measurements $\delta l = l' - l$ and investigate the relative error, $\delta l/l'$ of th measured line. The various elements of the analysis may be expressed as th relative errors

$$R_i = (\delta l/l')_i \quad (13.5$$

The Influence of Tilt upon Linear Measurements (R_1)

In Fig. 13.7, the point i is the isocentre which is the origin of the system c (U, V) coordinates, the U-axis being the principal line and the V-axis being th isometric parallel. Because of the conformal property of the isocentre, the lin $ia = r_a$ coincides with $ia' = r'_A$. Thus the tilt displacement of the point a to a' i the difference in the length of these lines, or

$$\delta r'_A = r_A - r'_A \quad (13.6$$

Similarly the displacement of b to b' occurs along the line ibb' and

$$\delta r'_B = r_B - r'_B \quad (13.7$$

In general the displacement is likely to be different at the two ends of the line s that the lines ab and $a'b'$ differ in orientation. However, this difference is smal for nearly vertical aerial photographs, so that no significant errors will b introduced if the two lines ab and $a'b'$ are taken as being parallel. Consequentl the difference in length, δl, depends only upon the positions of the termina

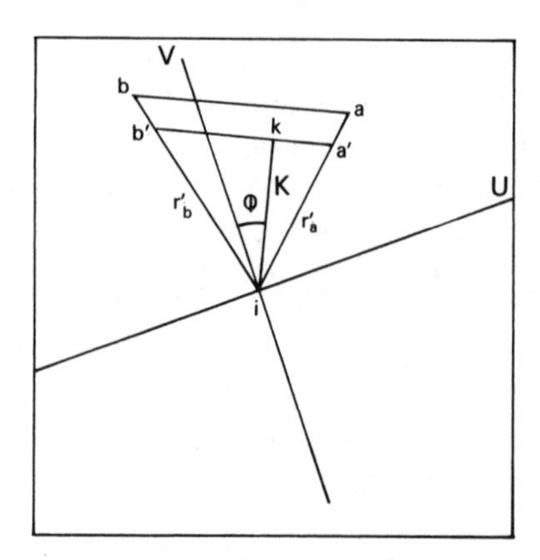

FIG. 13.7. Demina's analysis of the influence of camera tilt upon the representation of the line *ab*.

TABLE 13.3. *Percentage values for the relative error in length, R_1' for vertical aerial photography of size 228 × 228 mm and $f = 152{\cdot}4$ mm for different values of K and θ*

Distance from isocentre, K (mm)	Tilt (θ) 1°	2°	3°
10	0·1	0·1	0·2
20	0·1	0·3	0·4
30	0·2	0·4	0·6
40	0·3	0·5	0·8
50	0·3	0·7	1·0
60	0·4	0·8	1·2
70	0·4	1·0	1·6
80	0·5	1·1	1·7
90	0·6	1·2	1·8

points with respect to the isocentre of the photograph. Demina has shown that if the line ik is drawn to intersect $a'b'$ at right angles and $ik = K$, if this line makes the angle φ with the isometric parallel, we have

$$(\delta l/l')_\theta = \{[(l'_A + l'_B)\cdot\cos\varphi - K\cdot\sin\theta]/f\}\cdot\sin\theta \tag{13.8}$$

where $l'_A = ka'$ and $l'_B = kb'$ and θ is the angle of tilt.

Equation (13.8) attains its maximum value for a given amount of tilt where $\cos\varphi = 1$. This is where $a'b'$ coincides with the principal line, or is aligned in the direction of maximum tilt.

After evaluating the possible effects of variation in φ, K and θ, and introducing some approximations, we obtain the simple solution

$$R_1 = (\delta l/l')_\theta = 0{\cdot}6K\cdot\theta/f \tag{13.9}$$

In this equation, $\sin\theta \approx \theta$ radians for small angles of tilt and the quantity 0·6 is the mean for $\cos\varphi$ between $\varphi = 0$ and $\varphi = \pi/2$. We should note that since f forms the denominator of the fractions in both (13·8) and (13·9), the longer the focal length of the camera lens, the smaller is the relative error due to tilt.

The Influence of Surface Relief upon Linear Measurements (R_2)

Consider a truly vertical photograph upon which we wish to measure the length of the line $ab = l$. The ground point A has height h_A and that of B is h_B. We have used the height of B to determine the scale of the photograph, or

$$1/S_B = f/(H - h_B) \tag{13.10}$$

Therefore we imagine that the image point b is fixed, but a has been displaced to a_0 by an amount corresponding to the difference in height $h_A - h_B$.

It is the line $a_0b = l$ which we measure on the photograph. From equations (13.2) and (13.3) the displacement $aa_0 = \delta r$ and is directed along the straight line naa_0.

From equation (13.3)

$$\delta r = r \cdot \delta h / f \cdot S_B \tag{13.11}$$

where δh is the height difference $h_A - h_B$. This may also be expressed in the form

$$\delta h = l \cdot S_B \cdot \tan \beta \tag{13.12}$$

in which β is the angle of elevation or depression which might be observed from B to A on the ground. Then

$$\delta r = l' \cdot \tan \beta \cdot r \cdot \sin \gamma / f \tag{13.13}$$

where γ is the angle at both n and a in Fig. 13.8.

Thus the relative error in measurement resulting from ground relief may be expressed as

$$R_2 = (\delta l / l')_\beta = r \cdot \tan \beta \cdot \sin \gamma / f \tag{13.14}$$

The value of $\tan \beta$ may be determined from the heights h_A and h_B which have been measured on the ground or interpolated from a contour map or measured by parallax bar on the pair of photographs which show the whole of the line best. It is unlikely that β is often more than 10°; indeed it is reasonable to suggest that $\beta < 5°$ for all except mountainous terrain. For an angle as small as this we may apply the approximation that $\tan \beta \approx \beta$ radians. As in equation (13.9) we adopt the mean value of 0·6 for $\sin \gamma$. Consequently,

$$R_2 = 0{\cdot}6r \cdot \beta / f \tag{13.15}$$

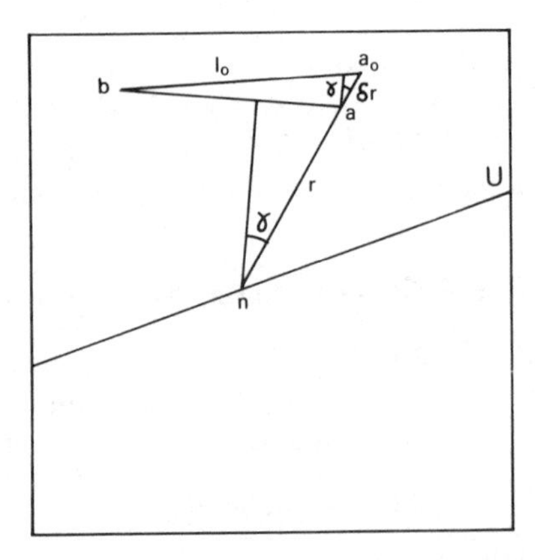

FIG. 13.8. Demina's analysis of the influence of surface relief upon the representation of the line *ab*.

TABLE 13.4. *Percentage values for* $R_2 = (\delta l/l')_\beta$ *for different values of r and* β, *calculated for wide angle photography, with* $f = 152{\cdot}4$ *mm*

Distance from nadir homologue, r (mm)	Vertical Angle (β)		
	1°	2°	3°
10	0·1	0·2	0·3
20	0·1	0·4	0·7
30	0·2	0·6	1·0
40	0·3	0·8	1·4
50	0·3	1·0	1·7
60	0·4	1·2	2·0
70	0·5	1·4	2·4
80	0·5	1·6	2·7
90	0·6	1·7	3·1

In other words we obtain the same result for R_2 as we found for R_1 with the exception that we are dealing with the angle of ground slope rather than the angle of aircraft tilt. In other words, we find that *uniform* slope across the working area of a photograph is geometrically equivalent to the same amount of tilt in that photograph. Photogrammetrists have known about this property for many decades and have exploited it in rectification.

The Influence of an Error in Flying Height (R_3)

From the description of the method of determining the scale of a photograph by means of linear measurement it is clear that a small error in the flying height, δH, is equivalent to a small error in determining the scale of the photograph. Thus, in Fig. 13.9, $a'b'$ of length l' represents the line AB

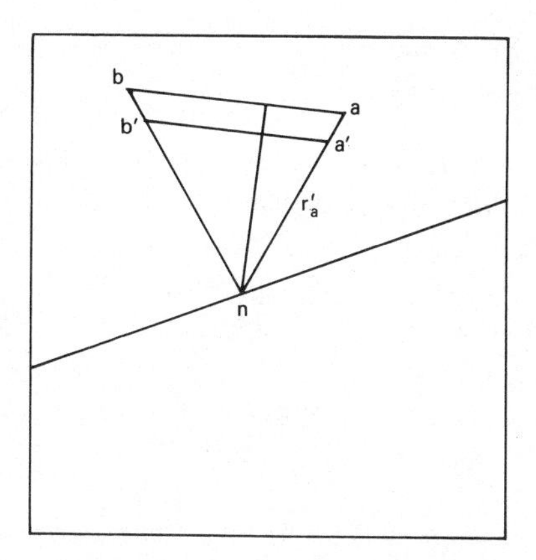

FIG. 13.9. Demina's analysis of the influence of a change (or uncertainty) in flying height upon the representation of a line *ab*.

photographed from flying height H, whereas the correct representation of the line is ab, photographed from the height $H + \delta H$. It follows that

$$l = r_1 \cdot \sin \gamma_1 + r_2 \cdot \sin \gamma_2 \tag{13.16}$$

Because l is parallel to l', from the symmetry of the figure,

$$\delta l = \delta r_{\delta H_1} \cdot \sin \gamma_1 + \delta r_{\delta H_2} \cdot \sin \gamma_2 \tag{13.17}$$

Therefore

$$\delta l_{\delta H} = \delta H \cdot l/H \tag{13.18}$$

so that

$$R_3 = (\delta l/l')_{\delta H} = \delta H/H \tag{13.19}$$

The Combined Influence of Tilt, Relief and Scale

Three different sources of error have been examined and the corresponding relative errors R_1, R_2 and R_3 have been derived in equations (13.9), (13.15) and (13.19). These may be combined to give an estimate of the relative error in the form

$$R = \{R_1^2 + R_2^2 + R_3^2\}^{1/2} \tag{13.20}$$

Substituting the right-hand sides of the three equations and putting $K = r$ in (13.9), the average relative error in linear measurement may be written

$$R = 1/f\{0{\cdot}4r^2(\theta^2 + \beta^2) + (\delta H^2 \cdot 1/S^2)\}^{1/2} \tag{13.21}$$

As a numerical example of the use of this equation, consider the example of wide-angle photography of scale 1/10,000 which satisfies the conditions that $\theta < 3°$ and $\beta = 5°$. It is further specified that $\delta H = 150$ metres. Then for $r = 50$ mm.

$$R = 1/152{\cdot}4\{1000(0{\cdot}00274 + 0{\cdot}00761) + 0{\cdot}0002\}^{1/2}$$
$$= 0{\cdot}021 \text{ or } 2{\cdot}1\%$$

In this example the value for δH is taken to be 10% of the flying height, which is larger than would normally be expected in practice. However, the effect of the term in δH is extremely small and is negligible in all except extreme cases. The most important term in the equation is that for r^2. This may be demonstrated by solving (13.21) for different values of r. It shows that for the conditions of photography specified above, R increases from 0·4% at a distance of 10 mm from the centre of the photograph to 3·8% at 90 mm from the principal point. The increase in relative error is linear, as illustrated in Fig. 13.10.

These calculations suggest a useful practical technique for applying the foregoing theoretical analysis to actual measurements. Given a strip, or block of photography of known focal length and approximate scale with standard

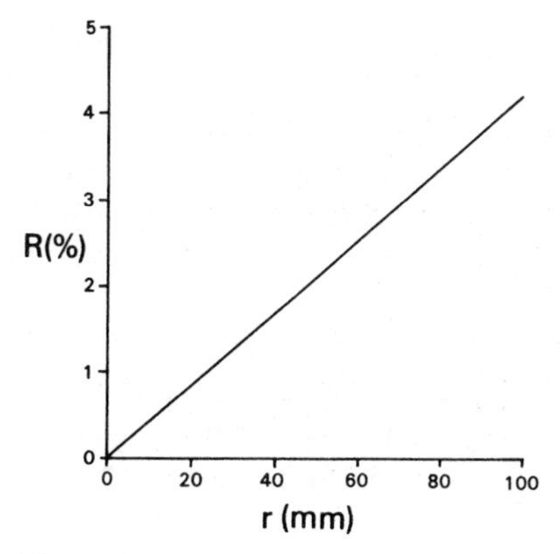

FIG. 13.10. The combined effect of camera tilt, surface relief and variation in flying height related to radial distance from the centre of the photograph.

60% and 20% overlaps, we assume limiting values of θ and β as above and calculate the combined relative error at the edge of the working area. Then by interpolation on a graph such as Fig. 13.10, we may select the values of r corresponding to $(\delta l/l')$ to 0·5%, 1%, 1·5%, etc. These may now be plotted as concentric circles on a transparent overlay, as in Fig. 13.11, which may be placed over each photograph on which measurements are intended. The relative error which is likely to be experienced in any part of the working area may be estimated by inspection.

Certain conditions of photography provide smaller values of the variables and therefore reduce the magnitude of the errors. Some survey cameras have gyro-stabilised mounts. These do not wholly eliminate tilt but may keep the camera within a few minutes of arc of the truly vertical position, which means that the value of θ^2 in equation (13.21) becomes practically negligible.

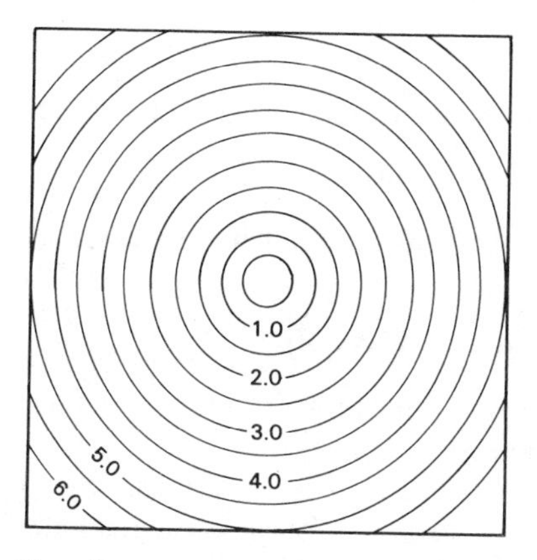

FIG. 13.11. A correction diagram constructed from the conditions illustrated in Fig. 13.10 to be used as an overlay on the photograph.

Similarly a *rectified print* is one in which $\theta = 0°$ by virtue of fitting the photographic images to ground control through adjustment of an optical projector (or rectifier). Photography of extensive flat land, such as marshes or flood plains gives $\beta = 0°$. The methods of rectification is particularly suitable for mapping flat lands so that in such cases both $\theta = 0°$ and $\beta = 0°$. Moreover, the scale adjustments which are introduced to fit the photographic images to ground control also eliminate most of the uncertainty of H so that the R_3 term is also zero. It follows that, under these conditions, virtually all the terms on the right-hand side of equation (13.21) vanish and we are left with only the technical errors of measurement which would be present if the measurements had been made on large-scale maps instead of photographs.

The Influence of Tilt and Relief upon Area Measurement

Demina examines the case of a square parcel whose sides are parallel to the principal line and the isometric parallel on the tilted photograph. She defines the error in length measured in the direction of the principal line as δl_v and that in the direction of the isometric parallel as δl_H. Then, from equation (13.8), she determines the maximum value for R_1, which occurs in the direction of the principal line (i.e. where $\varphi = 90°$) and finds

$$\delta l_v / l = 2r \cdot \theta / f \tag{13.22}$$

If the area of this parcel on a truly vertical photograph is

$$A = l^2 \tag{13.23}$$

the corresponding area on the tilted photograph is

$$A' = (l + \delta l_v)(l + \delta l_h) \tag{13.24}$$

Multiplying through and neglecting second-order terms, we obtain

$$A' = A(1 + \delta l_v / l' + \delta l_H / h') \tag{13.25}$$

The relative error in the measurement of area is therefore

$$\delta A / A = \delta l_v / l' + \delta l_H / l' \tag{13.26}$$

Substituting the values from the right-hand sides of (13.23) and (13.22), we obtain

$$(\delta A / A)_\theta = 3r \cdot \theta / f \tag{13.27}$$

where r is the mean value of the radii drawn from the centre of the photograph to the sides of the parcel.

We commented earlier that the longer the focal length of the camera lens, the smaller are the relative errors in linear measurement. Obviously the same applies to area measurement, and we see in Table 13.5 that the percentage relative errors in area are approximately one-half as large on wide-angle photography as for the same conditions on super-wide-angle photographs.

TABLE 13.5. *Percentage values for the relative errors in area measurement, $\delta A/A$, which result from the measurement of a square parcel at different distances, r, from the centre of the photograph, for tilts of 1° and 2° and for focal lengths, $f = 88$ mm (super wide angle) and $f = 152{\cdot}4$ mm (wide-angle) photography*

	$\delta A/A$ (%)			
	$f = 88$ mm		$f = 152{\cdot}4$ mm	
r (mm)	$\theta = 1°$	$\theta = 2°$	$\theta = 1°$	$\theta = 2°$
10	0·6	1·2	0·3	0·7
20	1·2	2·4	0·7	1·4
30	1·8	3·6	1·0	2·1
40	2·4	4·8	1·4	2·7
50	3·0	5·9	1·7	3·4
60	3·6	7·1	2·1	4·1
70	4·2	1·3	2·4	4·8
80	4·8	9·5	2·7	5·5
90	5·4	10·7	3·1	6·2

Clearly there are advantages in using longer focal length photography where this is available. This may not be of much use to the civilian user, who is more or less obliged to make measurements on wide-angle photographs. However, it is extremely important in military applications for which reconnaissance photographs taken by cameras having longer focal length lenses are also available.

The advantages of centrality and symmetry should also be stressed. The increase in relative error with increasing r corresponds to that found in linear measurement and points to the use of the central part of each photograph as much as possible. However, a large parcel which is both centrally placed and symmetrical about the principal point may be measured with apparently small errors. This is because the distortions in one part of the parcel compensate for those in another. If we were to subdivide such a parcel into a collection of smaller components, we would find that each of the small parcels would have relatively larger errors because they are not symmetrically disposed.

Demina's investigation of the errors arising from variation in surface relief is of the distortion of the same square parcel as that for the study of tilt errors. The remaining variables are the values of the ground heights of the corners of the parcel and the uncertainty in flying height which, as before, affects the scale. From equation (13.25), for a square parcel on an untilted photograph,

$$l_v = l_H = l$$

and

$$\delta A = 2A \cdot \delta l / l' \tag{13.28}$$

Consequently the relative error in determining the area is composed of

$$\delta A/A = 2\delta l_\beta/l' \tag{13.29}$$

and

$$\delta A/A = 2\cdot\delta l_{\delta H}/l' \tag{13.30}$$

Substituting (13.31) in (13.14) and (13.32) in (13.19) we obtain

$$(\delta A/A)_\beta = 2r\cdot\beta/f \tag{13.31}$$

and

$$(\delta A/A)_{\delta H} = 2\cdot\delta H/H \tag{13.32}$$

Combining equations (13.27), (13.31) and (13.32) as before, we obtain for the relative error of area measurement

$$\delta A/A = 2/r\{r^2(2{\cdot}2\theta^2 + 0{\cdot}4\beta^2) + \delta H^2\cdot 1/S^2\}^{1/2} \tag{13.33}$$

The same special conditions of photography and landscape which influence the accuracy of linear measurements also affect area measurements. There is, however, the important additional factor that the accuracy of area measurement depends only upon displacements of the parcel perimeter. The result is unaffected by any irregularities of the terrain within the parcel itself. This means that area measurements made directly on aerial photographs are particularly appropriate for the study and measurement of such objects as islands and lakes. Obviously, in these examples, the shoreline is the perimeter of the parcel and since, within the confines of the single aerial photograph, the lake or sea surface is practically a horizontal plane, there should be no displacement due to relief on the perimeter of the parcel.

There are, it seems, few experimental studies to support the theoretical work by Demina. Some work has been done by Smirnov and Chernyaeva (1964) and Smirnov (1967), which has attempted to examine the different error components arising in area measurement using aerial photographs. Their experimental work comprised measurement of more than 200 parcels by planimeter of photographs from six different sorties comprising different kinds of surface relief and also using both normal-angle and wide-angle photography. Moreover, as is common in much Russian work, we are left in some doubt about some of the data and therefore the validity of the results. For example, this work shows some striking improvements in the results obtained when the measurements were made on enlargements of the photographs compared with those made on the original contact prints. We are told by Smirnov (1967) that these enlargements were made in a photogrammetric rectifier, but we are not told by how much they were enlarged nor if tilt displacements were also removed in the process. Thus the bald statement that the percentage error for measurement of parcels of size 50–100 ha was 2·6% on the contact print but only 1·2% on the enlargement is not particularly informative if we do not know the scale of the original photography and therefore the dimensions of such parcels upon it, let alone these dimensions on the final enlargements.

Otherwise this work confirms most of the ideas which have been derived theoretically. For example it demonstrated that normal-angle photography ($f = 200$ mm for the standard Russian 180 × 180 mm format camera) is superior to wide-angle photography ($f = 100$ mm), that parcels measured near the centre of the photograph have smaller errors than those located near the edges, and that parcels on photographs of flat land are measured more accurately than those on photographs of mountainous terrain. However, it would have been much more worthwhile if they had produced more specific results. There is clearly enormous scope for further experimental work in this branch of cartometry.

SPACE IMAGERY AND EARTH CURVATURE

We have argued that, from the cartometric point of view, the effects of earth curvature are negligible on any aerial photography which has been taken from a flying height of 10 km or less. Where the photography has been acquired from a space vehicle, however, the flying height is very much greater – usually in excess of 150 km. Consequently the resulting images need to be transformed from the projection created by their representation before they may be used for mapping or measurement. Figure 13.12 illustrates the different variables affecting a photograph taken from space. Obviously these are the same kinds of error which have to be corrected on larger-scale photography, but now we have additional complications resulting from earth curvature and characterised by the replacement of a plane, or straight line in section, by the spheroidal or spherical surface which is elliptical or circular in section. Table 13.6, due to Colvocoresses (1972), estimates the values of these errors as they were expected to affect Landsat-1 imagery.

TABLE 13.6. *Summary of maximum displacements in space imagery calculated for Landsat-1 by Colvocoresse*

Source of displacement	Maximum uncorrected displacement (m)	Maximum displacement after processing and correction	
		(m)	(mm at 1/1,000,000)
Earth curvature	200	50	0·05
Atmospheric refraction	0·34	–	–
Tilt (1°)	440	53	0·05
Relief (for $h = 1000$ m)	160	160	0·16
Map projection (U.T.M. for imagery of the U.S.A.)	42	42	0·04
Apparent combined error	840	300	0·30

The amount of correction may be reduced by the much smaller scale of imagery to the extent that it may be virtually ignored. Thus the annoying correction for relief displacement which, we have seen are particularly awkward to apply in the use of conventional aerial photography, more or less vanishes when the flying height is many hundred times greater than any variations in ground height.

In any investigation of the corrections to be applied to space imagery it is necessary to distinguish between the conventional perspective images obtained from an ordinary camera system and those created line-by-line as a lens or mirror sweeps to and fro within a scanning sensor. The perspective images obtained from space include the photographs taken by conventional cameras on film during manned missions, together with the pictures obtained automatically from return beam vidicon (R.B.V.) cameras operating on certain of the Landsat vehicles. From the civilian point of view, the only space photography which most closely resembles conventional aerial photography has so far been taken on two different missions. In 1983 a Zeiss 30/23 metric camera was carried on the European Space Agency *Spacelab-I* launched on 28 November 1983. Some preliminary results of work done with this photography was described by Dowman *et al.* (1985). The second was the NASA/Itek LFC or *large-format camera* which was first used on Space Shuttle Mission 41-G in October 1982. The hardware of the L.F.C. and the results have been briefly described by Doyle (1985). Apart from the large-format negative, which measures 230 × 460 mm (9 × 18 in.), this camera corresponds to a typical aerial camera with a lens of focal length 305 mm (12 in.). It follows that from the operational height of just below 240 km, the scale of photography is approximately 1/800,000. Therefore the ground area covered by the single photograph measures approximately 178 × 356 km or a total area of about 63,500 km^2. The Zeiss camera used in Spacelab-I was a more or less standard aerial camera taking 228 × 228 mm format negatives on film and also having a lens of focal length 305 mm. The orbiting height of Spacelab was marginally heigher so that the scale of photography is nominally 1/820,000. At this scale the amount of relief displacement is very small. For example, a feature of height 1,000 m situated in one of the corners of a photograph is only displaced by 1·5 mm and small variations in height near the middle of the photograph are virtually undetectable. Such a vertical photograph represents a perspective azimuthal projection of the terrestrial surface. The plane of projection is tangential to the curved spherical surface at the nadir point, as shown in Fig. 13.12. The R.B.V. output from Landsat-1 through Landsat-3 also satisfies the perspective geometry of the conventional photograph illustrated here.

Where the images have been transmitted from an unmanned satellite as signals received by a ground station, the necessary corrections are most effectively made by digital processing of the resulting tapes. This applies particularly to the multispectral scanner (M.S.S.) output from the Landsat

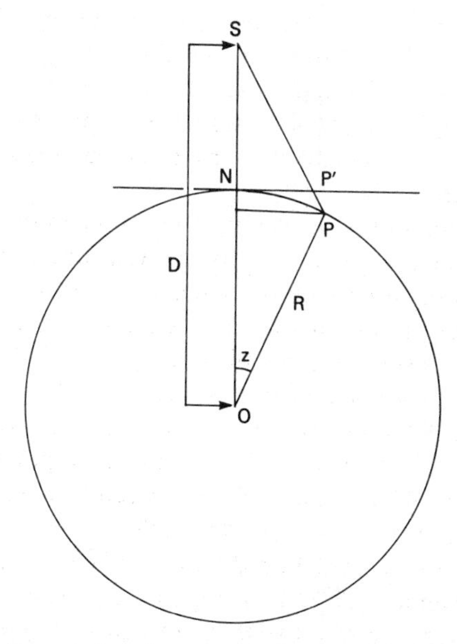

FIG. 13.12. The influence of earth curvature regarded as a perspective azimuthal projection in which the plane of projection is tangential to the spherical surface at the nadir image, N, of the sensor in the space vehicle, S, and the point P on the spherical surface projects as P' on the plane. It can be determined from the proportional triangles that the mapped distance $NP' = \rho$ of P' measured from the centre of the image, N, may be calculated from $\rho = [(D + R)R \cdot \sin z]/[D - R \cos z]$. This was the method suggested by both Volkov (1964) and Dumitrescu (1966) for mapping from space photographs in the early days of artificial satellites.

systems which was illustrated by Fig. 2.9, p. 29. The geometry of the multispectral scanner is more complicated than the simple instantaneous perspective image of ordinary photography because the satellite moves further in its orbit and the earth rotates beneath the scanner during the longer period needed to collect the image data.

The various predicted or measured motions of the satellite may be used to calculate a series of system or *bulk corrections* to be applied to a particular scene. There are 14 geometrical corrections which may be applied to the raw data, illustrated in Fig. 13.13. The result is a corrected image which may be regarded as being a parallel or orthogonal projection to the tangent plane. According to Colvocoresses (1974) the result is a form of pseudocylindrical projection "of rather curious characteristics rather than the perspective azimuthal projection of the ordinary photograph". The next step is to transform this into a projection which is more suitable for mapping purposes. At the time when the first Landsat images were being processed in 1972, it was assumed that since the U.T.M. projection was widely used for topographical mapping, this would be the logical choice for use with maps made directly from Landsat images and that transformation from the bulk-processed image to

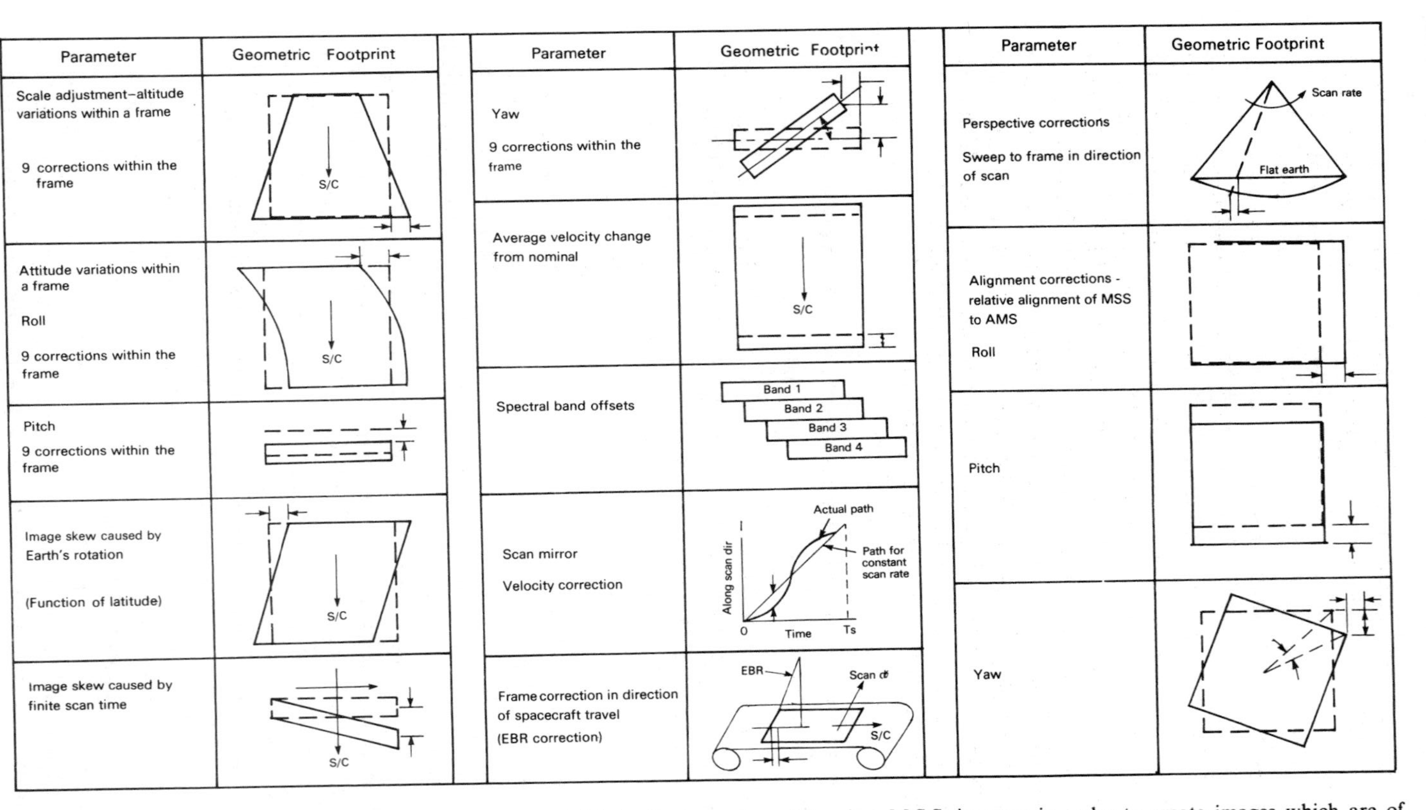

FIG. 13.13. The different corrections which have to be applied in bulk processing of Landsat M.S.S. imagery in order to create images which are of approximately constant scale.

U.T.M. would be sufficient. However, there are certain objections to the use of the U.T.M. in this work. For example, the effect of earth rotation means that the track of the satellite is not a great circle, but a more complex curve. Since this is inclined to the central meridian of each U.T.M. zone, the satellite track moves from one zone to another during the course of a single orbit. Where the junction between two zones occurs the scale errors of the projection are approximately 1/1,000. Landsat imagery has smaller residual scale errors after bulk processing and therefore some better system is required. A further minor objection to the use of the U.T.M. is that this ends in latitudes 80° north and south so that it cannot be used for representing the polar caps.

In order to overcome these difficulties the U.S. Geological Survey investigated the suitability of the oblique Mercator projection for continuous representation of M.S.S. imagery. There are several ways of deriving the oblique Mercator projection for the spheroid. None of these is ideal because either the scale along the centre line is not constant, or, if this condition be satisfied, the projection is not rigorously conformal. The version of the oblique Mercator adopted for use with Landsat imagery is that originally devised by Hotine and described by Brazier (1947) for use in topographical mapping of Malaysia and other countries whose longer axes are inclined to the meridians and for which, therefore, the Transverse Mercator is less suitable. Hotine called this the *rectified skew orthomorphic projection*, but in these new applications the Americans have renamed it the *Hotine Oblique Mercator*, or H.O.M. projection. There are three distinct improvements resulting from the use of this as the basis for the *precision* or *screen-corrected* output which is the final output from the processing system. For example, the projection is based upon north–south strips which are inclined at a small angle away from the polar axis, thereby matching the ground track of the satellite more closely.

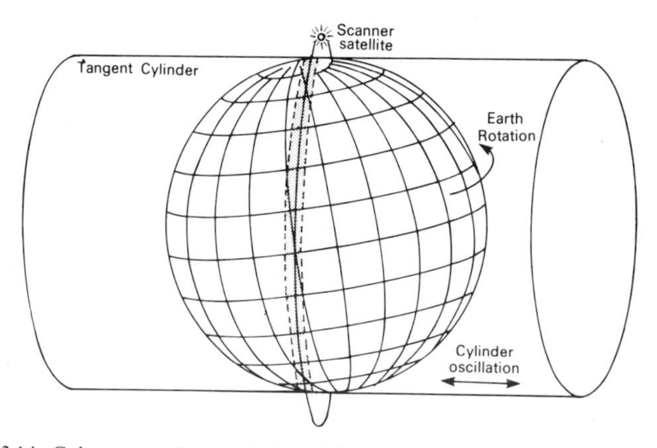

Fig. 13.14. Colvocoresses' concept of an oblique cylindrical projection surface which might take into account the earth's rotation during a Landsat orbit. This idea was subsequently developed by Snyder as the Space Oblique Mercator projection.

Notwithstanding the improvement, the H.O.M. projection is still not wholly suitable for continuous mapping of the earth within the width of the swath scanner by the M.S.S. or the thematic mapper on the later Landsat sorties. The discrepancies may not be detected in the single frame, but they become apparent in mapping long strips for each orbit and, in particular, matching them with the next strip formed by a later orbit. In the planning stages for Landsat-3, which was launched in 1978, Colvocoresses (1974) investigated the desirable criteria for a projection which would take into consideration the special requirements of mapping from a satellite orbiting about a rotating earth, for which ordinary map projections only provide a static and therefore partial solution. His call for ideas was taken up by Snyder (1978, 1982), who devised the *Space Oblique Mercator* or S.O.M. projection, which is rigorously conformal and in which the scale errors do not exceed 1/10,000.

Sooner or later it becomes necessary to refer Landsat M.S.S. imagery based upon the H.O.M. or S.O.M. projections to the U.T.M., if only to facilitate comparison with existing topographical maps which are based upon this projection. Some scale distortion is inevitable when transforming from one system to the other at the U.T.M. zone boundaries as before. The maximum scale distortion occurs at each zone boundary at the equator. The magnitude of the displacement which occurs here in reconciling the H.O.M. or S.O.M. projection with the U.T.M. corresponds to about 204 metres on the ground. At 1/1,000,000, which is the scale at which most Landsat imagery is initially printed as a photograph, this discrepancy is 0·2 mm, which is just detectable. Although this is small enough to be ignored on the single photograph, a mosaic created from several S.O.M. or H.O.M. images cannot be expected to fit the U.T.M. without measurable error.

14

Conventional Linear Measurements and their Reduction

"What do you consider the *largest* map that would be really useful?"
"About six inches to the mile."
"Only six inches!" exclaimed Mein Herr, "We very soon got to six yards to the mile. Then we tried a hundred yards to the mile. And then came the grandest idea of all! We actually made a map of the country, on the scale of a mile to the mile!"
"Have you used it much?" I enquired.
"It has never been spread out, yet", said Mein Herr: "the farmers objected: they said it would cover the whole country, and shut out the sunlight! So we now use the country itself, as its own map, and I assure you it does nearly as well."

(Lewis Carroll, *Sylvie and Bruno Concluded*, 1893)

We use the words *conventional* and *classical* to mean linear measurements made by dividers, chartwheel, pieces of thread and card. In other words, this category includes all the methods described except the probabilistic or longimeter methods described in the next chapter, and measurements made by digitiser or scanner which are considered in Chapter 21.

The conventional methods have been described in Chapter 3. Here we examine the precision of the different methods, their suitability for different kinds of work and the methods of reducing the crude measurements of distance in ways which take account both of the methods used and the scales of the maps on which they have been made.

THE PRECISION OF CONVENTIONAL LINEAR MEASUREMENTS

Any attempt to compare the precision of different cartometric measurements requires two sets of data:

- repeated measurements of the same line by the same method in order to establish its mean square error;
- measurement of the same line using different techniques.

Ideally the second set ought to comprise a series of samples from the first set. Although this seems to be the obvious procedure to adopt in making any comparative study of instrument precision, it is seldom realised. Usually the

TABLE 14.1. *Results of comparative testing of the precision of methods of measuring distance mode by Volkov (1950b). Standard error, in millimetres, at the optimum setting of the instrument*

Instrument	Length of line (mm)	Standard error of the single observation (mm)
Dividers (12 cm legs)	Up to 100	±0·3
Dividers (8 cm legs)	60–80	±0·25 – ±0·3
Spring-bow	40–50	±0·25
Beam compass	1000	±0·25
Steel scale	1000	±0·10
Chartwheel	—	1/50–1/100

available data turn out to be members of one set or the other, rarely of both. Thus we find that there are remarkably few distance measurements which are sufficiently comprehensive and complete to serve as adequate data.

From this rather small number of suitable test data we first describe the work done in this field by Volkov (1950), who made a series of measurements of

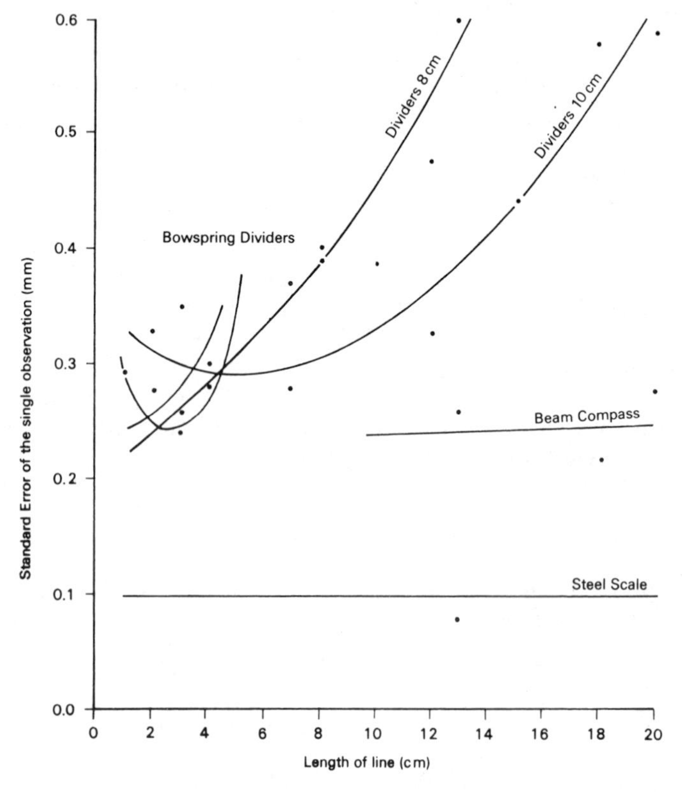

FIG. 14.1. The standard errors for measurement of different lengths of line using different instruments. (Source: Volkov, 1950b).

straight lines within the length range 2–20 cm. The detailed results are illustrated in Fig. 14.1. The following characteristics are evident:

Measurement by Scale

Measurement of the length of a line by direct comparison with a steel scale is evidently the most precise method of measurement. Moreover, within the range of distances tested, it is independent of the length of line. In fact, as we have already established in Chapter 6, the root mean square error of location of a point with respect to a scale graduation is about ±0.1 mm, and because there are two terminal points to a line, the combined reading error is

$$\pm 0{\cdot}1\sqrt{2} = \pm 0{\cdot}14\,\text{mm}.$$

This is the only error affecting the measurement, so that we may conclude that the precision is independent of distance within the working range of a single steel scale.

Measurement by Beam Compass

Measurement of the length of a line by beam compass. This technique is confined to lines of length 100 mm or longer. The principal error component is, again, in the location of the terminal points, but it has now increased to about

$$\pm 0{\cdot}17\sqrt{2} = \pm 0{\cdot}24\,\text{mm}.$$

Because other small instrumental errors (flexibility of the beam being the most significant?), the standard error increases slightly with distance.

Measurement by Dividers

Figure 14.1 indicates that some of the curves for dividers are characteristically U-shaped. This is because each type and size of dividers has an optimum setting and there is some loss in precision when the legs of the dividers are set too far apart or too close together. This is evidently because of the variable angular penetration of the map surface by the points as illustrated in Chapter 3. It does not happen to measurements made by beam compass because the points are perpendicular to the plane of the map irrespective of their setting. It follows that each type and size of dividers has its optimum setting. It can be seen that ordinary dividers having legs of length 10 cm are best for making measurements of length 6–9 cm, and that small spring-bow dividers are best for steps of 2–4 cm.

Measurement by Chartwheel

Volkov dismisses the use of the chartwheel as being too inaccurate for most purposes, and he assigns to it relative errors of order 1/50 to 1/100. This

TABLE 14.2. *Chartwheel measurements made by Kishimoto (1968). Mean distance and the standard error of the single observation expressed in millimetres*

Map	Line no. 1	2	3	4	5
1/2,500	–	–	–	–	609·7 ± 1·28
1/5,000	–	–	–	–	304·9 ± 1·28
1/10,000	1051·2 ± 1·49	560·2 ± 1·31	426·4 ± 1·36	371·7 ± 1·53	151·9 ± 1·35
1/25,000	417·5 ± 1·39	223·9 ± 1·30	166·3 ± 1·17	148·6 ± 1·50	–
1/50,000	207·7 ± 1·41	110·7 ± 1·25	80·2 ± 1·47	73·9 ± 0·97	–
1/100,000	103·0 ± 1·08	55·0 ± 1·10	40·6 ± 1·35	36·5 ± 1·05	–

Line No. 1: River Doubs, 10·4 km
Line No. 2: Highway between Montfaucon and St Brais, 5·5 km
Line No. 3: Valle di Duragno, 4·3 km
Line No. 4: Swiss-French frontier, 3·7 km
Line No. 5: Shoreline of Oberer Chatzensee, 1·5 km

corresponds approximately to standard errors which are about ten times greater than those for dividers measurements. Similar opinions are echoed in the statements of others – for example, Richardson (1961), Håkanson (1978), Galloway and Bahr (1979). Nevertheless none of these writers offers any quantitative information about the method. The most comprehensive set of chartwheel measurements are those conducted by Kishimoto (1968), who used this as the principal method of measuring the lengths of different features on maps. Since she was unaware of criticism of the method by others, she used her chartwheel measurements as the standard against which the other results might be compared. Altogether Kishimoto, and her students, made 445 separate measurements of five different kinds of feature. Most of the experiments were repeated on maps at four different scales. These results are summarised in Table 14.2. Inspection of the table indicates that the standard error of the single observation remains remarkably constant at about ± 1·3 mm. There is no indication that it varies according to the character of the line traced, nor the scale of the map measured.

In 1977 the present author investigated the *relative error*, as he called it, to be obtained from chartwheel and dividers measurements.

This parameter differs from the well-known *coefficient of variation* in that the sizes of the samples are taken into consideration by application of Student's t-test. The coefficient of variation of a distribution having mean x and standard deviation σ may be written as the percentage

$$f = 100\sigma/x \qquad (14.1)$$

For a sample of measurements, the standard error of the mean is the value used for σ. We may determine the *confidence limits* or *fiducial limits* for the assumed population mean, μ, of the measurements. This may be determined from

$$x - s_M \cdot t < \mu < x + s_M \cdot t \qquad (14.2)$$

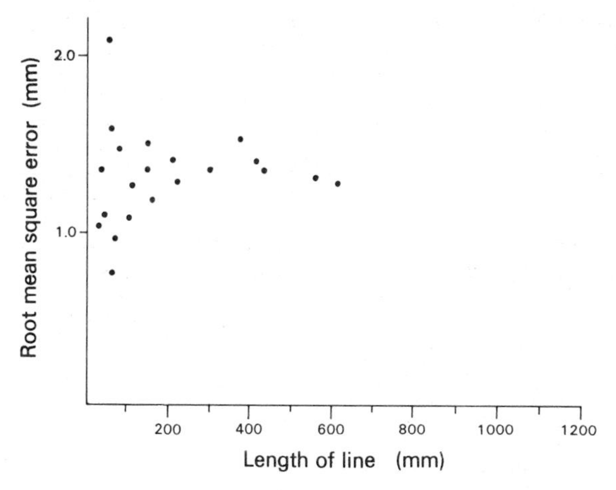

FIG. 14.2. Mean square errors resulting from the measurements by Kishimoto using a chartwheel, these being the results tabulated in Table 14.2. (Source: Maling, 1968).

where s_M is the standard error of the mean of a sample of n measurements and t is Student's value appropriate to the selected significance level and v, the number of degrees of freedom. We demonstrate this by means of the following example.

The length of a feature shown on a map is measured five times, giving a mean of 426.4 mm and the standard error of the single observation is $\pm 1{\cdot}356$ mm. Then the C.V. would be determined as

$$1{\cdot}356 \times 100/426{\cdot}4 = 0{\cdot}32\%$$

In the present example $v = 5 - 1 = 4$. If we accept the 95% probability level, from Table 6.4, p. 102, $t = 2{\cdot}776$. It follows that the term on each side of (14.2) is

$$s_M \cdot t = (1{\cdot}356 \times 2{\cdot}776)/2{\cdot}2361 = 1{\cdot}683$$

It follows that the confidence range is from 424·72 to 42·08 mm. The relative error is this range divided by the length of the line and expressed as a percentage. Therefore

$$f = 3{\cdot}366 \times 100/426{\cdot}4 = 0{\cdot}79\%$$

The range is about four times the standard error of the single observation, but it is adjusted for the size of sample, which is important in dealing with a large number of differently sized samples, many of which are only three or four measurements. For example, if this result had been based upon 25 separate measurements the value of f would have been reduced to less than one-half, or 0·32%, corresponding to the coefficient of variation.

In Kretschmer (1977), the present author determined values of f for the chartwheel measurements by Kishimoto and compared log f with log l, this

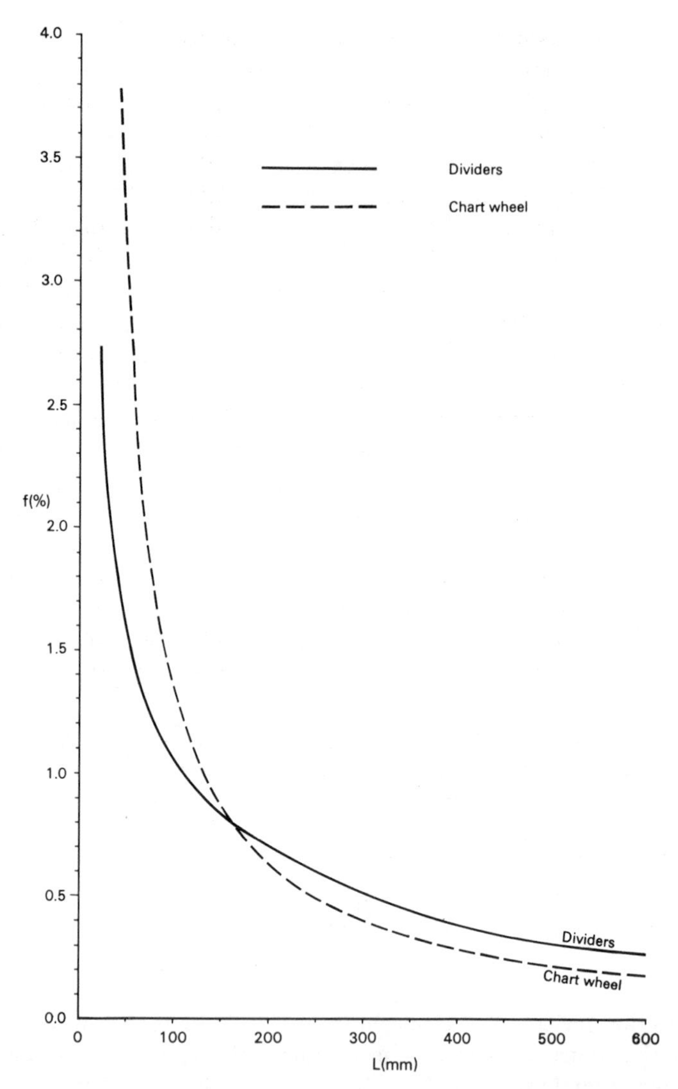

FIG. 14.3. Comparison of the relative error curves for Kishimoto's chartwheel and the Yorkshire Coast measurements by dividers. (Source: Maling, in Kretschmer, 1977).

being the transformation to the data giving the best fit to the data from the twelve empirical distributions examined. The curve, illustrated in Fig. 14.3, indicates that the relative error may be quite large for short distances, but for chartwheel measurements it has already fallen to 2% for a line of length 50 mm and it has been further reduced to 1% or less for distances in excess of 130 mm.

The agreement of most of Kishimoto's results with the illustrated curve is good. However, results such as those made by her students, and also by my students for the Yorkshire coast, do not fit the same distribution because the standard errors for these are much larger. For example, the measurements of the length of the River Doubs at 1/25,000 scale which was measured as a class exercise for Kishimoto's students produced the result $419{\cdot}0 \pm 5{\cdot}32$ mm, and this standard error is nearly five times greater than that obtained by Kishimoto from her own measurements listed in Table 14·2.

Using the same technique the author obtained comparative data for dividers. This is also shown in Fig. 14.3. The curve for chartwheel measurements intersects that for dividers measurements at a point corresponding to the relative error, $f = 0{\cdot}8\%$ and length of line 160 mm. Rounding off these figures we may use the following guiding principle for making linear measurements:

If the length of the line to be measured is greater than 150 mm (say 6 inches), then a chartwheel should give the *experienced operator* results with a relative error less than 1% on 95% of all occasions.

If the length of the line is less than 150 mm, and *for beginners*, measurements by dividers are to be preferred. Nevertheless, using dividers we cannot except the errors to be less than 2% if the line is shorter than 25 mm (or 1 inch).

Although this comprises somewhat meagre data upon which to base an important conclusion, it is virtually all that was available to the author in 1977. At that time he presented the agreement that, since the standard errors of chartwheel measurements are remarkably consistent and appear to be independent of the length of line, it is likely that an important component of user standard error is in consistently setting the same part of the measuring wheel over the terminal point of the line, and that the other important source of error is estimating the scale reading.

The author therefore argues that the loss in precision resulting from difficulties on manipulating a chartwheel on extremely irregular lines has been much exaggerated; otherwise we should detect an increase in standard error with decrease in map scale.

Other Methods

Volkov limits his tests to dividers and chartwheel. Only Kishimoto has attempted to determine the precision of the two other conventional methods of measuring distance.

Dividers Measurements by Unequal Steps

This is the method using dividers corresponding to the card method described on pages 38–39. Kishimoto limited her experiments to the measurement of only one of the lines; No. 4 in Table 14.2, this being a portion of the Franco-Swiss frontier. It is easy enough to make a theoretical estimate of the precision of the card method. Since this comprises N separate representations of lines

TABLE 14.3. *Kishimoto's measurements of the length of part of the Franco-Swiss frontier using dividers with unequal spacing (distances in millimetres)*

Map scale	No. of measurements	Length	Standard error of the single observation
1/10,000	10	366·2	±2·02
1/25,000	10	147·1	±1·13
1/25,000	55	146·1	±2·16
1/50,000	10	72·7	±0·65
1/100,000	10	35·5	±0·47

marked on a card, the theoretical precision is $\sqrt{N}$ times the graphical error of marking the card. From Chapter 9, plotting precision is of the order ±0·2 mm. Therefore the precision of measurement by card may be taken to be $\pm 0{\cdot}2\sqrt{N}$, but there does not seem to be any data to support this conclusion.

Distance Measurement by Thread

Kishimoto used thread to measure the length of only one line; that of the road from Montfaucon to St Brais in the Swiss Jura, which was the least irregular of the test lines used by her. Despite the various difficulties in manipulation which were described in Chapter 3, she did obtain tolerably good results using the method. The standard errors are given in Table 14.4. It suggests that the method does not seem to be significantly less precise than the use of a chartwheel, but this can hardly be called a thorough test of the method. From their measurements of three special test lengths of the Australian coast at 1/250,000, Galloway and Bahr (1979) found that "measurement by fine wire proved surprisingly consistent with a mean accuracy equivalent to a dividers intercept of 0·7 km". This evidently corresponds to dividers with $d = 2{\cdot}8$ mm.

CORRECTIONS TO DIVIDERS MEASUREMENTS

It was shown in Chapter 3 that measurement of an irregular line by dividers corresponds to making a series of chord measurements. Since the length of a

TABLE 14.4. *The length of the Montfaucon–St Brais road determined by Kishimoto using thread*

Scale	No. of observations	Distance (mm)	Standard error of single observation (mm)
1/10,000	10	558·95	±0·55
1/25,000	10	223·28	±0·86
1/50,000	10	111·74	±1·05

hord is always less than its arc, it follows that a measurement by dividers is lways shorter than the length of the line. It is therefore desirable to investigate hether measurements by dividers can be corrected to allow for their eparation. This subject has been examined more or less independently in both estern and eastern Europe. Because there is comparatively little overlap etween the methods of treatment we may deal with each approach separately.

The Western European Contribution; L. F. Richardson and his Followers

The distinguished physicist and mathematician L. F. Richardson died in 1953. During the later years of this life he had been preoccupied with the nathematical study of the incidence of wars, and his work *The Statistics of Deadly Quarrels* (eds. Wright and Lienaur, 1960) was published seven years after his death. An appendix to this work, entitled *The Problem of Contiguity*, eferred to here as Richardson (1961), appeared separately in *The General Systems Yearbook* for 1961, and is of particular interest in the present context. Like much of Richardson's other work – for example his pioneer studies on numerical methods of whether forecasting – it has continued to exercise a considerable influence through the intervening decades. In an investigation of some geometrical properties relating to the proximity of states, Richardson became intrigued by the lack of agreement between the published statistics for he lengths of the frontiers between certain European nations. He quotes some figures relating to these, which are just as striking as some of the discrepancies n coastline lengths recorded in Chapter 5. Consequently he made his own brief study of how the lengths of irregular boundaries may be measured on naps.

He found, like others before and after him, that the best results were to be obtained by walking a pair of dividers along a map of a frontier

> so as to count the number of equal sides of a polygon, the corners of which lie on the frontier...

as he describes the procedure.

He measured various features on atlas maps, notably the west coast of Great Britain from Land's End to Duncansby Head, the coastlines of South Africa and Australia and the land frontiers of Germany (in 1900) and of Portugal. As if to emphasise his description of the procedure as fitting a polygon to an irregular boundary, Richardson became preoccupied with the general, but evidently insoluble, problem of how an irregular line may be subdivided into exactly n equal parts so that the sum of the steps made by dividers is the measurement of the line without having to make an additional final measurement of length d' as is described in equation (3.1). His first four conclusions relate to this mathematical problem. On the general subject of measurements and their reduction he writes

(5) The total polygon length, including the estimated fraction, usually depends slightly on the point that is taken as the start.

(6) It is doubtful whether the tidal polygonal length of a sea coast tends to any limit as the side of the polygon tends to zero.

(7) To speak simply of the "length" of a coast is therefore to make an unwarranted assumption. When a man says that he "walked 10 miles along the coast", he usually means that he walked 10 miles near the coast.

We have already introduced the concept that an empirical curve appears to have infinite length in our introduction to the work of Steinhaus and Perkal in Chapter 2. Clearly Richardson had reached virtually the same conclusion independently of the work done in Poland. Indeed, bearing in mind that this work was published posthumously, Richardson may be considered to have identified the paradox of length some years before Steinhaus. Richardson continues:

(9) Although the phenomena mentioned in (5) and (6) above are peculiar and disconcerting, yet coexisting with them are some broad average regularities, which are summarised by the useful empirical formula

$$L \propto D^{-\alpha} \tag{14.3}$$

where L is the total polygonal length, D is the length of the side of the polygon and α is a positive constant, characteristic of the frontier.

(10) The constant α may be expected to have some positive correlation with one's immediate visual perception of the irregularity of the frontier. At one extreme, $\alpha = 0{\cdot}00$ for a frontier that looks straight on the map. For the other extreme the west coast of Britain was selected, because it looks one of the most irregular in the World; for which α was found to be $0{\cdot}25$. Three other frontiers which, judging by their appearance on the map were more like the average of the World in irregularity, gave $\alpha = 0{\cdot}15$ for the land frontier of Germany in about A.D. 1899; $\alpha = 0{\cdot}14$ for the

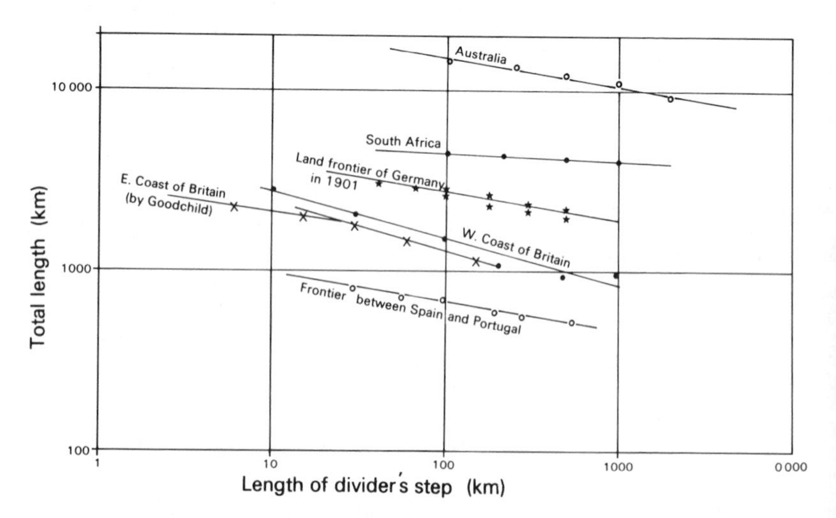

FIG. 14.4. The relationship between L and D obtained by Richardson for the different coastlines measured by him on atlas maps. The lines for the east coast of Britain are those obtained by Goodchild using the same technique. (Sources: Richardson, 1961; Goodchild, 1980).

land frontier between Spain and Portugal; and $\alpha = 0{\cdot}13$ for the Australian coast. A coast selected as looking one of the smoothest in the atlas, was that of South Africa, and for it, $\alpha = 0{\cdot}02$.

(11) The relation $L \propto D^{-\alpha}$ is in marked contrast with the ordinary behaviour of smooth curves for which

$$L = a + bl^2 + cl^4 + dl^6 + \cdots \tag{14.4}$$

where a, b, c, d are constants. This property is used in "the deferred approach to the limit".

Figure 14.4 illustrates the relationship between L and D for the measurements made by Richardson.

Other writers have made measurements which may be added to this list. For example, Goodchild (1980) has measured the length of the east coast of Britain on a map of approximate scale 1/150,000, using six different dividers settings at 5·0, 2·0, 1·0, 0·5, 0·2 and 0·1 inches (127, 50·8, 25·4, 14·7, 5·1 and 2·5 mm, respectively). He found that this series of measurements resolved themselves into two separate lines: with slopes $-0{\cdot}31$ and $-0{\cdot}15$ respectively, which meet at approximately $d = 25$ mm or 1 inch. This is illustrated in Fig. 14.4. Unfortunately he does not provide sufficient information about these measurements so that we cannot pursue this interesting conclusion here.

Galloway and Bahr (1979), whose work on the length of the coastline of Australia has already been examined in earlier chapters, obtained the measurements, shown in Table 14.5, of the mainland coast on maps of four different scales. The three right-hand columns give the coefficients for the logarithmic regression

$$\log L = m \cdot \log D + b \tag{14.5}$$

together with the correlation coefficient, r, derived from the curve-fitting procedure. Thus for the three measurements of the mainland coast of Australia at scale 1/2,500,000, Galloway and Bahr obtained the expression

$$\log L = 4{\cdot}4 - 0{\cdot}11 \log D$$

All the results quoted so far have been based upon the use of extremely large values for d, and often the measurements have been made on very small-scale maps. In order to demonstrate that the principles apply equally to the use of topographical maps and dividers set to small values for d, we provide the

TABLE 14.5. *Coastal lengths of mainland Australia, measured with different intercepts by dividers on maps of varying scales and quality (from Galloway and Bahr, 1979)*

	$D = 500$ km		$D = 1000$ km		$D = 1500$ km				
Scale	d (mm)	L(mm)	d (mm)	L(km)	d (mm)	L(km)	b	m	r
1/15M	33·3	11,800	66·7	10,600	100	9,800	4·525	−0·167	−0·998
1/6M	83·5	11,990	167	10,730	250·5	10,630	4·386	−0·115	−0·954
1/5M	100	11,910	200	10,690	300	10,470	4·401	−0·112	−0·975
1/2·5M	200	12,010	400	10,830	600	10,650	4·383	−0·114	−0·970

TABLE 14.6. *Application of the log–log expression to the Yorkshire coast measurements*

Map scale	L (from Table 4.2)	b	m	r
1/63,360	179·3	2·25355	−0·0269	0·990
1/250,000	176·6	2·24687	−0·0273	0·996
1/625,000	175·6	2·24453	−0·0370	0·962
1/1,000,000	177·8	2·24995	−0·0465	0·980

corresponding values for the Yorkshire coast, using the data given in Table 5.2, p. 68.

In fitting these data to the log–log relationship we have included only those map scales for which there are available five separate measurements.

With the development of methods of measuring distance using digitisers having cartographic format and precision, a certain amount of additional work has been carried out in this field, notably by the present writer, published in Kretschmer (1977) and by Muller (1986). Richardson's expression was further investigated by Beckett (1977), who used the equation

$$l_M/L = \beta \cdot S^{\gamma} \tag{14.6}$$

to relate the "true" length of a line, L, to the measured length l_M which has been measured on a map of scale $1/S$. β and γ are both constants which are supposedly independent of the length of a line having uniform sinuosity. For a line without sinuosity we may assume that $\beta = 1$ and $\gamma = 0$.

Beckett investigated two sets of data: his own chartwheel measurements of ten roads from maps ranging in scale from 1/2,500 through 1/2,000,000, and the Yorkshire coast measurements.

Since the true length of a feature is unknown, Beckett used a bivariate least-square regression analysis of log L_M on log L using the ratios between the measurements of the line at the largest map scale $1/S_M$ and that at another, smaller scale $1/S_N$.

Thus, given a series of measurements l_M, l_{N1}, $l_{N2} \ldots$ made at different map scales $1/S_M$, $1/S_{N1}$, $1/S_{N2} \ldots$,

$$\log(l_N/l_M) = \gamma \cdot \log(S_N/S_M) + \beta \tag{14.7}$$

From his analysis of the five different roads measured by chartwheel, Beckett obtained a value for $\gamma = 0{\cdot}014$. He then indicated, amongst other tentative conclusions, that

$$l_M/L = (1/S_M)^{0{\cdot}014}$$

so that an estimate of "true length" may be derived from a measurement of the feature made on a large-scale map. However, this begs the question concerning the definition of what is a large-scale map. Beckett's work was criticised by Gardiner (1977), who argued that the method of analysis depends for its utility upon the existence of statistically significant correlation between the variables of map scale and measured distance. This was not satisfied by the first set of data, but the correlation coefficients for the Yorkshire coast data are significant at $p < 0{\cdot}01$. However these produced values of γ in the range $-0{\cdot}032$ to $+0{\cdot}051$, and Gardiner was unable to demonstrate whether $-0{\cdot}032 < \gamma < 0{\cdot}051$ differs significantly from zero, or that the differences in significance between the two sets of data reflect differing procedures in cartographic generalisation for roads and coasts rather than the dependence on differences between measurements made by chartwheel and by dividers.

Mandelbrot and Statistical Self-similarity

Mandelbrot (1967) also starts from the assumption that a sea coast may be measured by walking a pair of dividers along it. He describes each step to be of length G and refers to the length of a line measured in this fashion to be $L(G)$. From a consideration of Richardson's result in (14.3) he introduces the equation

$$L(G) = M \cdot G^{1-D} \tag{14.8}$$

where M is a positive constant and D is called the *exponent of similarity*. Note that Mandelbrot's use of D is not the same as that of Richardson. Clearly $D \equiv 1 - \alpha$ so that the exponent of similarity varies between $D = 1$, for a straight line, though $D = 1{\cdot}02$ (South African coast), $D = 1{\cdot}13$ (Australian coast), $D = 1{\cdot}25$ for the irregular western coast of Britain to a maximum value of $D = 2{\cdot}0$. Geographically this would correspond to a lake having such a complicated shoreline that this would occupy its entire area. In practice we may determine D as follows.

Suppose that the length of a line has been measured by two different settings of dividers, d_0 and d_1. The length l_0 is found to contain n_0 steps and l_1 contains n_1 steps. The two estimates of length are $l_0 = n_0 d_0$ and $l_1 = n_1 d_1$. Then

$$D = \log(n_1/n_0)/\log(l_M/l_0) \tag{14.9}$$

Self-similarity is defined as the property of certain curves where each part of the curve is indistinguishable from the whole, or that the form of the curve is invariant with respect to scale. A corollary to this property is that it should not be possible to determine the scale of the curve from its form. Figure 14.5 shows how the simple pattern of straight line elements depicting an irregular curve may be composed of similar elements as a much smaller scale.

If the line behaves as a smooth curve of fixed length, when the sampling interval is halved the number of such intervals will be doubled, and D will be equal to unity, the conventional dimensionality of a line. When a curve is self-similar it is characterised by D, which therefore possesses many properties of a dimension, though it is usually a fraction greater than the dimensions

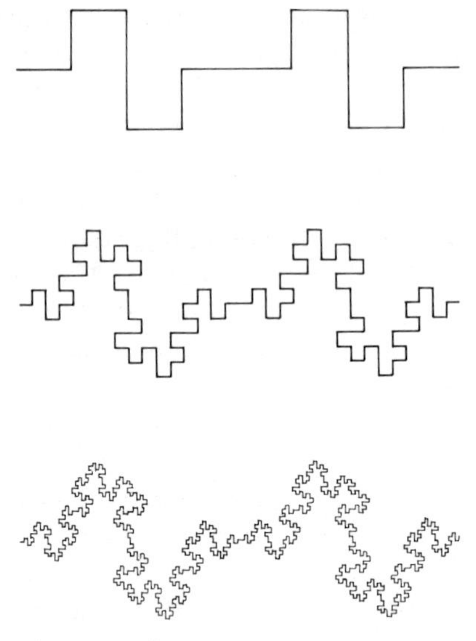

FIG. 14.5. Mandelbrot's concept of a statistical self-similarity applied to a simple linear pattern (top line) which is systematically made more complex in the two following illustrations (middle and bottom) without losing the original character of the line. (Source: Muller, 1987).

commonly attributed to curves. But for an irregular line on a map depicting some natural feature or boundary, the number of intervals corresponding to d will tend to more than double, and D will be greater than unity. The limit to D is provided by lines that grow with scale so as to eventually fill the space and thus have a dimensionality of 2. Mandelbrot coined the term *fractal* to refer to curves and surfaces with n-integer dimensionality. Richardson's measurements, as shown in Fig. 14.4, plot the length of the line against the sampling interval, D, that is $n_1 l_1$ against l_1. Rearranging the definition of D above we have

$$\log n_1 l_1 = a + (1 - D) \log l_1 \tag{14.10}$$

where a is a constant.

Although Richardson's data evidently support the existence of self-similarity, in reality the relation of this to the natural landscape, or maps of it, seems to be limited. Although an irregular line created by self-similarity may seem to be a convincing representation of an irregular coastline to the layman, or even to a mathematician, its appearance is quite unconvincing to the geologist or geomorphologist. Thus the notion that statistical self-similarity bears any relationship to the natural landscape is firmly rejected by Scheidegger (1970), Goodchild (1980) and Håkonson (1978). It is commonly claimed that the surface generated from Mandelbrot's equations has the appearance of a lunar, or other planetary, landscape, but this is surely because we are still somewhat unfamiliar with the detailed composition of these landforms and the processes which have given rise to them. No doubt when we know more about the principles of selenomorphology, Mandelbrot's statistically generated surfaces will appear as unconvincing as are his representations of the terrestrial surface today.

However self-similarity is only aspect of the fractal approach, and it would be unwise to reject the concept entirely. Goodchild has argued that the performance of D and a variable over various scales is a useful index of the nature of a real line shown on a map. This is an approach which was investigated by the present author nearly 30 years ago, in Maling (1962, 1963), who investigated several different coefficients as a means of defining the nature of a coastline having known geomorphological characteristics. In that investigation the use of them had all the appearances of a blind alley, but here it leads neatly into the consideration of the methods of reduction which have been developed in the Eastern Europe, knowledge of which has been slow to filter into the scientific literature of western countries.

The Eastern Contribution to Linear Measurement: N. M. Volkov and His Followers

The alternative approach to the subject was first studied by Volkov (1950), with subsequent modifications of the theory by Chernyaeva (1958), Malovichko (1951, 1958), Znamenshchikov (1963) and Frolov (1964). The history of this aspect of Russian cartometry goes back to the second half of the nineteenth century when Tillo, Strel'bitsky and Shokal'sky measured the lengths of the main rivers of Russia. Shokal'sky recognised the need to make some correction to allow for the separation of the dividers, but his method was derived from making visual comparison between portions of the line with standard curves of known sinuosity and length. This proved to be a somewhat unreliable method of correction, primarily because an irregular river outline cannot be matched to a standard curve for more than an extremely short distance.

It was the work by Volkov, using Shokal'sky's data for certain Asiatic Russian rivers, which put the analysis upon a more secure footing.

Volkov's Correction

Volkov showed that if the total distance measured is plotted against the separation of the dividers used for that measurement, these points lie close to a smooth curve. This is illustrated for some of the Yorkshire coast measurements in Fig. 14.6. If the equation of the curve is known this can be solved for the condition that $x = 0$ to give a corresponding point where the curve meets the ordinate. Since this corresponds in Fig. 14.6 to the condition that $d = 0$, we have determined the length of line which we might expect to measure if we were able to use a pair of dividers whose points were infinitely close to one another.

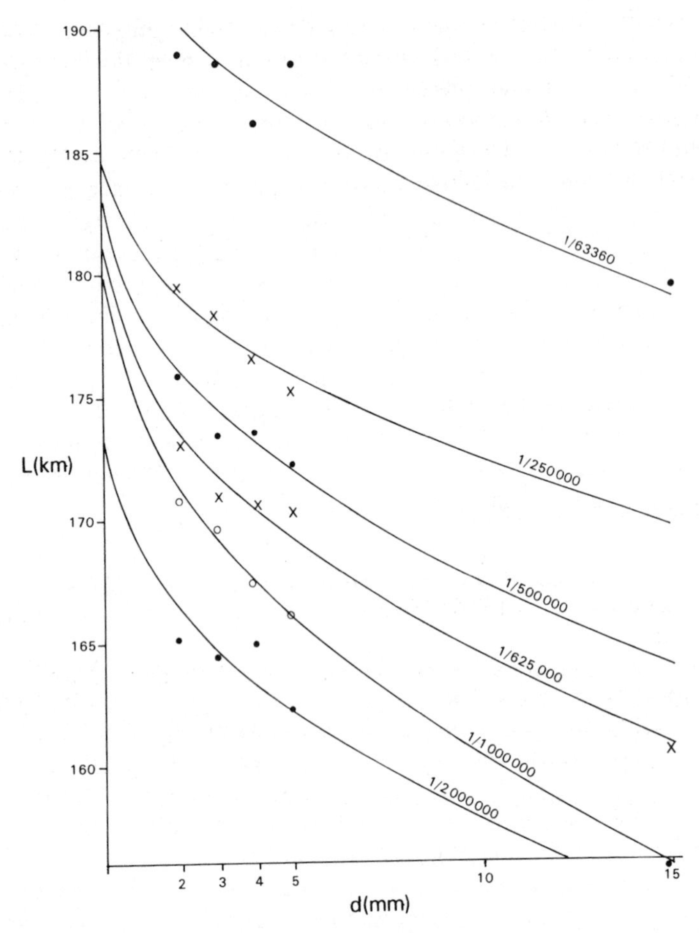

FIG. 14.6. Application to Volkov's fitting of parabolae to the measurements by dividers of the Yorkshire coast. The data are those tabulated in Table 5.2, p. 68. Curves have been calculated for each of the map scales used in the work from 1/63,360 to 1/2,000,000.

Volkov considered this curve to be a parabola of the form

$$l_0 = l_d + \beta \cdot \sqrt{d} \tag{14.11}$$

where l_0 is the length of line corrected for zero separation of the divider points, l_d is the length of the line measured with the dividers set to d units of separation, β is a constant for the line measured and evidently represents some measure of its sinuosity. Numerical results for equation (14.11) from a succession of measurements such as

$$(d_1, l_1), (d_2, l_2) \cdots (d_n, l_n)$$

may be obtained from the procedure for regression analysis which is available on some scientific pocket calculators. For example, using the Texas Instruments TI 54, TI 55-II and similar models it is sufficient to enter $x = \sqrt{d}$ and $y = l$ to find l_0 and β to solve (14.11).

As an example we take the Yorkshire coast measurements at the scale 1/1,000,000 which are as follows:

$x = d$(mm)	$y = 1$ (mm or km)
2	170·51
3	169·54
4	167·52
5	166·09
15	155·91

From these data we find that

$$y = 179{\cdot}653 - 6{\cdot}1032\sqrt{d}$$

so that where $d = 0, l_0 = 179{\cdot}653$. We also find that the correlation coefficient, $r = -0{\cdot}9982$.

There is an even simpler solution because a parabola may be described by the coordinates of two points lying upon it. In this case, therefore, all we require is two measurements of the same feature made on the same map with dividers set to different distances. If we call these two settings d_1 and d_2 respectively, and further specify that $d_1 < d_2$, then $l_1 > l_2$ and

$$l_0 = l_1 + k(l_1 - l_2) \tag{14.12}$$

where

$$k = \sqrt{d_1}/(\sqrt{d_2} - \sqrt{d_2}) \tag{14.13}$$

Although Volkov's method of correction appears to be based upon plausible theory, and it represents an interesting method of overcoming the

TABLE 14.7. *The length of the road (metres) between Montfaucon and St Brais (data from Kishimoto, 1968)*

Dividers setting d (mm)	Map scale 1/10,000	1/25,000	1/50,000	1/100,000
2	–	–	5535·0 (10)	–
4	–	5569·0 (10)	5474·0 (10)	–
8	–	5504·0 (10)	–	–
10	5573·0 (10)	–	–	–
20	5496·0 (10)	–	–	–
Chartwheel	5601·7 (30)	5596·3 (20)	5537·5 (20)	5495·0 (20)
Thread	5589·5 (10)	5582·0 (10)	5587·0 (10)	–

approximations inherent in the use of dividers, it is not always possible t
make this theory fit the facts. We may demonstrate this from the measure
ments made by Kishimoto (1968) for the Montfaucon–St Brais roa
(Table 14.7). If we solve equations (14.12) and (14.13) for the three values of l
we find that these values, which are listed in Table 14.7, exceed th
measurements made by chartwheel by appreciable amount, corresponding (a
shown in Table 14.9, p. 297) to distances of + 15·7 mm, + 5·2 mm an
+ 3·2 mm on the three maps for which dividers measurements are available
Without prejudice to the fact that the chartwheel results may be less precise
and do not in any sense represent absolute values for these distances, it i
inconceivable that a discrepancy in excess of 5 mm can be attributed to error
in manipulating a chartwheel when it has been used to measure a feature whic
has only gentle sinuosities. Therefore we must query the validity of Volkov'
correction when applied to this sort of line. Similar criticism of the Volko
correction has already appeared in the Russian literature.

Malovichko's Correction

Malovichko (1951) was the first to criticise Volkov's method on the ground
that it tended to overcorrect measurements which he had made of a series o
geometrical figures (semicircles and sine curves) whose lengths could b
determined independently by calculation. He proposed a different expression
namely

$$l_0 = l_1 + n_1/6(2a_1 - a_1) \qquad (14.14$$

where l_0 is the corrected length of a line measured by dividers set to d_1 and d
giving the distances l_1 and l_2 respectively; n_1 and n_2 are the number of step
needed to make the measurements and a_1, a_2 are coefficients to be determine
from l and n. Because these coefficients are awkward to calculate, Malovichk
(1958) later modified (14.14) to

$$L = l_1 + \tfrac{1}{3}(l_1 - l_2) - \tfrac{1}{3}k(l_1 - l_2) \qquad (14.15$$

where

$$k = (0{\cdot}5d_2 - d_1)/(d_2 - d_1) \qquad (14.16$$

The coefficient k may be tabulated as shown in Table 14.8 for some typica
ratios between d_1 and d_2. It will be observed that if one setting of the dividers i
double the other, then $k = 0$, the second term on the right-hand side o

TABLE 14.8. *The relationship between the ratio d_2:d_1 and the coefficient k for use in Malovichko's correction formula*

d_2:d_1	1·5	2·0	2·5	3·0
k	− 0·50	0·0	0·17	0·25

equation (14.15) vanishes and the positive to be applied to l_1 correction is reduced to the easily remembered *one-third of the difference between the two measurements.*

Znamenshchikov made further contributions at about this time. However, it is almost impossible to give a proper appraisal of his contribution because of the difficulty in locating copies of his published work. Both Malovichko and Znamenshchikov worked in Novosibirsk and published most of their cartometric material in either the *Works of the Novosibirsk Institute of Engineering Geodesy, Air Survey and Cartography* (Trudy NIIGAiK) or a volume entitled *Questions about Cartography* (Voprosy Kartografii) which was also published by the Novosibirsk Institute. Both of these publications had only small print-runs, so that it seems that the majority of copies were absorbed by libraries within the U.S.S.R. The author has failed to trace any copies of them in the U.K., and it is only because Malovichko published one of his papers in the Moscow periodical Geodeziya i Kartografi-

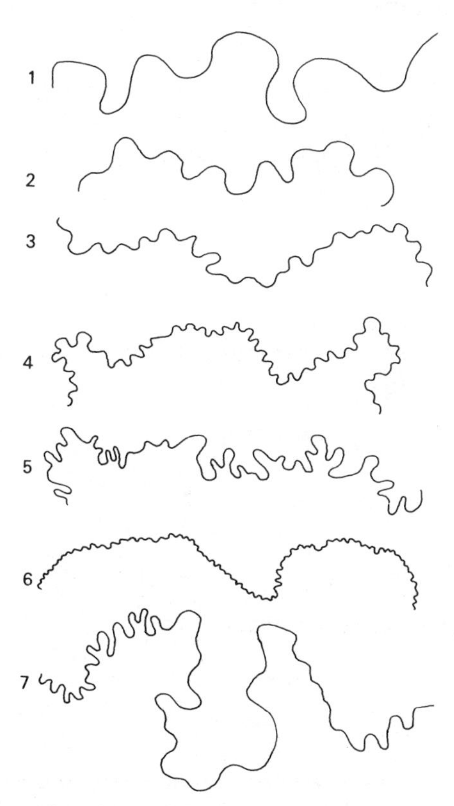

FIG. 14.7. The approximate appearance of the bent wires used by Chernyaeva in her experiments into the corrections to be applied to measurements by dividers.

ya, or *Geodesy and Cartography*, that he reached the wider international audience. The present writer has seen the Avtoreferat, which is the publicly distributed summary by Znamenshchikov of his doctoral thesis, and this is sufficient to whet one's appetite to see more of this work. The works by Chernyaeva and Frolov to be described next were published in Leningrad University periodicals, and these have extensive western circulation.

Chernyaeva's Correction

Chernyaeva (1958) carried out an elegant series of laboratory experiments using images of pieces of wire of known length to represent the lines which she had bent into sinuosities of predetermined magnitude. The pieces of wire were laid upon a sheet of photographic paper to reproduce their images for measurement. Figure 14.7 illustrates the different kinds of line studied. She demonstrated that both the Volkov and Malovichko formulae are liable to produce erroneous corrections for certain kinds of lines. Her own equations are based upon the expression

$$l_c = l_1(l_1 - l_2)k \cdot t \qquad (14.17)$$

where l_c is the corrected length determined from a measured length l_1 made with dividers set to a separation of d_1 units. She varied the value of b between $+2{\cdot}0$ and $+0{\cdot}5$ for increasing sinuosity of line and proposed a correction of the form

$$l_c = l_1 + (l_1 - l_2)l \cdot t \qquad (14.18)$$

where, as usual, l_1 and l_2 are the pair of measurements made with dividers set to d_1 and d_2 units respectively.

For practical use she proposed four different coefficients, k:
For $b = 2{\cdot}0$ (the smoothest curve)

$$k = d_1^2/(d_2^2 - d_1^2) \qquad (14.19)$$

For $b = 1{\cdot}5$

$$k = d_1^{1\cdot5}/(d_2^{1\cdot5} - d_1^{1\cdot5}) \qquad (14.20)$$

For $b = 1{\cdot}0$

$$k = d_1/(d_2 - d_1) \qquad (14.21)$$

For $b = 0{\cdot}5$ (the most irregular curve)

$$k = \sqrt{d_1}/(\sqrt{d_2} - \sqrt{d_1}) \qquad (14.22)$$

The last equation is, of course, identical with the Volkov correction k in (14.13).

The coefficient, t, is a further function of the separation of the dividers and of the number of steps, n_1 and n_2 required to measure the lengths l_1 and l_2 respectively. It provides a measure of sinuosity, thereby overcoming sheer

TABLE 14.9. *Correction of Kishimoto's measurements of the Montfaucon–St Brais road by the Volkov and Chernyaeva methods*

Map scale	Distance by chartwheel	Corrected distances			Differences		
		Volkov	Chernyaeva $b=2{\cdot}0$	Chernyaeva $b=1{\cdot}5$	δ_1	δ_2	δ_3
1/10,000	5601·7	5758·9	5598·7	5616·0	15·7	0·3	1·4
1/25,000	5596·3	5725·9	5590·7	5605·2	5·2	0·2	0·4
1/50,000	5537·5	5682·3	5555·3	5568·9	2·9	0·4	0·6

guesswork based upon visual inspection of the line. For her two conditions of "average" sinuosity, i.e. for $b = 1{\cdot}5$ and $b = 1{\cdot}0$, Chernyaeva has derived the following expressions
For $b = 1{\cdot}5$

$$t = \{[n_1/n_2(d_2/d_1)]^3\}^{1/2} \tag{14.23}$$

For $b = 1{\cdot}0$

$$t = n_1/[n_2(d_2/d_1)] \tag{14.24}$$

If we employ these corrections to the Montfaucon–St Brais measurement we obtain the values which are given in Table 14.9. In this table the three columns labelled δ_1, d_2 and δ_3 represent the differences (in millimetres) between the corrected distances based upon the dividers measurements and the results for the same map scale obtained by chartwheel. The difference δ_1 corresponds to the use of the Volkov correction, δ_2 corresponds to the Chernyaeva correction using $b = 2{\cdot}0$, and δ_3 is that for $b = 1{\cdot}5$. It must be emphasised that the chartwheel results are not to be construed as a standard, for we are comparing more precise dividers measurements against the less precise chartwheel. However, these figures do illustrate the way in which the Volkov correction produces an overestimate of the length, and how closely the Chernyaeva correction for a smooth curve matches the independent measurements.

Applying the Malovichko correction to the same measurements we find that for both 1/10,000 and 1/25,000 scales the corrected distances are the same obtained from the Chernyaeva correction for $b = 2{\cdot}0$.

Frolov's Correction

Another form of correction was proposed by Frolov (1964). He started from the premise that an irregular line could be imagined as being composed of a multitude of circular arc elements, and he established an integrated arc–chord relationship to convert from two linear measurements, l_1 and l_2 made from the same map with dividers set to d_1 and d_2 respectively. As before, we assume that

TABLE 14.10. *The corrections to be applied in the determination of* Δ *according to Frolov's correction*

$t = d_2/d_1$	K_1	K_2	K_3	$t = d_2/d_1$	K_1	K_2	K_3
1·25	1·778	2·346	8·397	2·3	0·233	0·171	0·489
1·33	1·286	1·505	4·996	2·4	0·210	0·152	0·432
1·5	0·8	0·795	2·448	2·5	0·190	0·135	0·386
1·67	0·559	0·896	1·476	2·6	0·174	0·122	0·346
1·8	0·446	0·376	1·098	2·7	0·159	0·110	0·313
1·9	0·383	0·311	0·901	2·8	0·146	0·100	0·285
2·0	0·333	0·263	0·756	2·9	0·135	0·092	0·260
2·1	0·293	0·225	0·645	3·0	0·125	0·084	0·239
2·2	0·260	0·195	0·558				

$d_2 > d_1$ so that $l_1 > l_2$. Putting the ratio

$$t = d_2/d_1 \tag{14.25}$$

we determine a series of terms K_1, K_2, K_3 and use these to determine a correction Δ. After some rather complicated algebra these terms are determined from

$$K_1 = 1/(t^2 - 1) \tag{14.26}$$

$$K_2{}^1 = 0{\cdot}585((t^3 - 1)/(t^2 - 1)](1/t^2 - 1)^{3/2} \tag{14.27}$$

and

$$K_3 = 1{\cdot}7\, t^2(1/t^2 - 1)^2 \tag{14.28}$$

We may now determine Δ from

$$\Delta = K_1(1 - l_2/l_1) + K_2(1 - l_2/l_1)^{3/2} - K_3(1 - l_2/l_1)^2 \tag{14.29}$$

and

$$l_0 = l_1 + \Delta l_1 \tag{14.30}$$

Since t and therefore the three K corrections depend wholly upon the ratio between the dividers used, the common values for these may be tabulated once and for all.

In order to test the merits of the different methods of determining the limiting distances the Yorkshire coast measurements have been subjected to a large number of different corrections. All the principal methods described in this section have been used, and these have been carried for for every possible combination of dividers settings d_1 and d_2. Therefore it has been possible to treat statistically with the various determinations of limiting distance. These results may be tabulated as shown in Table 14.11, which indicates a fair degree of consistency between the limiting distances derived from all except the Volkov method, which gives consistently larger values for l_0. The reader is invited to test which of the differences between these limiting distances are statistically significant using the method outlined on pages 120–122.

TABLE 14.11. *Yorkshire coast measurements by dividers; summary of limiting distances based upon the different methods of reduction proposed by Volkov, Malovichko, Chernyaeva and Frolov. Limiting distance and standard error of the single observation in km. Figure in parentheses indicates the number of separate determinations made by each method for each map scale*

Method	Map scale 1/25,000 (1)	1/63,360 (10)	1/250,000 (10)	1/625,000 (10)	1/1,000,000 (10)
Volkov	200·68	195·87 ± 3·47	186·16 ± 2·53	178·73 ± 4·01	179·37 ± 2·16
Malovichko					
Chernyaeva	192·20	189·10 ± 1·51	179·52 ± 1·55	173·06 ± 1·40	171·05 ± 1·52
$b = 1·0$	194·75	191·09 ± 2·14	181·79 ± 2·31	174·29 ± 1·94	173·41 ± 1·21
$b = 1·5$	192·85	189·7 ± 1·81	180·03 ± 1·68	173·00 ± 1·48	171·46 ± 1·36
$b = 2·0$	191·96	189·25 ± 1·19	179·65 ± 3·39	172·34 ± 1·38	170·45 ± 1·51
Frolov	192·55	189·49 ± 1·64	179·93 ± 1·53	173·13 ± 1·53	171·60 ± 1·27

THE INFLUENCE OF MAP SCALE: THE REDUCED LENGTH

Ever since we introduced Penck's measurements of the coastline of Istria, in Table 5.1, we have been preoccupied with the effect of map scale upon measured distance. Quite simply the larger the scale of the map the greater is the measured length of the line. We therefore need to investigate whether it is possible to treat with map scale in a fashion similar to that adopted with dividers measurements to determine the *limiting distance.*

Volkov (1950b) recognised the relationship between distance and map scale, and he introduced the concept of *reduced length* to be the solution of this expression, for the condition that this represents the distance which would be measured if it were possible to do this on a map of scale 1:1. Equations (5.1) and (5.2) were those proposed by him, and correspond to the assumption that the relationship between distance and scale is a parabola. Thus to obtain the reduced length for a particular feature it is sufficient to measure it on two maps having different scales.

For example, the Yorkshire coast values for scale 1/25,000 and 1/625,000 are as follows:

Scale $1/N_1 = 1/25{,}000$, limiting distance $l_2 = 192·55$ (by Frolov method)
Scale $1/N_2 = 1/625{,}000$, limiting distance $l_2 = 173·13$ (by Frolov method)

$$\sqrt{N_1} = 158·114, \sqrt{N_2} = 790·569$$
$$t = \sqrt{N_1}/(\sqrt{N_2} - \sqrt{N_1}) = 158·114/632·455 = 0·25$$
$$l_1 - l_2 = 19·42$$
$$t(l_1 - l_2) = 4·855$$
$$L_0 = 192·55 + 4·855 = 19·405.\text{km}$$

Nevertheless this kind of solution does not appear to provide a complete answer. In an example for which there are many different solutions the

TABLE 14.12. *The limiting distances for the Yorkshire coast determined from the Frolov formula and used to calculate the reduced lengths in Table 14.3*

Scale	Limiting distance	Scale	Limiting distance
1/25,000	192·55	1/1,000,000	171·60
1/63,360	189·49	1/2,000,000	166·63
1/250,000	179·93	1/2,500,000	165·07
1/500,000	176·68	1/6,000,000	159·48
1/625,000	173·13		

TABLE 14.13. *The reduced lengths for the Yorkshire coast, based upon Volkov's correction*

	$1/N_2$							
$1/N_1$	63,360	250,000	500,000	625,000	1 *M*	2 *M*	2·5M	6 *M*
25,000	197·72	198·39	197·12	197·41	196·49	195·81	195·60	194·83
63,360	–	199·18	196·57	197·13	195·51	194·44	194·11	192·93
250,000	–	–	187·78	191·63	188·26	187·20	186·80	185·17
500,000	–	–	–	206·75	188·94	186·73	186·07	183·66
625,000	–	–	–	–	178·91	181·37	181·19	186·14
1,000,000	–	–	–	–	–	183·60	182·84	179·96
2,000,000	–	–	–	–	–	–	179·85	176·40
2,500,000	–	–	–	–	–	–	–	175·25

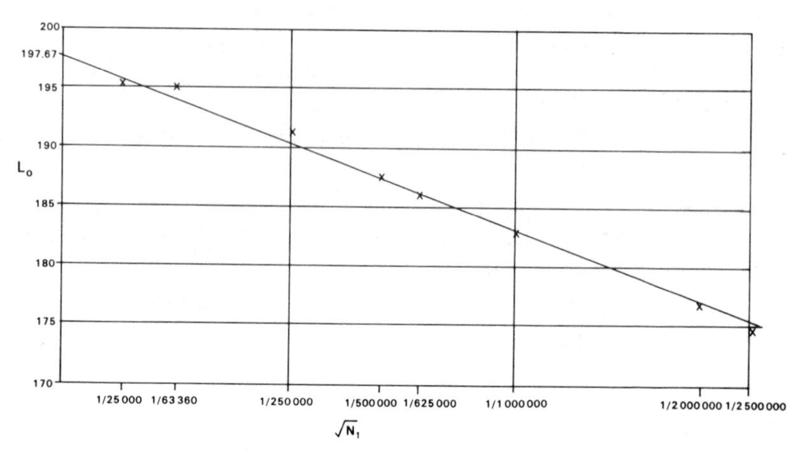

FIG. 14.8. The relationship between $\sqrt{N_1}$ and L_0 for reduced dividers measurements for the coastline of Yorkshire.

calculated reduced lengths may lie within an appreciable range of values. We may demonstrate this with reference to the Yorkshire coast experiment because nine different map scales were used and consequently there are 36 possible solutions. In order to standardise the data we use the limiting distances calculated for the different map scales corrected by the Frolov method. These values are shown in Table 14.12. Table 14.13 lists the results to be obtained from these data. We find that the range in the reduced lengths is from 206·75 km to 175·25 km, and even if we exclude the first of these values as being an unlikely result, the range 199·18–175·25 = 23·9 km or more than 12% of the length of the line. If we now compare the calculated values for reduced length with $\sqrt{N_1}$, for the larger map scale in each determination, we find there is a linear relationship between $\sqrt{N_1}$ and L_0 as illustrated in Fig. 14.8. Regression analysis demonstrates that this line is expressed by

$$y = 197{\cdot}67 - 0{\cdot}01456x$$

so that the line intersects the ordinate at the reduced length $L_0 = 197{\cdot}67$ km.

Håkonson's Methods of Reduction

A third approach to the method of determining a reduced length is that described by Håkonson (1978). This is sufficiently different from the other methods to need a separate description.

Because of the important limitation that its application is restricted to use with *closed curves*, i.e. the whole coastline of an island or the shoreline of a lake, it does not have the general applicability of the methods already described, although it is no doubt adequate for limnological work.

We have seen in Chapter 3 that Håkonson used a probabilistic method of measuring distance. It will therefore suffice to remark that there is no reason why the methods of reduction should not be applied to conventional measurements. Håkonson starts with the three quantities

- l, the measured length of the closed curve;
- $1/S$, the scale of the map used for measurement;
- A, the area of the feature enclosed by the line of length l.

It is the last of these measures, representing an important element of his analysis, which restricts the application of the method to closed curves bounding features of known area, A.

From these quantities he derives a measure of generalisation which he calls the *shore line development*, F,

$$F = l/2\sqrt{(\pi \cdot A)} \tag{14.31}$$

and from it he derives the measure, NF, the *normalised shore development*.

He now proceeds to establish a general relation between S, l, F and A. He obviously realised the importance of the concept of a map of scale 1:1, for he

TABLE 14.14. *Håkonson's measurements for Lake Vänern, Lake Vättern and Lake Mäleren, with comparative values for the reduced length*

Scale	Measured lengths		
	L. Vänern	Lake Mäleren	Lake Vätter
1/10,000	2,007	646	1,429
1/50,000	1,875	592	1,375
1/60,000	1,893	–	–
1/100,000	–	564	1,251
1/180,000	1,642	–	–
1/250,000	1,608	507	1,118
1/500,000	1,325	460	981
1/1,000,000	1,012	349	734
Reduced lengths			
Håkonson	1,943	642	1,413
Volkov	2,036	667	1,518

puts $F = NF$ for the map scale $S = 1$. For all other scales he found that the best empirical fit was

$$NF = F[K_2 - \log(s + a)]/(K_2 - K_1) \tag{14.32}$$

where: $a = 10^5 \cdot \log A$

$$K_1 = \log(1 + 10^5 \cdot \log A)$$

$$K_2 = \log(60{,}000{,}000 + 10^5 \cdot \log A)$$

In these expressions there are two empirical constants; the *scale constant* ($= 6{,}000{,}000$) and the *area constant* ($= 10^5$). The value of NF varies between 1 and approximately 10. For a circle, $NF = 1$ and for Lake Mäleren, described by Håkonson as "one of the most irregular lake basins in the world", $NF = 9{\cdot}95$.

Håkonson now introduces the *normalised shore length*, as he calls it, at scale 1:1 as

$$l_n = l(K_2 - K_1)/[K_2 - \log(s + a)]$$

In order to study the accuracy of the method, Håkonson derived all possible values for NF and used the mean of them, $\overline{NF}$, to find the percentage value of z

$$z = 100(NF - \overline{NF})/NF \tag{14.33}$$

Because of the limitation of his method to enclosed curves we cannot compare Håkonson's correction, especially the normalised shore length, for any of the data which we have hitherto used. However, we may apply the Volkov correction to Håkonson's own data and thereby obtain some idea of how his results compare with those analysed here. Håkonson measured the shores and areas of 12 Swedish lakes ranging in area from 5,893 km^2 to only 1·1 km^2. For the majority of these lakes he measured the length of the shore on five or six different map scales. The largest scale used by him was invariably 1/10,000; the smallest was 1/500,000 or 1/1,000,000 for all except the smallest examples. Table 14.14 provides measurements for the three largest lakes which he studied. The corresponding reduced lengths determined from equations (4.1) and (4.2) are also given in Table 14.14 as the Volkov reduced lengths. The inadequacy of the Håkonson reduction to the scale 1:1 is demonstrated here because the reduced lengths are shorter than the distances measured at the scale 1/10,000.

15

Probabilistic Methods of Distance Measurement

The mathematical ability evinced by BUFFON may well excite surprise; that one whose life was devoted to other branches of science should have had the sagacity to discern the true mathematical principles involved in a question of so entirely novel a character, and reduce them correctly to calculation by means of the integral calculus, thereby opening up an entirely new region of enquiry to his successors, must move us to admiration for a mind so rarely gifted.

(M. W. Crofton, *Philosophical Transactions of the Royal Society*, 1868)

In 1777 Georges Louis LeClerc, Compte de Buffon, described and explained the mathematical theory of a device which is now known as *Buffon's needle.* The nature of the original game and its application to probabilistic methods of measuring the length of a line on a map has already been mentioned in Chapter 3. Here we must consider the theory in greater detail in order to justify its cartometric or stereological uses.

"Clean tile" was originally a game in which players threw a stick upon a bare floor and wagered whether the stick would fall cleanly upon a single tile or plank or intersect one or more of the lines formed by the edges of the tiles or floorboards. In making the investigation of the probability of obtaining a "clean tile", Buffon created the branch of statistics known as *local* or *geometrical probability.*

Consider, first, the case where the stick is shorter than the distance between the cracks in the floorboards, or, in algebraic terms, $l < d$. If we designate each trial as X, we claim a "success" if the needle falls across one line and put $X = 1$. If the needle does not intersect a line we put $X = 0$ to represent a "failure". After k throws, or applications of the needle, the sum

$$\sum X = n \tag{15.1}$$

represents the total number of successes. Thus the relative frequency of the number of successes is n/k and, if k is large, this represents the probability that the needle intersects a line and

$$p = n \cdot k \tag{15.2}$$

Figure 15.1 illustrates one such success. If the centre of the needle falls at a distance x from one of the lines and it makes an angle θ with that line, the

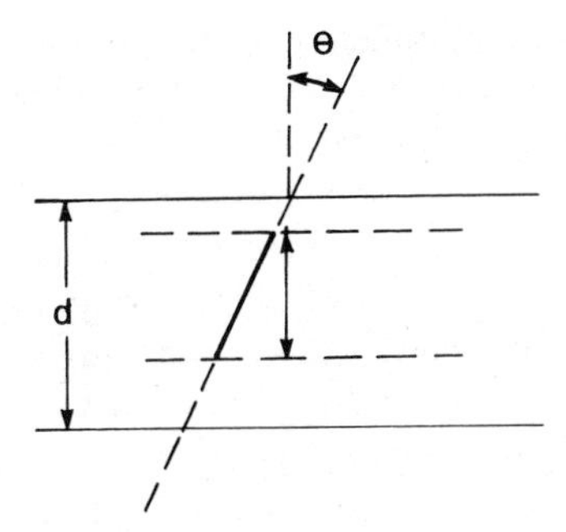

FIG. 15.1. The theory of Buffon's needle applying to a short stick of length l (thick line).

needle will intersect it if

$$x < (l \cdot \sin\theta)/2 \qquad (15.3)$$

Assuming that all the values of x from 0 to $d/2$ are equally likely, the expectation the needle will intersect a line is

$$E(X) = l \cdot \sin\theta / d \qquad (15.4)$$

Hence the probability for all values of θ is

$$\left[\int_o^\pi (l \cdot \sin\theta \cdot \mathrm{d}\theta)/d\right] \Big/ \int_o^\pi \mathrm{d}\theta = 2l/\pi \cdot d \qquad (15.5)$$

and since the player throws the needle k times and records n successes,

$$p = 2l/\pi \cdot d \qquad (15.6)$$

TABLE 15.1. *Determinations of the value of π using Buffon's needle. In the right-hand column there are the corresponding determinations of the length of the needle, l, assuming the customary value for π*

Experimenter	Needle length (units of d)	No. of throws (k)	No. of successes (n)	Estimate of π	l (calculated) (units of d)
Wolf, 1850	0·80	5,000	2,532	3·1596	0·795
Smith, 1855	0·6	3,204	1,218·5	3·1553	0·597
DeMorgan, c. 1860	1·0	600	382·5	3·137	1·001
Fox, 1884	0·75	1,030	489	3·1595	0·746
Lazzerini, 1901	0·83	3,408	1,808	3·1415929	0·833
Reina, 1925	0·5419	2,520	859	3·1795	0·535
Gridgeman, c. 1960	0·7857	2	1	3·143	0·785
Kahan, 1961	0·5387	3,000	1,042	3·046	0·546
Kahan, 1961	0·2693	3,000	514	3·075	0·269
Kahan, 1961	0·2693	3,000	527	3·002	0·276
Kahan, 1961	0·808	3,000	1,586	3·153	0·83
Kahan, 1961	0·404	3,000	782	3·185	0·409
Kahan, 1961	0·404	3,000	804	3·098	0·421

In order to demonstrate experimentally the validity of this result, mathematicians have been accustomed to use equation (15.6) to find a value for π. Table 15.1 is a summary of the results obtained by various investigators. However, Mantel (1953) has observed that nearly all the results which had been recorded were closer to the accepted value for π than might reasonably be expected, and suggested rather cynically that only those experiments which had provided good results had been published. Alternatively, the experiment had been terminated when the results obtained were exceptionally good. A typical example is Lazzerini's experiment. If this had been terminated one throw sooner or later, one half of the decimal places in the tabulated result would have been lost. An important exception to this criticism is the last set of determinations published by Kahan (1961) which were carried out by students and visitors to the Regent Street Polytechnic on the occasion of the college's Open Day in 1959 and represented a carefully monitored experiment.

LAPLACE'S EXTENSION TO THE THEORY OF BUFFON'S NEEDLE

An extension to the theory was made by Laplace in 1812, who proposed the addition of a second family of parallel lines perpendicular to the first set and forming an orthogonal square or rectangular grid. If the sides of the grid are a and b, and $l < a$, $l < b$, the procedure can be repeated. This time it can be shown, e.g. by Solomon (1978), that the probability that the needle will intersect the perimeter of one of the rectangles is

$$n/k = p = \{2l(a+b) - l^2\}/\pi \cdot a \cdot b \tag{15.7}$$

In the special case of a square grid, where $a = b = 1$,

$$n/k = \{4l - l^2\}/\pi \tag{15.8}$$

It might be argued that if A represents the event that the needle intersects the family of lines separated at distance a, and if B represents the event that the needle intersects the family of lines separated by distance b, the combination of the two probabilities $p(A)$, and $p(B)$ should make the method twice as efficient as the use of Buffon's needle with only a single family of lines. However, Schuster (1974) has pointed out that this argument is only valid if A and B are independent events. This is not true, because it can be shown that the numbers of successes n_A and n_B are negatively correlated and therefore dependent. Working with $d = 2l$ as the relationship between the square grid and the needle, Schuster was able to show that 100 independent observations made with respect to a square grid contain approximately the same information about π as 222 observations made on a family of parallel lines. For $d = l$, one throw of the needle to a square or rectangular grid contains at least 3·2 times the statistical information about the value of π as one throw on one set of lines. The implication is that, as l is increased with respect to d, so the efficiency of the measurement system increases.

Since the theory of Buffon's needle involves the simple division of an event X into either success or failure, the binomial distribution applies to any continued set of trials and the calculation of variance and other parameters. For a single trial the variance of X may be written as

$$\mathrm{Var}(X) = p(1-p) \text{ as in equation (6.23) for } n = 1.$$

It follows that the coefficient of variation is

$$\mathrm{C.V}(X) = \sqrt{(1/p - 1)} \tag{15.9}$$

and since we know p from (15.6),

$$\mathrm{C.V.}(X) = \sqrt{(\pi d/2l - 1)} \tag{15.10}$$

In the special case where $d = 1$ this simplifies to

$$\begin{aligned} \mathrm{C.V.}(X) &= \sqrt{(\pi/2 - 1)} \\ &= 0{\cdot}7555 \end{aligned} \tag{15.11}$$

It can also be shown that this is the minimum value for the coefficient of variation which can be obtained.

For the Laplace extension of the problem,

$$\mathrm{C.V.}(X) = \sqrt{\{\pi ab/(2l(a+b) - l^2) - 1\}} \tag{15.12}$$

which is minimised for $d = k = l$ to give the value

$$\begin{aligned} \mathrm{C.V.}(X) &= \sqrt{(\pi/3 - 1)} \\ &= 0{\cdot}2172 \end{aligned} \tag{15.13}$$

This result also shows that for a rectangle of given area, ab, where a and b are both greater than l, the lowest coefficient of variation is obtained by using the most elongated rectangle.

Buffon's Needle where $d > l$

Thus far we have assumed that $l < d$, so that one trial can only yield $X = 0$ or $X = 1$. With a longer needle the possibility of multiple intersections arises. The theory of the long needle has been investigated by Mantel (1953), Diaconis (1976) and others. Their work has been summarised by Solomon (1978).

This time we start from the expected number of intersections for a square grid, which Mantel writes as

$$E(X) = 4l/\pi \tag{15.14}$$

so that π may be determined from the expression

$$\pi = 4l/E \tag{15.15}$$

Assuming k trials ($k = 1, 2, \ldots, n$) with c_k intersections of the grid at the kth trial, the average number of intersections per trial is

$$\bar{c} = c_k/n \tag{15.16}$$

so that the estimate of π becomes

$$\pi = 4l/\bar{c}$$

Mantel estimates the variance in the determination of π for both the square and parallel systems of grid, and concludes that the precision in estimating π from the use of long needle on a square or rectangular grid is about 60 times greater than the precision obtained using the original Buffon needle technique with a needle of length $l = d$.

CARTOMETRIC APPLICATIONS OF THE THEORIES OF BUFFON AND LAPLACE

The arguments presented in the preceding section have been verified using the classic mathematical preoccupation with the experimental determination of the value of π. Suppose that we invert the problem and use the method to find the length of the needle. Rearrangement of equation (15.6) gives

$$l = (\pi/2)\cdot p\cdot d \tag{15.17}$$

To make the problem even more realistic we assume that the needle is a short straight line fixed in a plane, as on a map or microscope slide, and that the pattern of lines corresponding to the grid can be placed at random over this surface. After k applications of the grid we have counted a total of n intersections of the needle with the parallel lines. Since p is the relative frequency n/k as demonstrated by equation (15.2), we have

$$l = (\pi/2)(n/k)d \tag{15.18}$$

Using the values listed in the first four columns of Table 15.1 we may calculate

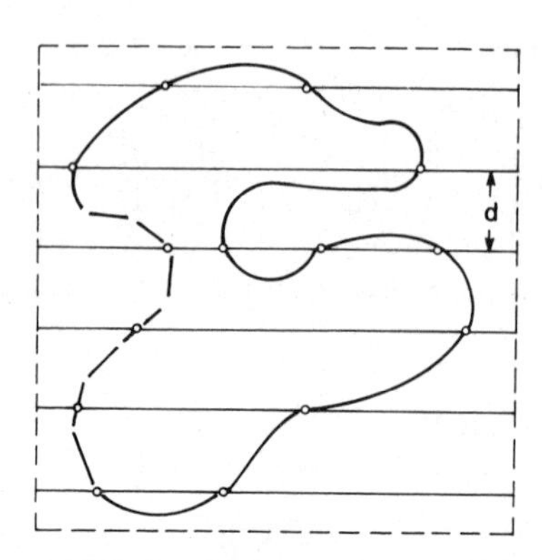

FIG. 15.2. Measurement of the length of the perimeter of an enclosed figure by the application of Buffon's needle of length l on parallel lines of separation d.

the values for the length of the short straight line, l. The results are given in the right-hand column of Table 15.1. Evidently the method works for a short straight line, but this could be measured with far less effort simply by placing a scale alongside it. Is there any possibility of using the same method to measure more irregular lines. Figure 15.2 illustrates an object bounded by a convex perimeter, which might be the outline of an organ or cell as seen under a microscope, or an island or lake shown on a map.

We imagine the perimeter to be composed of a number of short straight lines, δl. Then the number of throws must be

$$k = l/\delta l \tag{15.19}$$

and the mean number of intersections in the sample of k applications is

$$\begin{aligned} c &= n \cdot p(c) \\ &= (l/\delta l)(2\delta l/\pi d) \\ &= 2l/\pi d \end{aligned} \tag{15.20}$$

from which it follows that the length of the whole outline of the parcel is

$$l = \tfrac{1}{2}\pi d c \tag{15.21}$$

The corresponding value for a rectangular grid having sides a and b is

$$c = 2l/\pi\{a^{-1} - 2l(a \cdot b)^{-1} + b^{-1}\} \tag{15.22}$$

The most immediate practical use of Buffon's needle expressed in this form is in quantitative microscopy, where the linear element L_A, or total linear extent of a perimeter per unit area of a section revealed on a microscope slide, may have quantitative significance in petrology (e.g. Tomkieff, 1945) or metallurgy (e.g. Smith and Guttman, 1953).

Steinhaus' Longimeter

Although Steinhaus presents a closely reasoned derivation of the theory of measurement using his longimeter, the method is little more than a small modification of Buffon's needle. The only really original contribution appears to have been recognition of the order of the measurement. Steinhaus starts from a consideration of *Crofton's theorem*, which states that

> The measure of the number of random lines which need a given closed convex plane boundary, is the length of the boundary
>
> (Crofton, 1868)

This may be written in the form

$$\text{Length of } A = \frac{1}{2}\sum_{k=1}^{\infty} k|A_k| \tag{15.23}$$

where the arc A is intersected by A_k lines cutting A exactly in k places. It follows from (15.23) that since the number of intersections is proportional to the length of the arc, a repetition of the measurement increases the accuracy in the sense that it diminishes the stochastic error.

A measurement of the length of order n occurs with the aid of the Steinhaus longimeter just as in the classical Buffon experiment, with the one difference that for every straight longimeter no more than n points of the intersection of the line with the measured curve are counted. If the straight lines intersects the curve less than n times, or n times, then all of the points of intersection of the line and curve are counted. If, however, any line intersects the curve more than n times, it is accepted that this line intersects the measured curve only n times. As Steinhaus himself commented:

> The computation of the length of order m by means of the transparent sheet is perfectly simple: we have to count the intersections of every line L_i with A; if their number a_i exceeds m we have to replace a_i with m: practically this means that we shall stop counting when m is reached.

The longimeter comprises a transparent sheet with a family of equidistant parallels $L_i(i = \cdots -2, -1, 0, 1, 2, \ldots)$

The arc A to be measured cuts L_i in a_i points and

$$s = \sum a_i \tag{15.24}$$

where $\sum a_i$ is the number of intersections which the line makes with the parallels. Turning the sheet through an angle $\pi \cdot k/m(k = 0, 1, \ldots, m-1)$ we get s_k intersections and the grand total is

$$N = \sum s_k \tag{15.25}$$

Calling d the distance between the parallels or

$$d = L_i - L_{i+1} \tag{15.26}$$

we get the expression

$$L_{(a)} = Nd\pi/2m \tag{15.27}$$

as an approximation of the length of A. In this equation, m is effectively the number of applications. Steinhaus believed that, in order to overcome the paradox of length, we only have to stop the summation of series (15.23) at the mth term and write

$$\text{Length of order } m \text{, of } A = \frac{1}{2}\sum_{k=1}^{m} k|A_k| \tag{15.28}$$

The length of mth order is approximately the classical length in the sense that it tends towards the classical length when N increases to infinity. The length of order m serves for all empirical curves and is easy to measure. It would appear therefore that the problem is solved, and that the length of order m instead of classical length can be introduced conventionally.

In order to demonstrate the relationship between m, the order of measure-

TABLE 15.2. *The length of the Yorkshire coast at 1/1,000,000 measured by Steinhaus' longimeter*

Order (m)	Sample no. (1)	(2)	(3)	(4)	(5)	(6)	(7)	(8)	(9)	(10)
1	150·8	135·1	147·6	160·2	201·1	207·3	229·3	153·9	175·9	204·2
2	138·2	150·8	174·4	179·1	193·2	216·8	224·6	147·7	138·2	166·5
3	145·6	155·0	191·6	170·7	175·9	220·9	215·7	160·2	157·1	141·4
4	164·1	165·7	184·6	161·0	184·6	211·3	202·6	163·4	175·9	141·4
5	177·2	177·2	188·5	162·7	192·9	193·5	187·9	175·3	174·7	152·2
6	185·3	167·6	194·2	169·1	190·6	181·7	175·4	184·3	182·2	163·4
7	184·5	167·4	190·7	177·7	182·2	173·7	175·0	189·4	173·7	173·2
8	175·9	159·8	184·2	185·3	181·2	168·5	182·6	194·4	179·5	180·2
9	170·0	164·1	178·0	184·0	186·7	168·2	183·3	190·9	181·5	177·7
10	164·0	159·9	173·1	188·8	184·4	173·4	180·3	193·8	183·5	172·1

ment and the result, the following experiment has been carried out. The line measured is that of the Yorkshire coast on a map of scale 1/1,000,000. From Table 4.3, the length of the line on the I.M.W. sheet measured by dividers with $d = 2$ mm is 170·51 km (or mm). A longimeter comprising a family of parallel straight lines at separation $d = 2$ mm. The measurement was repeated 100 times, in ten samples of ten measurements each. The length of order m was determined for each sample in the sequence $m = 1 \cdots 10$. The results are tabulated as shown in Table 15.2.

Since the length of the line, of order 1, depends upon the result of the first measurement of the series and that of order 2 depends upon the first two measurements, it is not surprising that the lower-order measurements show the greatest fluctuations and are characterised in Table 15.3 by the larger values for the standard error. We see that the length of the line increases with

TABLE 15.3. *Mean and standard error of the mean for each order of measurement (recorded in the rows of Table 15.2)*

Order	Mean length (km or mm)	Standard error of the mean (km or mm)
1	176·54	±10·05
2	172·95	±9·79
3	173·31	±8·74
4	175·46	±6·61
5	178·26	±4·17
6	179·36	±3·22
7	178·75	±2·42
8	179·16	±3·00
9	178·44	±2·73
10	177·33	±3·40

increasing order, corresponding to the increased refinement in measurement resulting from a larger number of measurements. At the same time the standard error decreases, but for orders 8 to 10 there is a small fluctuation in both the measured length and the standard error. Steinhaus suggested the use of $m = 6$ for $d = 2$ mm, which gives results nearly as good as that for $m = 7$. It is not clear whether he hit on his recommended values by chance, or empirically, but this small experiment suggests that this is somewhere near the value of m which we need. We see that, for $m = 7$, the distance measured by longimeter is 8·24 mm longer than that measured by dividers. However it has been shown in Table 14.11 that at this scale the limiting distance determined by the Volkov method is 179·37 mm, so that in this case the result is only about 0·5 mm different from the limiting distance.

Perkal's Longimeter

Perkal describes the concept of length of order ε, this being a real number which corresponds physically to the radius of a circle. If this circle is placed repeatedly over an irregular line its circumference creates an envelope of arcs which lie at the distance ε either side of the line. Perkal calls this figure the *ε-aureole*. The ε-aureole of a curve X is the set of all points on the plane for which the distance from the curve is not greater than ε. In set notation

$$A_\varepsilon(X) = E_x[d(x, X) \leqslant \varepsilon] \tag{15.29}$$

where $d(x, X)$ is the distance from a point x to the nearest element of the set X.

The set $A_\varepsilon(X)$ may be regarded as a function of ε and X. It is monotonic increasing and continuous with respect to both arguments. The set $A_\varepsilon(X)$ may be regarded as defining the area within the ε-aureole, $a_\varepsilon(X)$ which is therefore also a continuous, monotonic increasing function of ε and X.

Line Convexity

It is necessary to distinguish between comparatively simple sinuous curves and irregular jagged curves when both are, strictly speaking, empirical curves. Perkal developed the concept of ε-convexity for this purpose and, in so doing, made an important contribution to the study of generalisation of lines such as is necessary in all forms of cartographic work. He defines ε-convexity as the condition when we may draw, at any point on either side of a line, the tangent circle having the point of tangency as the only point of contact between the line and the circle. It follows that an arc is ε-convex if every point on it has a radius of curvature of not less than $\varepsilon/2$. In other words we may imagine the line as an undulating, but smooth road surface upon which the circle of radius ε rides, like a wheel, always making contact at one place only. A line which is not ε-convex is a more irregular surface upon which the wheel makes more than one contact. The idea of replacing such irregularities with a circular arc of radius ε is fundamental to Perkal's concept of generalisation. For curves of 2ε convexity (with the ends separated by a distance of at least 2ε) the length of order ε is equal to the classical length. For other curves the length of order ε is equal to or less than the classical length.

Measurement by Perkal's Longimeter

In the simplest form, as illustrated by Fig. 15.3, the ε-aureole has the appearance of a sausage. The area $a_\varepsilon(X)$ of $A_\varepsilon(X)$ comprises three separate elements. First there is the strip extending ε units either side of the line X. If the line X is straight, the area of this element is that of a rectangle of length L_ε and width 2ε. The two other elements are the semicircles formed at the ends of the line X, corresponding to the rounded ends of the sausage. These combine to form a circle of radius ε, so that the area of the two ends is $\pi\varepsilon^2$. In order to measure the length of the line X we measure the area $a_\varepsilon(X)$, then we subtract the ends of the ε-aureole and divide the remainder by the width of the ε-aureole. Thus

$$L_\varepsilon(X) = \{a_\varepsilon(X) - \pi\varepsilon^2\}/2\varepsilon \qquad (15.30)$$

It should be evident that this equation is suitable for open-ended curves. If the line (X) is a continuous feature, like the outline of a lake or pond, the ends of the sausage are not subtracted. In short, therefore, the probabilistic method of linear measurement has become a probabilistic method of area measurement. Perkal's longimeter is therefore a form of overlay similar to that used for area measurement by direct point-counting methods. The area of the set is equal to the expected number of these points multiplied by the area of the unit figure of which each point is centre. Since the direct methods of area measurement depend upon the use of a systematic sampling frame of points, the design of the longimeter must be a regular network of points. As we have seen in Chapters 3 and 7, and as we shall see later in Chapter 18, the points must be aligned along a square grid formed by lines which are a units apart, or a triangular grid of separation b units. Perkal follows the traditional arguments outlined in

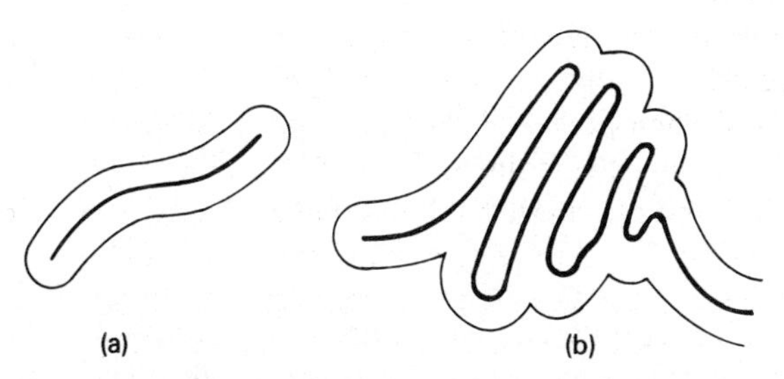

FIG. 15.3. The ε-aureole for (a) a simple, ε-convex line and (b) a more irregular line which is not ε-convex. (Source: Perkal, 1958).

Chapter 18 and accepts the view that the error of measurement by point-counting is directly proportional to the length of the perimeter and inversely proportional to the area of this feature. Moreover, the denser is the network of points, the more efficient is the method of measurement.

Direct measurement by point-counting comprises application of the dot grid k times on the feature, and the number of points falling in the area are counted. Using Perkal's notation we may record the number of points counted, n_1 in the first application, n_2 in the second and in the kth it is n_k points. For a square pattern of dot grid the area of the set is

$$a_\varepsilon(X) = (a^2/k)\sum n_k \tag{15.31}$$

and for the triangular, or hexagonal, pattern it is

$$a_\varepsilon(X) = [(b^2 \cdot \sqrt{3})/2k]\sum n_i \tag{15.32}$$

From equation (15.28) it follows that any point P of the grid will fall on $A_\varepsilon(X)$ if and only if a circle of radius ε centred at P meets the arc X. It is therefore sufficient to replace the dot grid overlay of points used for area measurement by a network of circles each of radius ε. It is not a necessary condition that the circumferences of the circles should be tangential to one another. However, it obviously simplifies the arithmetic if this condition is satisfied because we may replace a by ε^2 in (15.31) and b by $\sqrt{3\varepsilon^2/2}$ in (15.32).

If we number the circles of the ε-longimeter $1, 2, \ldots, r$ and the event Z_i is dependent on placing the ith circle on the arc X, then we may assign to x_i a random variable which either has the value 1 when the event Z_i occurs, or the value 0 when it does not, which is the occasion when the ith circle does not fall on the arc X. In other words we have returned to use of the familiar concept of "success" and "failure" which we have already encountered several times. By $x = x_1 + x_2 + \cdots x_r$, we mean the number of circles of the longimeter which fall on the arc X in one application. In other words $x \equiv \Sigma n_i$. If the occurrences of Z_i were independent of each other, and if the probability P of the occurrence of Z_i were independent of i, the random variable x would have a binomial distribution with a mean rp equal to the anticipated number of circles of the ε-longimeter falling on the arc X, with variance rpq, where $q = 1 - p$. But the events Z_i are not independent, and are correlated. Moreover if one circle of the longimeter falls on the arc X, then the probability that neighbouring circles of the longimeter also fall on this arc is greater than the probability that a random circle of the longimeter will fall on the arc X. The correlation between the events Z_i depends on the shape of the arc X. This increases the variance of the random variable x, making it greater than normal. On the other hand, the probability of the occurrence of Z_i is dependent upon i. Most people prefer to use the centre of the longimeter, so that when it is placed over the line X that line will pass near the centre of the overlay. For that reason the circles lying in the centre of the longimeter have a greater probability of coinciding with part of the arc X than do circles situated near the edges of the overlay. This reduces the variance of the random variable x to less than normal, and the size of this change depends upon the personal choice of the operator about which part of the longimeter is applied to the arc X. Since the two influences cancel one another to a certain extent, one can argue that the random variable x has variance rpq. The variance of the random variable N which is the summation over k of the independent random variable x (the combined number of circles falling on the arc in k independent applications of the longimeter) is therefore approximately equal to $krpq$, and with a large longimeter, where r is great in comparison to the anticipated value of the variable x, q is close to unity, and the variance in the random variable N is equal to krp/N. From this we should expect that the variance in the length of order ε of the arc X is of the order of the measured length, and the mean square deviation is of the order of the root of this length. Therefore the greater the length (of order ε) of the arc, the greater the precision (percentwise) in measuring with the longimeter. If, for example, the length of order ε of some curve is equal to 20, then it is expected that the probable average error of this length is of the order of 4–5, that is, 20–25%. If the length is equal to about 100, then the probable average error is of the order of 10, that is, about 10%. If ε is taken so that the ratio

$$\eta = \varepsilon / L_{\varepsilon}(X) \tag{15.33}$$

is not greater than 10%, then in view of the fact that $\varepsilon \geqslant 1$ and $L_{\varepsilon}(X) \geqslant 10$, the probable average error would become not greater than 30% of the length of order ε.

In practice measurement with the longimeter proved to be far more precise than would appear from the above considerations. Evidently the influence of

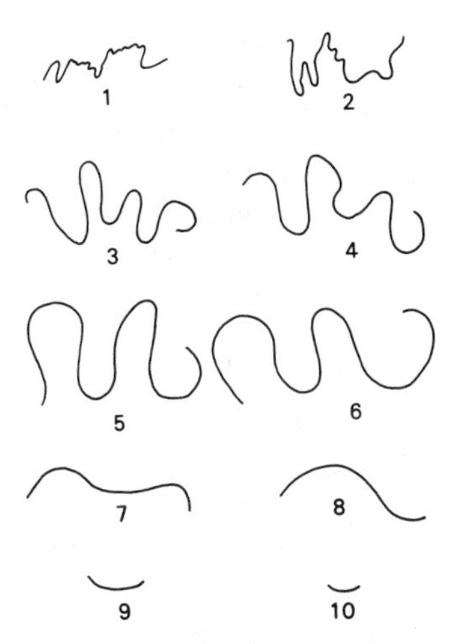

FIG. 15.4. The ten lines used by Perkal for his series of experimental measurements. The lengths of the lines reproduced in this figure do not correspond to those recorded by him. (Source: Perkal, 1958).

the different roles played by the central and extreme circles of the longimeter to diminish the variance is far stronger than the influence among the events Z_i. The result of this is that in practice the variance is remarkably less than normal.

Practical Tests of the Perkal Longimeter

Perkal describes a series of test measurements made by longimeter on the collection of ten different lines illustrated in Fig. 15.4. The purpose of these tests was, first, to examine evidence for systematic error; secondly to study the random errors of measurement of different lines using different longimeters; and, finally, to discover the time needed to make the measurements. The measurements were repeated many times and by three different operators. In one case a line was measured 200 times.

Systematic Errors

Line No. 2 in Fig. 15.4 was measured 60 times by counting the number of circles of the 3 mm longimeter touching the line when it was laid at random over the line. This experiment was carried out by three operators, *A*, *B* and *C*. Table 15.4 shows the number of circles touching or intersecting the line, together with the mean, the variance and mean square deviation. It can be seen that there are no evident systematic differences between the mean counts by operators *A* and *B* or between *B* and *C*, but there are differences in precision demonstrated by the variances. We see that there is an appreciable difference between the variances obtained for operators *A* and *B*, but little between *A* and *C*.

TABLE 15.4. *The results of Perkal's experiment to discover systematic errors in measurement of one line (No. 2) with one longimeter (3 mm) and three different operators*

Number of circles meeting line No. 2	Frequency		
	Operator *A*	Operator *B*	Operator *C*
12	–	1	2
13	2	5	8
14	9	13	24
15	29	12	15
16	15	23	10
17	4	4	1
18	–	1	–
19	1	1	–
Mean	15·23	15·20	15·43
Variance	1·02	1·73	1·15
Mean square deviation	0·13	0·17	0·14

Random Errors

Random errors appear to depend to a large extent upon the operator. For example, the variance for operator A was 20% less than that for B. In Table 15.5 we see the results of measuring one line using one longimeter. The length of order 3 mm we obtain using the sum of six measurements ($k=6$). Therefore the variance of order 3 mm will be the sum of the variances of six random independent variables each of variance s^2, or $6s^2$. This is a constant coefficient which does not affect the variation. The mean square deviations of these lengths calculated from six applications and the same deviations expressed as percentages of the length of the line of the arc are also given. In the beginning of the paragraph we calculated that in the case of normal variation the mean square deviation should be of the order of $\sqrt{90} \approx 9{\cdot}5$, therefore the empirical variation of length is actually subnormal.

The random error of measurement (mean square deviation) can be estimated on the basis of one measurement made up from six applications of a longimeter with $\varepsilon = 3$ mm. The error will actually be estimated with less precision, but for comparative purposes it will be sufficient. Ten curves are shown in Fig. 15.6. Double measurements of length were made on each of these, or order 3 mm, 5 mm and 8 mm with the appropriate square longimeter. The results are presented in Table 15.6. From this table we can investigate how the measured length decreases with increasing ε, just as the length of a line measured by dividers decreases as d increases. In the last line of Table 15.6 are given the lengths $L(X)$ of these lines measured by dividers (called the *incremental method*, by Perkal). Perkal states that the size of the unit step, d, was determined by measuring 100 such steps along a straight line, this being the same method of calibration advocated by the writer in Chapter 5, p. 68. However, Perkal evidently did not appreciate the importance of recording the size of the dividers separation, d. Therefore we must estimate this.

TABLE 15.5. *The results of Perkal's experiment to investigate the random errors of measurement by longimeter. The line is No. 2 and it has been measured using the 3 mm longimeter*

	Operators		
	A	*B*	*C*
Average length of order 3 mm of line No. 2	86·7	86·5	81·9
Mean square deviation of the length obtained from six applications	2·47	3·22	2·62
Deviation expressed as a percentage	2·9	3·7	3·2

TABLE 15.6. *Comparison of the mean lengths of Perkal's ten curves using three different orders ε, and two different operators. These results are compared with dividers measurements*

Order	Operator	Line 1	2	3	4	5	6	7	8	9	10
$L_3(X)$	*A*	69	87	145	135	165	158	64	58	26	10
	B	67	86	145	138	175	166	65	63	20	10
$L_5(X)$	*A*	61	79	132	132	168	163	65	56	20	11
	B	54	75	121	128	174	171	61	60	20	13
$L_\theta(X)$	*A*	56	63	111	107	156	159	66	57	21	11
	B	51	69	106	111	158	162	66	58	20	9
Dividers		76	105	151	138	176	171	65	58	22	10

Lines with small irregularities produce similar measured lengths, whether this be done by longimeter or dividers, this corresponding to Perkal's claim that for 2ε-convex lines the length to order ε approximates to the classical length. The first three lines have lengths of order 3 mm, which are shorter than those measured by dividers, suggesting that the missing value for $d > 3$ mm. The lengths of order 5 mm of these curves are shorter than the lengths of order 3 mm, and still shorter are the lengths of 8 mm order. This agrees with what we already know about the limiting distance of a line which has been measured using dividers at different settings, *d*. The next line (4) has a length of 3 mm order equal to the length established by dividers. On the other hand its length of 5 mm order is somewhat smaller and the length of 8 mm order is considerably smaller. The next two lines (5 and 6) have the lengths of order 3 mm and 5 mm equal to the dividers measurement, but the length of order 8 mm is smaller. Finally the measurement of the last four lines (7, 8, 9 and 10) show lengths of order 3 mm, 5 mm and 8 mm, all similar to the length measured by dividers. Lines 9 and 10 are especially short. The ratio $\eta = \varepsilon/L(X)$ for these

TABLE 15.7. *The results of Perkal's experimental measurements showing the ratios η_ε and W_ε obtained for different lines using different longimeters*

k	ε	Line 1	2	3	4	5	6	7	8	9	10
6	η_3	4·4	3·5	2·1	2·2	1·8	1·9	4·7	5·0	13·0	30·0
	W_3	3·5	2·8	1·7	1·7	3·3	1·7	3·7	3·9	6·3	12·2
10	η_5	8·2	6·3	3·8	3·8	3·0	3·1	7·7	8·9	25·0	40·0
	W_5	3·6	3·6	3·6	2·1	3·1	1·8	2·2	5·8	9·5	15·5
16	η_8	15·3	13·0	7·1	7·5	5·1	5·0	12·1	15·0	38·1	72·8
	W_8	7·9	5·7	3·0	4·2	3·6	2·1	5·5	6·0	13·2	18·2

lines approaches 75% (when $\varepsilon = 8$ mm). The remaining lines have lengths exceeding 60 mm, and for these the ratio η is decidedly smaller, although it occasionally exceeds 10%, which was earlier suggested to be as the upper acceptable limit. Despite this the random errors inherent in the results of measurements are not large. Table 15.7 consists of the ratios η_ε and the *indices of variation*, that is the percentage error in the length of order ε, W_ε. The number of applications of the longimeter necessary for one measurement of length of order ε are given in the column headed k. Actually, if the entire measurement were repeated n times (that is the longimeter were applied to the line nk times) the average error of length or order ε obtained from these n measurements would become $\sqrt{n}$ times smaller.

In those cases where the ratio η does not exceed 10% the error in measurement does not exceed 6% and in only two cases does it exceed 4%. It is further seen that when ε increases, in general, the error becomes greater and η also increases.

The lengths of lines 1, 2 and 6 measured by Steinhaus' longimeter are given in Table 15.8 for comparison. This table presents the averages obtained for 50 applications of the longimeter for each result. These averages contain random errors more or less of the order resulting from single measurements with the ε-longimeter. For example, the average error (from 50 applications of Steinhaus' longimeter) of the length of order 1 of arc No. 1 is established at 4.5%.

Finally Perkal considers the times required to make the measurements of the various lines. The measurement of length of order 3, 5, or 8 mm of any arc in Fig. 15.4 required 2–4 minutes. Such a measurement (consisting of 6, 10 or 16 applications of the longimeter) results in an error of about 4%. To obtain the length of order n of such an arc with the same precision by Steinhaus' method would require 50 applications, which would take 15 minutes, or four times longer than is required by the ε-longimeter. The dividers method (a double measurement and calibration of the dividers) needed more than 10 minutes. The precision of this method depends on the shape of the curve. For lines having small curvature (e.g. line 6) the dividers method gives very precise results (the error is about 0·3% of the length), but for more irregular lines the precision is much less. For example line No. 1 has an error of about 4%.

TABLE 15.8. *Comparative values for the lengths of three of Perkal's lines measured using Steinhaus' longimeter*

	Steinhaus order (m)							
Line	1	2	3	4	5	6	7	8
1	44	62	74	78	79	80	80	81
2	47	71	90	100	108	110	111	111
6	77	126	153	163	68	168	168	168

We now turn to a few additional measurements made by the author during the early 1960s. The majority of these were made with a square pattern longimeter scribed at $\varepsilon = 5$ mm. This comprised an array of 22 × 20 circles which were subsequently reproduced photographically on glass plates to provide other longimeters at $\varepsilon = 0{\cdot}5$ mm, 1 mm, 1·5 mm, 2 mm, 2·5 mm, 3 mm, 3·75 mm and 5 mm. The number of applications of each longimeter was 2ε as recommended by Perkal, but, for $\varepsilon = 3{\cdot}75$, k is not an integer. Therefore the longimeter was applies 15 times ($= 4\varepsilon$) and the total number of circles counted, n_k was divided by 2. The author found that the two smallest sizes of longimeter were most inconvenient for practical use because of the difficulty of counting accurately the number of extremely small circles meeting the line to be measured. Moreover, the size of the longimeter at extreme reduction was only that of a postage stamp. It was therefore necessary to measure a line such as the Yorkshire coast in a considerable number of very short lines. Perkal's own figures for η relating to his lines 9 and 10 indicate the unreliability of measurements of such lines.

The Yorkshire Coast

The original measurements of the length of the Yorkshire coast were made in this fashion. Table 15.9 lists the results.

The limiting distance for each map scale has been calculated using Volkov's method, treating ε as if it were identical to d. For comparison we also show the limiting distances based upon the dividers measurements which were given in Table 13.11, p. 299. This does not imply that the Volkov method of correction is to be preferred to any of the others, it is merely used to attempt to compare the results of measuring the same line in different ways. It should be observed that the limiting distance obtained by longimeter measurements is greater than the result obtained by dividers.

It is difficult to account for some of the sudden jumps in the measured distances such as that of 10 km in the measurement made at 1/1,000,000 between the 2·5 mm and 3·0 mm longimeters. It is such erratic results which causes some operators to lose confidence in this method of measurement.

TABLE 15.9. *The length of the Yorkshire coast measured by D. H. Maling using square pattern longimeters*

	Length of order ε						Volkov Limiting Distance	
Map Scale	1·5 mm	2 mm	2·5 mm	3 mm	3·75 mm	5 mm	Longimeter	Dividers
1/250,000	184·0	190·25	191·5	187·5	183·75	177·5	200·92	186·16
1/625,000	179·4	176·3	181·9	177·5	174·4	173·1	187·77	178·73
1/1,000,000	175·0	179·0	178·0	168·0	168·75	168·0	190·86	179·37

Measurement of Straight Lines and Circular Arcs

Another experiment, more closely allied to that carried out by Perkal, was introduced as a class exercise for students studying for the postgraduate Diploma in Cartography about 1963. Some of this work was subsequently analysed by A. O. Hughes as an unpublished diploma project.

The object of one exercise was to study the variance between operators and between longimeters for a succession of measurements made of simple rectifiable curves. In other words the attempt was made to exclude any noise created by measuring empirical curves like some of those of Perkal, or actual lines on maps like the Yorkshire coast. The lines measured were straight lines, or circular arcs constructed on drafting plastic by coordinatograph. Table 15.10 shows the results for two such experiments.

An important result shown by an analysis of variance carried out by Alun Hughes is that systematic variation between longimeters was, in our experiments, much greater than the variation between operators. This is difficult to explain for, as demonstrated by Perkal, the differences between the measurements made by different operators would be expected, but there is no theoretical basis for the variation shown by the longimeters.

In the 30 years which have elapsed since publicity to the Polish work was given by the present author (Maling, 1962) and the initial enthusiasm shown by Tobler and Nystuen (1966) for it, this method seems to have dwindled so that although books on quantitative geography, e.g. Cole and King (1968), Chorley and Haggett (1969), Taylor (1977) pay lip service to this work, it essentially remains a curiosity. The application of it to cartography as an "objective means of generalisation", as Perkal (1958b) puts it, was examined by the present author in Maling (1963), who still believes that it can fulfil a useful role here. It is also interesting that the work of both Steinhaus and Perkal seem

TABLE 15.10. *Experimental measurements of straight line of nominal length 50 mm and circular arc of length 106 mm made by six different operators using seven different square pattern longimeters*

Operator	Longimeter $\varepsilon = 1{\cdot}0$	1·5	2·0	2·5	3·0	3·75	5·0	Operator mean
Line *AB* (straight) ≈ 50·0 mm								
A	49	49	49	54	49	51·5	48	49·9
B	47	48	51	47	48	50·5	48	48·5
C	50	49	51	49	44	50·5	47	48·7
D	49	51	51	51	50	51	50	50·4
E	49	50	49	51	47	50	50	49·4
Longimeter means	48·8	49·4	50·2	50·4	47·6	50·7	48·6	49·4
Line *EF* (circular arc) ≈ 106 mm								
B	98	97	107	102	100	104	–	101·3
C	97	97	110	106	104	107	–	103·3
D	102	102	107	101	100	104·5	–	102·8
E	93	104	102	103	100	100·5	–	100·4
Longimeter means	97·5	1000·0	106·5	103·0	101·0	103·8	–	101·9

to be virtually unknown in stereology, where the unmodified Buffon method of measurement is still preferred.

THE PARADOX OF LENGTH AND THE CONCEPT OF ORDER IN DISTANCE MEASUREMENT

We have already seen in Chapter 3, page 46, that Steinhaus' major preoccupation was to find a method of measuring the lengths of empirical curves. The probabilistic methods of measurement which have resulted from the investigations of both Steinhaus and Perkal may be specified to be of a particular *order*. This is a most important distinction between those longimeter methods and the so-called classical methods already described in Chapter 13. Steinhaus introduced the paradox of length through his classic description of the property of empirical curves, in which he made specific reference to the cartometric application of measuring the length of the left bank of the River Vistula. In Chapter 3 we briefly examined the mathematical concept of rectification and demonstrated how the length of a rectifiable arc may be calculated for a few elementary examples. By contrast, empirical curves are generally not rectifiable. The series of lengths obtained by repeated measures with finer and finer instruments does not ordinarily converge to a finite value. The greater the accuracy with which an empirical curve is measured the longer it becomes. Perkal has drawn attention to the well-known phenomenon that apparently smooth-edged objects, such as the edge of a razor blade or a smooth-edged leaf, turn out to have quite irregular edges when these are studied under greater and greater magnification. He has written:

> Can these small discrepancies, seen only through the microscope, cause an essential difference in the length of the arc? Yes, if the length of the arc is not a continuous functional. Arbitrarily near to the arc A, one may draw another arc B [for example indented as shown in Fig. 15.5] which is much longer than the arc A; and so, near the approximate edge of a razor blade seen by the naked eye there is a much longer complicated curve representing the same edge as seen under the microscope, and if we look at this edge through more and more powerful microscopes we can see more and more complicated curves with increasing length. One cannot even talk of the real, or final shape of the edge of the blade. Moreover one cannot talk of the real length of the edge of the razor blade.

Skellam (1972) carries the argument a stage further.

> Consider, for example, the following quotation from Stamp (*Britain's Structure and Scenery*). "It has recently been calculated that the coast of England and Wales alone is some 2751 miles in length." To some, no doubt, this is an informative statement, but for those who are not acquainted with the operational procedure which yields the number 2571 it has no meaning. To sail round the coast is one thing and to run a tape measure round every boulder is another.
>
> We no longer believe that things have their own true proper names, but we commit a comparable mistake when we assume that spatially extended objects necessarily possess an intrinsic mathematical property called "length", and the error is perpetuated by the structural defects of our language. Strictly speaking, it is not the object but its mathematical model which has the length of the arc? Yes, if the length of the arc is not a continuous functional. Arbitrarily near to an instrumental procedure, number and counting included.
>
> Because we are free to choose the meaning of such terms as "length of coastline", there is far more flexibility in modelling than is commonly realised.

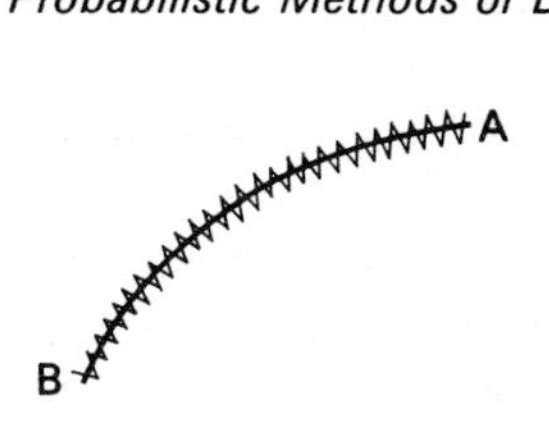

FIG. 15.5. Perkal's concept of an empirical curve.

It therefore seems that we must draw the conclusion that the "true length" of an object is an illusion, and that, consequently, that measurement of length is a singularly profitless occupation. Nevertheless we know that it has considerable value in many kinds of technical or scientific applications, from carpentry to geodetic surveying. It is obviously useful in quantitative map use, for example at the relatively crude level of measuring the distance between two points A to B in order to estimate the time which will be needed to make the journey between them. As the cartometric requirement for a reliable measure of distance or length of line becomes more precise, so the map user becomes increasingly aware of the influence of the paradox of length, and recognises that at best the results of measurement appear to be approximations.

However, the inference that at the molecular level the length approaches infinity is false, since a variable can increase and yet have a finite upper bound. By way of a familiar example, Bibby (1972) quotes the sum of the terms

$$S(a) = 1 + a + a^2 + a^3 + \cdots + a^n \tag{15.34}$$

The more terms one takes in this sum the larger it gets, yet assuming that $a < 1$ it never exceeds the finite upper bound of $1/(1 - a)$. Similarly, more refined measurement may increase the apparent length of a river or coastline, but this does not mean that it can increase *without limit*. The correct interpretation is that as measurement becomes more refined all curves tend to a finite length. The only possible exceptions are curves which have discontinuous derivatives *at an infinite number of points*.

In this attempt to explain and criticise Steinhaus' work, Perkal noted that

> This approximate length is useful in general for the comparison of the length of curved lines in geographical maps or on several maps of various generalisations. It can be admitted that in applying the length of the indicated order, for example the fifth, Penck would not have obtained such variations in the measurement of the length of the coast of Istria.

However Perkal follows this with the telling criticism that

> Scientists want to measure the true length, the so-called classical length of the empirical curve, and want to believe that such a length exists. It seems that it is easier for the scientist to digest generalisation of a curve rather than to introduce a new abstract measurement, such as the length of order *m*.

This does provide at least a handful of straws which may be clutched in following some of the procedures outlined here. After all, the paradox of length has quite a lot to do with generalisation of lines and their representation at different scales.

We have seen that, in the opinion of both Steinhaus and Perkal, the only practical, pragmatic solution to the paradox is to recognise an *order* of measurement; to identify what irregularities have geographical or practical significance and conduct the measurements so that these may be taken into consideration. Moreover, to repeat a point made at the outset of this work, the majority of cartometric measurements are made for comparative purposes.

A further, and important, practical point is that order is related to *resolution* in the sense of the use of this word in optics. Any practical cartographic or cartometric process has a limiting resolution. Thus the chartwheel, the points of dividers, the linkages in the coordinatograph drives of a photogrammetric plotter, or the movements of the plotting head of an automatic plotter all have a limiting resolution which may arise from difficulties in manipulation, mechanical backlash or small optical imperfections. All of these have their particular limits, ranging from the rather inaccurate way in which most users have to move a chartwheel round awkward corners to the limiting resolution of dividers which is presumably about one-half of the diameter of the prick mark made as these touch the paper. The present author, in Maling (1962), has remarked upon the relationship between ε and the diameter of the floating mark in a photogrammetric plotter, and the practical use of this to generalise linear detail in plotting and drawing linear features. Provided that the different measurements have been made to similar order, we obtain results which may be compared. In this field of measurement we may select a particular order of measurement, according to either the definition of Steinhaus, which we have seen is a limit to the number of applications and total counted, or that of Perkal which is the circle of radius ε controlling the width of the ε-aureole. By restricting ourselves to a particular numerical value for order, or changing order to accompany changes in map scale, we may make a succession of measurements of different features which may, now, be reasonably compared. The alternative solution, as we have seen, is that for determination of limiting distances for all measurements made at a particular map scale, and then to determine the reduced distance from a succession of limiting distances relating to each map scale at which the measurements were made.

16

Geometrical Methods of Area Measurement

Also you must note, there are diverse fashions of landes, and therefore diversly to be measured. And some manner of lande lieth in suche sondrie fourmes, that it must needes be measured, not in the whole, but in divers parcelles, every parte by itself. Also where a pece of Lande is to be devided into divers partes, of which eche one muste bee measured by hym selfe, then ye ought vigilauntly to consider, into how many parcells, and into how many and what manner fashions they must be divided, that ye may measure every parte accordying to his fourme or fashion.

(Valentine Leigh, *The Moste Profitable and Commendable Science of Surveying of Landes Tenementes, and Hereditamentes*, 1577)

Some of the methods to be described in this chapter were already known and used in Britain in the middle sixteenth century. The work entitled: *This boke sheweth the maner of measuring of all maner of lande, as well as woodlande, as of lande in the fielde*, appeared in 1537. It had been written by Richard Benese, a Canon of the Augustinian Priory of Merton, and it is regarded as being the first English contribution to the subject and symptomatic of a new awareness of the need for greater accuracy in the description and measurement of land.

More than two centuries earlier there had been enacted the statute known as *Extenta Manerii* (possibly 4 Ed.III or 1276) which dealt, for the most part, with legal and fiscal matters such as customary dues, rents and tenures. According to this statute the Crown required a quantitative statement about the extent of meadow, arable, pasture and woodland, the respective areas of which were to be set down. Jones (1979) has recognised an improvement in the quality of manorial surveys dating from the early fourteenth century and which he attributes to the working of this statute. However, Taylor (1947) argued that relatively crude estimation of land areas persisted for at least another century, and does not consider that the improvement took place until early Tudor times. She has written:

It must be emphasised that a series of rapid changes in technique took place from the late fifteenth century onwards, consequent upon three related developments. These were advances in and wider dissemination of, arithmetic and geometry; the invention of simple instruments for surveying; and the addition of a 'plat' or plan to the written part of a survey. As a natural result of these changes, the mathematical part of the survey increased in importance, while the legal and judicial parts fell into the background. This shifting emphasis was, of course, stimulated by social and economic changes. The growing proportion of land rented or leased by the arce, the wholesale transfers of land ownership consequent upon political upheavals, coupled with rising land values, all alike pointed to the advantages of more accurate mensuration.

Richard Benese provided tables for computing land values in which the range of these extended from 3s.4d. per acre to £6.3s.4d, these presumably represented the extremes which he thought likely to occur. Ralph Agas in 1596 (*A Preparative to Platting of Landes and Tenementes for Surveighs*) and George Atwell in 1658 (*The Faithful Surveyor*) indicate some land values as high as £9 per acre. Professor Taylor considers that, once land values had risen to this extent, estimates made to the nearest half-acre and rough measurements resulting in errors of the order of 20–25% made by untrained men could no longer be tolerated. Thus an appreciation had developed about the significance of the areas of individual fields and woods, and there was no great intellectual hurdle to be jumped in applying this idea to describe an entire estate or a parish. However, nearly a century had to elapse from the work of Benese to the measurement of larger units of land, such as a whole countries. It was not until 1636, when Gerard Malynes (*Consuetudo, Vel, Lex Mercatoria*) published the areas of the countries of the world, the ratio of land to sea and of uninhabited to inhabited land. According to Schmiedeberg (1906) and Proudfoot (1946), it is assumed that he used geometrical methods to obtain these figures, but there is no firm evidence to confirm this.

The principal applications of the geometrical methods of area measurement have changed remarkably little since Tudor times, and they are still most commonly used for estate surveys and as an aid to management and planning. In many countries such surveys are also needed in order to register title to land. They also have important civil engineering applications, especially where there is need for volumetric calculations of earthworks for mass haul, cut-and-fill and associated information, and they are especially useful where the ground detail of roads and field boundaries approximate to straight lines and many of the parcels correspond to simple geometrical figures where the parcels are irregular, and are bounded by sinuous lines, it is necessary to make allowance for the small areas lying between these and the sides of the geometrical figures into which the parcel has been subdivided for the purpose of executing the survey. The measurement and computation of these *marginal elements* adds significantly to the workload, particularly if the user does not have access to a computer.

The way in which a typical estate survey is carried out on the ground has also changed little over the centuries, comprising in the most part the old-fashioned but extremely effective method of *chain surveying*. The geometrical methods of area measurement are particularly appropriate for use with a chain survey, because the subdivision of the area to be surveyed is a network of triangles, all the sides of which are measured. Consequently the data are available from the field measurements to compute the area of each triangle forming the framework of the survey. The method of measuring and plotting boundaries and other detail is by making *offset measurements* along lines which are perpendicular to the sides of the triangles. These offsets may also be used to measure the areas of the marginal parts of parcels. Later innovations in surveying technique, for example the development of tachymetry and other methods of optical distance measurement, followed much later by the methods of electronic distance measurement, have encouraged the use of traversing and radiation for plotting detail at a large scale. Since the normal method of fixing position from such data is to calculate the grid coordinates of each survey station, the appropriate method of determining areas has been to use the methods of coordinate geometry and calculate the area of entire polygons directly from the coordinates of the points which have been established along the perimeter. Such methods are especially well-suited for computer processing, and since the coordinates of points on a map or plan can also be measured and recorded by digitiser, the methods can be used for wider range of applications than their traditional restriction to large-scale surveys. Before digital computers became readily available the geometrical methods of area measurement, particularly those which are based upon the multiplication and addition of successive coordinates, combined with the use of such techniques as Simpson's rule and the trapezoidal rule to determine the marginal areas, were considered to be too slow and clumsy for all except for simple plots. Thus Monkhouse and Wilkinson (1952) dismiss Simpson's rule with the comment that "It is doubtful whether the mathematical tedium of this method is worth the slightly more accurate result." Dickinson (1969) holds a similar opinion and claims that he "has measured many hundreds of acres without ever having recourse to it." However, it is unlikely if any of them ever had to make an estate survey or compute earthwork volumes. In the days before automatic data processing the geometrical methods of area measurement were neither convenient nor suitable for the kinds of measurements typically required by geographers, such as evaluation of the extent of land-use categories on medium- or small-scale maps. The sort of objections which these authors raised in their criticism of these methods indicate that they were inclined to treat all methods of area measurement as having equal utility, irrespective of the size of parcel, the irregularity of its outline or the convenience of the method. Otherwise they would not have included the geometrical methods in books intended primarily for geography students. We must therefore restate our thesis *that not every cartometric method is equally suitable for every use.* In the days before digital computing invaded almost every facet of professional work, the methods to be described in this chapter could have been safely ignored by the geography student; but today they cannot. The pendulum of utility has now paradoxically swung in the opposite direction, and the determination of area from the coordinates of points along the boundary of a parcel is a suitable method of determining area for anyone who has access to a suitable digitiser or vector cartographic data already in machine-readable form. Using modern methods of data collection and storage there is no reason why they should not be used to measure the area enclosed by a coastline or other irregular boundary, even if this comprises many hundreds of thousands of separate coordinated points.

THE AREA FORMULAE OF PLANE GEOMETRY

The formulae which follow are presented without proof, for all of them are well known and they are listed here to remind the reader how each of the measurements and calculations are made.

The Area of a Triangle

Figure 16.1(a) illustrates a triangle ABC in which one side is designated the *base*, b, and the line perpendicular to it which passes through the opposite vertex is called the *height*, h. Then, the area,

$$A = \tfrac{1}{2} b \cdot h \tag{16.1}$$

If any of the angles of the triangle are also known, alternative solutions are

$$A = \tfrac{1}{2} a \cdot b \cdot \sin C \tag{16.2}$$

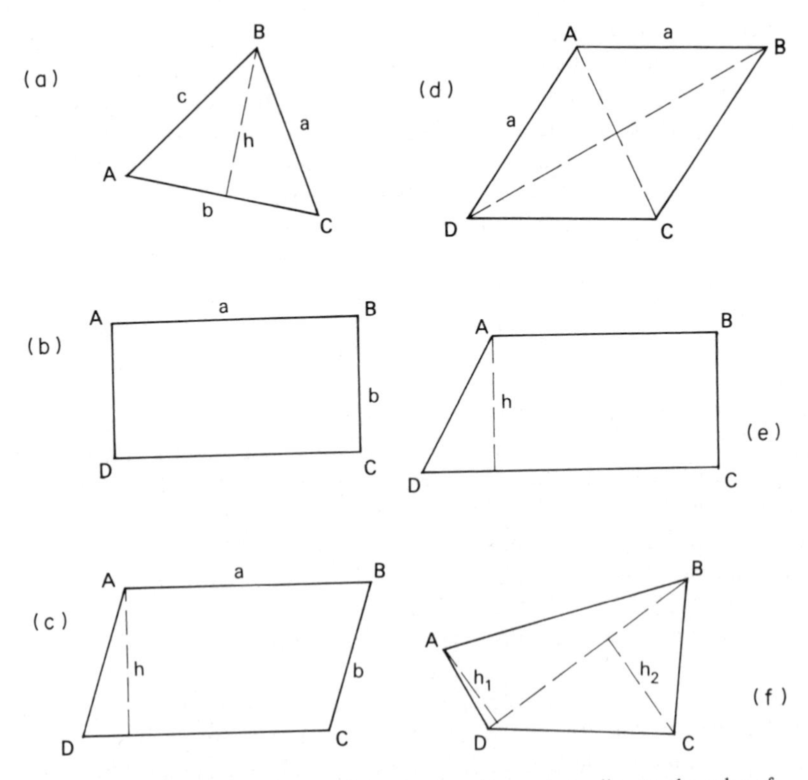

FIG. 16.1. Measurement of area of simple plane figures according to the rules of Euclidean geometry: (a) the triangle ABC; (b) the rectangle $ABCD$; (c) the parallelogram $ABCD$; (d) the rhombus $ABCD$; (e) the trapezoid, $ABCD$; (f) any quadrilateral $ABCD$.

$$A = \tfrac{1}{2}a \cdot c \cdot \sin B \tag{16.3}$$

$$A = \tfrac{1}{2}b \cdot c \cdot \sin A \tag{16.4}$$

For a triangle in which the perpendicular height had not been determined, and the angles have not been measured, but the lengths of all three sides are known, the *half-perimeter formula* may be used

$$A = \sqrt{\{s(s-a)(s-b)(s-c)\}} \tag{16.5}$$

where

$$s = \tfrac{1}{2}(a+b+c) \tag{16.6}$$

or is one-half of the length of the perimeter of the figure.

The Area of a Rectangle

In Fig. 16.1(b) the rectangle $ABCD$ has the sides $AB = a$ and $BC = b$. Quite simply.

$$A = a \cdot b \tag{16.7}$$

The Area of a Parallelogram

In Fig. 16.1(c) the figure has side $AB = a$ and $BC = b$. Also h is the length of the perpendicular to one pair of the parallel sides. Then

$$A = a \cdot h \tag{16.8}$$

Also

$$A = a \cdot b \cdot \sin B \tag{16.9}$$

The Area of a Rhombus

The length of each side of the figure is equal to a. In Fig. 16.1(d) the lengths of the diagonals, AC and BD are also known,

$$A = \tfrac{1}{2}(AC \cdot BD) \cdot \sin B \tag{16.10}$$

Also, as a special case of (16.9)

$$A = a^2 \cdot \sin B \tag{16.11}$$

The Area of a Trapezoid

In Fig. 16.1(e) the lengths of the pair of parallel sides AB and CD are known; so, too, is the perpendicular distance, h, between them. Then

$$A = \tfrac{1}{2}h(AB + CD) \tag{16.12}$$

The Area of Any Quadrilateral

The length of the diagonal *BD* is known; so, too are the length of the two perpendiculars h_1 and h_2 joining the diagonal *BD* to *A* and *DC* respectively.

$$A = \tfrac{1}{2}BD(h_1 + h_2) \tag{16.13}$$

If the area of the figure is to be determined from measurement on a map or plan, it is usually preferable to employ the formulae which make use of linear measurements only, for these can be measured more accurately than angular measurements. An exception to this rule occurs when special equipment intended to measure angles is used. The Coradi Digimeter Model DMP, to be described later, is a case in point, because this measures polar coordinates. Similarly, if angular measurements have been made in the field, there is little objection to the use of them and equations such as (16.2)–(16.4) and (16.9) may be employed.

DETERMINATION OF THE AREA OF THE FRAMEWORK OF A PARCEL

The geometrical method of determining the area of the framework of a parcel comprises subdivision of the parcel into a network of simple geometrical figures, measuring certain elements of these figures, calculating the individual areas of them and finally combining the results. Unless the parcel to be

TABLE 16.1. *Calculation of the area of the polygon ABCDEFGH from the areas of the six component triangles*

Triangle	b(m)	h(m)	A(m^2)	A(ha)	Mean area (ha)
ABH	846	401	169,623·0	16·962	
	799	424	169,388·0	16·939	16·942
	429	789	169,240·5	16·924	
BHG	846	500	211,500·0	21·150	
	923	458	211,367·0	21·137	21·149
	505	838	211,595·0	21·159	
BDG	884	696	307,632·0	30·763	
	750	821	307,875·0	30·787	30·777
	923	667	307,820·5	30·782	
BCD	581	496	144,088·0	14·409	
	750	385	144,375·0	14·437	14·428
	497	581	144,378·5	14·438	
DEG	435	860	187,050·0	18·705	
	890	421	187,345·0	18·735	18·712
	884	423	186,966·0	18·697	
EFG	890	266	118,370·0	11·837	
	781	303	118,321·5	11·832	11·827
	308	767	118,118·0	11·812	
				Total area	113·835 ha

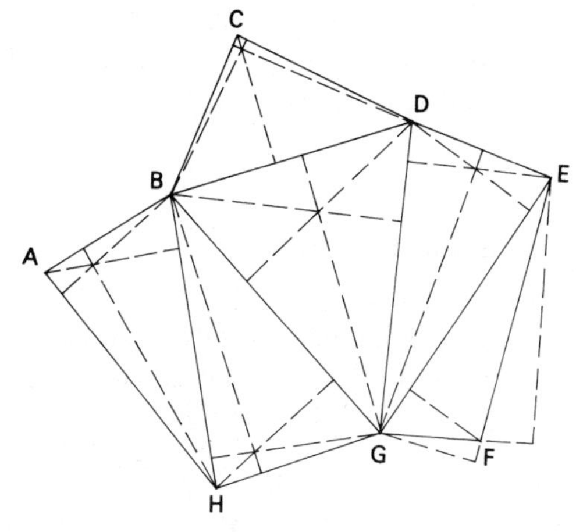

FIG. 16.2. The subdivision of the polygon *ABCDEFGH* into six triangles for measurement of its area. The broken lines are the perpendiculars to the sides of the triangles.

measured corresponds closely to a particular form of quadrilateral, it is generally more convenient to subdivide it into a series of triangles, as illustrated, for example, by Fig. 16.2. We have seen this is the normal practice in chain surveying but it applies equally where the user does not have access to the original measurements and the dimensions of the figures are scaled from a map or plan. The polygon *ABCDEFGH* has eight points on the perimeter, and it can be divided into six triangles. Table 16.1 gives all three solutions of

TABLE 16.2. *Calculation of the area of the polygon ABCDEFGH from the areas of the six component triangles using the half-perimeter formula*

Triangle	Sides (m)	s	$(s-a)$	$(s-b)$	$(s-c)$	$A\,(\mathrm{m}^2)$	A (ha)
ABH	846						
	799						
	429	1,037	191	238	608	169,295	16·929
BHG	846						
	923						
	505	1,137	291	214	632	211,539	21·154
BDG	884						
	750						
	923	1,279	395	529	355	308,185	30·818
BDC	581						
	750						
	497	914	333	164	417	144,273	14·427
DEG	435						
	890						
	884	1,105	669	215	221	187,015	18·701
EFG	890						
	781						
	308	989	99	209	681	118,278	11·828
						Total area	113·857 ha

equation (16.1) for each components figure. The half-perimeter formula (16.5) has also been used to calculate their areas and the results of this solution are given in Table 16.2. The difference between the two results is 0·002 ha, or 0·02% of the total area.

ERRORS OF THE GEOMETRICAL METHOD

Even a simple parcel such as that illustrated in Fig. 16.2 may be subdivided in several different ways. Therefore different side lengths may be used in the calculations and, as shown in Table 16.1, slightly different results are obtained for the area of each component triangle. Since the accuracy of the total area of the parcel may vary for this reason, as well as those errors of measurement which we have already examined, it is desirable to investigate the theoretical accuracy of the method and attempt to answer three questions:

- What is the best form of geometrical figure?
- Does the shape of the figure affect the accuracy of the results?
- Does the number of component figures affect the accuracy of the results?

The Form of the Figure

We assume that the parcel has been subdivided into a network of triangles, as in the example studied, and that the area of each triangle has been calculated from equation (16.1). We need to establish the relationship between the standard error of the calculated area, as well as those for the measured base and height of each triangle.

Equation (16.1) may be expressed in natural logarithms as

$$\ln A = \ln b + \ln h - \ln 2 \tag{16.14}$$

Differentiating this expression we find that

$$\mathrm{d}A/A = \mathrm{d}b/b + \mathrm{d}h/h \tag{16.15}$$

The terms $\mathrm{d}A$, $\mathrm{d}b$ and $\mathrm{d}h$ represent infinitely small increments in area and side length, which we may regard more realisticallly as corresponding to small standard errors of these variables. If s_A, s_B and s_H are the standard errors in area, base length and triangle height respectively, we have

$$(s_A/A)^2 = (s_B/b)^2 + (s_H/h)^2 \tag{16.16}$$

The same kind of argument may be used to determine the relative precision of the area calculated from a rectangle, parallelogram or rhombus in which the area is also derived from two linear measurements. For the trapezoid, the half-perimeter formula and for the measurement of the area of any unspecified quadrilateral three measurements, are required. Therefore the corresponding expression will contain an additional term on the right-hand side.

It was established in Chapter 14, page 279, that the precision of measurement of the length of a straight line made by steel scale on a piece of paper is uniform and independent of the length of line (within the practical constraint that the measured line is not longer than the scale). Thus we may assume that

$$s_B = s_H = s \tag{16.17}$$

and substitute this in equation (16.16). However, it should be stressed that the base of a triangle is usually determined more accurately than is its height, because measurement of h must be preceded by location of this line. Since by definition the line representing h is perpendicular to the base, its length may also be influenced by any small angular errors arising from faulty construction. This additional error can be shown to be smallest in an Isosceles triangle.

Making the appropriate substitutions in (16.16) we obtain

$$s_A/A = (s/b \cdot h)\sqrt{(b^2 + h^2)} \tag{16.18}$$

For a triangle, $b \cdot h = 2A$. Therefore for the area of a triangle we find

$$s_A = \tfrac{1}{2} s \sqrt{(b^2 + h^2)} \tag{16.19}$$

and if $b \approx h$,

$$s_A = s \cdot \sqrt{A} \tag{16.20}$$

For a rectangle, $A = b \cdot h$. Therefore (16.18) becomes

$$s_A = s\sqrt{(b^2 + h^2)} \tag{16.21}$$

and for a square or rhombus in which $b = h$

$$s_A = s \cdot \sqrt{2A} \tag{16.22}$$

Comparison of equations (16.19) with (16.21), and (16.20) with (16.22), demonstrates that theoretically the area of a triangle may be determined with greater precision than the area of a quadrilateral.

Where three linear measurements are required, such as for the trapezoid,

$$A = \tfrac{1}{2}(b_1 + b_2) \cdot h$$

as in (16.12), then

$$s_A^2 = \tfrac{1}{4}\{h^2 s_{B_1}^2 + h^2 s_{B_2}^2 + (b_1 + b_2)^2 \cdot s_H^2\} \tag{16.23}$$

Where $b_1 \approx b_2 \approx h$, $s_{B_1} \approx s_{B_2} \approx s_H \approx s_1$, we have

$$s_A = s\sqrt{(3A/2)} \tag{16.24}$$

or the area of a trapezoid is also determined with less precision than that of a triangle.

The Shape of the Triangle

In order to study the influence of the shape of a triangle upon the calculated area, we use the term K to represent the shape of the figure. This is established as

$$K = h/b \tag{16.25}$$

Substituting this in equation (16.19)

$$s_A = \tfrac{1}{2} s \cdot b \sqrt{(1 + K^2)} \tag{16.26}$$

But, in this case,

$$A = \tfrac{1}{2} K \cdot b^2 \tag{16.27}$$

and

$$b = \sqrt{(2A/K)} \tag{16.28}$$

Therefore

$$s_A = s \cdot \sqrt{A} \cdot \sqrt{(1 + K^2/2K)} \tag{16.29}$$

Comparison of equations (16.20) and (16.29) indicates that the area of an elongated triangle contains errors which are $(1 + K^2)/2K$ times greater than the area of a triangle in which $b = h$ and $K = 1$.

Can one therefore determine for each value of K the condition when

$$\sqrt{\{(1 + K^2)/2K\}} = \text{minimum?}$$

In order to answer this question we differentiate this expression and make the derivatives equal to zero. In other words,

$$\mathrm{d}Q/\mathrm{d}K = \{4K^2 - 2(1 + K^2)\}/8K \cdot Q = 0 \tag{16.30}$$

where

$$Q = \sqrt{\{(1 + K^2)/2K\}}$$

The solution is $K = 1$, so that the best results obtain for those triangles in which $h \approx b$. Inspection of Table 16.1 shows that the triangle BDG, in which $b = BD$, is the only case where K approaches unity. By contrast the most ill-conditioned triangle in the figure is EFG for, where $b = EG$, $K = 0.299$.

Knowing the optimum shape of triangle we may make a sensible decision about how best to subdivide an irregular parcel into its component triangles; and to decide which side of each triangle ought to be chosen as the base.

However, we have reached this conclusion from the assumption that $s_B \approx s_H$, or that both lines have been measured with equal precision. Often this is not so, because the sides of the triangles have been measured in the field whereas their heights have been scaled from the map or plan after the survey has been plotted. Consider, for example, the extremely ill-conditioned triangle in which the base, $b = 500$ metres, measured in the field and having a relative error of 1/2,000. In the same triangle, $h = 1000$ metres, but this has been measured from a map of scale 1/10,000 and is only known to the nearest metre. Then the relative error in the measurement of h is 1/10,000 and, from equation (16.16)

$$(s_A/A)^2 = (1/1000)^2 + (1/2000)^2$$

so that

$$s_A/A = 1/625$$

In this example the short length of the base has a smaller influence upon the relative error of the calculated area than had the measured length of the height of the triangle. On the other hand, if the same triangle had been wholly on the plan, so that b is also measurd to the nearest metre then

$$(s_A/A)^2 = (1/1000)^2 + (1/50)^2$$

and therefore

$$(s_A/A) \approx 1/50$$

In this example the error in the measurement of the shorter side has absorbed the whole of the relative error in the measurement of the height of the triangle, and therefore it is the larger relative error which determines the accuracy of the calculation.

The Number of Triangles

The final question which this theoretical investigation should answer is whether the number of component triangles in a parcel affects the accuracy of the measurement of its area.

Let the polygon, of area A, be subdivided into n triangles, each having $b \approx h$ and having individual areas $A_1, A_2, \ldots, A_n$.

Provided that the same linear distances are not used for measuring the areas of adjacent triangles, i.e. assuming that the area of each component figure has been derived independently,

$$s_A = (s_{A_1}^2 + s_{A_2}^2 + \cdots s_{A_n}^2)^{1/2} \tag{16.31}$$

but according to (16.20)

$$s_{A_i} = s \cdot \sqrt{A_i} \tag{16.32}$$

Therefore

$$\begin{aligned} s_A &= s(A_1 + A_2 + \cdots + A_n)^{1/2} \\ &= s \cdot \sqrt{A} \end{aligned} \tag{16.33}$$

as in equation (16.20).

We would obtain the same result if the number of component triangles was n_1, n_2 or any other number. Consequently the number of component triangles does not have any influence upon the accuracy of the measurement, *provided that the base/height ratio of each triangle is approximately equal to unity and the measurements have all been made independently of one another.*

DETERMINATION OF AREA FROM COORDINATE DIFFERENCES

The next stage is to consider how area may be calculated from the coordinates of the points on the perimeter of an n-sided polygon. Figure 16.3 illustrates the same eight-sided polygon $ABCDEFGH$ upon which are superimposed the Cartesian axes OX and OY. We adopt the normal graph convention that x

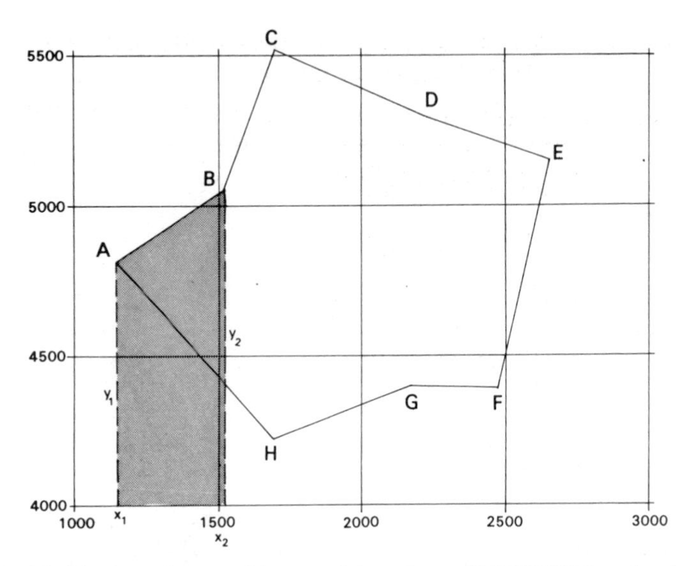

FIG. 16.3. The determination of the area of the polygon *ABCDEFGH* from the plane Cartesian coordinates of the perimeter points. The grid lines are shown at intervals of 500 metres. The shaded figure is the trapezoid formed by the line *AB* of the perimeter with the grid lines.

represents the grid easting and y is the grid northing. Inspection of the figure indicates that it may be imagined as being composed of the areas of eight trapezoids with bases $x_1, x_2, \ldots, x_8$ and with heights $(y_2 - y_1)$, $(y_3 - y_2) \cdots$ $(y_8 - y_1)$. Modifying equation (16.12) for the present purpose, the area of the figure may be expressed in the general form

$$2A = \sum(x_i - x_{i+1})(y_{i+1} - y_i) \tag{16.34}$$

Then

$$\begin{aligned} 2A &= \sum(x_i y_{i+1} - x_i y_i + x_{i+1} y_{i+1} - x_{i+1} y_i) \\ &= \sum(x_i y_{i+1} - x_{i+1} y_i) + \sum(x_i y_{i\mp} + x_i y_i) \end{aligned} \tag{16.35}$$

Since the second sum in this equation is equal to zero, we obtain

$$2A = \sum x_i y_{i+1} - \sum x_{i+1} y_i \tag{16.36}$$

The equations which follow may be derived from these two basic formulae (16.35) and (16.36). In equation (16.35)

$$\sum x_{i+1} y_{i+1} = \sum x_i y_i$$

so we may rewrite the equation as

$$\begin{aligned} 2A &= \sum(x_i y_{i+1} - x_i y_i - x_{i+1} y_i + x_i y_i) \\ &= \sum x_i(y_{i+1} - y_i) - y_i(x_{i+1} - x_i) \end{aligned} \tag{16.37}$$

Putting $\delta x_i = x_{i+1} - x_i$ and $\delta y_i = y_{i+1} - y_i$ equation (16.37) may be rewritten

TABLE 16.3. *Computation of the area of the polygon ABCDEFGH using the grid coordinates of the eight points on the perimeter*

Part 1: Coordinates of the eight points

Point	Easting (x) (m)	Northing (y) (m)
A	1,160	4,809
B	1,518	5,046
C	1,698	5,509
D	2,231	5,278
E	2,647	5,149
F	2,478	4,387
G	2,173	4,396
H	1,700	4,220

Part 2: Computation of area from equation (16.39)

x_i	y_{i-1}	y_{i+1}	$(y_{i-1} - y_{i+1})$	$x_i(y_{i-1} - y_{i+1})$
1,160	4,220	5,046	−826	−958,160
1,518	4,809	5,509	−700	−106,260
1,698	5,046	5,278	−232	−393,936
2,231	5,509	5,149	360	803,160
2,647	5,278	4,387	891	2,358,477
2,478	5,149	4,396	753	1,865,934
2,173	4,387	4,220	167	362,891
1,700	4,396	4,809	−413	−702,100
				$2A = 2{,}273{,}666$
				$A = 1{,}136{,}833\ \text{m}^2$
				$= 113.68$ ha

Part 3: Alternative solution based upon equation (16.40)

y_i	x_{i-1}	x_{i+1}	$(x_{i+1} - x_{i-1})$	$y_i(x_{i+1} - x_{i-1})$
4,809	1,700	1,518	−182	−875,238
5,046	1,160	1,698	538	2,714,748
5,509	1,518	2,231	713	3,927,917
5,278	1,698	2,647	949	5,008,822
5,149	2,231	2,478	247	1,271,803
4,387	2,647	2,173	−474	−2,079,438
4,396	2,478	1,700	−778	−3,420,088
4,220	2,173	1,160	−1,013	−4,274,860
				$2A = 2{,}273{,}666$
				$A = 1{,}136{,}833\ \text{m}^2$
				$= 113{\cdot}68$ ha

as

$$2A = \sum x_i \delta y_i - \sum y_i \delta x_i \tag{16.38}$$

Moreover, since

$$\sum x_i y_{i-1} = \sum x_{i+1} y_i$$

we obtain, from equation (16.36)

$$2A = \sum x_i (y_{i-1} - y_{i+1}) \tag{16.39}$$

Using similar reasoning, but reckoning the bases of the trapezoids to be $y_1, y_2, \ldots, y_n$ and their heights to be $(x_2 - x_1), (x_3 - x_2) \cdots (x_n - x_{n-1})$, we may obtain the alternative expression

$$2A = \sum y_i (x_{i-1} - x_{i+1}) \tag{16.40}$$

As an example of the use of equation (16.39) we compute the area of the parcel *ABCDEFGH* which has already been done using direct solutions of the areas of the component triangles. The computation may be laid out in the form of Table 16.3.

THE HERRINGBONE METHOD

The coordinate differences in equations (16.39) and (16.40) may be replaced by the sums of the increments. In other words,

$$2A = \sum x_i (\delta y_{i-1} + \delta y_i) \tag{16.41}$$

Similarly

$$2A = -\sum y_i (\delta x_{i-1} + \delta x_i) \tag{16.42}$$

From equation (16.41) we may derive the expression

$$2A = \sum x_i \delta y_i + \sum x_{i+1} \delta y_i \tag{16.43}$$

From equation (16.43)

$$2A = \sum \delta y_i (x_i + x_{i+1})$$

but

$$x_i + x_{i+1} = 2x_i + \delta x_i$$

Therefore

$$\begin{aligned} 2A &= \sum \delta y_i (2x_i + \delta x_i) \\ &= \sum \delta y_i \delta x_i + 2\sum x_i \delta y_i \end{aligned}$$

or

$$-A = \tfrac{1}{2} \sum \delta x_i \delta y_i + \sum x_i \delta y_i \tag{16.44}$$

Similarly

$$A = \tfrac{1}{2}\left(\sum x_i y_{i-1}\right) - \tfrac{1}{2}\left(\sum y_i x_{i-1}\right) \tag{16.45}$$

The last is the best known solution, sometimes known as the *herringbone method* from the form of the cross-multiplications in parentheses in (16.45). In the example already studied we determine

$$\sum x_i y_{i-1} = +76{,}790{,}188$$

$$\sum y_i x_{i-1} = -74{,}516{,}522$$

Therefore

$$2A = 2{,}273{,}666$$

$$A = 1{,}136{,}833\ \mathrm{m}^2$$
$$= 113{\cdot}683\ \mathrm{ha}$$

Determination of Area from Polar Coordinates

It is similarly possible to determine the area of a parcel from the coordinates of points on the perimeter which have been measured in polar coordinates. Figure 16.4 illustrates the method with reference to a simple quadrilateral, *ABCD*, whose four points have been recorded in (r, θ) polar coordinates measured from the origin *O*. From equations (16.2)–(16.6) we may express the area of a triangle in terms of two sides and the included angle. Thus the triangle *ADO* comprises two measured radii vectors r_A and r_D and the included angle is $\theta_D - \theta_A$, which is the difference between the two vectorial angles measured from some initial line *OP*.

We may see that in this simple example we may determine the area of the entire figure from the sum of the areas of triangles *ODA* and *ODC*. Moreover, we may determine the part of the figure lying outside the parcel by means of the two triangles *OAB* and *OBC*. Consequently

$$\text{Area } ABCD = (\text{area } OAD + \text{area } ODC) - (\text{area } OAB + \text{area } OBC) \tag{16.46}$$

In terms of the measured coordinates

$$\begin{aligned} \text{Area } ADO &= \tfrac{1}{2}[r_A \cdot r_D \cdot \sin(\theta_D - \theta_A)] \\ \text{Area } ODC &= \tfrac{1}{2}[r_D \cdot r_C \cdot \sin(\theta_C - \theta_D)] \\ \text{Area } OAB &= \tfrac{1}{2}[r_A \cdot r_B \cdot \sin(\theta_B - \theta_A)] \\ \text{Area } OBC &= \tfrac{1}{2}[r_B \cdot r_C \cdot \sin(\theta_C - \theta_B)] \end{aligned} \tag{16.47}$$

Such a system of equations may be extended to more complicated figures, but we do not devote space to this here. The only reason for treating briefly with the use of polar coordinates here is to indicate the method employed in the Coradi *Digimeter Model DMB*, which was one of the first, and most successful, instruments to be developed for measurement of area by digital methods. The equipment comprises a measuring head, the geometrical centre of which represents the origin of the polar coordinates, *O*. The measuring mark is mounted on the end of a steel bar which may be moved to different distances, *r*, and different azimuth angles, θ, about the origin. The polar coordinates are

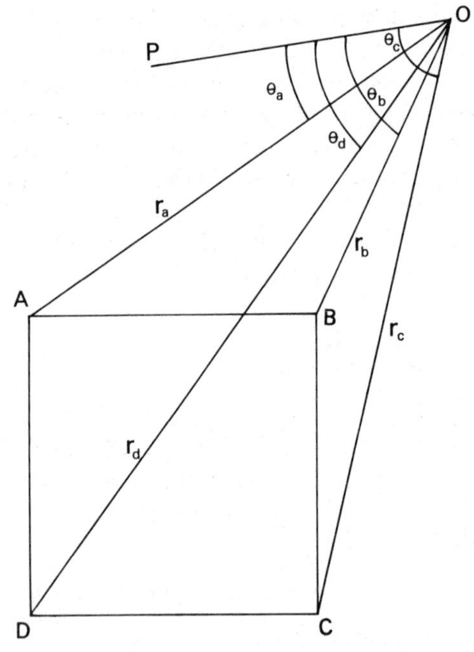

FIG. 16.4. The area of a quadrilateral $ABCD$ expressed in terms of the polar coordinates of the corner points.

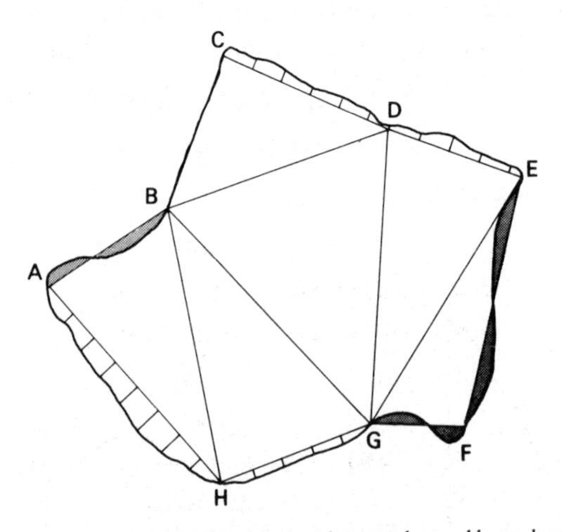

FIG. 16.5. The polygon $ABCDEFGH$ showing the actual parcel boundary and how the marginal element may be measured. The lines AB, EF and FG measured on the ground or a map have been chosen as give-and-take lines which separate equal (shaded) portions inside and outside the parcel. The lines CD, DE, GH and HA lie within the parcel and equidistantly spaced offsets have been used to locate the boundary and to calculate the areas of these marginal elements. The line BC matches the parcel boundary so closely that no special treatment of it is required.

recorded on tape or disc by pressing a footswitch every time the measuring mark occupies a required point. In use, the measuring head is placed upon the map in a position where the mark can be moved round the perimeter points of a parcel. An ordered sequence of such points are measured separately and individually, taking these in either clockwise or anticlockwise order round the parcel. It is necessary to remeasure the coordinates of the starting point after completion of the circuit. In this respect the operation of the Digimeter Model DMB is similar to a planimeter for, indeed, it is a member of Category II of planimeters, to be discussed in Chapter 17.

MEASUREMENT OF THE MARGINAL ELEMENTS

If the true boundary of a parcel does not concide with any of the straight lines of the geometrical framework, a portion of the parcel will lie between this and the true boundary as shown in Fig. 16.5. If the framework lies wholly within the parcel and such marginal elements are ignored, the calculated area of the parcel is smaller than its true area.

MEASUREMENT OF AREA FROM OFFSETS

We have already seen that the method of locating points of ground detail and boundaries in chain surveying is by means of offsets; using linear measurements made from the main framework of the survey at right angles to its component lines. From the point of view of area measurement it is useful for the offsets to be measured at equidistant intervals along the main lines of the framework.

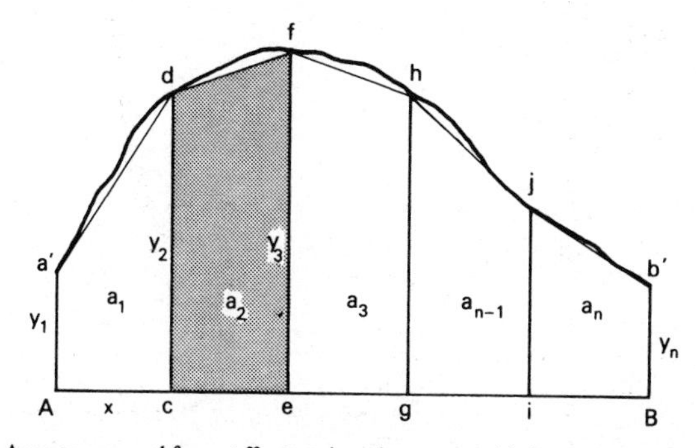

FIG. 16.6. Area measured from offsets using the trapezoidal rule. It is assumed that the parcel boundary is represented by a straight line over the short distance between the offsets. Since the offset lines are parallel to one another, the unit figure, such as *cdef*, is a trapezoid.

The Trapezoidal Rule

The geometry of this method is illustrated in Fig. 16.6, which shows a series of offsets measured from the line AB, namely cd, ef, gh, ij, etc. Examination of any single figure, such as $cdfe$, formed by these lines indicates that if the portion of the parcel boundary, df, is short enough to be regarded as a straight line, the resulting quadrilateral is a trapezoid. From equation (16.12), a_2, the area of the small element $cdfe$, is

$$a_2 = \tfrac{1}{2}ce(cd + ef)$$

Similarly, the area of the small trapezoid $efgh$ is

$$a_3 = \tfrac{1}{2}eg(ef + gh)$$

Consequently the area of the whole of the marginal elements $Aa'f \cdots b'B$ is

$$A = a_1 + a_2 + a_3 \cdots + a_n \qquad (16.48)$$

If we denote $Aa' = y_1$, $cd = y_2$, $ef = y_3$ etc. to $Bb' = y_n$ and, moreover, the offsets are equidistantly spaced along the survey line so that $Ac = ce = eg \cdots = x$, the whole computation may be summarised as the *trapezoidal rule.* This may be written in the form

$$A = \tfrac{1}{2}x(y_1 + 2Y + y_n) \qquad (16.49)$$

where

$$Y = \sum y_i \qquad (16.50)$$

Equation (16.47) may also be written in the form

$$A = x\{(y_1 + y_n/2) + Y\} \qquad (16.51)$$

If, moreover, the two end points of the line, A and B, are situated on the true boundary of the parcel, $Aa' = y_1 = Bb' = y_n = 0$ so that equation (16.51) simplifies to

$$A = x \cdot Y \qquad (16.52)$$

The Mean Ordinate Rule

This is another simplification which also assumes the boundary to be composed of short straight lines. If the parcel boundary is roughly parallel to the survey line, the difference between the lengths of successive offsets is small and the pairs of values, such as cd and ef may be replaced by their mean value, m_2.

The sum now comprises the means of successive pairs, i.e.

$$M = \sum m_i = m_1 + m_2 + m_3 + \cdots + m_n \qquad (16.53)$$

and

$$A = x \cdot M \qquad (16.54)$$

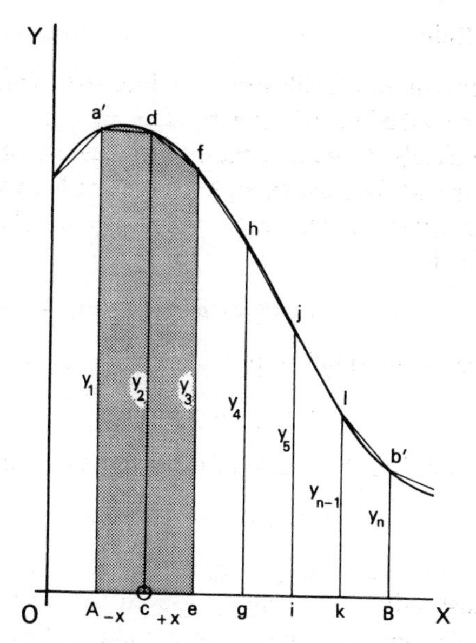

FIG. 16.7. The geometry of Simpson's rule. The area of the element $Aa'fe$ is determined by integration of the expression defining the area of the parabola passing through the points A', d and f.

Simpson's Rule

Thus far it has been assumed that the boundary of the parcel approximates to a succession of short straight lines between successive offsets. This may be justified if the distance, x, between the offsets is made small enough. This is demonstrated by the worked example in the next section where we compare the results obtained from offsets made at intervals of 100 m, 50 m and 20 m along the chain line. However, there is a limit to this process of subdivision because the method becomes too time-consuming to be used in field survey although it remains a practical possibility when the measurements are all being made from a map or plan. Another means of overcoming this difficulty is to replace the straight line df with a curve. Simpson's rule provides one solution of this sort. Fig. 16.7 shows a curve which is a more accurate representation of the boundary than the rectilinear elements shown in Fig. 16.6. The notation used for the points on the curve and along the X-axis are the same as those used in Fig. 16.6. Thus the ordinates corresponding to the terminal offsets are Aa' and Bb' respectively, and a succession of equidistant offsets cd, ef, gh, ij and kl have also been measured. We make the assumption that the curve passing through the points a', d, f, h, j, l and b' is a parabola of the form

$$y = a + bx + cx^2 \tag{16.55}$$

In order to define such a line in terms of (x, y) Cartesian coordinates it is necessary to define the origin and axes of the coordinate system. We therefore consider the point C to be the origin of the axes. The X-axis coincides with the line AB and is positive in the direction of B. Therefore the Y-axis is parallel to the offsets measured from this line and is positive in the direction of the parcel boundary.

If the separation of the offsets is x_1, the coordinates of the point A is $(-x_1, 0)$ and that of e is $(+x_1, 0)$. Since equation (16.54) is satisfied by all the points on the boundary $a', d, f \ldots$, we may now express the lengths of the three offsets algebraically as

$$\begin{aligned} Aa' &= a - bx_1 + cx_1 = y_1 \\ cd &= a = y_2 \\ ef &= a + bx_1 + cx_1^2 = y_3 \end{aligned} \tag{16.56}$$

Adding the expressions for y_1 and y_3,

$$y_1 + y_3 = 2(a + cx_1^2)$$

Therefore

$$\begin{aligned} 2cx_1^2 &= y_1 + y_3 - 2a \\ &= y_1 + y_3 - 2y_2 \end{aligned}$$

and

$$cx_1^2 = \tfrac{1}{2}(y_1 + y_3 + 2y_2) \tag{16.57}$$

The area enclosed between the points $Aa'fe$ may be found by integrating equation (16.53) between the limits $-x_1$ and $+x_1$, i.e.

$$A = \int_{-x_1}^{+x_1} (a + bx + cx^2)\,\mathrm{d}x \tag{16.58}$$

$$= [ax + \tfrac{1}{2}bx^2 + \tfrac{1}{3}cx^3] \tag{16.59}$$

$$\begin{aligned} &= 2ax_1 + \tfrac{2}{3}cx_1^3 \\ &= 2x_1(a + \tfrac{1}{3}cx_1^2) \end{aligned}$$

Therefore

$$\begin{aligned} A &= 2x_1\{y_2 + \tfrac{1}{6}(y_1 + y_3 - 2y_2)\} \\ &= 2x_1\{(4y_2 + y_1 + y_3)/6\} \end{aligned} \tag{16.60}$$

$$= x_1\{(y_1 + 4y_2 + y_3)/3\} \tag{16.61}$$

Similarly the area of the figure $efji$ may be shown to be

$$A = x_1\{(y_3 + 4y_4 + y_5)/3\}$$

where y_4 is the offset gh and y_5 is the offset ij.

Combining the two results, and generalising the expression for a figure comprising n offsets,

$$A = x_1/3(y_1 + y_n) + 2\sum y_i + 4\sum y_i' \tag{16.62}$$

provided that n is an odd number. In equation (16.62), y represents the odd-numbered ordinates and the summation is from $1 = 3$ through $1 = n - 2$. The symbol y' denotes the even-numbered ordinates and the summation is from $i = 2$ through $i = n - 1$.

If n is odd, there is an even number of strips such as *cdfe* and Simpson's rule may be stated as follows:

> *The area is equal to the sum of the first and last ordinate plus twice the sum of the remaining odd ordinates plus four times the sum of all the even ordinates, multiplied by one-third of their common distance apart.*

Simpson's rule was first described during the eighteenth century as a method of integration. It has been much used ever since for two important cartometric applications: to determine the areas of irregular parcels according to the methods described above, and in the volumetric determination of amount of material to be excavated or added (cut-and-fill) in civil engineering applications. On the whole the method has served the civil engineer well, for it is a better approximation of a curved boundary than is the succession of straight lines assumed by the trapezoidal rule. Amongst other methods which have been described, there are the *Formule de Poncelet* and *Formule de Nicolosi*, which have been described in the older French textbooks on surveying (e.g. Carrier, 1947). We do not consider these methods here.

Simpson's rule suffers from two important disadvantages. The first is the way in which a discontinuity occurs at the junction of two parabolic arcs where these represent an irregular boundary. This error may be important if the boundary has inflection points as is illustrated by the discrepancies between the boundary and the two parabolical at the point *i*. The second source of error occurs when the number of offsets is even, so that there is an odd number of strips. Then the solution is not unique because two different solutions are possible depending upon which elements of the line are fitted to the parabolae. This is illustrated by Fig. 16.8.

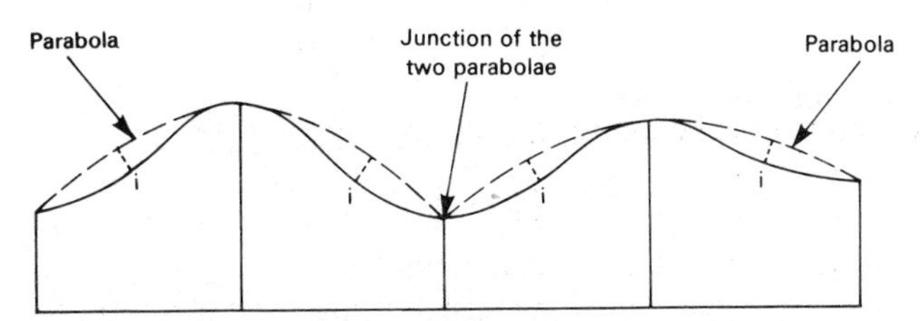

FIG. 16.8. Discrepancies between the parabolae fitted to a sinuous curve (full line) showing maximum differences at the inflection points, *i*, and discontinuities at the junctions between the parabolae. (Source: Ahmed, 1982).

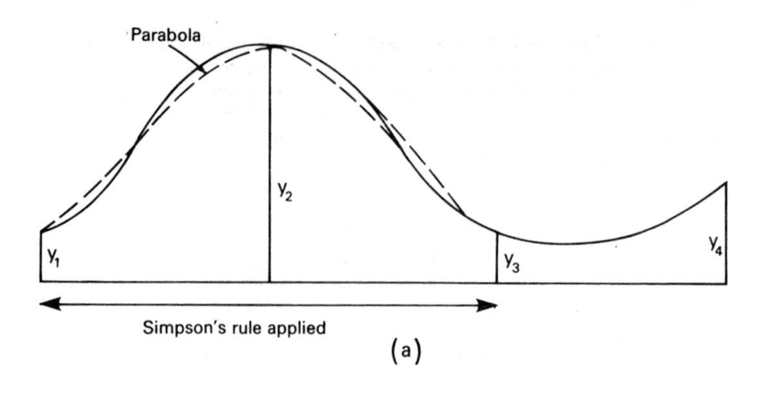

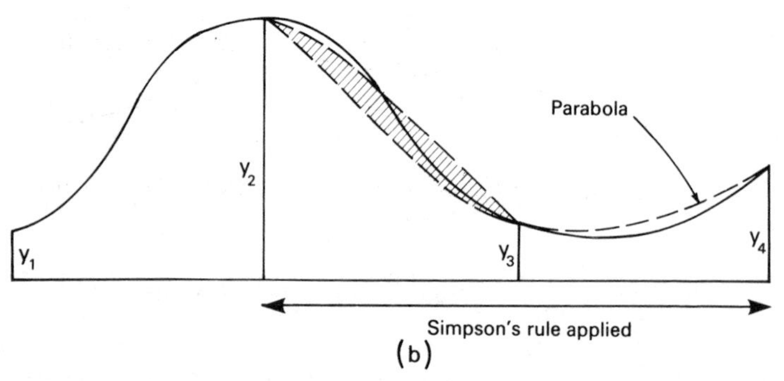

FIG. 16.9. Alternate parabolae fitted to data points. In (a) the parabola is fitted to the ordinates y_1, y_2 and y_3. In (b) the parabola is fitted to the ordinates y_2, y_3 and y_4. The shaded area indicates the discrepancy between the 2 curves between y_2 and y_3. (Source: Ahmed, 1982).

Ahmed's Method

Fouad Ahmed (1982) has examined the problem of measuring the areas of marginal elements anew, and has produced a solution for fitting a polynomial curve of the form

$$f(x) = a_0 + a_1x + a_2x^2 + a_3x^3 \tag{16.63}$$

through the data points represented by the ends of the offsets, *def*, etc. His final expression, which is here presented without proof, is

$$A = x/24\{a(y_1 + y_n) + b(y_2 + y_{n-1}) + c(y_3 + y_{n-2}) + d\Sigma y_i\} \tag{16.64}$$

In this expression, x is the distance between the offsets which are successively of length $y_1, y_2 \ldots y_n$. The numerical values for the constants a, b, c and d are integers which depend upon n. The values for these are listed in Table 16.4. We now compare the various techniques with reference to a specific example.

TABLE 16.4. *Numerical values for the constants a–d used in Fouad Ahmed's polynomial solution for the measurement of marginal areas from offsets*

No. of offsets n	Constant a	b	c	d
3	8	32	–	–
4	9	27	–	–
5	9	28	22	–
6	9	28	23	–
7 or more	9	28	23	24

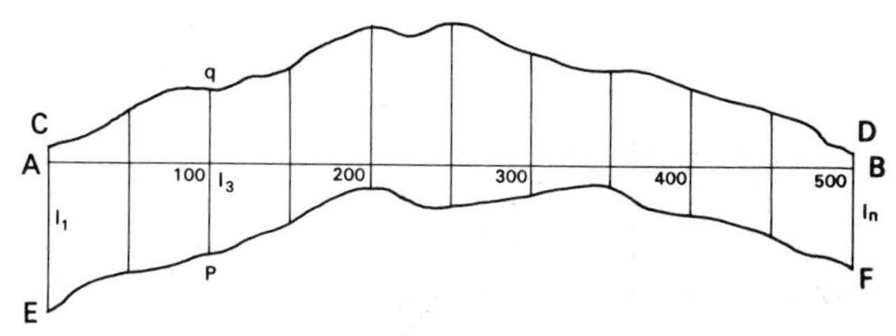

FIG. 16.10. A narrow parcel *CDFE*, measured by means of offsets at 50-metre intervals along the line *AB*.

An Example of Calculating the Areas of Marginal Elements by Different Methods

Figure 16.10 illustrates an elongated parcel, *CDEF*, which has been plotted using equidistantly spaced offsets from the straight line *AB* which is one side of the framework of a large scheme. The length of *AB* is 500 metres, so that if the offsets were measured at 50-metre intervals along the line, each boundary *CD* and *EF*, would be defined by 11 such lines.

TABLE 16.5. *Lengths of the offsets illustrated in Fig. 16.10 (in metres)*

Distance from *A*	Offsets to Line *CD*	Offsets to Line *EF*	Sum of offsets
0	10	90	100
50	32	65	97
100	40·75	55·75	96·5
150	58	33	91
200	82·5	15	97·5
250	82·5	27·5	110
300	66·25	19·25	85·5
350	54·5	13·25	67·75
400	43·75	29·5	73·25
450	31·25	45	76·25
500	8	63·75	71·75

Solution by the Trapezoidal Rule

From equation (16.51)

Figure $ACDB$: Area $= 50\{(10 + 8)/2 + 491{\cdot}5\} = 25{,}025\,\text{m}^2$

Figure $ABFE$: Area $= 50\{(90 + 63{\cdot}75)/2 + 303{\cdot}25\} = 19{,}006\,\text{m}^2$

Solution by the Mean Ordinate Rule

From equations (16.53) and (16.54)

Figure $ACDB$: Area $= 50 \times 500{\cdot}5 = 25{,}025\,\text{m}^2$

Figure $ABFE$: Area $= 50 \times 380{\cdot}125 = 19{,}006\,\text{m}^2$

Solution by Simpson's Rule

From equation (16.62)

Figure $ACDB$: Area $= 50/3\{18 + 2 \times 233{\cdot}25 + 4 \times 258{\cdot}25\}$
$= 25{,}292\,\text{m}^2$

Figure $ABFE$: Area $= 50/3\{153{\cdot}75 + 2 \times 119{\cdot}5 + 4 \times 183{\cdot}75\}$
$= 18{,}796\,\text{m}^2$

Solution by Ahmed's Method

From equation (16.64)

Figure $ACDB$: Area $= 50/24\{9 \times 18 + 28 \times 63{\cdot}25 + 23 \times 84{\cdot}5$
$+ 24 \times 343{\cdot}75\} = 25{,}263\,\text{m}^2$

Figure $ABFE$: Area $= 50/24\{9 \times 153{\cdot}75 + 28 \times 110 + 23 \times 85{\cdot}25$
$+ 24 \times 10\} = 18{,}784\,\text{m}^2$

It is also instructive to use the same data to investigate the effect of reducing the spacing between the offsets. To do this we reduce the value of x to 20 metres and also increase it to 100 metres (Table 16.6). From these data for $x = 20\,\text{m}$ the reader may verify that the area of the figure $ACDB$ is

25,010 m^2 by both the mean ordinate and trapezoidal rules
25,009 m^2 by Ahmed's method.

Because $n = 26$, Simpson's rule is not applicable to the whole figure.

The results of the different solutions may be summarised in Table 16.7, which demonstrates the obvious advantage of defining an irregular boundary in detail, by using numerous closely spaced offsets rather than attempting to

TABLE 16.6. *The lengths of offsets measured from the line AB to the boundary CD taken at intervals of 20 metres (in metres)*

Distance from A	Line CD	Distance from A	Line CD
0	10	260	83
20	15	280	73·25
40	27	300	66·25
60	38	320	58
80	46·75	340	54·5
100	40·75	360	55
120	47	380	50·5
140	53·25	400	43·75
160	63	420	38·75
180	73	440	34·25
200	82·5	460	28·25
220	75·75	480	12
240	82	500	8

measure the area with a few widely spaced offsets. However, the method used is also important. The difference in measured area between the solutions for $x = 20$ metres and $x = 100$ metres is 3% for the trapezoidal rule but it is only 0.9% for Ahmed's method, demonstrating that greater approximation is present in the use of the trapezoidal rule.

THE TOTAL AREA OF THE PARCEL

We now consider the whole figure $CDFE$ as a single parcel which is intersected by one survey line AB, from which various offsets were measured as tabulated in Tables 16.5 and 16.6.

Obviously the total area of the parcel is the sum of the areas of the two figures $ACDB$ and $ABFE$. Combining the results obtained for the 50-metre offsets solution, the total area of the parcel is

By the trapezoidal and mean ordinate rule: 44,031 m^2

TABLE 16.7. *Comparison of different measurements of the figure ACDB (in metres)*

	Distance between offsets		
Method	20 metres	50 metres	100 metres
Trapezoidal rule	25,010	25,025	24,225
Mean ordinate rule	25,010	25,025	24,225
Simpson's rule	–	25,292	–
Ahmed's method	25,009	25,263	24,789
No. of offsets, n	26	11	6

By Simpson's rule:	44,088
By Ahmed's method:	44,047

Nearly all the elementary textbooks on surveying describe measurement of the area of such a parcel in one step.

Consider the parcel to be the figure *CDFE* as illustrated in Fig. 16.10 showing parallel lines such as *pq* which are not referred to the line *AB* but which correspond to the total length of both offsets. For example $pq = 96{\cdot}5$ metres $= l_3$ and all the other lines $l_1, l_2 \ldots l_n$ have the lengths recorded in the righ-hand column of Table 16.7.

The equations used to find the area of the parcel are easily derived from (16.50), (16.53), (16.62) and (16.64), usually by substituting l_i for y_i. Thus

Trapezoidal rule: $A = x\{(l_1 + l_2)/2 + \Sigma l_i$ (16.65)

Simpson's rule: $A = x/3(l_1 + l_n) + 2\Sigma l_i + 4\Sigma l_i'$ (16.66)

As in (16.62) l_i are the odd-numbered and l_i' are the even-numbered lines.

Ahmed's method: $$A = x/24\{a(l_1 + l_n) + b(l_2 + l_{n-1}) + c(l_3 + l_{n-2}) + d\Sigma l_i\} \quad (16.67)$$

The method now approximates to measurement by means of linear measurements, l_i, along equidistant parallel lines. The techniques employed in this kind of measurement are considered in some detail in Chapter 20.

17

Area Measurement by Planimeter

Jacob Amsler, as I have been told by the late Prof. Hesse, was at that time a student at Königsberg, where the late Prof. Franz Neumann encouraged students to work at the lathe and otherwise use their hands. He thus was enabled to make his first instrument with his own hands in about 1854.... Many thousands of them have been manufactured by Amsler at his works in Schaffhausen, and though in England many are sold with the name of an English firm engraved on them, practically all have come from Schaffhausen.

(O. Henrici, *British Association for the Advancement of Science, Report*, 1894)

Henrici (1894) described three different types of instrument:

Type I, or *orthogonal planimeters*
Type II, or *polar coordinate planimeters*
Type III, *planimeters of the Amsler type*

The Type I planimeters are now of historical interest only. Some of the earliest investigations into instrumental methods of area measurement, made early in the nineteenth century, were devoted to trying to construct instruments of this sort. We do not consider them further in this book, but refer the interested reader to Kneissl (1963), as well as Henrici's classic paper, for an account of how some of them were intended to work.

The Type II instrument is essentially a small coordinatograph which is used to measure the radius vector and vectorial angle of each point on the perimeter of a parcel. It was seen in Chapter 16 that it is possible to calculate the area of a parcel by means of polar coordinate differences. However, these are tedious to measure and compute without the use of a digital computer. Consequently it was not until digital data capture and processing arrived during the 1960s that the method had any real practical application. The Coradi *Polar Digimeter Model DMB* is typical of the instruments produced at this time which exploited the advantages of automatic data processing.

The Type III instruments have been, and remain, the most important to cartometry ever since the first Amsler instrument was constructed. Most modern planimeters still closely resemble the original design. During more than a century there have been small modifications which make the instrument more accurate in use or more versatile in its applications.

The simple planimeter operates by the movement of a measuring wheel which is in physical contact with the plane surface of the map or drawing, but in the family of disc-planimeters the wheel operates in contact with the machined surface of a metal disc or plate.

DESCRIPTION OF THE INSTRUMENTS

The Polar Planimeter

All Type III planimeters function on the principle that an index mark, or *tracing point*, is moved round the perimeter of the parcel to be measured and during this operation the movements of the measuring wheel are recorded upon a scale. Measurement of the parcel has been completed when the whole perimeter has been traced. Fig. 17.1 illustrates the main components of a typical polar planimeter. There are two rigid straight arms which are hinged at the *pivot*, *B*. The *pole arm*, *PB*, is anchored at the point *P* by means of a pin, spike, or by the sheer mass of the steel block which represents the pole. The length of this arm, R', is constant. The second arm, known as the *tracing arm*, *BC*, contains the tracing mark at the point *C*. This is either a steel point which can be set to ride just above the surface of the map, or it is a lens mounted within a cylindrical barrel and having an opaque mark, engraved on the under-surface of the lens which rides flat on the map. In modern instruments the arms may be detached from one another at the pivot. Consequently the pole arm may be placed either to the left or right of the tracing arm, as is illustrated in Fig. 17.2. These may be referred to as *pole left*, in which the pole lies to the left of the parcel to be measured and *pole right*, which is the alternative setting. An instrument in which the poles can be reversed is known as a *compensating* polar planimeter. The term is used because some of the systematic errors which arise from small irregularities in manufacture and assembly of the instrument may be removed by using the planimeter in these two positions. There are two ways in which the concept of pole position may be

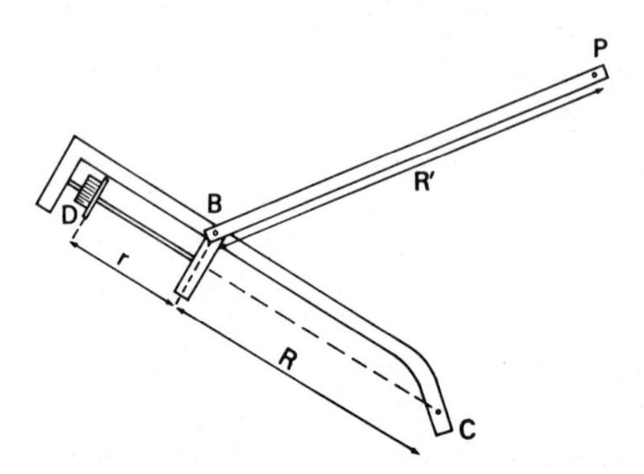

FIG. 17.1. The principal components of a polar planimeter.

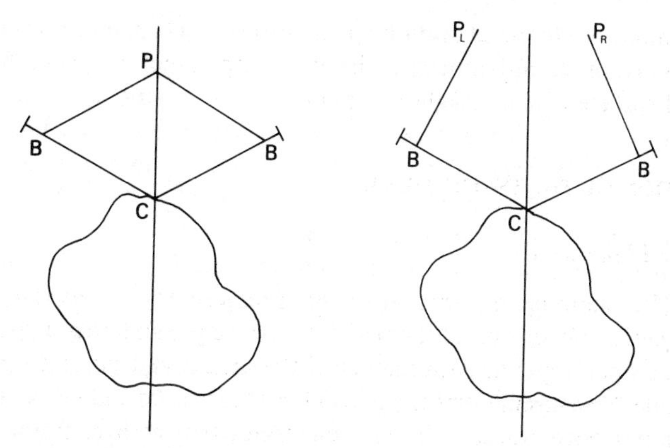

FIG. 17.2. The pole position of a planimeter: (a) represents the "same-pole" condition where the instrument is used both left and right of the same pole; (b) represents the effect of changing the position of the pole with respect to the parcel to be measured.

interpreted. In Fig. 17.2(a) the pole P remains the same point throughout, but the relative positions of the arms and the measuring wheel are reversed about an axis from the pole to the centre of the parcel. This is known as the "same-pole" condition. In Fig. 17.2(b) two entirely different positions for the pole, P_L and P_R have been adopted. This is obviously the "different-poles" condition.

The *measuring wheel* is mounted on part of the tracing arm in the vicinity of the pivot. For convenience in design the position of the measuring wheel may be offset to one side of the rod comprising the tracing arm, but provided that the axle of the wheel is parallel to the axis of the tracing arm, and the end of the tracing arm is bent outwards to create the straight line CD, parallel to this axis, the theory of the instrument in unaffected. Another wheel may be mounted in a plane parallel to the axle of the measuring wheel. This provides stability to the

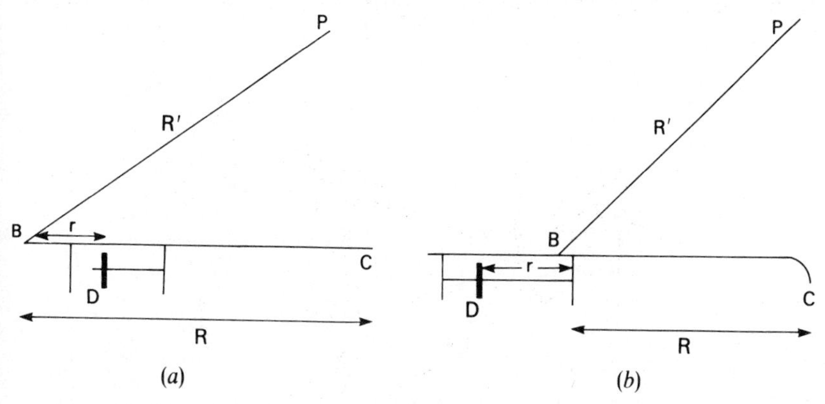

FIG. 17.3. The relationship of the pivot, B, of a polar planimeter to the measuring wheel, D.

frame containing the pivot and measuring wheel but it plays no other part in the operation of the instrument. It should be noted that the pivot, *B*, and the measuring wheel may occupy different relative positions. It is therefore necessary to distinguish between the *inner-pivot* instruments illustrated in Fig. 17.3(a) and the *outer pivot* instrument illustrated in Fig. 17.3(b).

The movements of the measuring wheel are read on two scales. The disc scale is turned by the rotation of a worm drive on the axle of the measuring wheel. This scale measures revolutions of the wheel on a scale numbered 0 through 9, so that the total range of the scale represents ten complete rotations of the wheel. The second scale is a drum which is an integral part of the measuring wheel, again subdivided into the numbered divisions 0 through 9. Each of these divisions is further subdivided into ten parts. Consequently there are 100 marked divisions on the complete circumference of the drum. Adjacent to this scale is a vernier which permits further subdivision of the drum scale to one-tenth of each marked division so that one unit on the vernier corresponds to 1/1,000 of the circumference of the measuring wheel or 1/10,000 of the total scale capacity of the instrument. A reading of the two scales therefore produces a four-digit number, e.g. 1584, as illustrated in Fig. 17.4. The first digit, 1, is read from the disc scale, the second digit, 5, is a numbered division of the drum. The third digit, 8, is determined by counting the number of lines from that marked 5 on the drum to the zero line of the vernier. The final digit, 4, is read from the coincidence of one of the vernier lines with the divisions of the drum. Since the final digit is a vernier reading, we describe the whole scale reading as being in *vernier units* (*v.u.*). It will be seen later that the determination of the area corresponding to one vernier unit is an important stage in checking and calibrating a planimeter before making any measurements.

The difference between one scale reading, made before tracing round the parcel, v_1, and the second reading made on completion of tracing, v_2, is a linear measurement of the movements of the wheel. This is

$$\delta v = v_2 - v_1 \tag{17.1}$$

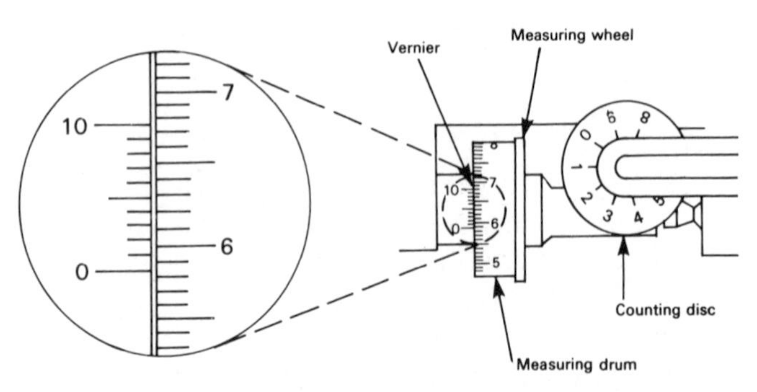

FIG. 17.4. The method of reading the scales of a polar planimeter.

and we describe by l the linear distance on the map which has been recorded on the scale by δv. During the tracing process the wheel rolls and slides over the surface of the map. The sliding movement does not produce any alteration in scale reading. It follows, therefore, that the distance l which is recorded by the scale is less than the actual track, s, followed by the wheel. The fact that both kinds of movement occur is critical to an understanding of how the instrument operates.

The scales and recording devices on ordinary polar planimeters have changed little since the earliest days, apart from enclosing them in a metal capsule as a protection from dust and fingertips. However, the use of two measuring wheels and scales on the Russian MIIZ planimeter deserves mention. This has the advantage that two scale readings are always available, thereby providing both a check upon gross errors of scale reading and some measure of compensation for systematic errors.

A change in the length of the tracing arm, R, alters the scale of the movements made by the measuring wheel. This property may therefore be used to set a *variable-arm* planimeter to read directly in the required units of measurement, so that one vernier unit corresponds to a simple fraction of the acre or hectare represented on the map in use. The actual setting to be made to the length of the tracing arm depends upon the dimensions of the particular instrument and this needs to be determined by calibration. *Fixed-arm* planimeters are designed to measure in only one set of units. For example, many fixed-arm instruments made for the British and North American markets work on the single conversion that 1,000 vernier units = 1 disc scale division = 10 square inches, or 1 v.u. = 0·01 square inches. At one time the relationship between the pivot and measuring wheel illustrated in Fig. 17.3 distinguished the variable-arm instruments (Fig. 17.3a) from the fixed-arm instruments (Fig. 17.3b), but this is no longer necessarily true of modern instruments.

The Polar Disc Planimeter

The polar disc planimeter was first introduced about 20 years after the invention of the polar planimeter. Since that time the disc instruments have been regarded as the superior instruments because they consistently produce more precise measurements. However, they are also appreciably more expensive. In these instruments the measuring wheel moves upon the plane surface of a metal disc and is therefore free from the uncertainties which may arise if it has to traverse minor irregularities of the paper surface, for example as the wheel of an ordinary planimeter moves off the edge of the map sheet. However, the real advantage of the disc instruments is the greater precision of the scale.

As shown in Fig. 17.5 the pole is P, represented by a massive cylindrical block approximately 15 cm in diameter. It is held in position by its own weight,

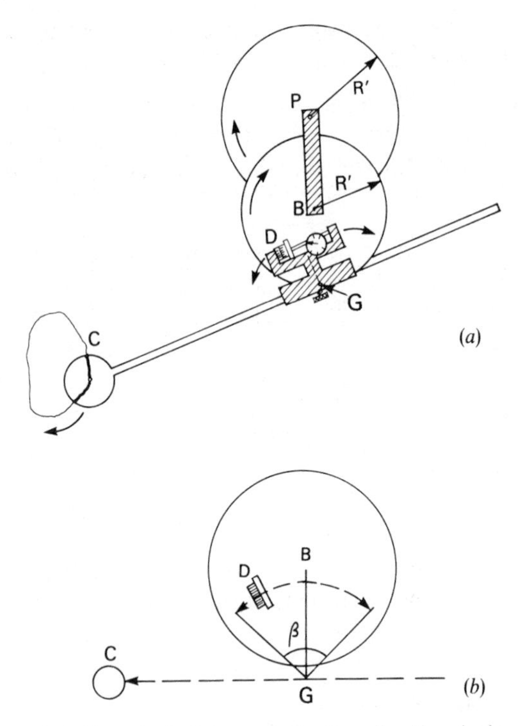

FIG. 17.5. The principal components of a polar disc planimeter.

but its massive size is not only for this reason. The upper rim of the cylinder is cut as a gear. When the pole arm is attached to the centre of the block this gear impinges upon another which is part of the vertical axis through the disc. The pole arm corresponds to the distance *PB* and is of fixed length. The axle of the disc is located on the axis of the pole arm, partly overlapping the cylindrical block of the pole unit. When the pole arm is turned clockwise about the pole, the gears cause the disc to rotate in a clockwise direction also. Moreover, the gears increase the rotation of the disc, the amount being expressed by the ratio R'_1/R'_2, where R'_1 is the radius of the pole block and R'_2 is that of the gear on the disc axle. In the Coradi Polar Disc Planimeter $R'_1 \approx 7{\cdot}5$ cm and $R'_2 \approx 6{\cdot}4$ cm, so that as the pole arm is moved through the angle α, the equivalent angular movement created by rotation of the disc is about $1{\cdot}17\,\alpha$.

The measuring wheel, *D*, is connected to the tracing arm by means of a *Y*-shaped bracket. This is constrained to move across the disc as the pivot is turned through the angle β. The distance of the measuring wheel from the pivot, and the diameter of the disc, are so arranged that the whole movement of tracing an arc involves turning the tracing arm through the angle $\beta = 90°$. If no other movement occurs as β changes, the measuring wheel slides across the surface of the disc without producing any alteration in the scale reading.

The results of these changes to the traditional pattern of polar planimeter area may be summarised as follows:

1 The total movement of the wheel is restricted to the surface of the disc. Apart from the advantages of uniformity of surface and absence of map edges, this makes the instrument more economical in the space needed to measure a large parcel.
2 The run of the measuring wheel is much increased by the exaggerated movements of the disc during its rotation. Therefore the difference between the scale readings obtained by polar disc planimeter is much greater than on the ordinary polar planimeter. This, in turn, means that the factor k is smaller.
3 The longer run of the scale means that its range must be increased. Thus the main scale is five times longer, being subdivided into 50 units, each representing one complete rotation of the measuring wheel. For this reason it is necessary to read five-digit numbers for each scale position, these lying within the range 00,001 through 50,000 units.

Operation, adjustment and calibration of the polar disc planimeter is similar to that described later for the ordinary polar planimeter.

The Automatic Reading Planimeter

A further development of the polar disc planimeter has been the introduction of automatic scale reading, thereby allowing measurement to proceed without frequent pauses to read the scales and paving the way to the use of automatic data processing in area measurement. This is now a highly desirable element in the maintenance of national registers of land for legal, administrative, fiscal and planning purposes. The Stanley–Cintel *automatic reading planimeter* was one of the first instruments in this field, developed during the 1950s to meet OS requirements. The instrument is essentially no more than a polar disc planimeter in which the measuring wheel carries a transparent disc on which are reproduced radiating lines. A beam of light passes through the disc and interruption of the beam by one of these lines activates a counter, which can output the results in digital form on paper or magnetic tap for subsequent computer processing.

The prototype Stanley–Cintel equipment was first tried by the Areas Section of the Large Scales Division of the OS in 1959, and a later version of it began to replace the existing equipment a few years later. Since the middle 1960s it has been the only method used to collect all area information which is required for certain of the large-scale maps. The methods of collecting the data and its subsequent processing in the early years were described by McKay (1966).

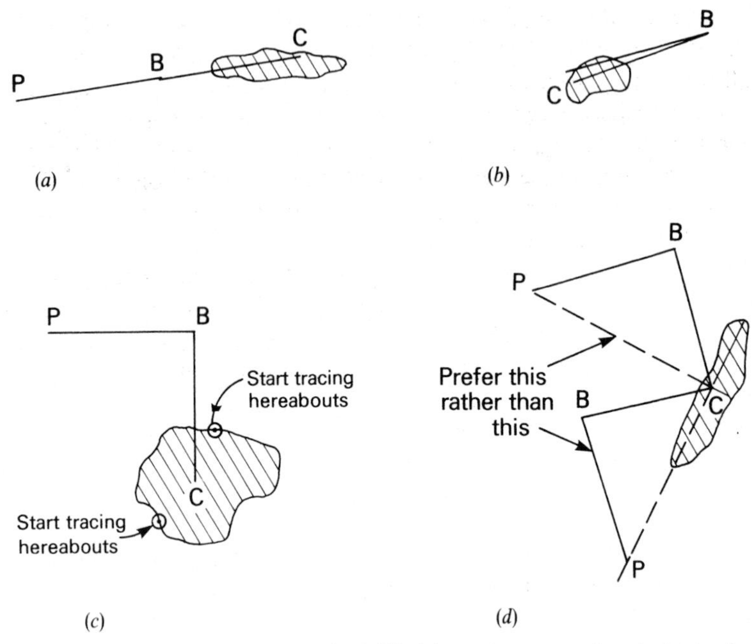

FIG. 17.6. The main conditions to be fulfilled in setting up a polar planimeter for operation in the pole-outside mode: (a) and (b) represent the two limiting conditions which render the instrument inoperable because the measuring mark cannot be moved further away from or closer to the planimeter pole; (c) and (d) represent the relationship between pole, pivot and measuring mark which should produce the optimum conditions for measurement of a symmetrical parcel (c) or an asymmetrical parcel (d).

Operation of a Polar Planimeter

There are two distinct modes of operation of a polar planimeter. If the parcel to be measured is small compared with the operating range of the instrument, the pole is placed outside the perimeter of the figure to be traced. This is known as the *pole-outside* mode of operation, and is the method normally used in cartometric work. For a larger parcel, which can only be traced in its entirety when the pole is located within the perimeter of the parcel, the instrument is operated in the *pole-inside* mode. From the point of view of understanding the theory of the polar planimeter, the pole-inside mode is easier to visualise, but practical use of it does give rise to certain difficulties in understanding the scale readings. Moreover the optimum sizes of parcels which can be measured in this mode (in many instruments $> 400\,cm^2$) is much larger than commonly encountered in cartometric work. This is owing to the various kinds of controls which need to be introduced into area measurement in order to reduce the instrumental and tracing errors created by the equipment and methods used, and the cartographic errors owing to the map projection in use, and paper distortion. It is a recurrent theme of this book that area measurement must be controlled by reference to the mathematical base of the map, either to the grid on a large- or medium-scale map or the graticule on a small- or atlas-scale map. In order to make the most efficient use of these controls, especially to reduce the unpredictable effects of paper distortion, the dimensions of these figures ought not to exceed 100–200 cm^2. This is a much smaller area than can be measured by planimeter in the pole-inside mode. It follows that there are strong arguments favouring only the pole-outside mode of operation for all cartometric work.

We start from the assumption that the map or drawing to be measured has already been mounted on a drawing board or table, and that the edge of the map lies some inches inwards from the edge of the drawing board so that the planimeter will not move across the edge of the working surface.

1 Place the instrument upon the drawing board in a position where it is possible to trace round the parcel without stretching the working range of the instrument to its limits. It is desirable that the measuring wheel should not cross the edge of the paper on which the map is printed during the tracing process. If draughting tape has been used to mount the map on the board, it is important that the measuring wheel should never make contact with it, or be required to ride over it. If a parcel lies so close to the edge of the map that movement of the wheel across the edge is unavoidable, it is necessary to attach paper of similar thickness to the board. This must be butt-jointed to the edge of the map, for there must be no overlap of changes in the thickness of the paper at such joints.
2 Choose a position for the pole by placing the tracing point near the centre of the parcel and the two arms set approximately perpendicular to one another. This is illustrated in Fig. 17.6(c). If the parcel is extremely elongated, as in Fig. 17.6(d), it is necessary to locate positions for the pole with respect to the shorter axis through the parcel rather than the longer axis. If the parcel is still too large to measure after these precautions have been taken, it is necessary to subdivide the parcel into smaller figures and measure these separately. However, if the grid or graticule is being used to control the measurements, such difficulties ought not to arise.
3 Select a point on the perimeter of the parcel which may serve as the starting and finishing point for each tracing. Ideally this should also be located in a position where the two arms of the planimeter are at right angles, to one another and the tracing arm is tangential to the perimeter of the parcel when the tracing point is set there. There are cogent reasons for making this choice, which will be explained later when we consider the causes of error in planimeter operation. The starting point ought to be marked for positive identification.
4 Record the scale reading.
5 Trace the perimeter of the parcel in the clockwise direction until the tracing point has returned to the starting point. Read the scales again. The

difference δv_1, between the second and first readings, is the area of the parcel expressed in vernier units.

6 Trace the perimeter of the parcel from the same starting point in the anticlockwise direction until the tracing point has returned to the starting point. Read the scales for a third time. The difference between the third and second reading, δv_2, provides a second measure of the area of the parcel. If suitable tolerances have been imposed for a particular kind of work, the difference between δv_1 and δv_1 should lie within that tolerance.

7 Disconnect the pole arm from the tracing mechanism, and either reverse the relative positions of pole, pivot and tracing point, as illustrated in Fig. 17.2(a), or physically move the pole to a different position on the other side of the parcel as in Fig. 17.2(b). There seems to be no definitely preferred method of making this change, although the writer prefers to move the pole to an entirely different position. Test, as before, that the whole parcel can be traced with the pole in the new position. If necessary select a new point from which tracing can start which again satisfies the condition that the two arms are perpendicular to one another when the tracing point is situated there. Then repeat stages 4, 5 and 6 obtain another pair of area measurements related to the second position of the pole.

The positioning of the pole relative to parcel is largely governed by the behaviour of the planimeter when its arms are move far apart or close together. There are two limits to the working range of a polar planimeter.

- the pole arm and tracing arm may form a straight line before the furthest extremity of the parcel has been reached (Fig. 17.6(a));
- the two arms may fold back towards one another to make an extremely acute angle at the pivot (Fig. 17.6(b)).

In the first case the pole is situated too far from the parcel and the tracing point cannot reach across it, for obviously it is impossible to move the point further from the pole than the distance $R + R'$. In the second case difficulties usually begin when the two arms make an angle of about 15° to one another, when some components mounted on the tracing arm may interfere with the movements of the pole arm. Then the instrument no longer moves freely. This condition only arises when the pole is located too close to the perimeter of the parcel.

The result of following the procedure outlined above is to obtain six scale readings and four different values for the area of the parcel in two different pole positions. In other words:

Pole right	*Pole left*
Initial scale reading, v_1	Initial scale reading, v_4
1st tracing (clockwise)	1st tracing (clockwise)
Second scale reading, v_2	Second scale reading, v_5
2nd tracing (anticlockwise)	2nd tracing (anticlockwise)
Third scale reading, v_3	Third scale reading, v_6

Then

$$\begin{aligned} \delta v_1 &= v_2 - v_1 \qquad & \delta v_3 &= v_5 - v_4 \\ \delta v_2 &= v_2 - v_3 \qquad & \delta v_4 &= v_5 - v_6 \end{aligned} \tag{17.2}$$

Moreover,

$$v_1 \approx v_3 \quad \text{and} \quad v_4 \approx v_6$$

which are checks against gross errors even if no working tolerances have been specified. Putting

$$\bar{v} = (\delta v_1 + \delta v_2 + \delta v_3 + \delta v_4)/4 \tag{17.3}$$

we have established a mean value for the differences in scale reading based upon the four different combinations of pole position and direction of tracing. A conversion factor, k, is taken from the calibration card kept with the instrument, or has been determined from a new calibration by the operator. Then the area of the parcel, A, is determined from

$$A = \bar{v} \cdot k \tag{17.4}$$

and this is expressed in the units of measurement which have been used to define k.

The sequence of measurements outlined above should be regarded as being *the minimum observation routine* needed to measure the area of a parcel by polar planimeter set in the pole-outside mode of operation. Just as the observation routine normally used with a theodolite to measure horizontal angles is intended to cancel small systematic errors, so, too, this routine of measurement by polar planimeter is intended to improve the quality of the measurements by means of sequential repetition. More precise results may be obtained by increasing the number of observations; for example by making two or more circuits of the parcel perimeter between successive scale readings, by using more than two pole positions etc.

GEOMETRICAL THEORY OF THE POLAR PLANIMETER

It has already been noted that the measuring wheel moves by rolling and sliding over the surface of the map, and that, as a result, the distance l recorded by revolutions of the measuring wheel is shorter than the length of the track, s,

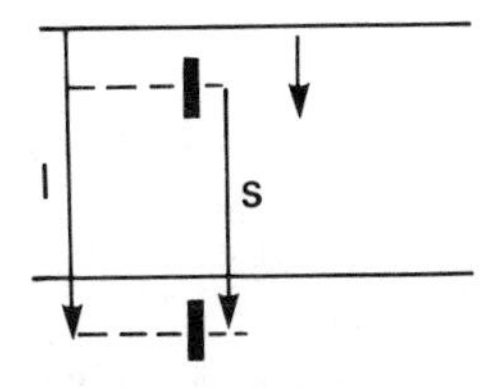

FIG. 17.7. The movement of the measuring wheel: (1) rotation: $l = s$.

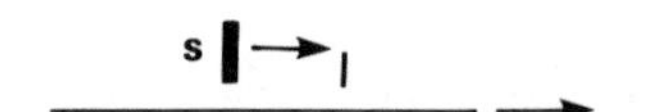

FIG. 17.8. The movement of the measuring wheel: (2) sliding: $l = 0$.

which the wheel actually makes over the paper. Four different kinds of movement may be recognised.

1 When the wheel moves in the direction perpendicular to its axle, as illustrated in Fig. 17.7, the wheel rolls throughout and traces the arc of length l which corresponds exactly to the track s.
2 When the wheel is moved in the direction of the axle, as illustrated by Fig. 17.8, the wheel slides across the paper without turning. Therefore the scale reading is unchanged and $l = 0$ irrespective of the distance s tracked by the wheel.
3 When the wheel is moved in some other direction so that it moves neither wholly in the direction of the axle nor at right angles to it, the wheel traces along the line KK' by a mixture of rolling and sliding. It may be seen from Fig. 17.9 that s is the hypotenuse of the right-angled triangle (KQK', and it is the side KQ which represents the change in the readings, l. Therefore

$$l = s \cdot \sin \varphi = r \cdot \cos \lambda \tag{17.5}$$

where φ and λ are the angles illustrated in Fig. 17.9.
4 As the tracing arm rotates about the pivot, B, the wheel turns and slides along the axis $s = KK'$, recording the length of the recorded run as

$$l = r \cdot \beta \tag{17.6}$$

The Theory of Measurement in the Pole-inside Mode

Notwithstanding the fact that the pole-inside mode of operation is far less used in cartometry than is the pole-outside mode, it is desirable to consider the mechanics of this mode of operation in any comprehensive study of the polar planimeter. Moreover, it is probably easier to visualise, and certainly to illustrate, the theory of the instrument for this mode of operation.

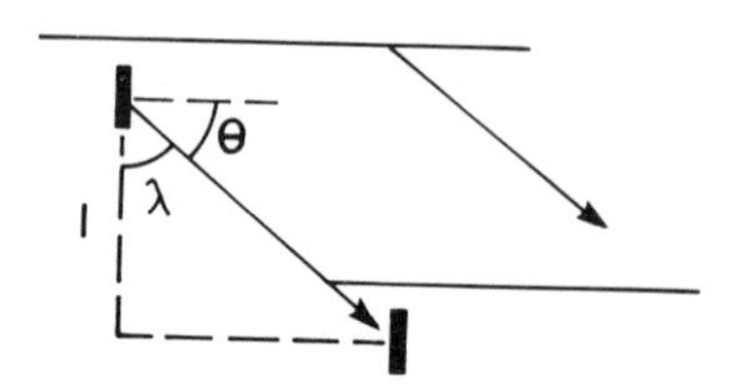

FIG. 17.9. The movement of the measuring wheel: (3) combined rotation and sliding, $s > l > 0$.

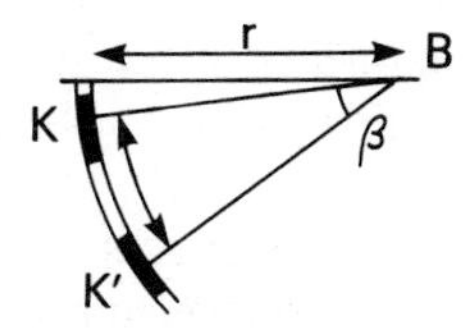

FIG. 17.10. The movement of the measuring wheel resulting from rotation of the tracing arm about the pivot.

In Fig. 17.11, P is the pole of the planimeter and the tracing point is placed initially on the parcel perimeter at c. As the tracing point is moved in a clockwise direction from the starting point, it occupies successively the positions, $c_1, c_2 \dots c_n$. Within the first element of the perimeter, defined by the movement of the tracing point from c to c_1, the planimeter sweeps through the small area of the parcel, $A_1 = Pbcc_1b_1$. The tracing point now moves onwards to c_2 and the planimeter sweeps through the next element of area, A_2. This process is repeated until the tracing point returns to the point c and the planimeter has passed through all the small elements and

$$A = A_1 + A_2 + A_3 \cdots + A_n \tag{17.7}$$

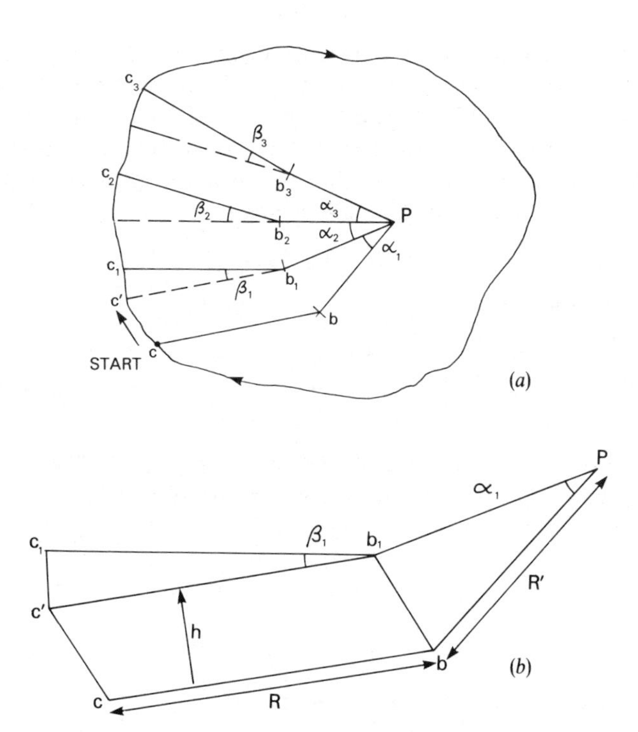

FIG. 17.11. The geometry of operating a polar planimeter in the pole-inside mode.

We now consider the components of the single element of area A_1. An enlarged version of this portion of the whole parcel is illustrated in Fig. 17.11(b).

The total movement of the planimeter as it moves through this part of the parcel may be resolved into the displacement of the tracing arm from the position bc to the position b_1c' followed by the rotation of the tracing arm about the pivot, moving into the position b_1c_1.

Since the total area of A_1 is the figure $Pbcc'c_1b_1$, it is composed of three elements:

- the area of the sector Pbb_1
- the area of the parallelogram $bcc'b_1$,
- the area of the sector $b_1c'c_1$.

Combining these, using the sides and angles illustrated in Fig. 17.11,

$$A_1 = \tfrac{1}{2}R' \cdot bb_1 + R \cdot h + \tfrac{1}{2}R \cdot c'c_1 \tag{17.8}$$

From Fig. 17.11 it can be seen that in either case

$$bb_1 = R' \cdot \alpha_1$$

and

$$c'c_1 = R \cdot \beta_1$$

The value of h corresponds to the arc l recorded by the measuring wheel during the movement of the tracing point from c to c', whilst the total track of the wheel is KK', as in Fig. 17.9. Note that in the case of the inner-pivot planimeter (Fig. 17.11a), the wheel turns in the anticlockwise direction as the tracing arm rotates about the pivot at b_1 so that the distance measured represents a reduction in the scale reading. Consequently

$$l_1 = h - KK' \tag{17.9}$$

but from equation (17.6)

$$KK' = r \cdot \beta_1$$

Therefore

$$h = l_1 + r \cdot \beta_1 \tag{17.10}$$

In the case of the outer-pivot planimeter, as shown in Fig. 17.11(b), the measuring wheel moves forward during this operation so that

$$l_1 = h + KK'$$

$$= h + r \cdot \beta_1$$

and

$$h = l_1 - r \cdot \beta_1 \tag{17.11}$$

Substituting these values for the terms in (17.8)

$$A_1 = \tfrac{1}{2}R'^2\alpha_1 + R \cdot l_1 + R \cdot r \cdot \beta_1 + \tfrac{1}{2}R^2 \cdot \beta_1 \tag{17.12}$$

Similarly the small element of area A_2 comprises

$$A_2 = \tfrac{1}{2}R'^2 \cdot \alpha_2 + R \cdot l_2 + R \cdot r \cdot \beta_2 + \tfrac{1}{2}R^2 \cdot \beta_2 \tag{17.13}$$

with similar expressions for all the individual elements A_i which make up the parcel. Combining all such expressions as in equation (17.7)

$$\begin{aligned} A = {} & \tfrac{1}{2}R'^2(\alpha_1 + \alpha_2 + \cdots + \alpha_n) + R(l_1 + l_2 + \cdots + l_n) \\ & + R \cdot r(\beta_1 + \beta_2 + \cdots + \beta_n) + \tfrac{1}{2}R^2(\beta_1 + \beta_2 + \cdots + \beta_n) \end{aligned} \tag{17.14}$$

When the parcel is traced in the pole-inside mode, the pole-arm of the planimeter rotates about P through a complete circle. Therefore

$$\alpha_1 + \alpha_2 + \alpha_3 + \cdots + \alpha_n = 2\pi \tag{17.15}$$

Similarly, if all the angles β were superimposed upon the single position of the pivot, b, we would also find that the tracing arm also describes a complete circle about the pivot. Consequently

$$\beta_1 + \beta_2 + \beta_3 + \cdots + \beta_n = 2\pi \tag{17.16}$$

Moreover the arc l_i is the amount by which the measuring wheel rotates as each element of area is measured. Designating l' as the sum of all the movements of the wheel, or

$$l' = l_1 + l_2 + \cdots + l_n \tag{17.17}$$

Substituting these expressions in (17.14)

$$\begin{aligned} A &= \pi R'^2 + R \cdot l' + 2\pi R \cdot r + \pi R^2 \\ &= Rl' + \pi(R'^2 + 2R \cdot r + R^2) \end{aligned} \tag{17.18}$$

All the quantities in the second term on the right-hand side of (17.18) are constants, being the dimensions for a particular instrument, or for a particular setting of the tracing arm, R. We call this term the *planimeter constant*. It has two forms

$$C = \pi(R'^2 + 2R \cdot r + R^2) \tag{17.19}$$

for the inner-pivot instrument, or

$$C = \pi(R'^2 - 2R \cdot r + R^2) \tag{17.20}$$

for the outer-pivot instrument. Substituting one or other of these expressions in (17.18) we have

$$A = R \cdot l' + C \tag{17.21}$$

Since the term l in equation (17.1) corresponds to l' in (17.21), we may also write

$$A = R \cdot t \cdot \bar{v} + C \tag{17.22}$$

In this equation t is the length of the arc of the circumference of the measuring wheel which corresponds to 1 vernier unit. Thus in an instrument having a measuring wheel of diameter 20 mm, the circumference of it is 62·83 mm and the value of $t = 1/1{,}000 \times 62{\cdot}8 = 0{\cdot}063$ mm and is a constant. The value of R is only constant for a fixed-arm planimeter. Since the value for R may be changed through more than 100 mm in a variable-arm planimeter, it is better to consider here the special case of the fixed-arm instrument which is intended to read in metric units, namely 1 *v.u.* = 10 mm². We require that

$$k = R \cdot t = 10 \tag{17.23}$$

In order to satisfy (17.23) the length of the tracing arm should be made 150·15 mm. The constant k represents the unit of area which corresponds to 1 *v.u.* on the planimeter scale. It is a measure of area because, from (17·23), it is the area of the narrow rectangle of length R and width t.

Before the significance of the second, C, can be appreciated, it is desirable to consider the theory of planimeter operation in the pole-outside mode.

Theory of Measurement in the Pole-outside Mode

The diagram which corresponds to Fig. 17·11(b) is Fig. 17·12, which, for simplicity, shows only the geometry of the outer-pivot instrument. The diagram is labelled in the same way as Fig. 17.11, so that the early stages of the argument refer to the same sides and angles used earlier. Indeed the entire presentation of the argument is identical to that given in the preceding section up to the definition of the total area of the parcel in equation (17.17). For the pole-outside more of operation, it is clear from Fig. 17.12 that on completion

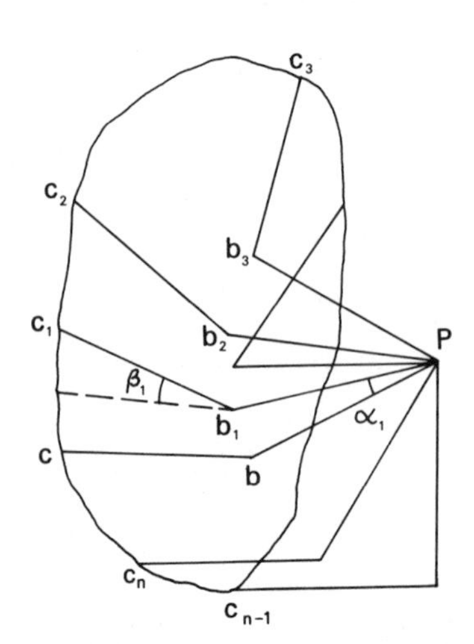

FIG. 17.12. The geometry of operation of a planimeter in the pole-outside mode.

of tracing the perimeter of the parcel,

$$\alpha_1 + \alpha_2 + \cdots \alpha_n = 0 \tag{17.24}$$

and, similarly,

$$\beta_1 + \beta_2 + \cdots \beta_n = 0 \tag{17.25}$$

Therefore practically the whole of the right-hand side of equation (17.17) vanishes, leaving the final expression

$$A = R \cdot l' \tag{17.26}$$

or, as before,

$$\begin{aligned} A &= R \cdot t \cdot \bar{v} \\ &= k \cdot \bar{v} \end{aligned} \tag{17.27}$$

Since the planimeter constant, C, has vanished, this constant is evidently only significant when working in the pole-inside mode of operation.

The Zero Circle of The Polar Planimeter

It is instructive to set a polar planimeter in the centre of a large sheet of paper, place the arms as illustrated in Fig. 17.13, so that they intersect at an angle slightly greater than a right angle, and trace the circumference of the circle whose radius is CP. If the tracing point can be directed accurately, the measuring wheel will always slide at right angles to its axis. Consequently the perimeter of a large circle may be traced without recording any movement of the measuring wheel. Not surprisingly this circle is called the *zero circle, null circle, neutral circle* or *datum circle.* Figure 17.13 illustrates the circumference of the zero circle, as this is established in both varieties of instrument. The conventions used in Fig. 17.13(a) and (b) are those used earlier in Fig. 17.2 and Fig. 17.11. Putting ρ as the radius of the zero circle, we obtain, from

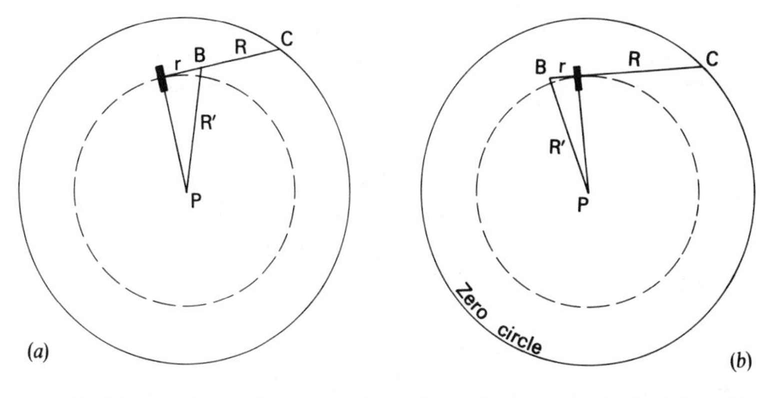

FIG. 17.13. The nature of the zero circle and its use in measuring in the pole-inside mode.

Fig. 17.13(a),

$$\rho^2 = (R - r)^2 + AK^2 \tag{17.28}$$

But

$$AK^2 = R'^2 - r^2$$

Therefore

$$\begin{aligned}\rho^2 &= R^2 + 2R\cdot r + r^2 + R'^2 - r^2 \\ &= \sqrt{(R'^2 + 2R\cdot r + R^2)}\end{aligned} \tag{17.29}$$

or, in the case of the design of planimeter illustrated by Fig. 17.13(b):

$$\rho^2 = \sqrt{(R'^2 - 2R\cdot r + R^2)} \tag{17.30}$$

Interpreting these results in the light of equations (17.19) and (17.20), we see that in each case

$$C = \pi\rho^2 \tag{17.31}$$

or the *planimeter constant is the area of the zero circle*, expressed in the units used to measure the arm lengths of the planimeter. It should be noted that some authors derive the planimeter constant as the radius of the zero circle rather than its area. However, it is an incomplete solution and the user still needs to determine the area of the zero circle in order to calculate the areas of any parcels measured in this mode.

The planimeter constant is usually determined by the manufacturer and is recorded on a label attached to the lid of the box in which the instrument is stored. This value for C is usually derived from the measurements of R', r and R made during assembly of the planimeter. It is less easy for the user to make sufficiently accurate measurements of these components, particularly that for r. Therefore it is customary for the user to obtain the value for C by calibration measurements.

Vernier Readings Made in The Pole-inside Mode

Since the zero circle represents a circumference along which no change takes place, the readings of other sizes of parcel measured in the pole-inside mode must be made with reference to this datum. The appropriate formula is

$$A = C + k(v_2 - v_1) \tag{17.32}$$

where the measurement is restricted to only two readings of the vernier made at the start (v_1) and completion (v_2) of one tracing in the clockwise direction. As before, multiple measurements would be appropriate, using both directions of tracing as well as different pole positions.

CALIBRATION OF POLAR PLANIMETER

From the examination of the theory of the polar planimeter it has been seen that two constants are required for use with the instrument, k and C. The constant k is the conversion factor needed to convert the scale readings from vernier units into suitable measures of area. The constant C represents the area of the zero circle, and is only needed if the planimeter is to be used in the pole-inside mode. One method of calibration, which is similar to the techniques used in other branches of cartometry, is to measure a geometrical figure of known area, such as a block of grid squares, or even a suitably sized graticule quadrangle, the area of which has been determined using the methods described in Chapter 11. Another method is to use a special jig provided by the manufacturer of the instrument.

The Calibration Jig

This is also called a *checking ruler* or *calibration ruler*, and it is used to determine k. It is a small steel strip, illustrated in Fig. 17.14, having a fine needle, A, near one end with a small hole, B, near the other. The distance AB between these points represents the radius of a circle which can be traced by the steel ruler as it rotates about the needle points at A. The distance AB is designed to represent a circle having the area of an integer number of square

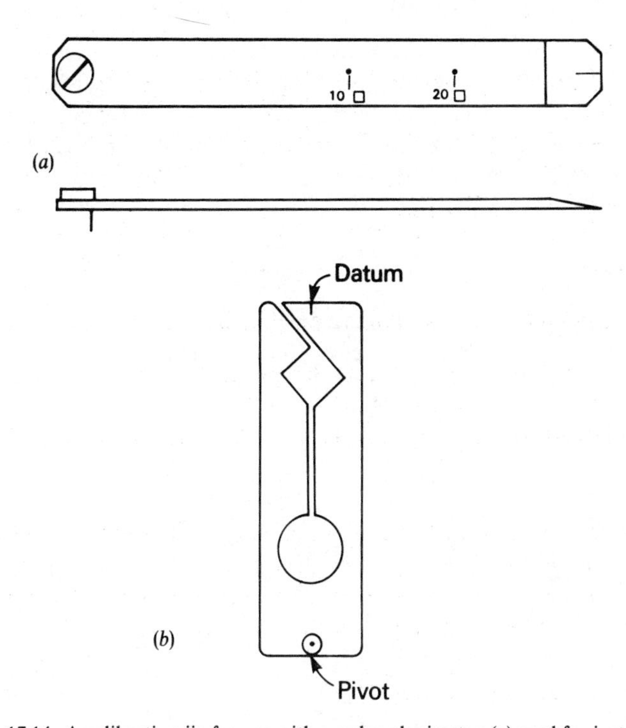

FIG. 17.14. A calibration jig for use with a polar planimeter: (a) used for instruments having a steel point as the tracing mark; (b) used for instruments in which the measuring mark is at the centre of the reading lens.

centimetres or square inches. This can be verified by plotting the points A and B carefully, and measuring the distance between them by steel scale. Some examples have the precise area, as determined by the manufacturer, engraved upon the surface. Some jigs contain more than one hole, thereby allowing calibration to be carried out for more than one size of circle. The jig described, and illustrated in Fig. 17.16(a), is intended for use with a planimeter having a tracing point. Planimeters with tracing lenses require a more elaborately shaped end to the jig, as illustrated in Fig. 17.14(b), which will hold the barrel of the tracing lens in only one position but allow the lens to rotate within the holder. The technique is to use the jig as a template with which to trace a circular parcel of known size. It is placed on the drawing board in a suitable position with respect to the arms of the planimeter and to its pole. The needle point forming the centre of the circle is pressed gently into the drawing board. A mark is made with a fine needle against the engraved line which serves as the datum for tracing. It is now necessary only to read the scale, rotate the jig about its own centre through a complete circle, ensuring that the tracing point remains riding in its own pin hole.

Determination of the Conversion Factor

By repeating a succession of readings, these being the usual combinations of clockwise and anticlockwise tracings from different pole positions, a succession of different values for v are obtained. From these, we determine a mean value $\bar{v}$ together with the standard error. If A' is the known area of the circle described by the jig, from (17.04), obviously

$$k = A'/\bar{v} \tag{17.33}$$

The procedure and calculations may be demonstrated as follows:

> The planimeter to be calibrated was a fixed-arm polar planimeter, model Nr 641, manufactured by Röst of Vienna and designed to measure in metric units.

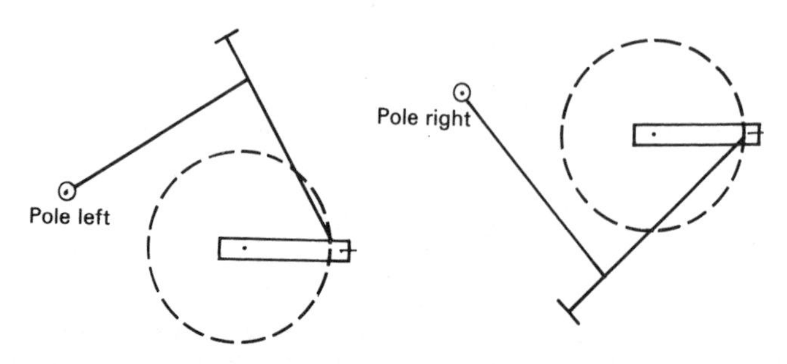

FIG. 17.15. Calibration jig in use.

1 The calibration jig supplied by the manufacturer was placed on a sheet of paper, and prick marks made by the needle point at *A* and by putting a fine needle through the small hole at *B*. The jig was removed and the distance between the two points measured by steel scale (using the method described on page 107). A sample of five measurements yielded a mean length 56.56 $\pm$ 0·05 mm. Therefore the area of the circle made by this jig is 10050 mm^2 (100·50 cm^2) and not the exact 10000 mm^2 which might be expected from cursory examination of it.
2 The planimeter was set on the drawing board for operation in the pole outside mode with the pole set about 26 cm distant from the pole of the jig, as illustrated in Fig. 17.15.
3 The circular parcel was described by moving it about its own pole, first in the clockwise and secondly in the anticlockwise directions. The results of these measurements are given in the left-hand column of Table 17.1.
4 After completion of the three pole-right readings the pivot was disconnected and the part of the planimeter comprising the tracing arm and measuring unit was placed to the right of the pole arm. The same position for the planimeter pole was maintained. The tracing and reading procedure was repeated, these being the "pole-left" observations recorded in the right-hand column of Table 17.1.

In addition to the method described in detail above, the following variations may also be used in a calibration routine:

1 Using the "same pole" the parcel defined by the jig was traced twice between readings:

$$\bar{v} = 1{,}002{\cdot}4 \pm 0{\cdot}25 \text{ v.u.}, \quad k = 10{\cdot}03 \text{ mm}^2$$

2 Using "different poles" the parcel was traced once in each direction

$$\bar{v} = 1{,}004{\cdot}0 \pm 1{\cdot}41 \text{ v.u.}, \quad k = 10{\cdot}01 \text{ mm}^2$$

3 Using "different poles" the parcel was traced twice as before

$$\bar{v} = 1{,}003{\cdot}75 \pm 2{\cdot}06 \text{ v.u.}, \quad k = 10{\cdot}01 \text{ mm}^2.$$

TABLE 17.1. *Calibration of fixed-arm planimeter Röst 18112. Single tracing in the "same-pole" position*

Pole right				Pole left			
v_1	0421			v_4	1153		
		δv_{12}	1003			δ_{45}	1002
v_2	1424			v_5	2155		
		δv_{23}	1003			δv_{56}	1001
v_3	0421			v_6	1154		

$\bar{v} = 1002{\cdot}25 \pm 0{\cdot}96$ v.u.
$k = 10{,}050/1{,}002{\cdot}25 = 10{\cdot}03$ mm^2 or $0{\cdot}1003$ cm^2

These three variations in the calibration procedure yield results which are so close to one another there is no significant difference between them. Consequently it is not possible to recommend any single method as being superior to the others, although evidently double tracing in the "same pole" configuration gives the most precise results. On the other hand the use of "different poles" is clearly the least precise, but the reason for this is that we changed the position of the pole in order to ensure that the second set of measurements is wholly independent of the first. This, after all, is the object of making repeated measurements both in calibration and for obtaining new data. Far more important than the choice of method is the precision with which the area of the circle described by the jig can be obtained. If it had been assumed that the area of the circle was 10,000 mm^2 exactly, the difference in the calculated value for k would differ by 6 units in the second decimal place (9·97 instead of 10·03, where k is expressed in square millimetres). Since a small error in k is a source of systematic error, it follows that careful measurement of the radius traced by the jig is more important than worrying about which routine of planimeter measurements is to be preferred. This being so, it is necessary to comment on how the test measurement ought to be made for the kind of jig illustrated in Fig. 17.14(b) where there is no hole to serve as an accurate mark. In this case the best solution is to make a prick-mark on the paper first, place the index of the tracing lens over this while it is sitting snugly within the part of the jig which holds it, and finally press the pole of the jig into the paper.

Determination of the Planimeter Constant, *C*.

The typical calibration jig supplied with a polar planimeter is only about 60 mm in length. Consequently it can only be used to simulate a parcel which can be measured in the pole-outside mode and this size of parcel (approximately 113 cm^2) is far too small to accept the planimeter in the pole-inside mode. Determination of C may be attempted by measuring the areas of a series of circles of known size which are most easily plotted by beam compass. In order to remove any uncertainty about the meaning of the scale readings it is wise to construct a series of three concentric circles, one having a radius greater than ρ, another less than ρ and the third approximating to ρ. The individual parcels are now measured in the pole-inside mode, using the various combinations of pole position and tracing direction which have already been recommended. A further improvement of the method, which can be used if the planimeter has a tracing point rather than a lens, together with a point at the pole rather than a block of machined steel, is to construct a plastic jig, similar in appearance to the metal calibration jig illustrated in Fig. 17.14(a). This may be made from a strip of glass-clear vinyl or polycarbonate plastic of gauge 0·05 mm or thicker. An old plastic ruler might serve for this. Holes are pricked or drilled in it, at one end to accommodate the planimeter pole and the other corresponding to the position of the tracifng point on circles of different radii.

Knowing the distances between the holes, corresponding to AB on the metal jig, it is possible to calculate the area of each circle traced. The method is not as precise as the use of a steel jig for calibration if only because there may be small twists and bends affecting the plastic strip as this is moved round the drawing board.

Adjustments to the Polar Planimeter

An alteration in the length of the tracing arm, R, changes the scale at which the instrument operates. Since t is constant, it follows that if R is altered in equation (17.23), k changes too. This modification may be used for several purposes, but in each case the object of the adjustment is to make k the simplest possible conversion factor so that a minimum of mental arithmetic is required in its use. A good example of the need to adjust the length of the tracing arm is to set a suitable value for R which will allow the user to work directly in both metric and British standard units. The Coradi Cora-Senior planimeters which are usually sold by retailers in North America and the British Commonwealth have only been calibrated to measure in British standard units. For example, the instrument reads 1 v.u. $= 0{\cdot}01$ inches^2 when the tracing arm is set to read 108·6 mm. One obvious way of obtaining the results of measurements in square centimetres instead of square inches is to multiply all the results by 6·45245, this being the required conversion factor. It is easy enough to do nowadays, using a pocket calculator, but only a few years ago it would have been necessary to use a slide rule and lose some significant figures in the process. Therefore an important alternative was to reset the length of the tracing arm to the scale value $R = 168{\cdot}2$, when 1 v.u. $= 0{\cdot}10\,\text{cm}^2$. The no arithmetic is needed other than to insert the decimal point between the third and fourth digits, which converts the answers into square centimetres.

The second application of adjustments to the length of the tracing arm is to derive a value for k which allows the operator to read ground area directly from the planimeter. It is therefore possible, for example, to set the instrument to read areas in acres directly from a map of scale 1/10,560, in hectares from a map of scale 1/25,000 or in square metres at 1/1,250.

The third application of adjustment to the length of the tracing arm is to obtain the measurement of area from a map or drawing, the paper of which has been deformed by changes in the humidity of the environment, or from maps which have been stored tightly rolled for a long time. In some of the older literature relating to area measurement these are quaintly described as "shrunk plans". In order to justify making this adjustment it must be assumed that deformation has been regular in all directions, and that it is sufficiently uniform over a large part of the map. The same technique is sometimes used as a general, but unspecified, correction if the results of check measurements reveal an unexplained discrepancy.

Thus far we have considered the length of the tracing arm to be the distance

R as this was originally defined in Fig. 17.1, and as it has appeared in many subsequent diagrams. When this quantity was used in the equations relating to the theory of the polar planimeter it was tacitly assumed that the three linear variables, R', R and r were all expressed in the same units. However, the scale engraved on the tracing arm of a variable-arm planimeter is not standardised, but differs from one manufacturer and model to another. For example, some of them are graduated in millimetres, but many more are graduated in units of half-millimetres. Some of the scales measure distance outwards from the tracing point or the index mark on the tracing lens, but others have their origin which is some centimetres distant from these points. The way in which the units are numbered along the scale also varies from one instrument to another. It follows that the scale on the tracing arm is not the distance R. However, calculations which make use of the readings on this scale, which we will denote by R, all bear a similar relationship to this distance. We do not need to know the exact relationship between R and R because we are going to compare the readings made with the planimeter set to R_1 with those made at the setting R_2. This is acceptable with the single instrument, but it is essential to realise that the R settings calculated for use with one instrument cannot be used for another.

One method of calculating the adjustment of the tracing arm as a proportion between two other settings is that recommended for use with Coradi planimeters. Expressed in the notation already used

$$R = R_1 - [(R_1 - R_2)(k_1 - k)]/(k_1 - k_2) \qquad (17.34)$$

In this equation it is required to determine a new setting R which corresponds to the constant k, whilst we have at our disposal the calibration data for the setting R_1, giving the constant k_1, and the setting R_2 giving k_2. It is also necessary to specify that $R_1 > R_2$ so that $k_1 > k_2$.

We examine the use of this expression in the determination of the setting R which is needed to make a Cora-Senior planimeter, calibrated only for British Standard units, read directly in metric units.

1 The calibration jig supplied with the instrument has been measured in advance of this work. This has shown that the radius of the circle described by the jig was $9{\cdot}045 \pm 0{\cdot}028$ cm. Therefore the area of the calibration circle is $257{\cdot}02\ \text{cm}^2$ or $39{\cdot}858\ \text{in}^2$.
2 The tracing arm was set to 175·4, which is the maximum value for R which can be set in the instrument. The standard calibration procedure was carried out to find a mean scale value of 2467·1 v.u. Therefore 1 v.u. $=$ $257{\cdot}02/2467{\cdot}1 = 0{\cdot}1042\ \text{cm}^2$.
3 The tracing arm was set to 108·6. The result of an identical calibration made at this setting was a mean scale value of 3980·3 v.u. Therefore 1 v.u. $= 0{\cdot}0646\ \text{cm}^2$.

4 For the solution of equation (17.34),

$$R_1 = 175{\cdot}4, \quad k_1 = 0{\cdot}1042$$
$$R_2 = 108{\cdot}6, \quad k_2 = 0{\cdot}0646$$

and we must seek a solution for $k = 1, 0{\cdot}1$ or $0{\cdot}01$. It will be found that only $k = 0{\cdot}01$ represents a realistic value, so we insert this in (17.34). Then

$$R = 175{\cdot}4 - (66{\cdot}8 - 0{\cdot}0042)/0{\cdot}0396$$
$$= 168{\cdot}3$$

5 This value was set on the tracing arm and the calibration procedure was repeated. The resulting mean scale value was 2567.4 v.u. so that $k =$ 0.1001.

6 Putting $R_2 = 168{\cdot}3, k_2 = 0{\cdot}1001$ in equation (17.34) a second solution of the equation gave a revised setting of the tracing arm as 168·2. Calibration was repeated yet again to provide the final value for $k = 0{\cdot}100\,\text{cm}^2$.

Adjustments to Make the Planimeter Read Directly in Ground Units

The value of k which corresponds to a particular setting of the tracing arm may be calculated to give a simple fraction for units of ground area measured on a map of given scale $1/S$. This is done, as in all map scale manipulations, in two stages: first by multiplying map distances by the scale denominator, S, and secondly by dividing this result by a factor to bring it into units measured on the ground. Since k represents the area of a narrow rectangle with sides R and t, both sides must be multiplied by S. Thus, from (17.23)

$$k = RS \cdot tS$$
$$= R \cdot t \cdot S^2 \qquad (17.35)$$

Obviously the right-hand side of this equation expresses area on the ground in those units by which R and t have been measured. For most map scales, S^2 is an exceedingly large number and therefore the right-hand side of equation (17.35) must be divided by the conversion factor from square millimetres into hectares or square kilometres.

Having determined k for the initial setting, the adjustment to R follows, using either equation (17.34) or (17.35) to make the conversion. The following example will serve to demonstrate the method. From the earlier calculations, where $R = 168{\cdot}2$, $1\text{ v.u.} = 0{\cdot}1\,\text{cm}^2$. Also at the scale 1/25,000, $0{\cdot}1\,\text{cm}^2 =$ 0·625 ha. Therefore using the planimeter with this value of R, on a map of 1/25,000, 1 v.u. = 0·625 ha. Obviously a more suitable setting of the tracing would be that allowing area to be read directly in units of 0·5 ha. From equation (17.35)

$$R = (k/k_1) \cdot R_1$$
$$= (0{\cdot}5/0{\cdot}625) \cdot 168{\cdot}2$$
$$= 134{\cdot}6$$

Checking this by the customary calibration outlined above, the mean scale value is 3209·5 v.u. and $k = 0{\cdot}08008$. At 1/25,000, 1 square centimetre corresponds to 6·25 ha and 0·8008 square centimetres = 0·5005 ha, which is practically the required result. Now putting $R = (0{\cdot}5/0{\cdot}5005) \times 134{\cdot}6$, we find the final setting 134·5 which provides the correct setting to obtain the factor $k = 0{\cdot}5$ ha.

Transfer From One Map Scale to Another

The final stage is to investigate what is the relationship between the values of k determined for different map scales. For example, if the factor k_1 has been determined for use with a map of scale $1/S_1$, what is the value of the factor k_2 to be used to make a similar conversion using measurements made on a map of scale $1/S_2$? From (17.35)

$$k_1 = R \cdot t \cdot S_1^2 \quad \text{and} \quad k_2 = R \cdot t \cdot S_2^2$$

Dividing the second expression by the first, we have

$$k_2/k_1 = R \cdot t \cdot S_2^2 / R \cdot t \cdot S_1^2$$
$$k_2 = (S_2^2/S_1^2) \cdot k_1 \tag{17.36}$$

For example, on a map of scale 1/25,000 1 v.u. corresponds to the area 0·5 ha. What is the value of 1 v.u. at this setting of the tracing arm if the planimeter is used to measure area on a map of scale 1/10,000?

$$k_2 = (10{,}000^2/25{,}000^2) \times 0{\cdot}5$$
$$= 0{\cdot}08 \text{ ha}$$

Adjustment For Paper Distortion

If the measurement of the area of a parcel of known size, such as a block of grid squares, produces a result which does not correspond to the calculated area of this figure, the discrepancy may be conveniently attributed to paper distortion. One method of making allowance for this discrepancy is to adjust the length of the tracing arm before embarking on a prolonged series of measurements to be made on the same map sheet. Suppose that the area of the parcel, determined by calculation, is A, but measurement of it by planimeter set to R_1 gives the area A_1, the correction to be made to the planimeter may be derived from the simple proportion

$$R = R_1 \cdot (A_1/A) \tag{17.37}$$

The same procedure may be used as a temporary form of calibration if no steel jig is available, but obviously any paper deformation contained in the grid will be transmitted to the values of R or k determined in this fashion.

Adjust or Calculate?

Since adjustment of the length of the tracing arm involves repetition of the calibration procedure, whether this be done by jig or measurement of the map grid, it is debatable how often an alteration should be made. Is it not more convenient to accept that the measurement contains certain known systematic errors to be removed by calculation?

Twenty years ago it was still considered preferable to make adjustments to a planimeter rather than solve each of the problems which have been described by calculation. For example, the multiplication of a large number of measured parcels each by the factor $k = 0{\cdot}0646$ introduces an additional stage into the data collecting process, and any arithmetical mistakes will be a secondary source of gross error. Now that pocket scientific calculators are so easily available, and so cheap, it has become more economical to solve such problems numerically. It is now far cheaper to buy a fixed-arm planimeter and a calculator rather than a variable-arm planimeter.

INDIRECT MEASUREMENT BY PLANIMETER

The foregoing solution, especially equation (17.37), indicates that there is yet another method of using a planimeter which avoids all the calibration and adjustment procedures. This is to use the instrument to make *indirect measurements* in a similar way to the techniques used for area measurement by linear sampling and point counting.

A test block of grid squares, or a graticule quadrangle, the area of which may be calculated, has area A_0. This block is measured by planimeter and the mean reading for the area in v.u., or $\bar{v}'$ has been obtained. The required parcel, which preferably is located within the test block, is also measured and the mean reading of $\bar{v}$ vernier units is also obtained. Obviously the ratio $\bar{v}/\bar{v}'$ is the proportion of the whole block or quadrangle occupied by the parcel. Since we know the area A_0 from an independent source, the required area of the parcel is

$$A = A_0(\bar{v}/\bar{v}') \tag{17.38}$$

Obviously A is expressed in the same units as A_0.

Although we only give scant attention to this technique here, indirect measurement is one of the major methods of cartometry, and indeed the whole of Chapter 19 is devoted to the consideration of it. The point is that its application to planimeter measurements is probably the least important of the indirect methods of measurement. The reason for this is that a planimeter measurement represents a complete measurement of the area of a parcel, whereas many of the point-counting and linear measurement methods do not. We argued in Chapter 3, and will demonstrate further in following chapters, that the methods of statistical sampling which may be used with these techniques convert incomplete measurements into useful results. However, the whole point of using indirect methods is that these are supposed to save some of the time which would be needed to make direct measurements. In a sense, therefore, making indirect measurements, by planimeter is like wearing both belt and bracers to keep one's trousers up; it distinguishes the pessimist from the optimist.

THE ACCURACY OF THE PLANIMETER

Much of the published work upon the accuracy of different methods of area measurement is based upon the comparison of the measurements made in a particular fashion with planimeter measurements of the same figures. Almost invariably the planimeter measurements have been taken as standard, and are therefore assumed to be without error. This is, of course, a colossal assumption to make, for all measurements contain some errors. It also means that there is no recognised standard against which to evaluate the accuracy of a planimeter. Therefore it is necessary to measure the precision of the instrument, by determining the standard error of samples of measurements representing repeated tests and supporting this with an analysis of the precision of the various stages in the measurement process. The outcome of crude or empirical testing is usually the comparison between standard error and parcel area. It is customary to try to fit a curve to the results having the form

$$S = a + b\sqrt{A} + cA \qquad (17.39)$$

in which s is the standard error of the planimeter measurements of parcels of area A, with a and b and c being coefficients to be determined from the curve-fitting process. This equation approximates to a parabola. If the standard

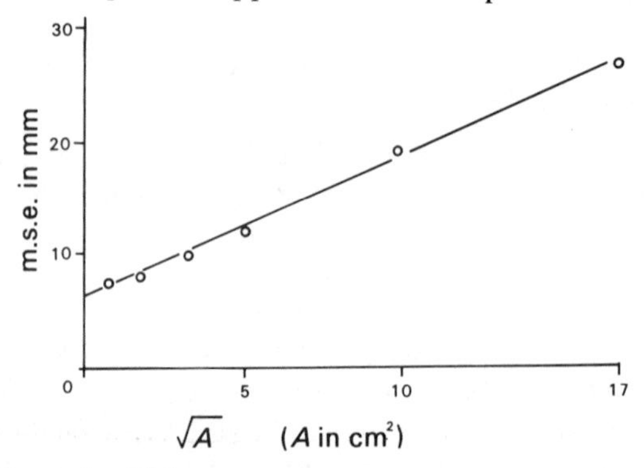

FIG. 17.16. The relationship between root mean square error and parcel size for typical tests of planimeter precision. Plot of root mean square error (in mm) against $\sqrt{A}$, where A is expressed in cm^2. (Source: Zill, 1955).

error is plotted against the square root of the parcel area, as in Fig. 17.16, the relationship is close to a straight line. The weakness of this approach is that it fails to isolate the different sources of error. These must be separated on the bases of analytical studies in order to estimate the theoretical contribution of each. The studies by Maslov and co-workers (1955, 1958, 1959), creates a picture of the magnitude of the different errors experienced, and their total agrees tolerably well with the results obtained from the testing and curve-fitting.

The Errors of The Polar Planimeter

Four separate sources of instrumental and operating error may be distinguished and allocated their individual standard errors:

- *Errors in reading the vernier.* Contained in the standard error s_1.
- *Error which arise in matching the tracing point against the point on the parcel perimeter which has been chosen to start tracing.* Represented by the standard error s_2.
- *Error which arise from the mechanical operation of the instrument.* Caused by friction, backlash and flexure of the various components. The nature of the map surface and the way in which the measuring wheel behaves as it passes over surfaces of different qualities should also be investigated in this context.These combine to form the standard error s_3.
- *Errors which occur during tracing.* Produce the standard error s_4.

The combination of these defines the standard error of measurement of a parcel by polar planimeter, or

$$s_a^2 = k^2(s_1^2 + s_2^2 + s_3^2 + s_4^2) + [(s_k/k)\cdot A]^2 \qquad (17.40)$$

The final term in this equation is an expression of the standard error of the calibration process.

Errors in Reading The Vernier

In any measurement process the limiting resolution of the scale is an important source of error. If an instrument has a vernier to assist reading the scale, this does not eliminate the problem, but merely transfers it to a portion of the scale which is one-tenth of what would be considered the limiting resolution using estimation. On a vernier scale a reading must serve for all measurements within the range $\pm\frac{1}{2}$ v.u., giving rise to a rectangular distribution having these limits. From the statistics of such a distribution as described, for example, by Lyon (1970), the theoretical value of the standard error ought to be

$$\begin{aligned} s_0 &= \pm\sqrt{2}(0{\cdot}5/\sqrt{3}) \\ &= \pm 0{\cdot}41 \text{ v.u.} \end{aligned}$$

This has been confirmed in practice, for example by Maslov, who collected a sample of 75 observations of random settings of a planimeter vernier made independently by 15 different operators. This gave rise to a standard error of the single observation $\pm 0{\cdot}43$ v.u. He also made 20 tracings each of a series of small parcels, using three different planimeters for another stage of the investigation to be described later. This gave a standard error of $\pm 0{\cdot}42$ v.u. for reading the vernier. Baer (1937) obtained a slightly lower value, $\pm 0{\cdot}36$ v.u. from his own work. Therefore we may put

$$s_0 = \pm 0{\cdot}4 \text{ v.u.}$$

as a likely value for the standard error.

Determination of the area of a parcel comprises the difference between two vernier readings, such as $v_2 - v_1$. Consequently the standard error s_1, must be determined from the combination of two standard errors of equal magnitude.

$$\begin{aligned} s_1 &= \pm s_0 \cdot \sqrt{2} \\ &= \pm 0{\cdot}57 \text{ v.u.} \end{aligned}$$

We have already seen that it is common practice to repeat tracing several times. In the example of making three vernier readings separated by one clockwise and one anticlockwise tracing, and repeating this sequence for two pole positions, there are altogether, six scale readings and four tracings. Then

$$\begin{aligned} s_1 &= \pm s_0(\sqrt{6/4}) \\ &= \pm (0{\cdot}4 \times 0{\cdot}612) \\ &= \pm 0{\cdot}24 \text{ v.u.} \end{aligned}$$

Where the measurement involves repeating the tracing operation, such as following the parcel perimeter three times between reading the vernier, we have, for one pole position

$$\bar{v} = \tfrac{1}{2}[(v_4 - v_1)/3 + (v_8 - v_5)/3] \qquad (17.41)$$

The intermediate scale readings, corresponding to v_2, v_3, v_6 and v_7 at the end of each circuit, have either not been made or are not used in the calculation. In this example there are four readings from six tracings so that

$$\begin{aligned} s_1 &= \pm s_0(\sqrt{4/6}) \\ &= \pm (0{\cdot}4 \times 0{\cdot}333) \\ &= \pm 0{\cdot}13 \text{ v.u.} \end{aligned}$$

Generalising these results we find that where there is a succession of n_1 readings which have been obtained from tracing the parcel n_2 times,

$$s_1 = \pm s_0 \cdot (\sqrt{n_1/n_2})(\text{v.u.}) \qquad (17.42)$$

The advantage of tracing the perimeter of a parcel several times is made abundantly clear. It is recommended that this be done for any parcel of area less than 10 cm^2. The extra time needed to repeat the tracing process is small compared with that required to set the planimeter in a suitable position, and to read the vernier before and after each measurement. For the present discussion it is assumed that a sequence of measurements comprise six scale readings and four circuits of each parcel. Then $\sqrt{n_1/n_2} = 0{\cdot}612$, which value will be incorporated into the equations which follow.

Errors Owing to Incorrect Positioning of the Tracing Point

The standard error s_2 is that resulting from lack of agreement between the position of the tracing point when it is supposed to occupy the position which has been chosen to start and finish tracing. It should be assumed that there is no detectable gross error in making this setting. Thus it is assumed that any instrumental causes, such as the parallax created by viewing the tip of the tracing point from different angles if this lies too far above the surface of the map, have already been eliminated. The size of the discrepancy which gives rise to s_2 is that corresponding to the limit of resolution for unaided vision, and may be taken as $d = 0{\cdot}08$ mm. The effect of this discrepancy upon the movements and scale readings of the planimeter may be resolved into the errors in two perpendicular directions – δ_1 which lies along the perimeter of the parcel and δ_2 which lies across it.

DISPLACEMENT δ_1

If the planimeter tracing point can be set on the parcel perimeter at a point which coincides with an arc element of the zero circle, small errors in positioning the point over the mark will have no effect upon the scale readings which are obtained. This is naturally achieved by finding a place on the parcel perimeter at which the planimeter arms are perpendicular, and choosing the starting point here.

According to Maslov (1955) the discrepancy δ_1 may be expressed by the equation

$$\delta_1 = [d \cdot \cot\theta (R + r)]/R \quad \text{v.u.} \tag{17.43}$$

where θ is the angle made at the pivot between the planimeter arms, and the distances R and r are the lengths of the arms (in millimetres to correspond with d) as illustrated in Fig. 17.1. For a planimeter in which $r = 30$ mm and $R =$ 150 mm, $\delta_1 = 0{\cdot}0096 \cot\theta$ v.u. If $\theta = 90°$, $\delta_1 = 0$, which confirms the advantage of setting the planimeter over the starting point with the arms at right angles. Solving equation (17.43) for other values of θ indicates that the discrepancy δ_1 soon becomes significant. For example where $\theta = 30°$, the value of δ_1 increases to 0·3 v.u., even if d is only 0·08 mm.

DISPLACEMENT δ_2

Although the longitudinal component of this error can be eliminated by correctly locating the planimeter, the transverse error cannot be corrected in this fashion. Therefore this component represents the whole of the content of the mean square error s_2. It can be shown that the mean square error, s_2, may be expressed as

$$s_2 = (1{\cdot}3r/R)(\sqrt{n_1}/n_2)$$
$$= \approx 0{\cdot}3(\sqrt{n_1}/n_2) \quad (17.44)$$

for typical planimeter dimensions. As before n_1 is the number of scale readings and n_2 is the number of times the perimeter of the parcel has been traced. If, therefore, we use the sequence outlined earlier, $s_2 \approx 0{\cdot}18$ v.u.

This standard error may be reduced somewhat if the planimeter is equipped with a tracing lens rather than a steel point. Since the circular index mark is engraved upon the flat under–surface of the cylindrical barrel of the lens, the possibility of introducing any parallax error is virtually eliminated. The index mark is usually a ring of about 0·4 mm diameter which it is easy to centre over a line of width 0·2 mm. It follows that, both in locating the starting point and tracing the perimeter of the parcel, the tracing lens usually provides more precise results. This has been amply confirmed by sets of comparative measurements made with point and lens by Gerke (1951), Kuhlmann (1951) and Zill (1956).

The Mechanical Errors of the Polar Planimeter

The mechanical errors of the instrument arise from such imperfections as friction in the bearing, inaccuracy in the dimensions or assembly of certain components leading to backlash, and incorrect positioning of the measuring wheel, characterised by eccentricity in the setting of its axle. The influence of the quality of the paper surface over which the wheel moves is also a component of this error. A major problem in evaluating the standard error which may be attributed to these causes is the difficulty in designing a suitable experiment which is free from errors created by other factors, notably that of tracing. Maslov overcame this difficulty in an ingenious fashion. He argued that if a straight line is traced by planimeter, its length, L, may be regarded as representing two sides of a square having area A, or

$$A = \tfrac{1}{2}L^2 \quad (17.45)$$

If the distance L is measured using a straight edge to guide the tracing point, a series of different square parcels may be simulated without introducing any tracing errors. For his tests, Maslov used three different polar planimeters, each set for the same value of k. The results of these measurements are given in Table 17.2. The areas of the parcels which are simulated by the linear

TABLE 17.2. *The derivation of the mean square errors of three different polar planimeters from scale-assisted linear measurements made by A. V. Maslov (1955)*

		Planimeters			Mean square error (in v.u.) due to various causes		
L(cm) (1)	A cm^2 (2)	1 (3)	2 (4)	3 (5)	Mean square error of the three instruments (6)	s_{12} (7)	s_3 (8)
5·66	8	0·87	0·78	0·63	0·77	0·65	0·41
8·95	20	0·81	0·81	0·81	0·81	0·65	0·48
12·65	40	1·18	1·52	0·87	1·22	0·65	1·03
17·90	80	1·46	1·51	0·71	1·28	0·65	1·10
28·30	200	1·36	1·84	1·00	1·44	0·65	1·28
40·00	400	1·96	1·73	1·27	1·68	0·65	1·55

measurements are given in column (2). Columns (3), (4) and (5) represent the standard errors for each of the three planimeters, and these are combined as a single value in column (6). Column (7) represents the standard errors s_1 and s_2 which are, of course, not related to the area of the parcel measured and have the constant value of 0·65 v.u. It may therefore be argued that the standard error arising from mechanical reasons s_3, is the difference between columns (6) and (7). There is a curvilinear relationship between the standard error s_3 and the size of the parcel A, which evidently has the form

$$s_3 = x \cdot A^y \tag{17.46}$$

and from analysis of the data presented in Table 17.2,

$$y \approx 0{\cdot}3 \text{ so that}$$
$$s_3 = 0{\cdot}225\sqrt[3]{A} \quad \text{v.u.} \tag{17.47}$$

Shape of Parcel

Thus far it has been assumed that the figures traced have been square. It is now desirable to introduce an additional factor which takes into consideration the shape of a parcel, at least to the extent of distinguishing between a square and a rectangle. Maslov considered the properties of a coefficient which is merely the ratio of the sides a and b of a rectangle, or

$$K' = a/b \tag{17.48}$$

and showed that the standard error s_3 is increased by a factor

$$E = \sqrt[3]{[(1+K')^2/4K']} \tag{17.49}$$

Table 17.3 gives some examples of E for different rectangular shapes. This indicates that if $K' < 3$, the influence of parcel shape may be safely ignored.

TABLE 17.3. *Maslov's coefficient E describing the asymmetry of rectangles*

$K' = a/b$	E	$K' = a/b$	E	$K' = a/b$	E
1	1	6	1·27	20	1·77
2	1·04	7	1·32	40	2·19
3	1·10	8	1·36	60	2·49
4	1·16	9	1·41	80	2·74
5	1·22	10	1·45	100	2·94

However although this represents an increase in the standard error, repeated tracing of the parcel has the opposite effect of reducing s_3 by a factor of $\sqrt{n_2}$. Consequently the addition of these two factors it modifies (17.47) to

$$s_3 = (0{\cdot}068/\sqrt{n_2})\cdot\sqrt{(1/k)}\cdot E\cdot\sqrt[3]{A} \tag{17.50}$$

The Influence of the Map Surface

It has been suggested that the quality of the paper surface over which the tracing wheel moves may also be a potential source of error, and this has been included under the same heading as the mechanical errors. It seems, at first sight, that the measuring wheel is more likely to slip and slide on the glossy surface of art paper or a glazed photographic print than on the rougher texture of drawing paper. It follows that if this supposition be correct, a measurement made upon a glossy surface ought to be less accurate than a similar measurement made upon a rough surface.

There appear to have been very few attempts to study this aspect of planimeter use. Baer (1937) quotes earlier work by Ketter, who investigated seven different kinds of paper and also the glossy surface of tracing linen and glass. Evidently Ketter found that the variations between the measurements made on glossy and rough paper did not differ by more than 0·1% of the parcel area, and concluded that paper texture has much less influence upon the results than did other sources of error. Baer's own investigations involved measuring the areas of circular parcels using a calibration jig of length 10 cm on five different types of paper. The pole distance was kept constant in both trials so that the route followed by the measuring wheel was always the same. His results are given in Table 17.4.

The greatest difference between the readings in all these trials is only 18 vernier units or 0·06% of the area measured. Therefore we conclude that the texture of the paper surface has an insignificant effect upon s_3. This is encouraging for those who wish to make measurements of area by planimeter on photographs. Potential users of the method may well have been deterred from using a planimeter because of the widespread belief that the surface is too slippery, rather than for any reasons relating to the variable geometry of the

TABLE 17.4. *Measurements of areas of circles, in vernier units, using a polar planimeter on surfaces of different textures made by Bear (1937)*

Surface	Trial 1	Trial 2
Lithographic paper	311,360	311,360·5
Calendered machine-made paper	311,364·5	311,363·5
Rough texture paper	311,368	311,367
Extra-rough texture paper	311,372	311,371
Watercolour paper	311,378	311,377·5

photographic image. However, it should be stressed that both of these investigations were made more than half a century ago, they are based on what we would nowadays regard as unacceptably small samples and long before some of the materials which are now commonly used in cartography had been invented. There is obviously scope for additional work to be done on this subject.

Tracing Errors

These generally arise from the failure to follow the parcel perimeter exactly during tracing; that the tracing point or index mark occasionally deviates from the line to one side or the other because of unsteadiness of the hand, momentary loss of concentration, imperfect vision, and parallax between point and line. Although the size of these deviations may be primarily owing to operator defects, other factors such as the weight of the planimeter and the angle between its arms also enter the theoretical argument. We have already noted that if a typical polar planimeter is being used close to its physical limits, so that the angle at the pivot approaches its extremes, the action becomes stiff and operation of the instrument is unreliable. Baer's work on the operation of planimeters with the angle between the arms set to fixed values confirms this.

Maslov (1955) investigated the effects of tracing errors, using two different sets of data collected by him in 1944 and 1950. These experiments were made by tracing a series of parcels of different sizes, 20 times each, using six different instruments. The standard errors obtained by him, and the calculated standard errors based upon the theory which has already been established, allowed him to tabulate the results in Table 17.5.

The mean square error for each determination is tabulated under the heading s_v, and is the combination of the four standard errors s_1, s_2, s_3 and s_4. Using the same argument as that applied in the evaluation of the mechanical error, the standard errors due to reading the vernier and setting the tracing point are independent of area and represent the constant $\pm 0{\cdot}65$ v.u. This is deducted from s (using equation 6.16, of course) to leave the standard errors tabulated under the heading s_{34}, which is therefore the combination of the

TABLE 17.5. *Mean square errors of measurements by planimeter for $k = 0{\cdot}09$ cm² (from Maslov, 1955)*

1944 experiments			1950 experiments		
Area (cm²)	Errors (v.u.)		Area (cm²)	Errors (v.u.)	
	s	s_{34}		s	s_{34}
8	0·97	0·72	2·1	0·81	0·48
20	1·07	0·85	5·3	0·87	0·58
40	1·23	1·04	12·5	1·03	0·80
80	1·59	1·45	50·3	1·40	1·24
200	2·16	2·06	113·0	1·83	1·71
400	2·71	2·63	201	2·03	1·92

mechanical and tracing errors. Examination of the relationship between $\log x$, $\log A$ and $\log s_{34}$, using the same technique as that described for the evaluation of the mechanical errors, yields a solution of the equation

$$s_{34} = 0{\cdot}347\, A^{0{\cdot}328} \tag{17.51}$$

or

$$s_{34} = 0{\cdot}345 \sqrt[3]{A} \tag{17.52}$$

and having eliminated the mechanical errors by subtracting the expression for s_3

$$s_4 = 0{\cdot}260 \sqrt[3]{A} \tag{17.53}$$

However, it is still necessary to consider the influence of a variation in k, of the asymmetry of the parcels and the effect of making repeated tracings. Combining all of these in the form which was established earlier, we obtain

$$s_4 = (0{\cdot}017/\sqrt{n_2}) \cdot 1/k \cdot E \cdot \sqrt[3]{A} \tag{17.54}$$

It will be seen later that this standard error is larger than the others which have been evaluated, which agrees with the conclusions of other writers. For example, Baer demonstrated that the tracing error approximated to

$$s_t = c \cdot k \cdot \sqrt{U} \tag{17.55}$$

where s_t is the standard error of tracing a perimeter of length U. The term k is that used earlier to denote the value of 1 vernier unit, and c is a coefficient to be determined experimentally. Since equation (17.55) is that for parabola, it follows that a graph of the standard error plotted against $\sqrt{U}$ approximates to a straight line, as is shown in Fig. 17.19, p. 392 which illustrates the results obtained by Zill (1956).

The Calibration Error

Only the pole-outside mode of operation is under serious consideration. Consequently this term refers only to the precision with which the factor k has been determined. The appropriate expression is

$$(s_k/k)^2 = (s_{A0}^2 + s_u^2 \cdot k^2)/A_0^2 \tag{17.56}$$

in which s_{A0} refers to the standard error of the area, A_0, of the circular parcel defined by the calibration jig, and depends almost wholly upon the precision with which the distance AB in Fig. 17.14 between the pole point and the small hole have been measured. Putting $d_1 \pm s_D$ as the distance and standard error of the single observation calculated from a series of measurements of this distance by steel scale, the area, A_0 and the standard error of the area of this parcel may be calculated

$$(s_{A0}/A_0)^2 = (2s_D/d_1)^2 \tag{17.57}$$

The term s^2 is made up from the same four error terms already described. Therefore

$$s^2 = [0{\cdot}4(\sqrt{n_1/n_2})]^2 + (0{\cdot}3\sqrt{n_1/n_2})^2 + [0{\cdot}043(1/\sqrt{n_2})\cdot 1/k \cdot E \cdot {}^3\sqrt{A_0}]^2 + [0{\cdot}015(1/k\sqrt{n_2})\cdot E \cdot {}^3\sqrt{A_0}]^2 \tag{17.58}$$

The variation in this standard error is demonstrated by the following measurements by Maslov, who measured a series of rectangular parcels formed by groups of grid squares. From these data we may draw the following conclusions.

1. The relative error in determining the value of the vernier unit decreases with increasing area of the figure used for calibration.
2. There is no particular advantage to be gained from using test parcels which are larger than 300 cm^2. Moreover, parcels of this size cannot be traced when the tracing arm has been set to the shorter distances on its scale. The gaps in Table 17.5 indicate combinations of R and A_0 for which measurement is impossible.
3. The relative error of determining the size of the vernier unit ought to be determined independently of the value of the vernier unit.
4. The size of the relative error indicates that the value of the vernier unit should be calculated to four significant figures.

Since the constant k is multiplied by v, an error in k will behave as a systematic error and increase in proportion to the size of the parcel. If the calibration has been done by measuring grid squares of area 200 cm^2 or larger, a suitable average value for the standard error will be approximately

$$s_k/k = 0{\cdot}0009A \tag{17.59}$$

corresponding to the values in the fourth column of Table 17.6. If, however, calibration has been done using the typical metal jig, the area of the resulting circular parcel is much smaller so that the relative error is higher, corresponding approximately to the values in the third column in Table 17.6.

TABLE 17.6. *Relative mean square error (s_k/k) for planimeter calibration*

		Area of calibration figure, A_0(cm²)			
k(cm²)	R(cm)	100 $E=1$	200 $E=1{\cdot}04$	300 $E=1{\cdot}10$	400 $E=1{\cdot}16$
0·10	17·7	0·00127	0·00095	0·00084	0·00081
0·09	15·0	0·00126	0·00094	0·00083	0·00080
0·08	13·3	0·00125	0·00093	0·00083	0·00080
0·07	11·7	0·00123	0·00092	0·00082	0·00079
0·06	10·0	0·00122	0·00092	0·00082	0·00079
0·05	8·3	0·00121	0·00091	—	—
0·04	6·7	0·00120	—	—	—

The Combined Standard Error

It is now possible to combine the results of the separate stages in this analysis to form a single equation for the mean square error of measurement by planimeter. These are combined in the sense of equation (17.42), i.e. for a mean square error expressed in vernier units

$$s=[0{\cdot}4(\sqrt{n_1/n_2})]^2+[0{\cdot}3((\sqrt{n_1/n_2}))]^2 +[0{\cdot}068(1/\sqrt{n_2})\cdot(1/\sqrt{k})\cdot E\cdot\sqrt[3]{A}]^2 =+[0{\cdot}023(1/\sqrt{n_2})\cdot(1/k)\cdot E\cdot\sqrt[3]{A}]^2+0{\cdot}0009A^2 \quad (17.60)$$

Putting as before $\sqrt{n_1/n_2}=0{\cdot}612$ and $1/\sqrt{n_2}=0{\cdot}5$,

$$s=0{\cdot}918+0{\cdot}034\cdot 1/\sqrt{k}\cdot E\cdot\sqrt[3]{A}+0{\cdot}011\cdot 1/k\cdot E\cdot\sqrt[3]{A} +0{\cdot}0009A^2 \quad (17.61)$$

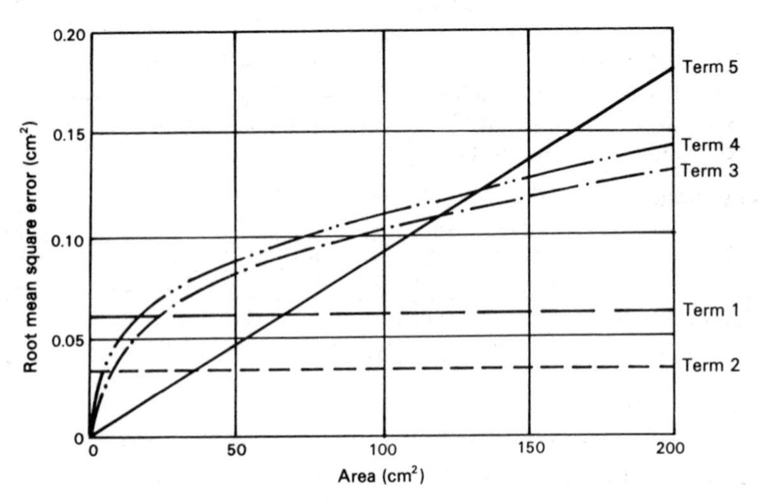

FIG. 17.17. The combined influence of the different sources of error upon the precision of the polar planimeter. (Source: Maslov, 1955).

The influence of each term has been plotted as a series of five graphs in Fig. 17.17. These were calculated on the basis that $k = 0{\cdot}1\,\text{cm}^2$, $E = 1{\cdot}0$ and $n_2 = 1$. The following conclusions may be drawn:

- For parcels smaller than about $10\,\text{cm}^2$ in area the shape is unimportant because the size of the third and fourth terms is less than those of the first and second. By reducing the value of k the error of measuring small parcels is significantly reduced. This is because the first term, which predominates, may be reduced in direct proportion to k. The third term in the equation is also reduced, but this is only in proportion to $\sqrt{k}$.
- An important way of reducing the standard error is to increase the number of circuits traced. This has already been mentioned as being the obvious method of increasing the scale reading, and numerical examples were given earlier to indicate how useful this technique may be.
- The polar planimeter is an unsuitable instrument for measuring the areas of very small parcels ($A < 3\,\text{cm}^2$). Figure 17.17 indicates that the mean square error of measurement, even after two tracings of each parcel, may be larger than the areas of the parcels themselves. It follows that for very small parcels, other methods of area measurement, such as point-counting methods, are to be preferred. This conclusion was also confirmed by some rather crude time-and-motion studies conducts by the author in 1968, which demonstrated that measurement of very small parcels by planimeter is more time-consuming than by point-counting methods.

Maslov simplified equation (17.61) to an expression more readily comparable with the empirical formulae derived by other workers in this field and based upon (17.41).

$$s_a = 0{\cdot}7 + 0{\cdot}3\sqrt{A} + 0{\cdot}003A \tag{17.62}$$

However, this shortened form of working formula is satisfactory only for evaluating measurements of parcels which are smaller than $200\,\text{cm}^2$.

This result can be compared immediately with the following equations derived by other workers in this field:

Volkov (1946):

$$s_A = 0{\cdot}68 + 0{\cdot}028\sqrt{A} \tag{17.63}$$

Zill (1955):

$$s_A = 0{\cdot}04 + 0{\cdot}0188\sqrt{A} - 0{\cdot}00034A \tag{17.64}$$

Lüdemann (1927)

$$s_A = 0{\cdot}027\sqrt{A} - 0{\cdot}00151A \tag{17.65}$$

Montigel (1924):

$$s_A = 0{\cdot}000986A_0 + 0{\cdot}00164\sqrt{A \cdot A_0} + 0{\cdot}0000072A \cdot A_0 \tag{17.66}$$

(where $A_0 = 56{\cdot}77\,\text{cm}^2$)

Table 17.7 compares some of these results. They have been examined by

TABLE 17.7. *Mean square errors of area measurement by polar planimeter according to different authorities. Mean square errors expressed in square centimetres*

Area (cm^2)	Lüdemann (1927)	OS tol-erance (÷6)	Volkov (1946)	Maslov (1955)	Zill (1956)	Montigel (1924)
0·5	0·018	0·020	0·074	0·073	0·053	0·11
1·0	0·025	0·023	0·077	0·076	0·058	0·12
2·0	0·035	0·033	0·080	0·080	0·066	0·12
5·0	0·053	0·053	0·088	0·088	0·080	0·14
10·0	0·070	0·075	0·100	0·100	0·096	0·16
20·0	0·091	0·105	0·110	0·110	0·117	0·19
50·0	0·115	0·167	0·130	0·140	0·156	0·25
100·0	0·119	0·230	0·160	0·190	0·194	0·34

Kneissl (1963), who concluded that the expression

$$s_A = 0{\cdot}02\sqrt{A} \tag{17.67}$$

is a fair approximation to most of these determinations.

In addition to the results obtained by testing, a standard error based upon the OS working tolerance for area measurement is also given. This tolerance was originally introduced by the Areas Section as a means of quality control in the days when the scale-and-trace method of measurement was still being used for all production work, and which was retained after the change to planimeter methods in the 1960s. This is the statement that two measurements of the same parcel of area A should agree within the limits $\sqrt{(2 \cdot A)}/100$. For a normal distribution it is assumed that such a tolerance is equivalent to the range of a large sample of measurements, that the range is limited by $\pm 3\sigma$, and therefore

$$s \approx \pm \text{tolerance}/6 \tag{17.68}$$

Comparison With Other Types of Planimeter

This chapter has emphasised the properties and operation of the simple polar planimeter, virtually to the exclusion of all other types. However, it is desirable to conclude the section upon the accuracy of these instruments with some statement of how the polar planimeter compares in accuracy with other designs.

Polar Disc Planimeter

There are three studies of the precision of the polar disc planimeter, which have been published by Idles (1951), Gerke (1951) and Zill (1955). The empirical equations derived by them are as follows:

Idler:

$$s_A = 0{\cdot}0108 + 0{\cdot}0049\sqrt{A} \quad (\text{cm}^2) \tag{17.69}$$

TABLE 17.8. *Precision of the polar disc planimeter. Mean square error (cm²)*

Area (cm²)	Idler	Gerke	Zill
1	0·016	0·011	0·024
2	0·018	0·012	0·030
5	0·022	0·015	0·039
10	0·026	0·017	0·050
20	0·033	0·021	0·065
50	0·045	0·032	0·091
100	0·060	0·046	0·116
200	0·080	0·071	0·142

Gerke:

$$s_A = 0{\cdot}91 + 0{\cdot}0206\sqrt{A} + 0{\cdot}000162A \quad (\text{cm}^2) \tag{17.70}$$

Zill:

$$s_A = 0{\cdot}11 - 0{\cdot}00029A + 0{\cdot}0134\sqrt{A} \quad (\text{cm}^2) \tag{17.71}$$

Table 17.8 compares these three equations for the mean square errors of parcels within the range 1–200 cm².

Rolling Disc Planimeter

This instrument has not been described in this book, for it is seldom encountered in use for cartometric work in English-speaking countries. However, Zill has examined the precision of this instrument and his conclusions are included here for comparison with the others.

Zill has derived the expression for the precision of the rolling disc planimeter as

$$s_A = 0{\cdot}005 - 0{\cdot}00023A + 0{\cdot}01\sqrt{A} \tag{17.72}$$

which indicates that the instrument is marginally more precise than is the polar disc planimeter. Figure 17.18 illustrates the curves corresponding to the different instruments. This indicates that the instruments ought to be ranked in the following order of precision:

Least precise	Polar planimeter with tracing point
	Polar planimeter with tracing lens
	Polar disc planimeter
Most precise	Rolling disc planimeter

This result is not only the consequence of the improved precision with which the scales may be read on the disc instruments resulting from the exaggerated movements of the measuring wheel. Zill has also indicated that a similar order of precision is also maintained in tracing. Using the form of equation which has already been briefly mentioned on page 378 but relating to the mean

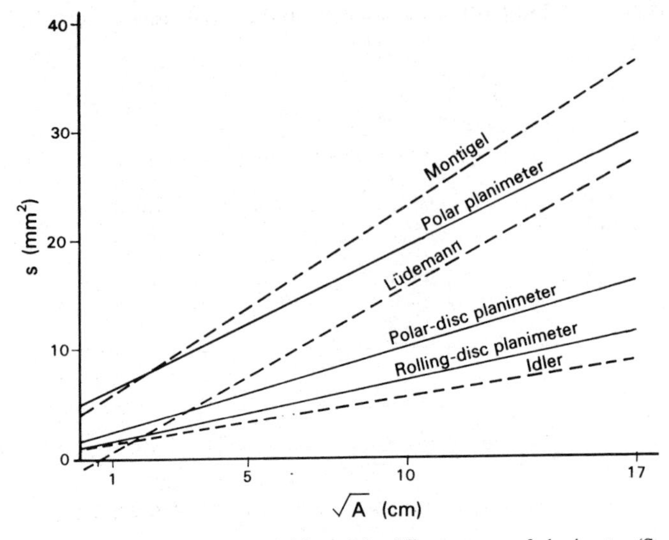

FIG. 17.18. A comparison of the precision of the different types of planimeter. (Source: Zill, 1955).

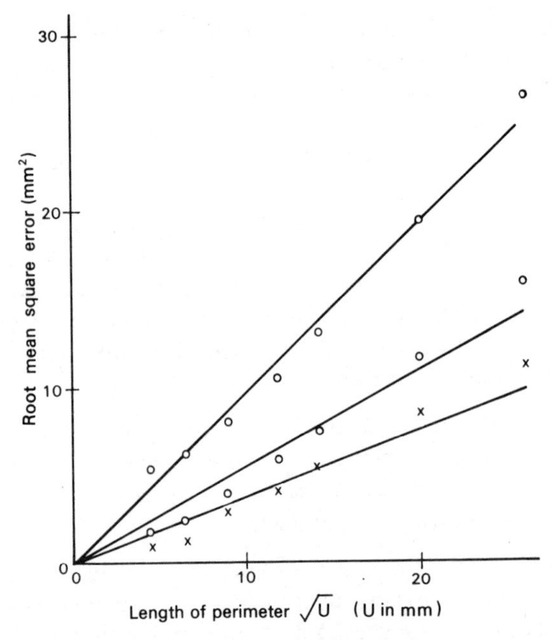

FIG. 17.19. The relationship between instrument precision and length of parcel perimeter. Upper line: Polar planimeter; middle line: polar disc planimeter; lower line: rolling disc planimeter. (Source: Zill, 1955).

square error of tracing the perimeter of a parcel of length U, Zill's three equations are:

Polar planimeter:	$s_A = 0{\cdot}8\sqrt{U}$	(17.73)
Polar disc planimeter:	$s_A = 0{\cdot}42\sqrt{U}$	(17.74)
Rolling disc planimeter:	$s_A = 0{\cdot}3\sqrt{U}$	(17.75)

In these equations s_A is expressed in square millimetres and U in millimetres. Figure 17.19 illustrates the relationship between these three expressions.

18

Direct Measurement of Area by Point-Counting Methods

If anyone can see well enough to distinguish one type of dot from another and is able to count, success is assured.

(Walter F. Wood, *Professional Geographer*, 1954)

We have already described in Chapter 3 the difference between those counting and measuring techniques used to make complete measurements of parcel area and those which only comprise partial measurements but which provide good results if they are treated as techniques in statistical sampling. The reader will recall the comparison which was made on pages 56–58 between the method of squares and the use of the dot grid, or point-counting methods, and that the author has also distinguished between direct and indirect methods of measurement by counting. The majority of the early descriptions of area measurement by counting relate to indirect methods of area measurement. This is the natural consequence of its use in inventory sampling where the output are a series of frequencies of occurrence of the various categories of the variable being studied and are treated as proportions of the total area. It follows that the area of each category of land use or land cover must subsequently be calculated from the total area of the region which must have been measured or calculated independently. A further and fundamentally important distinction between the methods is that using the direct method *one sample comprises one application of the grid over the parcel to be measured.* Using indirect measurement, *one sample comprises one point counted.* It follows that the two methods are quite different to one another despite the fact that both are based upon counting points. The present chapter is intended to study the direct methods in detail; the indirect methods are examined in Chapter 19. The various procedures and strategies used in both are summarised in Fig. 18.1.

Whereas area measurement by the method of squares is exceedingly old, the sampling methods are comparatively new. They evidently originated about 50 years ago as methods of two-dimensional sampling from maps to obtain data about agricultural land use and related variables. A little later they began to be used by foresters in North America for evaluation of timber stands, because the sampling methods were especially suited to working directly from vertical aerial photographs. By the time Yates (1949) wrote his classic work on

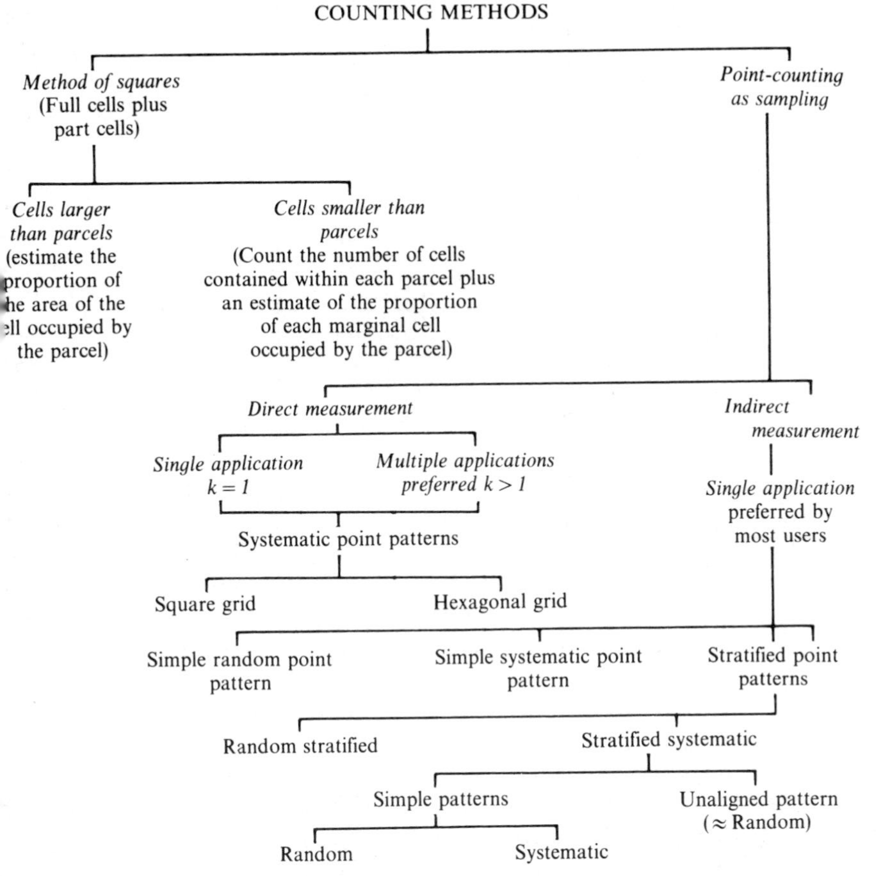

FIG. 18.1. Summary of the different methods of area measurement by counting points of cells.

sampling methods the use of a grid or point pattern superimposed over the map or aerial photograph was already well established as a convenient frame for sampling. The earliest reference to the use of point-counting as a direct method of determining the areas of individual parcels appears to have been a paper by Abell (1939) describing its forestry applications, but differing from the usual methods by virtue of the fact that his unit figure was a small rectangle.

A few years later Bryan (1943) described the *modified acreage grid*, which was a square pattern of dots separated by intervals of $\frac{1}{8}$ inch, or a density of 64 points per square inch. This is illustrated in Fig. 18.2, which shows how different sizes of square, having side lengths $\frac{1}{8}$, $\frac{1}{4}$ and 1 inch may be distinguished by choosing suitable combinations of dots and lines. Because different multiples of the unit area are quickly recognised, such a grid is easier

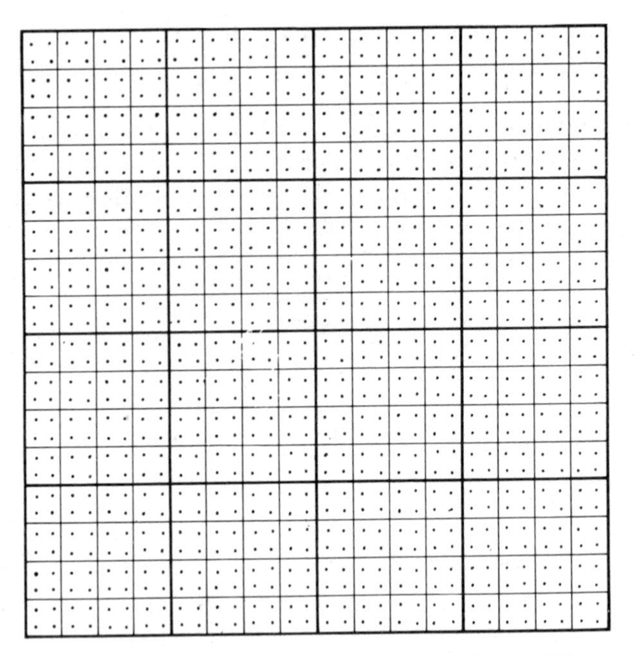

FIG. 18.2. The modified acreage grid. (Source: Bryan, 1943).

to use than an uniform pattern in which all the dots are of equal diameter or the lines are of uniform width.

After World War II, references to the methods become more frequent. In the American literature of this period we note the introduction of such terms as the *dot templet* by Wilson (1949), *dot grid* by Tryon *et al.* (1955) and the *dot planimeter* by Wood (1954) to describe virtually the same device as the modified acreage grid.

The essential character of the direct method is that the areas of parcels are measured individually and separately. The dot grid is placed over the map in a random position and the dots lying within the boundary of parcel i are counted. The grid is now moved into a different position, usually by shifting the overlay in both dimensions and through an arbitrary rotation. This procedure is repeated several times. After a series of k applications of the grid, yielding counts of $n_1, n_2, \ldots, n_k$ points, the arithmetic mean, $\bar{n}_i$, of the counts may be calculated. Then A_i, the area of parcel i, is

$$A_i = \bar{n}_i \cdot d_1^2 \tag{18.1}$$

for a square pattern, or

$$A_i = \bar{n}_i \cdot (\sqrt{3}/2) \cdot d_2^2 \tag{18.2}$$

for the hexagonal point pattern grid. It follows, moreover, that the standard error of the single random application is

$$s_i = \sqrt{[\sum(n_i - \bar{n}_i)^2/(k-1)]} \tag{18.3}$$

The sampling strategy advocated here is that each application of the overlay represents a single random observation which is functionally related to the area of the parcel. Hence the error of the area estimate is also an error of the dot count. This error may be estimated by drawing a sample as described. From the theory of sampling presented in Chapter 7, the estimates of the mean number of points, $\bar{n}_i$ contained within the parcel i, are normally distributed so that the standard errors and confidence intervals may be calculated in the customary fashion. However, the size of the sample is small. It is economical to use point-counting methods only if the time required to complete the measurement is less than the time needed to measure the area of a parcel in some other way. In most production work it is therefore necessary to restrict the size of the sample to about $k = 5$.

Although the square and hexagonal grid patterns are those most commonly used for direct measurement, other kinds of grid are used from time to time. There is no reason why a rectangular cell having sides l and m, or unit area $a = l \cdot m$ should not be used. Indeed Abell (1939) used rectangular grids whose dimensions were chosen as being small fractions (1/25 and 1/100) of the $1' \times 1'$ graticule appearing on the maps used by him. Bellhouse (1981) also derives the general theory of point-counting in terms of rectangular cells. Moreover, there is no reason why the dots to be used for sampling should not occupy some other position within each cell than its geometrical centre, as was implied on page 58. This, too, is illustrated by Bellhouse. The only condition which must be satisfied is that every point on the grid must occupy the same position within its cell.

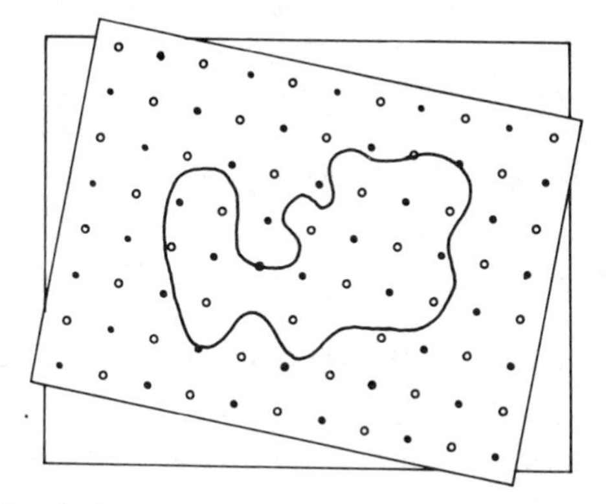

FIG. 18.3. The dot planimeter showing the arrangement of A-type symbols (block dots) and B-type symbols (circles). (Source: Wood, 1954).

TREATMENT OF MARGINAL POINTS

Sooner or later the occasion arises where the perimeter of the parcel appears to coincide with one of the dots on the overlay. Then a decision has to be made whether the point ought to be counted or not. In several early descriptions of the use of the dot grid it was supposed that a proportion of such points ought to be included in the count. Thus Bryan suggested that one-half of those points which *fell exactly* on the perimeter of the parcel ought to be counted. Wood went one stage further and used two different kinds of dot symbol on the overlay, as is shown in Fig. 18.3. Since these alternate in both rows and columns, there are equal numbers of each symbol on the overlay. Wood's rule is that where the perimeter *touches* an *A*-type symbol this point is counted, but where it touches a *B*-type symbol this is not counted.

Some later writers, such as Monkhouse and Wilkinson (1952) and Muehrcke (1978), followed Wood without reservation. Hiwatashi (1968) and Loetsch *et al.* (1973) both used Bryan's form of correction. Gierhart (1954) stated that all points touching the perimeter should be counted. By contrast, Bonnor (1975) made no special allowance for marginal points. The author believes that there is no strong evidence, one way or the other, to suggest that use of Wood's or Bryan's rules, or neglect of the marginal points, materially affects the results. The need for any rules depends upon the size of the dot drawn on the overlay, the width of the line comprising the parcel boundary on the map, and choosing which of the two criteria should be used for selection; whether exact coincidence or mere touching represents the rule for selection. Much of the uncertainty about whether a dot lies on one side of the boundary or the other is caused by the need to use dots which are large enough to be easy to see and count. Similarly the lines on the map should also be easily recognisable. It is more likely that dot and line will appear to coincide where there is the combination of a large dot and a thick line than would happen if the points and lines approximated to their Euclidean definitions. In most cases, however, it should be obvious to the observer whether a dot ought to be counted or not, and that truly doubtful cases are quite rare. This was confirmed by Yuill (1971), who introduced a subroutine in his simulation program to apply Wood's rule should a point in the overlay coincide with a parcel boundary. Because his points and lines satisfied their Euclidean definitions, he found that this subroutine was never used.

GROSS ERRORS IN COUNTING, AND SOME INSTRUMENTS TO REDUCE THEM

The risk of making gross errors is higher in counting than in making instrumental measurements. There are likely to be three kinds of mistake:

- Some of the points to be counted are missed and are excluded from the count; therefore the total recorded is smaller than it ought to be.

- Some of the points may be counted more than once; therefore the total is greater than it ought to be.
- The observer may forget the number of points counted before completing measurement of a parcel.

One way of overcoming them is to try to remove any external source of interruption which may cause the observer to lose concentration. Extreme measures may be necessary to avoid interruption, such as disconnecting the telephone and working behind locked doors. For the benefit of the reader who only supervises the work of others, it should be stressed that it is unreasonable to expect a student, draughtsman or junior technician working in a crowded, noisy laboratory or drawing office to produce consistently reliable work. Simple precautions against making gross errors include marking the map or the overlay as counting proceeds, keeping a record of every 10 or 100 points counted, or employing the hand-held mechanical counter used by attendants at a turnstile or a gate to count the number of people or sheep passing through. Various electrical and mechanical instruments have also been produced to facilitate counting under laboratory conditions. The less sophisticated type of counter suitable for cartometric work is exemplified by the *Markounter* and the *MK Area Calculator.*

The Markounter was developed in Britain by Scientifica and Cook Electronics in the 1960s. It was originally designed for biological work, for example to count bacterial colonies, but it has also been found to be extremely useful for cartometric work. The instrument comprises a veeder counter which is electrically operated by a press-to-make switch which is operated by a short arm containing a sleeve to hold a pencil.

It therefore operates when the device is held in the normal writing position and the pencil point is pressed against the overlay, the purpose being to make a mark at the same time as one point is counted. A similar device has also been described by Kenady (1961), who modified a typical ballpoint pen for this purpose.

The object of marking the dots as they are counted is to overcome the first two sources of gross error, for it is evident at a glance which points have been counted. The dot grid itself may be marked, but it is better to place a sheet of glass-clear plastic over the dot grid and use a wax pencil or fibre-tipped pen to make the marks. Working on a separate plastic sheet rather than the grid itself reduces the risk of spoiling the images of the dots through over-zealous cleaning.

The principal disadvantage of the Markounter is that counting necessitates deliberate pressure of the pencil point upon the overlay. When this process is repeated many hundreds or even thousands of times it becomes extremely tiring for the operator. The speed with which it can be used may be as high as 3 points per second for a small parcel in which all the dots can be reached without moving the hand from where it rests on the drawing board. However, in normal work the counting rate is about 1·3 points per second. Experiments carried out by the author have shown that the time needed to count dots with a Markounter is directly proportional to the number of dots up to a total of about 1,000 dots, or 12 minutes of continuous counting. Thereafter operator fatigue sets in, the movements of the pen become slower and the time needed to count the last 100 points is appreciably longer than that required to count the first 100.

The *MK* Area Calculator is an American instrument which was specifically designed for area measurement. It comprises an electrical counting mechanism like that of the Markounter, but in this instrument the circuit is made by the contact between the graphite "lead" of a propelling pencil and copper strips which are printed upon the matt surface of a sheet of transparent plastic. The copper strips form equidistant parallel straight lines which are interconnected so that the circuit is completed wherever they are touched by the pencil point. The under-surface of the plastic sheet contains a family of equidistant parallel straight lines printed in red so that these, combined with the copper strips, create a uniform square grid. Two different sizes of grid are available, the first having lines at intervals $d_1 = \frac{1}{8}$ inch, or 3·2 mm, having a density of 64 points per square inch. This will be referred to as the *MK* 64/in. grid. The second size of grid has separation $d_1 = \frac{1}{10}$ inch, or 2·5 mm, giving rise to the density of 100 points per square inch. This is the *MK* 100/in. grid.

The grid is placed over the map with the printed circuit uppermost. The operator draws the pencil across the grid using the red lines as a guide. Each time the pencil point makes contact with a copper strip the circuit is completed

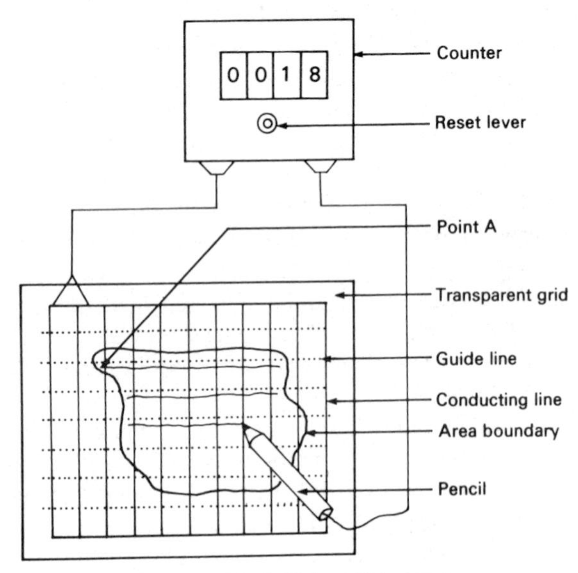

FIG. 18.4. The MK Area Calculator.

and the veeder counter is advanced one digit. The effect is to count the number of copper strips crossed in drawing a single straight line from one side of the parcel to the other. By following each red line which intersects the parcel, the total number of grid intersections contained within it are counted. At the same time a series of parallel lines are drawn in pencil on the grid to show which points have been included in the count. The system is illustrated in Fig. 18.4.

The advantage of the *MK* instrument over the Markounter is that counting and marking is carried out as a continuous drawing process. The pencil point is only raised and lowered where the red lines intersect the parcel perimeter. It is therefore a much less tiring instrument to use. There are, however, two important disadvantages.

The first is that the counting mechanism is not sufficiently sensitive to accept signals in quick succession. The instrument is reliable only if it is operated rather more slowly than a comfortable drawing speed. If the pencil is moved too quickly some points will not be recorded so that the measured areas are smaller than they should be. This disadvantage means that the operator must be continually aware of how the instrument is behaving. Rideout (1962) recommended that the user should make an independent visual count of the points passed during measurement, but the author has found that it was more convenient to listen to the clicks made by the veeder counter as this moved, and apply a correction immediately the instrument failed to respond to a contact.

The second defect of the instrument is the difficulty of cleaning the pencil lines from the grid. Although a special cleaning fluid is provided by the manufacturer for this purpose, this is not wholly successful, and gradually a thin layer of burnished graphite spreads across the plastic sheet. It follows that the grid becomes progressively less transparent so that it becomes increasingly difficult to see map detail through it. Moreover, a delay occurs while the residue of cleaning fluid evaporates from the plastic, for the grid cannot be used again until it is quite dry. A large number of experiments carried out by the author in 1968 showed that after about 3,500 individual measurements, and after being cleaned about 100 times, a grid had become so opaque that it could no longer be used efficiently. In production work, involving measurement of numerous small parcels, this would mean replacing a grid after measuring about ten O.S. 1/2,500 maps, which would add significantly to the cost of making routine measurements. When it was first available the cost of the *MK* Area Calculator was similar to that for the best-quality variable-arm polar planimeters made in Switzerland and Germany. These can be expected to operate for many years without having to replace essential components after measuring only a few maps.

SYSTEMATIC ERRORS AND CALIBRATION OF A DOT GRID

The main source of systematic error in this method of area measurement arises from variation in the dimensions of the dot grid, the cells of which do not correspond to the supposed value of d. Such discrepancies arise from errors in construction and drawing of the grid, or from subsequent deformation of the plastic base in use, cleaning and storage.

Although it might appear that an obvious method of calibrating a dot grid is to measure the sides of the unit cell to find d_1 or d_2, it is usually difficult to measure this distance with sufficient accuracy. This is because the diameters of the dots, widths of lines, or the copper strips on the *MK* grids have to be large enough to be legible, and it is therefore difficult to identify the centres of the symbols or lines with sufficient precision. For this reason calibration is more reliable if carried out across the whole grid, rather than between pairs of adjacent points.

The alternative approach is to use an indirect method of calibration; to make a series of test measurements of parcels of suitable size and known area, simply dividing the number of points counted by the known area, thereby finding the conversion factor equivalent to the area of the unit cell, a. One should use test areas which are appreciably larger than any parcels likely to be measured subsequently. The smaller the test areas, the greater is the risk of retaining some undetected systematic errors. Since the test parcel must be of known size, it is generally convenient to use a block of grid squares or a graticule quadrangle for this purpose. An alternative to using a map is to construct a series of simple geometrical figures of known size. The pattern map devised by McKay (1967) is a rather complicated example, comprising a large number of geometrical parcels of different shapes and sizes, the areas of which may be determined independently.

THE THEORETICAL ACCURACY OF POINT-COUNTING METHODS

Between 1965 and 1975 a series of independent investigations were made to attempt the evaluation of the accuracy of the point-counting methods by the following writers:

- Hiwatashi (1968) investigated the problem from the point of view of a forester. His approach was primarily statistical.
- Frolov and Maling (1969), whose work was subsequently modified by Lloyd (1976), attempted to analyse the theoretical problem of how points might fall within an enclosed figure represented by the parcel boundary.
- Yuill (1971), who worked for the Dartmouth College Remote Sensing Project, considered one of his goals was to extend the investigations of Frolov and Maling on an empirical basis using more irregular shapes. His work is of interest in two respects; first, because he used a computer program to simulate the measurement process, thus allowing him to use large test samples without the work becoming laborious. Secondly, he found that parcel shape was comparatively unimportant and concluded that most of the variability was a direct function of parcel size.

- Loetsch *et al.* (1973) devote a short section to this subject in their all-embracing and definitive work on the techniques and methods of forest inventory. Although the sample of measurements collected and analysed by them is small compared with the masses of data collected by the other authors listed here, there method of analysis is interesting as providing a link between the different investigations. Moreover their main results are remarkably close to one of the principal conclusions of Frolov and Maling.
- Zöhrer (1978) returned to this study later.
- Bonnor (1975) is another forester who treated the problem statistically, particularly with reference to what he called the *maximum allowable error*. Bonnor was also preoccupied with the influence of parcel shape upon the accuracy of the results.

Space will not permit presentation of a detailed analysis of the theoretical discussion initiated by Frolov and Maling. Only sufficient theory is given to indicate the framework upon which the principal practical conclusions are founded.

The Random Start

Most of these investigations start with the assumption that a square point pattern is laid at random over a parcel. The square cells which lie entirely within the perimeter of the parcel in all applications of the grid deserve little special comment, for it is the behaviour of the perimeter as it passes through the marginal cells which needs investigation. It is assumed, in creating the model illustrated in Fig. 18.5, that both the point pattern and the perimeter of the parcel satisfy the Euclidean definitions of having no size or width, and that the size of the unit cell is small enough to regard the element of the perimeter contained within the cell as being a straight line. In practice this corresponds to

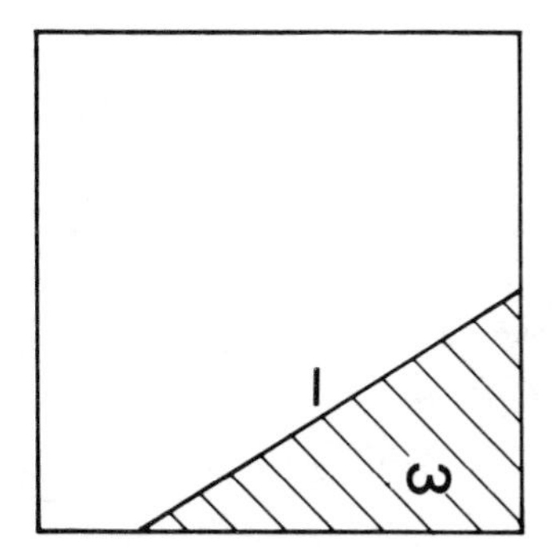

FIG. 18.5. The unit cell of a square dot grid which contains a portion of the parcel perimeter of length l and encloses a small area ω of the parcel. (Source: Frolov and Maling, 1969).

the concept of the square or rectangular pixel which plays such an important part in remote sensing and computer graphics. Thus we substitute a series of short straight lines, one lying in each of the marginal cells, for the true perimeter of the parcel. The smaller the size of the unit cell, the less will be the error arising from making this assumption. It follows that the short straight line can occur anywhere within the cell and orientated in any direction. Therefore it is necessary to devise a means of generating a random position for the line which can be used subsequently in the analysis. Having determined the limiting values for l, which is the length of the straight line element representing the part of the parcel perimeter within this cell, and the small area, ω, contained between this line and two or three sides of the cell (the shaded part of Fig. 18.5), average values for them are determined from integration of all possible random positions of the line l.

After some rather awkward algebra we obtain two constants,

$$\bar{l} = 0{\cdot}7979$$

and

$$\bar{\omega} = 0{\cdot}2127$$

These initial stages in the argument were also examined by Lloyd, who approached the subject from the standpoint of an electronics engineer investigating *visual scene analysis* (*VSA*). The subject is concerned with acquiring pictorial information from a scene using an input device such as a television camera, storing the information in a machine-readable form and *enhancing* the images or performing analyses to reveal certain characteristics of the original scene. In essence the part of VSA which is common to

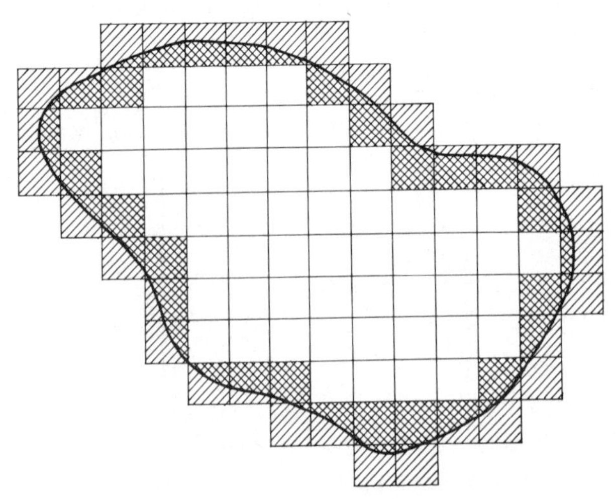

FIG. 18.6. A parcel superimposed upon a square grid showing the marginal cells by shading.

cartometry is the need to examine how an irregular boundary passes through a pattern of pixels in order to assign threshold illumination round the image. In order to measure the image illumination it is necessary to count the number of pixels contained within, and marginal to, the image. This is obviously equivalent to the cartometric applications of point-counting. The analysis of how the image boundary passes through the marginal pixels is identical to the theoretical study of the accuracy of point-counting methods of area measurement.

Lloyd rejected the Frolov–Maling model for randomness and proposed his own solution with slightly different results, viz.

$$\bar{l} = 0{\cdot}8621$$

and

$$\bar{\omega} = 0{\cdot}2469$$

The Components of the Standard Error

In each of the cells which coincide with the parcel perimeter there is either a positive or a negative error in measurement according to the place where the rectilinear element intersects the cell wall. It can be shown that the standard error for the whole parcel will be

$$s = k_1 \cdot \sqrt{n_p} \tag{18.4}$$

where $k_1 = 0{\cdot}2127$ is an integral solved by Frolov and Maling for the different positions where the line l intersects the base of the cell and n_p is the total number of marginal cells through which the perimeter passes. This quantity may be determined from

$$n_p = k_2 \cdot L \tag{18.5}$$

where L is the cumulative length of all the small rectilinear elements l_i which were substituted for the real perimeter. The coefficient k_2 may be expressed as

$$k_2 = 1/\bar{l} \tag{18.6}$$

so that the numerical value for k_2 is the constant 1·2533 or 1·16 depending upon which model is preferred. The length of the true perimeter of the parcel may be related to the area of the figure enclosed by all the straight line elements by the expression

$$L = k_3 \sqrt{A} \tag{18.7}$$

where k_3 describes the shape of the parcel. The three equations (18.4), (18.6) and (18.7) are now combined to form

$$s_A = k_1 (k_2 \cdot k_3)^{1/2} \cdot \sqrt[4]{A} \tag{18.8}$$

In the examination of the standard error of the measured parcel, Frolov and

Maling made use of the expression for the relative root mean square error. This has the form

$$f = s_A/\sqrt{A}$$
$$= k_1(k_2 \cdot k_3)^{1/2} \cdot A^{-3/4} \tag{18.9}$$

This equation is most conveniently expressed in logarithmic form by grouping the three k coefficients as

$$K = 2 + \log k_1 + \tfrac{1}{2}\log k_2 + \tfrac{1}{2}\log k_3 \tag{18.10}$$

Consequently,

$$\log f = K - 0{\cdot}75 \log A \tag{18.11}$$

and introducing the size of the grid squares, d_1,

$$\log f = K + 1{\cdot}5 \log d_1 - 0{\cdot}875 \log A \tag{18.12}$$

The Influence of Parcel Shape

Rewriting equation (18.7)

$$k_3 = L/\sqrt{A}$$

we have a coefficient which may be used to describe the shape of the parcel numerically. Thus $k_3 = 3{\cdot}5449$ for a circle, 4·0 for a square and 4·2426 for as rectangle with ratio of the sides 1:2. The quantities k_1 and k_2 are constants for any square grid and only k_3 varies from one parcel to another. It follows that the coefficient K is a quantity which changes value only with variation in the shape of the parcel. Table 18.1 gives some values for K appropriate to some geometrical figures. When these results are applied to equation (18.11) we obtain a series of graphs which indicate the influence of shape upon parcels of different sizes illustrated by Fig. 18.7. Some of these are quite small; much

TABLE 18.1. *The influence of parcel shape upon the coefficient K*

Figure	K (Frolov and Maling)	K (Lloyd)
Circle	1·652	1·699
Square	1·678	1·726
Rectangle with ratio of sides		
1:2	1·691	1·739
1:10	1·798	1·846
1:50	1·956	2·004
1:100	2·029	2·077
Equilateral triangle	1·706	1·754
Lune formed by two circular arcs with ratio 1:2 between the radii	1·716	1·764
Isosceles right-angled triangle	1·719	1·767

FIG. 18.7. The shapes of parcels derived from simple geometrical figures used in the study of the influence of parcel shape upon the relative mean square error. (Source: Frolov and Maling, 1969).

TABLE 18.2. *The difference $\delta f = f_1 - f_2$ determined from the Frolov–Maling constant, f_1, and that of Lloyd, f_2*

Parcel shape	Area (n)		
	10	100	1000
Square	−0·91	−0·16	−0·03
Rectangle			
1:2	−1·02	−0·18	−0·03
1:10	−1·31	−0·23	−0·04
1:50	−1·55	−0·28	−0·05
1:100	−2·22	−0·40	−0·07

smaller, indeed, than many other writers have supposed, though the comparative unimportance of shape was subsequently confirmed by Yuill (1971). Table 18.2 compares the results to be obtained from equation (18.11) using the alternative mathematical models. This table shows that for large parcels the differences are practically negligible, and that it only becomes significant for counts of 10 points or less.

The validity of equation (18.11) was tested by the author using the data collected in the experimental work with the *MK* Area Calculator on the pattern map. Because the largest frequency of parcel size on this map is the smallest size class (0–0·999 cm^2) and the distribution of parcels by size is an *L*-

TABLE 18.3. *Relative mean square errors of measurement of area on the pattern map obtained with the MK Area Calculator*

Class range (cm^2)	Mean count (n)	Root mean square error (s)	f(%)	Size of sample
0–0·999	5·08	0·943	15·9	30
1–1·999	14·31	1·205	7·7	30
2–2·999	24·57	1·476	5·5	15
3–3·999	34·79	1·541	4·2	15

shaped pattern, the author took a random sample from each of the four smallest size classes and used the mean values for these samples to fit a curve through them. Table 18.3 illustrates these results. The curve fitted to these data is

$$\log f = 1{\cdot}771 - 0{\cdot}75 \log n \tag{18.13}$$

Some years later, and unaware of our work, Loetsch *et al.* (1973) investigated the measurement of three irregular parcels, which they considered to be a fair representation of a portion of a typical forest map, using three different sizes of dot grid. Each sample comprised 20 measurements and Loetsch, too, determined the "standard error percent of the single observation" which is the same as the relative mean square error, f, in our terminology. Their solution was

$$\log f = 1{\cdot}782 - 0{\cdot}76 \log n \tag{18.14}$$

which represents a quite extraordinary degree of agreement with the Frolov–Maling results.

The Use of Different Sizes of Grid for Different Sizes of Parcel

We now concentrate attention upon the lower third of Fig. 18.8. For most purposes it is sufficient to demonstrate that the relative error of a series of measurements does not exceed some specified value such as 1% or 5%. By inspection, the corresponding point count can be read from Fig. 18.8 upon which the levels, $f = 2\%$, $1{\cdot}5\%$, 1% and $0{\cdot}5\%$ have been distinguished. Table 18.5 gives the values of n corresponding to these levels. The two determinations due to Hiwatashi are (*i*) derived from his data, recomputed by the

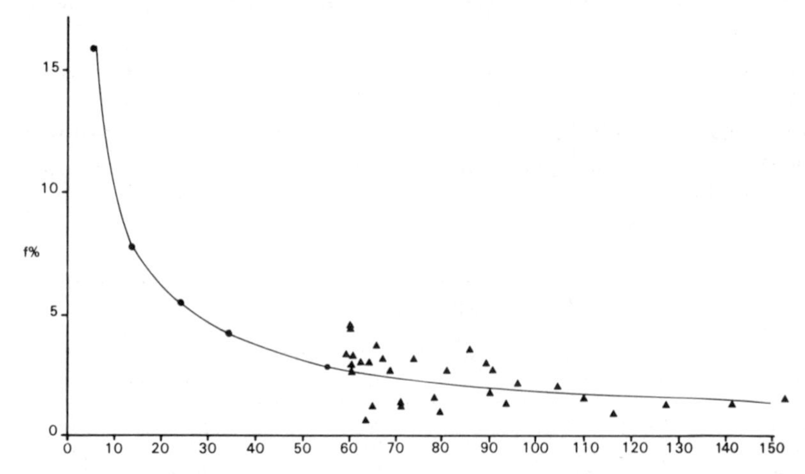

FIG. 18.8. The relative mean square error derived by Frolov and Maling (1969) for point-counting using a square MK/64 grid on parcels of the McKay pattern map.

TABLE 18.4. *Values of n corresponding to different values of f*

$f(\%)$	Hiwatashi (*i*)	Hiwatashi (*ii*)	Frolov and Maling; Loetsch *et al.*
2	32	50	90
1·5	49	–	120
1	88	150	230
0·5	240	–	570
0·2	–	–	1,800

present author to give the curve illustrated in Fig. 18.8, and (*ii*) from his own conclusions given in the text of his paper.

If we specify that f should not exceed 1·5%, the minimum number of points to be counted in a parcel should be about 120 and for $f = 2\%$, $n = 80$ points. Since this conclusion is based upon the number of points counted, and is not related to the separation of points on a dot grid, Frolov and Maling suggested that *it is possible to discard the conventional notion of using the same overlay to measure all the parcels on a map, irrespective of size, and use an overlay of density appropriate to the size of the parcel and an expected count of about n dots.* Table 18.5 indicates the recommended point densities to use with parcels of different sizes for $f = 2\%$.

Hiwatashi also came to the conclusion that since precision is dependent upon the number of points coinciding with a parcel, it is possible to calculate critical values for the number of points required. He found that a count of 150 points per parcel is needed to achieve a relative mean square error of 1%, but that a count of only 50 points is adequate to meet the 2% specification.

From his computer-simulated measurements, Yuill (1971) derived the equation

$$Y = [6{\cdot}278 - 0{\cdot}8534 \ln n]^2 \tag{18.15}$$

TABLE 18.5. *Recommended point separation for dot grids to measure parcels of different sizes, based upon the conclusion by Frolov and Maling that 2% relative mean square error corresponds to n = 80 points*

Area of parcel (cm^2)	d_1 (cm)	Point density (points/cm^2)
4	0·2	25
16	0·4	6
36	0·6	3
64	0·8	2
100	1·0	1

where Y is the expected range of error and n is the number of points used to determine the area of a parcel. He concluded that, using 250 points, there is a 99% probability that the area measured by dot grid would be within 2·45% of the true value, but for 100 points this value would have risen to 5·51%. Although this appears to be a rather large relative error this figure is based upon the *maximum* expected range of errors. The *average* expected error is only one-third of this value, or, for 100 points, it is 1·89%, which corresponds more closely with the figures determined by Frolov and Maling.

THE PRECISION OF POINT-COUNTING

Maslov (1955) regards the method of squares as being the normal use of a grid to measure area. He also regards this as only being suitable for measurement of very small parcels, for which, as we have seen, planimeter measurements are unreliable. He takes as the lower limit of area measurement by planimeter a parcel of area 1 cm^2, and he places the upper limit for area measurement by the method of squares as being a parcel of area 2 cm^2. As is customary in the use of this method, the measurement is made by means of a single application. He considers that the ideal ruling of the grid lines is at every 1 mm or 2 mm. Therefore the largest size of parcel measured by the finer grid should be represented by a count of about 200 squares.

Maslov investigated the accuracy of determining area by grid as being of similar order to the accuracy of the geometrical methods described in Chapter 16. He therefore writes as a first approximation, the expression

$$s = 0{\cdot}08\sqrt{A} \tag{18.16}$$

which relates the standard error to parcel area through the constant 0·08, which approximates to the mean square error in drawing which we have encountered in Chapters 9 and 16. Maslov investigated the precision of a square grid for which $d_1 = 1$ mm. He measured the areas of five small parcels ten times each, and from these determined the standard error. The results of this small experiment are given in Table 18.6. Like the instrumental and

TABLE 18.6. *Mean square error in the determination of area by the method of squares: Maslov's experiment*

Area of parcel (cm^2)	Square grid	
	Mean square error	Relative error
0·05	0·006	1/8
0·11	0·010	1/11
0·30	0·014	1/22
0·56	0·021	1/27
1·06	0·029	1/37

TABLE 18.7. *Expressions for the mean square error of point-counting methods of area measurement derived by Köppke (1967)*

d_1 (mm)	a (mm^2)	S_A (mm^2)
1·41	2	$0{\cdot}7 + 0{\cdot}08\sqrt{A}$
2·24	5	$1{\cdot}8 + 0{\cdot}12\sqrt{A}$
3·16	10	$3{\cdot}5 + 0{\cdot}14\sqrt{A}$
4·47	20	$7 + 0{\cdot}15\sqrt{A}$

tracing errors obtained from planimeter measurements, the relationship between standard error and parcel size approximates to a curve of the form

$$s = b \cdot A^x \tag{18.17}$$

which leads to the logarithmic expression

$$\log s + x \cdot \log A - \log b = v \tag{18.18}$$

From the data in Table 18.6, Maslov found

$$s = 0{\cdot}029A^{0{\cdot}55},$$

which may be rounded off to the simple expression

$$s = 0{\cdot}03\sqrt{A} \tag{18.19}$$

Köppke (1967) has made use of a similar equation incorporating the size of the unit cell. His version of the expression is

$$s_A = 0{\cdot}35a + \beta\sqrt{A} \tag{18.20}$$

where a is the area corresponding to one unit cell of the grid and the coefficient β is also dependent upon a. The equations in Table 18.7 apply to the different sizes of grid used by him. Like the corresponding planimeter errors, illustrated in Figs. 17·16 and 17·18, pp. 378–392, each graph in Fig. 18.9 is a straight line because the mean square error is plotted against the square root of the area. Moreover, we may superimpose the corresponding graphs describing the mean square errors of the polar planimeter and polar disc planimeter. It is seen that, according to Köppke,

- The closer the separation of the points in a dot grid, the higher is the precision of the measurements. Thus a grid for which the area of the unit cell, $a = 2\,\text{mm}^2$ provides more precise results than the more widely spaced grid in which the area of the unit cell $a = 5\,\text{mm}^2$.
- Measurements by a dot grid having $d_1 = 1{\cdot}4$ mm appear to be more precise than those made by polar disc planimeter. Dot grids of size $d_1 = 2{\cdot}2$ mm and 3·2 mm produce more precise measurements than a compensating polar planimeter.

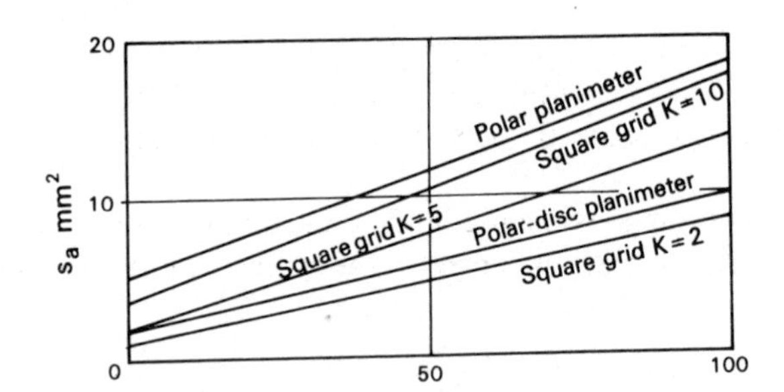

FIG. 18.9. The precision of measurement related to parcel size. (Source: Köppke, 1967).

If this were the complete answer, it would suffice to use a dot grid ruled as closely as is practicable for ease of counting, and to use this for any kind of area measurement and for any size of parcel. Before the errors of area measurement by direct point-counting methods had been studied in the early 1960s the user was often recommended to employ a closely-ruled grid "for greater accuracy", notwithstanding the increase in time which might be needed to execute the count if this advice were followed.

Although it is evident that a close ruled point pattern gives increased precision, we must realise that as d is reduced, the number of points to be counted per unit area increases according to geometrical progression. If we regard the counting rate as constant, the time needed to make a count also increases in geometrical progression. Consider the examples of measuring a parcel of area 4 cm^2 using a series of different dot grids ranging from $d_1 = 10$ mm to $d_1 = 1$ mm (Table 18.8). The times are based upon the constant counting rate of 1·3 points per second already established for the Markounter, and do not take into account any reduction in speed which occurs as the operator tires. The figures in the table demonstrate that a finely ruled dot grid is quite impracticable for measuring any but small areas. Even for a parcel as small as 4 cm^2, measurements using the two grids with the widest separation can hardly compare in speed with measurements made by planimeter, unless the user is willing to sacrifice the size of sample.

TABLE 18.8. *Time required to measure a parcel of area 4 cm^2 using dot grids of different densities and a constant counting rate of 1·3 points per second*

Point separation d_1	Number of points in parcel n	Time required to count the points in 1 application t_1 (seconds)	Time required to count the points in 5 applications t_5 (seconds)
10 mm	4	5	26
5	16	21	1 m 44
2	100	2 m 10	10 m 50
1	400	8 m 40	43 m 20

A variety of other attempts have been made to fit curves to these data by multiple regression and similar techniques. The following examples are of interest for comparison with the results given in equations (18.13), (18.14) and (18.19).

Zöhrer (1978) suggests the use of the following regression equation for the estimation of standard error percent of the estimate for the measurement of an area:

$$\log(S\%) = 1{\cdot}739 - 0{\cdot}755 \log n + 0{\cdot}457 \log q \qquad (18.21)$$

where n is the number of dots falling within the parcel and q is the *perimeter ratio*, i.e. the perimeter of the parcel divided by the perimeter of a circle of the same area (q varies from 1·0 for compact shapes to 4·0 for very irregular shapes).

After making repeated measurements by dot grids of individual parcels of a variety of shapes and sizes Yuill concluded that shape had no significant effect upon the precision of area measurement. Parcel size or number of dots used to measure the area was the sole significant factor in his regression equation:

$$\sqrt{S\%} = 3{\cdot}625 - 0{\cdot}493 \ln n \qquad (18.22)$$

where S is what he calls the *average expected error* corresponding to 1 standard deviation from the mean.

THE MAXIMUM ALLOWABLE ERROR AND GRID DENSITY

The principal contribution by Bonnor (1973) to the subject has been to explore methods of specifying a maximum allowable error, E, for the area estimate, and use this to select a dot grid of suitable density to meet the accuracy specifications which have been imposed.

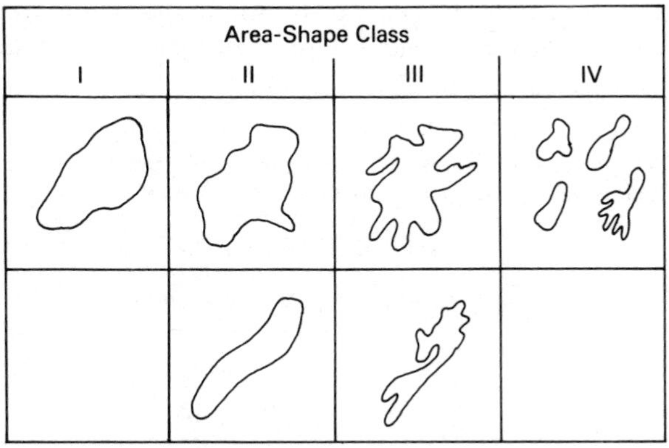

Area-Shape Class

I II III IV

FIG. 18.10. Area–shape categories. (Source: Bonnor, 1973).

Like many of his predecessors, Bonnor was preoccupied with the possible influence of parcel shape upon the accuracy of the results. He therefore devised a classification of the parcels into *area–shape categories*, assigning each parcel into one of four different categories illustrated in Fig. 18.10. Bonnor made a number of test measurements, using a variety of different densities of square grid and parcel sizes within the range 0·8 through 143 cm^2. The subsequent analysis is based upon the determination of the percentage error, e, of each parcel measured by grid using planimeter measurements of them to represent their "true" areas. A plot of e against n has the same L-shaped distribution which has already been encountered in making a study of the variation of f with n or A. As in these studies a straight line graph may be obtained by plotting $\log e$ against $\log n$.

From these data Bonnor has separated the results into three distinct area–shape categories and derived three expressions of the form

$$\log e = b - c \cdot \log n \tag{18.23}$$

viz.:

Area–shape category I

$$\log e = 1{\cdot}2458 - 0{\cdot}6367 \log n$$

Area–shape category II

$$\log e = 1{\cdot}5371 - 0{\cdot}6426 \log n$$

Area–shape category III

$$\log e = 1{\cdot}4440 - 0{\cdot}5814 \log n$$

He now proceeds to determine the upper limit of the prediction interval. Eventually Bonnor obtains three equations for different categories of shape, and from these he is able to identify the density of dot grid needed to meet a particular allowable error. This is given in Table 18.9. The gaps in the table indicate entries for which $d_1 < 1$ mm which would give rise to a dot grid which would be difficult to use. Because Bonnor has used a *maximum allowable error* rather than the relative mean square error, the values in this table are more pessimistic, needing the use of a more closely ruled grid for each class of parcel size than was specified by Frolov and Maling.

TABLE 18.9. *Recommended separation of grid points for measuring parcels of different sizes using Bonnor's method of analysis; based upon a maximum allowable error of 2%*

Area of parcel (cm^2)	Dot grid size, expressed in values of d_1 mm		
	Area–shape I	Area–shape II	Area–shape III
4	1·2	–	–
16	2·4	1·5	1·0
36	3·5	2·2	1·4
64	4.5	2·9	2·0
100	5·8	3·8	2·4

SAMPLING FOR DIRECT MEASUREMENT BY ONE APPLICATION ONLY

Bryan's account of the method contains the statement:

> When considerable accuracy is desired, several separate counts should be made, changing the position of the grid over the area each time. The average number of dots is used for computing acreage. Greater accuracy results with each additional count.

Wood held a similar view and it is, of course, the procedure advocated by the author from Chapter 4 onwards. Indeed the statistical theory developed in this chapter depends upon this method of obtaining a random sample of measurements. However, it is evident that many other writers on the subject supposed that one application of the dot grid would suffice. For example Gierhart (1954), Naylor (1956), Winkworth (1956), Rideout (1962) and Köppke (1967) all made tests or described production work using only one application of a dot grid.

Notwithstanding the fact that a sample of size $k = 1$ has little statistical use, the results of measurements made using a single application of the grid compare favourably with other methods of area measurement. It is therefore desirable to investigate if there is any reason why direct measurement by dot grid should not be confined to a single application of the grid. The measurements made by the author on the pattern map provide suitable data. For an investigation of this sort it is desirable to choose large enough parcels to show detectable errors. Thus, using the MK 64/in. grid, a parcel of about $3\,\text{cm}^2$ area is represented by 30 points. If any significant differences in the results are to be detected, this should be noticeable in a parcel as large as this, whereas in a smaller parcel for which $n < 10$ random variations in the count might influence the results. By limiting the analysis to parcels of area $3\,\text{cm}^2$ or greater, we have a population comprising data for more than 150 separate parcels.

Each parcel had been measured by five separate applications of the grid. These results may be divided into two samples for purposes of testing.

1 Sample I comprises the number of points counted in the *first* application of the grid.
2 Sample II comprises the mean number of points counted obtained from the remaining four applications of the grid.

It follows that for any parcel i we may determine the difference

$$\delta_i = n_i' - \bar{n}_i \qquad (18.24)$$

where n_i' is the count of points made in the first application of the grid and $\bar{n}_i$ is the mean of the count of points made in the second, third, fourth and fifth applications. The frequency distribution for δ_i determined from measurements with the MK 64/in. grid, is given in Table 18.10. Corresponding values for some measurements made with the MK 100/in. grid and a hexagonal dot grid having $d_2 \approx 3$ mm are given in Table18.10. Total number of parcels = 151; $\delta = 0{\cdot}13 \pm 0{\cdot}19$ points. The final column in Table 18.11 is the correlation

TABLE 18.10. *Difference between Sample I and Sample II for parcels larger than 3 cm² area on the pattern map. Measurements made by Maling with the MK 64/in grid*

δ_i	−6	−5	−4	−3	−2	−1	0	+1	+2	+3	+4	+5	+6
f	0	4	1	4	20	29	30	29	20	7	3	3	1

Total number of parcels = 151; $\delta = 0.13 \pm 0.19$ points

TABLE 18.11. *Comparison between Sample I and Sample II as measured using different grids*

Method	d (mm)	Size of sample	δ	s_m (in number of points counted)	r
MK 64/in	3·2	151	0·13	0·189	0·9995
MK 100/in	2·5	109	−0·11	0·321	0·999
Hexagonal	3·0	31	−0·07	0·362	0·9998

coefficient, r, derived from comparison of the corresponding values for n_i' and $\bar{n}_i$ for each of the parcels investigated. Inspection of these tables indicates that there is no need to undertake any more sophisticated statistical analyses to draw the conclusion that there is no significant difference between the two samples, irrespective of the grid used. Consequently a case has been made for restricting dot grid measurements to only one application. However, this conclusion ignores two important factors:

1 The precision of the results is enhanced by increasing the size of the sample.
2 There is a continuing need to check that gross errors have not occurred.

The first of these factors has already been described and recommended in earlier chapters. Provided we remember Weightman's homily on page 115 about overdoing this, repetition of measurements does serve to improve the results.

Elimination and checking for gross errors is wholly dependent upon having two or more measurements which may be compared.

There are 155 parcels of area 3 cm² or greater on the pattern map. In the case of the measurements made with the *MK* 64/in. grid, which represents a complete measurement of the entire map, four parcels (2·6%) have been excluded from the analysis in Table 18.10 because the values for δ indicate the presence of a gross error in counting. For the *MK* 100/in. measurements, made on only part of the pattern map, a further two parcels (1·8%) evidently contained gross errors in the first application. For the measurements made with the hexagonal dot grid and the Markounter, which is the much smaller sample of only 31 parcels, there are no gross errors. It has been emphasised that the only way of detecting a gross error is to repeat a measurement. Despite the use of mechanical aids to mark and count the points there is still the risk that an error in counting will occur. In a single application of a dot grid this check is absent. The author has made too many gross errors in his lifetime ever to regard the single observation with equanimity.

19

Indirect Measurement of Area by Point-Counting Methods

The equivalence of areal proportions to volumetric proportions was suspected and announced by Delesse in 1848.... Though Delesse used the relation to good advantage, he did not actually prove it. Nor did any other geologist. As a consequence it was always regarded with considerable skepticism. Those who placed any confidence in it might justify their creduility by pointing to an experiment, commonly too small to demonstrate anything at all, in which the bulk chemical composition calculated from the mean of a few modes of dubious quality agreed fairly well with an actual chemical analysis of unknown quality. Sometimes the procedure was reversed, and measured modes were compared with modes calculated from chemical analyses. Occasional tests of this kind could convince only those who had a powerful will to believe. And geology is by tradition an agnostic science.

(F. Chayes, *Petrographic Modal Analysis*, 1956)

We have seen that indirect measurement by point-counting is the alternative to the methods of measurement described in the previous chapter. Moreover it was the forerunner to the methods of direct measurement already studied, and it has been used in many disciplines other than cartometry. We shall see how different investigations into this method of area measurement converge to create a reasonably coherent picture of its precision and utility. The use of indirect measurement in the various fields of stereology has created a body of theory relating to the relative merits of random and systematic sampling, the desirable size of sample and the influence of parcel size and shape upon the results, which greatly extends the hitherto somewhat restricted view of the applications confining these to measurements on maps. It is only the terminology which changes from one subject to another. Thus the study of a *constituent* of a rock in petrographic *modal analysis* is equivalent to the study of the *α-phase constituent* of a polished surface in metallurgy, to a *profile* in biological morphometry and to the representation of a *parcel* on a map or a *polygon* on a remotely sensed image.

In direct measurement we measure and determine the area of each parcel separately. In indirect measurement it is common practice to sample several variables simultaneously. In the example which follows, of land-use categories shown on a map, the counting process consists of recording how many points on the overlay coincide with each category of land use. For example, the results of such a count are given in the second column of Table 19.1. The third column represents the areas of each land-use category determined by multiplying the proportion of points in each category by the area of the whole map.

The method has other applications: for example to measure by sampling the area occupied on a map or photograph of a single variable the distribution of which is much fragmented and measurement of which by individual parcels would be excessively slow. This is illustrated here by the measurements in Table 19.3, which comprise estimates of the total area of woodland on a topographical map of southern Ontario.

Yet another application exists in one-step estimation of land-use or land-cover categories from aerial photography. Here the term *one-step* means that no maps have been made specially, and therefore the boundaries defining each category of the variable have not been plotted. The use of indirect measurement techniques allows the estimation of the proportions of different categories of land-use for huge underdeveloped areas of the Third World, such as the work by DOS on land-use in Malawi described by Stobbs (1967, 1968). At the other extreme it has been used, for example by Emmott and Collins (1980), Emmott (1981), for small and intricate patterns of urban land-use in an industrial town in northwest England. We have already seen in Chapter 9 that similar methods are used to determine the areas of land-use or land-cover categories in order to evaluate the qualitative accuracy of a map compiled from a Landsat image, or from conventional aerial photography.

AN EXAMPLE OF INDIRECT MEASUREMENT OF LAND-USE CATEGORIES

Let us first consider area measurement by this means, using the example of Gregory (1963). It is measurement of the land-use categories shown on Sheet 13 (*Kirkby Stephen and Appleby*) of the Popular Edition of the One-inch to One-mile Ordnance Survey map series upon which were compiled the results

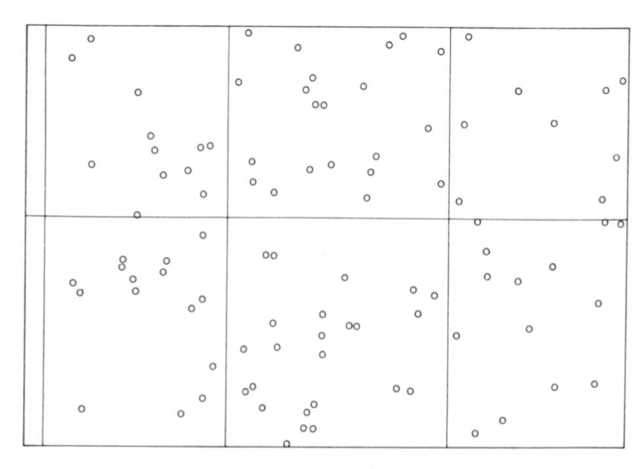

FIG. 19.1. Random point pattern used to measure the areas of land-use categories on the Kirkby Stephen and Appleby sheet of the Popular Edition One-inch to One-mile *O.S.* map. (Source: Gregory, 1963).

TABLE 19.1. *Frequency and areas of land-use categories on Sheet 13, Kirkby Stephen and Appleby, of the first Land Utilisation Survey of Great Britain.* (Data from Gregory, 1963)

Category	No. of points n	Area (km^2)	Standard error (km^2)	Standard error (%)	Relative error f (%)
Arable	8	110·3	37·3	2·7	33·9
Grassland	31	427·1	63·7	4·6	14·9
Woodland	6	82·8	32·7	2·4	39·4
Moorland	55	757·8	68·5	5·9	9·0
Total	100	1378			

of the first Land Utilisation Survey of Great Britain made in the 1930s. For this exercise four land-use categories were sampled simultaneously by recording the land-use at each of 100 points distributed at random over the map. The positions of these points had been located by extracting successive values from a table of random numbers and treating them as grid coordinates. The resulting point pattern is illustrated in Fig. 19.1.

The first obvious feature of this distribution is that, because the arrangement of points is random, we cannot use a direct method of measurement by point-counting. Quite simply, in a random distribution there cannot be a constant value, d, which is necessary to convert the count, n, into the area of a parcel through the equations (3.16), (17.1) or (17.2). Therefore it is necessary to use the proportion n/N of the points coinciding with one land-use category, n, to the total number of points counted in the whole area, N. This provides a fraction of the total area of the map which has been determined in some other way. In this example the total area of the map may be calculated independently as being $1378\,km^2$. It follows, for example, that the area of woodland determined from this measurement is

$$A_W = (6/100)\cdot 1378$$
$$= 82{\cdot}8\,km^2.$$

THE THEORY OF INDIRECT MEASUREMENT BY POINT-COUNTING

Consider a structure having only two constituents. In the cartometric model, B is a parcel situated on a map M. It is implied that B is contained within M so that we also have a quantity $M - B$ which are those parts of the map which, to employ that useful Scots word, lie outwith the parcel B. The probability that a point located at random in M will also lie in B is

$$p = A_B/A_M \tag{19.1}$$

where A_B is the area of the parcel B and A_M is the area of the map.

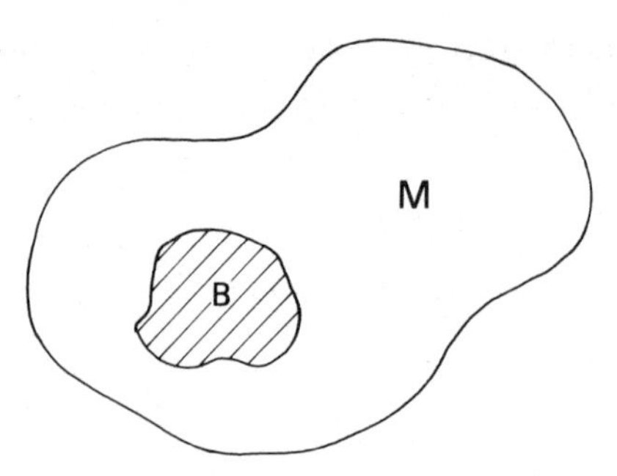

FIG. 19.2. The diagrammetic representation of a single parcel, B, situated upon a map, M.

Conversely the probability that a point does not lie in B is

$$q = A_{(M-B)}/A_M \tag{19.2}$$

We recall from Chapter 6 that since $p+q=1$, $q=(1-p)$.

The determination of the area of B is obtained by simple proportion. If an overlay is placed at random over the map so that M is covered by N points of which n points lie in B then

$$A_B : A_M = n : N \tag{19.3}$$

or

$$A_B = (n/N) \cdot A_M \tag{19.4}$$

It will be convenient to retain $p/(p+q)$ for the theoretical proportion and use

$$z = n/N \tag{19.5}$$

for the proportion actually measured. By definition $n < N$ and consequently $A_B < A_M$.

In a measurement system such as that described each point should be regarded as being an individual probe which is independent of the others. For each point tested we have to ask the question whether it lies in the parcel or not. Because the answer must be either p or q, we may argue that

> *The proportion z of N points lying within a randomly distributed constituent whose probability of occurrence at any one point is p, is a random variable having the binomial distribution*

The Validity of the Binomial Distribution

The essential requirement for the binomial distribution is that each point of the grid must have a chance p of being in the parcel, independently of the other points. Also the binomial implies a finite probability p^N of all N points being in the parcel and $(1-p)^N$ of no points lying in the parcel.

The reader is reminded of the brief introduction to this distribution which appears at the end of Chapter 6, and that when the variable is measured as a proportion, the mean μ and standard deviation, σ of the distribution are given by

$$\mu_p = p \tag{4.25}$$

$$\sigma_p = [p(1-p)/n]^{1/2} \tag{4.26}$$

Expressed in terms of area measurement, for example as employed by Barrett and Philbrook (1970) in forestry applications, the standard deviation for the measurement of parcel B is

$$\sigma_B = A_M[z(1-z)/N]^{1/2} \tag{19.6}$$

$$= [A_B(A_M - A_B)/N]^{1/2} \tag{19.7}$$

If, for example, substitute $n = 6$, $N = 100$, $z = 0{\cdot}06$, $1 - z = 0{\cdot}94$, $A_M = 1378$ and $A_B = 82{\cdot}8$ in equations (19.6) and (19.7), we obtain the standard error $\pm 32{\cdot}7\,\text{km}^2$ for the calculated area of woodland on the Kirkby Stephen and Appleby map.

One generally assumes that estimates of z or A_B are normally distributed, and that each dot represent a single random observation. Using the typical dot grids which have been described and illustrated in Chapters 4 and 18, N is large enough to satisfy assumptions about the estimates of z, but the assumption concerning dots representing random observations is only valid if the point pattern is truly random.

Since the indirect method of measurement depends upon the determination of the proportion z of those which lies in the parcel, and since the ratio is that of the smaller area to the larger area, the proportion of the total count that falls in the smaller area is to be regarded as being an unbiased estimate of that ratio. In other words an estimate of the true proportion p of that parcel is given by z. In general, however, $z \neq p$ and it is desirable to know the extent by which z may be expected to deviate from p. This may be done by studying the results of repeated applications of the sampling procedure. We have already stressed that one of the main factors distinguishing direct from indirect methods of measurement is that multiple applications are not needed to obtain area by indirect methods. However it is a useful method of evaluating the relative accuracy of each. If k sets are sampled by making repeated random applications of a dot grid, as in the direct method of measurement by point-counting, the values of z obtained from each sample will be scattered about the true value of p, and from these values it is possible to estimate the mean and standard deviation of the distribution of a hypothetical infinite population. The standard deviation of this distribution is the quality from which the accuracy of z may be assessed.

To estimate the standard deviation of the proportion z, determined from a count of n points, the following procedure was adopted by Gladman and Woodhead (1960).

A sample of k groups were obtained, and a value z determined from each group. The best estimate of the proportion is given by the mean

$$\bar{z} = (\Sigma z)/k \tag{19.8}$$

and the standard deviation of the distribution of z as estimated from k applications is:

$$s_n = [\Sigma(z - \bar{z})^2/(k-1)] \tag{19.9}$$

The standard error of the mean, $\bar{z}$ is:

$$\begin{aligned} s_z &= s_n/\sqrt{k} \\ &= \{\Sigma(z - \bar{z})^2/[k(k-1)]\} \end{aligned} \tag{19.10}$$

and this is clearly to be equated with the standard deviation σ of the distribution of z as estimated from nk points.

Thus for each count of nk points, two values of the standard deviation are determined, associated with nk N points respectively.

These conclusions were tested experimentally by Gladman and Woodhead (1960), first through a known binomial distribution and secondly through a series of measurements made by microscopic examination of polished steel surfaces.

The first of these experiments is of value to us in demonstrating the close agreement of the results to a true binomial distribution. This was obtained by determining the proportion of odd numbers, z, in k sets of n two-figure random numbers. In this case the value of p is known exactly, since there are equal numbers of odd and even numbers and $p = 0{\cdot}500$. Clearly, the results of the random-number experiment should conform to a binomial distribution with $p = 0{\cdot}05$. Table 19.2 indicates that this is the case, but there is the expected scatter in s. The scatter in this case is wholly comparable with that found in the point-counting experiments.

Their experimental measurements were made using a Swift automatic point counter and stage. In microscopy this device is equivalent to using a square pattern dot grid on a map or photograph, and therefore represents a systematic sampling strategy. The value of s of their actual measurements were compared with the theoretical values for σ. This showed that there was a close fit of their data to a straight line of unit slope and zero intercept. In short they had demonstrated that

> on the average, experimentally determined values of s tend to the theoretical value of σ and that it is reasonable to use the calculated σ to specify the accuracy of a point-counting experiment.

Indeed they were sufficiently confident to state later that

> we may regard it as proved that standard deviations calculated from the binomial approximation provide a satisfactory description of the accuracy obtained in point-counting for the specimens investigated. As these specimens cover a variety of structures and values of p ranging from 0·1 to 0·5, it seems reasonable to suppose that the application of this approximation is valid generally,

subject to the restrictions mentioned earlier, i.e. that the constituents shall be distributed at random, that the number of particles shall not be large compared with the number of grid points and that the structure shall be finely divided relative to the grid spacing. If these restrictions are not met it is likely that the binomial approximation will underestimate the accuracy of a determination.

At about the same time that Gladman and Woodhead were making their investigation, Hillard and Cahn (1961) were also investigating the mathematical theory of the procedures for volume-fraction analysis in quantitative metallography. They cover much the same ground in terms of the methods of measurement; using both point patterns and lines for both random and systematic sampling techniques. However, the significance of some of the variables introduced is more relevant to the study of the constituents of metals than to cartometric investigations. Apart from the fact that the later stages of their analysis are concerned with the fundamental theoretical problem of stereology, namely of showing that two-dimensional sampling may be applied to the study of three-dimensional distributions, they employ certain variables which we might not consider to be appropriate in land-use or similar studies. For example, they introduce a parameter which is, in effect, a dimensionless measure of the degree of dispersion in the size distribution of the constituents of a specimen. This may be significant in microscopic study of the composition and textures of metals, but may be less relevant to the study of land-use or land-cover shown on maps or aerial photographs. However, their study does offer some interesting conclusions. Of particular interest to us in the present context is their emphasis upon the superior precision of what they call a *coarse grid* used in systematic sampling. They define the coarse grid as being one in which the distance between adjacent points is greater than the largest diameter of any parcel to be measured on the map. This opens up a large and little-explored field of study in cartometry which, at first sight, appears to represent the opposite of what we intuitively regard as being the desirable procedure; namely to use the most closely divided systematic network of points which is compatible with the time required to count them.

TABLE 19.2. *Theoretical results derived by Gladman and Woodhead (1960) from a known binomial distribution, being the proportion of odd numbers in k sets of randomly selected numbers*

No. of sample n	No. of samples k	Total no. of points $N = n \cdot k$	Mean proportion (should be 0·50) z	Standard errors of proportion z			
				s_n	σ_n	s_z	σ_z
25	22	550	0·496	0·098	0·100	0·0210	0·0213
25	22	550	0·483	0·069	0·100	0·0147	0·0213
25	22	550	0·456	0·097	0·100	0·0207	0·0213
50	11	550	0·484	0·063	0·071	0·0190	0·0213
50	11	550	0·466	0·081	0·071	0·0244	0·0213
10	55	550	0·500	0·183	0·158	0·0246	0·0213
10	55	550	0·466	0·170	0·158	0·0229	0·0213

THE NATURE OF THE SAMPLE

It will be recalled that Chapters 7 and 8 were concerned with the subject of statistical sampling. One object of treating with the theory of spatial sampling in such detail was to create an adequate background to this kind of area measurement; especially the choice of a suitable form of dot grid and the size of the sample needed to obtain satisfactory results. At this stage we do not have to justify the theoretical reasons for making these decisions, for that was treated with in those chapters. Here we may concentrate upon the practical problems which may arise from the use of certain kinds of grid, and such additional problems as the methods of selecting grids to be used with aerial photography.

Sampling from Maps

We have already seen that the two major subdivisions of sampling strategy applied in two dimensions are the random point pattern and the systematic point pattern. Either of these may be combined with stratification if the nature of the distribution warrants this. Usually the point pattern is constructed on a transparent overlay which is applied to the map in the way already described for measurement of the areas of land-use categories on the Kirkby Stephen and Appleby map. The number of points represents the size of the sample, and since the validity of the binomial distribution has been established, an equation of the form of (7.14) and illustrated by Table 7.4, p. 118, is appropriate. Inspection of this table indicates that Gregory's choice of only 100 points to be sampled on the Kirkby Stephen and Appleby map represents rather low levels of probability and permissible error *if this is the only area to be measured*, but may be an acceptable size if this is only one of many land-use maps to be sampled. Because this is a random sampling strategy the use of (7.14) is statistically acceptable.

If we prefer to use a systematic sample, then it is usual to employ a transparent overlay like the dot grids already illustrated. This time the sample size given by equation (7.14) has no theoretical justification; nevertheless it serves as a useful guide to the number of points which should be sampled. The orientation of the dot grid to the map may be arbitrary but this does not, as we saw in Chapter 8, constitute a wholly acceptable definition of randomness. Therefore a systematic point pattern oriented parallel to the neat lines of the map will suffice. Indeed the overlay may be replaced by the map grid itself, and the sampling points are represented by the points of intersection of the grid lines. This has the considerable advantage that the sampling points may be easily and accurately recovered on maps of different scales and ages, provided they are all based upon the same grid system. The fact that the same sampling points may be occupied on maps of different ages is especially useful in measurement of *change* in land use. An example of this procedure is represented by the Second Land Utilisation Survey of Britain, for, as briefly mentioned in Coleman *et al.* (1974) and by Lawrence (1979), in this work the national grid intersections on the 1/25,000 OS maps were used as the sampling points. The method was originally chosen for area measurement on grounds of accuracy, speed and economy in both labour and equipment, suitability to land-use material and freedom from eyestrain. At the time it was first introduced as the production method of determining land-use areas it was criticised on the grounds that the spacing of the grid intersections (1 km or 40 mm on the 1/25,000 scale map) was far too great to provide reliable measurements. However, as we have seen, the work of Hillaird and Cahn (1961) indicates that a coarse grid may well be acceptable for systematic sampling. Besides, it is also possible to subdivide the printed grid further, for example into units of 500 × 500 m, and plot these on the map without much risk of introducing any systematic errors. If there are any changes in the dimensions of the map resulting from paper deformation the printed grid behaves in exactly the same way as the rest of the map detail, whereas a grid reproduced on some kind of overlay might not fit the map in the way intended.

Normally a region or administrative district to be measured is larger than the single large- or medium-scale map on which the thematic information has been compiled and plotted. For example the Ordnance Survey 1/10,560 (or 1/10,000) scale maps have a standard format of 5 × 5 km and the second series, or Pathfinder maps at 1/25,000 scale represent 20 × 10 km. Both are appreciably smaller areas than the single county, sometimes even the rural district. For this reason it is a common requirement to extend sampling over several adjoining sheets. In doing this it is important to avoid making the common mistake of duplicating the sample of points lying along the sheet edges. We are accustomed to thinking of the content of a map in terms of grid squares rather than the points which define these. Thus there are 25 grid squares on each map sheet of the Six-inches to One-mile series, and 200 of them on each 1/25,000 map. There are a total of 36 grid intersections on the Six-inch map and these are 231 of them on a 1/25,000 scale sheet. At the Six-inch scale 20 of these points are shared with adjoining sheets because they lie on the neat lines. Similarly there are 60 points on the 1/25,000 sheet which lie on the neat lines. In order to avoid duplication of sampling, a simple rule must be followed, such as counting only those marginal points lying on the northern and western neat lines.

Samplings from Vertical Aerial Photographs

A rather different procedure has to be followed in dealing with vertical aerial photography. This is primarily because of the overlap between successive photographs, which is intended to ensure continuity of stereoscopic photographic cover. It is an essential requirement of random sampling that all ground points have the equal chance of being selected. Therefore it is necessary to verify that the sampling strategy is both complete and unique, i.e. that there are no gaps in the photographic cover, but at the same time, no parts of the ground can be sampled more than once. Usually, therefore, the portion of the photograph to be sampled is a small rectangle which is not covered by the lateral overlap between strips, and which only appears on two successive photographs.

Another decision which must be made at the outset is whether the whole block, i.e. every photograph, is to be sampled, or whether multistage cluster sampling should be adopted. Using this as the sampling strategy, the first stage is to select individual photographs which are to be studied in detail at the second stage. Clearly a block of vertical aerial photography is especially convenient for this practice because it is easy to make either a random or systematic choice of the photographs to be used. Thus Zeimetz *et al.* (1976) carried out the following routine to ensure that the sampled areas were selected without bias. The first selection was carried out by selecting photos from alternate flight strips using the photo index sheets for this purpose. The decision whether to start with the first or second strip was determined by tossing a coin. A number from 1 to 10 was picked randomly to specify the first photo of the selected flight strips. Starting from this photo, every tenth photo in each alternate narrow was chosen. Hence the first photo was selected at random and the remainder were chosen systematically. The original aim was to obtain an area sample of at least 10% or more of the surface area of each county. The number of photos required for this coverage was based on two calculations. Each photo was assumed to represent a usable area of 8 square miles. The average size of the counties being studied was 646 square miles: thus about 80 photos per county would be needed. Photos from two different coverages overlapped about 65%. The effective area represented by each pair of photos was thus reduced to 5·2 square miles, which indicated that 13 photos were needed for 10% coverage. With this guideline the goal of 10% coverage was exceeded, and approximately 15% of the area was used for the second step, the point sampling within the chosen photographs.

In order to sample points from these photographs, the interpreter made use of:

- a random set of points, 20 per square mile, which had been selected from a table of random numbers;
- a piece of 9 × 9 in. graph paper with lines at intervals of 1/20 in. (1·27 mm).

Thus any point representing an area on the ground 83 ft square had an equal chance of being selected. The points marked on the graph paper were transferred to plastic templates. Five such templates were drawn up, each providing eight sample possibilities (four cardinal positions times the two sides). The templates each received a number I–IV and opposite sides were labelled *A* and *B*. The template corners were numbered 1 through 4 in a clockwise direction. The template to be used was placed on the air photo; the specified number was at the top right-hand corner of the template (or left-hand corner if side *B* had been selected). After the template has been placed on the most recent photo, the equivalent points were located on the older photo and the land use could be interpreted on each photo. Interpreters used magnification as necessary, and they interpreted all points monocularly. For some counties early cover was at 1/20,000 while the later cover was at 1/40,000. For these the template was placed on the earlier photograph and the equivalent points were located on the newer cover.

The method which has been described seems to be unduly complicated. More often the sampling is done in one step, using the central part of each photograph for sampling. Therefore the main preoccupation has been to avoid duplication of ground detail on successive photographs through overlap, or, in overcompensating for this, leaving gaps where the ground detail is not shown on any of the photographs. This involves definition of a small rectangular *effective working area* on each photograph which does not overlap with its neighbours. The typical dimensions are those illustrated in Fig. 19.3.

Sampling Random Point Patterns on Aerial Photographs

A truly random method of sampling was employed by the (former) Land Resources Division of the Directorate of Overseas Surveys, and has been described in some detail by Stobbs (1967, 1968), who developed its use for the Land Systems classification and land capability mapping in Malawi. It was later employed by Alford *et al.* (1974) for similar work in Nigeria.

At the outset the size of the sample had to be calculated. Stobbs' version of equation (7.14) was

$$N = (100 - P)(38{,}400/PE^2) \tag{19.11}$$

He argued that if the most critical land-use category covers about 5% of the total area of the unit and the sampling error is to be no greater than 5%, then

$$\begin{aligned} N &= (100 - 5)(38{,}400/5 \times 5^2) \\ &= 95(38{,}400/125) \\ &= 29{,}184 \text{ points.} \end{aligned}$$

Since there were 1000 photographs covering the study area, the sampling density was 29 randomly located points on each photograph. Stobbs now makes use of an "effective area" in the central portion of each photograph. The dimensions of this were calculated from knowledge of the average fore-and-aft and lateral overlap, and templates were constructed in transparent plastic which indicated the effective working area on the photograph. In order to ensure that the photographs are not all sampled by the same arrangement of points several templates were prepared, each having its unique distribution of random points. These templates were then chosen at random for use with particular photographs. The particular sampling points were then also selected at random within the confines of the selected area, using a table of random numbers to choose from 200 numbered squares on graph paper. For the Malawi investigation ten such random patterns were constructed, the use of these being varied at random between the photographs. In order to make a permanent record of the points sampled, as well as to speed up the preparatory work, the overlays were converted into templates, each having 18 pins mounted in the positions of the points to be sampled. It was then necessary only to position the template over a photograph with reference to the fiducial marks in order to mark all the sampling points simultaneously by stamping the point pattern on the photographic print. In later work the number of templates with different point patterns was increased to 30. Stobbs has recorded the following advantages of the method described:

- the sample is spread over the whole of the effective area;
- as the sampling points can be distributed at random it is possible to calculate a fully valid estimate of the sampling error;
- the sampling intensity can be determined beforehand according to the size of the sampling error considered to be acceptable;
- as no linear measurements are involved, small-scale changes within and between photographs do not affect the area estimates which, in the final analysis, are based upon accepted map areas;
- photo-interpretation is confined to the number of sampling points.

The Systematic Point Pattern

The following variations have been employed:

- Wilson (1949) chose the sampling frame of format 4·5 × 2·5 in. for use with 9 × 9 in. format photography in a block having 60% fore-and-aft overlap and 30% lateral overlap. Theoretically the size of the figure should be a square of 6·3 × 6·3 in. in the centre of the photo. However, as we have seen, operational photography cannot be maintained precisely at these standards, and over rough terrain variations in overlap are considerable. To minimise chances for the samples taken on several photos to overlap on common areas of terrain, it is advisable to confine the template pattern to the smaller area. Figure 19.3 illustrates the arrangement of the effective working area on the template.
- A similar procedure was adopted by Loesch *et al.* (1973), who adopted a grid of similar proportions located centrally on each photograph. Although the points formed a regular systematic grid, the actual points in this grid which were used for a particular photograph were chosen at random.

A similar procedure might be used with réseau photography, where the 1 cm point pattern formed by the small crosses may be used instead of constructing

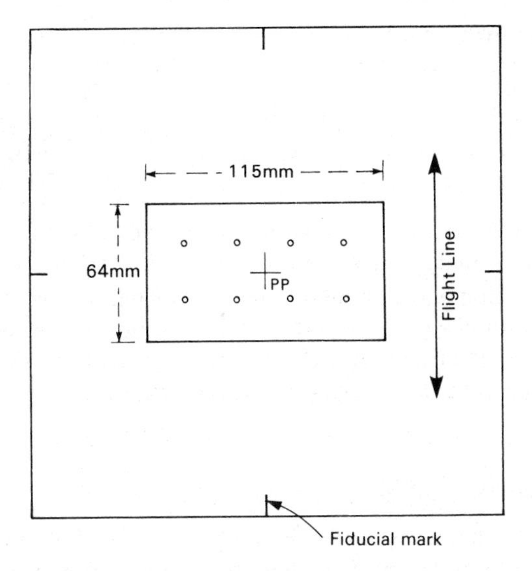

FIG. 19.3. The effective working area on a 9 × 9 in. vertical aerial photograph assuming 60% fore-and-aft overlap with 30% lateral overlap. (Source: Wilson, 1949).

a special grid overlay for each photograph. Although the method bears superficial similarity to using an existing grid on a map, it is important to stress that there are important differences. The réseau is a precise grid which is located with respect to the optical system of the camera; its principal point and the fiducial marks on the frame. Therefore the réseau pattern is unique to each photograph and bears no particular relationship to those patterns of crosses shown on adjacent photographs. Thus the réseau cannot be regarded as being a portion of a much larger grid system, whereas the map grid extends uniformly over a much larger portion of the earth's surface than does the single map sheet. Consequently the same map grid may be used to sample points on many adjacent map sheets as if it were one gigantic overlay. Moreover, if photographs of different ages are to be used, the ground positions corresponding to the réseau crosses are quite different on the different sets of photography.

The final example of the direct use of aerial photography is the work by Emmott and Collins, who developed a working procedure for the measurement of land-use changes from vertical aerial photography. Their study was specifically concerned with sampling directly from photography, and not using plotted land-use boundaries which had been first plotted on a map, or even outlined on the photographs during preliminary interpretation. Moreover the principal object of the study was to measure change, and therefore it was useful to locate the same sampling points on photographs of different ages and different scales.

CHOICE OF METHOD

There are several examples of comparative testing of the two methods of area measurement, each of which comprises using the same data, treating it first as if the measurement had been obtained from the direct method and secondly using the same data for indirect measurement. We describe two such tests:

- Measurement of a single variable show on a map, the distribution of which is much fragmented. The example is that used by Pielou (1974), illustrating the determination of the extent of woodland on part of a 1/50,000 scale map of southern Ontario.
- Measurement of urban land-use in a test area of part of Preston measured by Emmott and Collins (1980) and Emmott (1981).

Pielou's Measurements of Much-Fragmented Woodland

We have seen that the best way of combining the merits of systematic sampling with that of a random sample is to sample systematically several times over, using a coarse grid of sampling points that is relocated at random several times. This is essentially the solution described by Quenouille (1949), which was referred earlier and is the procedure adopted in direct measurement. It has also been used by Pielou (1974), who has additionally counted all the points occurring in a single application of the overlay on the map. Application of the regular grid of points gives not only different counts, n, but also the total N. It follows that a series of k values of z are obtained from k applications of the grid. Since several such estimates are obtained, one on each occasion that the grid is relocated, the sampling variance of the estimate, and hence its standard error, can be found.

Pielou's sampling frame comprised a rectangle of ground dimensions 5×10 km, or 10×20 cm on the map. Sampling was done using an overlay with a 1 cm grid, using the intersections of the grid lines as the points to be counted. This is a relatively coarse grid, the use of which evidently corresponds to the suggestions of Hillard and Cahn (1961) and runs counter to those of Köppke (1967) and others quoted in Chapter 18. Using the sampling procedure outlined for direct measurement Köppke drew ten samples from the map, each for the overlay in a different position, whereas for indirect measurement, only one such application of the overlay would have been needed.

Naturally the total number of points counted varies slightly from one application to another. The systematic pattern of points on the overlay fits the sampling frame exactly in one position, giving a total of 200 points. For all other positions and orientations of it there is variation of about ± 5 points within the sampling frame.

There are four different methods of determining the area of woodland.

1 We treat it as if it were an example of direct measurement by point-

counting. The mean number of points counted is $\bar{n} = 449/10 = 44{\cdot}9 \pm 4{\cdot}75$. Since $d = 1$ cm, $a = 1$ cm^2 and the measured area of woodland on the map is 44·9 cm^2. Converting this into ground units, the area of woodland is $11{\cdot}23 \pm 0{\cdot}37$ km^2.

We should note, however, that we are not using precisely the same technique which has been described earlier. The direct method of measurement of area by point-counting is normally used to measure one parcel at a time. In this case the method is being used to sample a large number of small irregular parcels simultaneously. It appears to be a valid application of the method of direct measurement. There is a hint in Husch (1971) that this may be done, but the author is not aware of the method having been exhaustively studied.

2 The method proposed by Pielou comprises determination of the estimate z for each application of the overlay. Since the sum $\sum n = 449$ and the sum $\sum N = 2{,}002$, it follows that the mean $\bar{z} = 449/2{,}002 = 0.224 \pm 0.002$. In other words, 22·4% of the area shown on this part of the map is woodland. Converting this into ground units by multiplying by the area of the sampling frame (50·0) we have the area of woodland equal to $11{\cdot}2 \pm 0{\cdot}35$ km^2. It can be seen that the area determined by this method agrees well with (1) above.

3 However, the recommended practice for indirect measurement has been based upon the single application of the point pattern. If we look at the areas determined from the individual values of $A_M{\cdot}z$ given in the right-

TABLE 19.3. *Measurement of the total area of woodland shown on part of the 1/50,000 sheet 31 C/7 West, Sydenham, Ontario, published in the National Topographical Series of Canada. Measurements from Pielou (1974)*

Application no.	n	N	Estimates of p: $z = n/N$	Estimates of p: $n/200$	Area (km^2) from $A_M{\cdot}z$
1	41	199	0·206	0·205	10·3
2	39	196	0·199	0·195	9·95
3	47	195	0·241	0·235	12·05
4	42	202	0·208	0·210	10·4
5	50	202	0·248	0·250	12·4
6	43	202	0·213	0·215	10·65
7	55	204	0·270	0·275	13·5
8	43	199	0·216	0·215	10·8
9	46	201	0·229	0·230	11·45
10	43	202	0·213	0·215	10·65
Sum	449	2002	2·24	2·25	112·15
Mean	44·9	200	0·224	0·225	11·215
s	±4·75	2·9	0·022	0·024	1·12
s_m	±1·50	0·9	0·007	0·007	0·354

hand column of Table 19.3, we see that although the mean area ($11{\cdot}215\,\text{km}^2$) agrees with that obtained in other ways, there is only one sample (No. 9) which approximates to this value.

4. There is yet another way of applying Pielou's method, which is an approximation but saves counting the total number of points within the sampling frame for each application of the grid. This is to assume that in a regular sampling frame of known size the total number of points contained within it is normally distributed about the mean value which corresponds to the count made when the grid on the overlay coincides with that on the map. In this case the number of points is 200, and we see from Table 19.3, column 3, that the mean $\bar{N}$ of the total counts is also 200. We proceed as outlined above, by drawing samples and counting n, the number of points coinciding with woodland. Then we determined for each sample the ratio $n/200$ as the proportion of woodland. These values are shown in the fifth column of Table 19.3 and they yield the mean $\bar{z} = 449/2{,}000 = 0{\cdot}2245 + 0{\cdot}0075$. Converting this result into ground units, we have as the measure of the area of woodland $11{\cdot}23 \pm 0{\cdot}0375\,\text{km}^2$, which agrees well with the other determinations.

Pielou's method reinforces the opinion that there are some advantages in using a comparatively coarse systematic grid, as proposed by Hilliard and Cahn (1960), which we have seen appears to be particularly suitable for use with much-fragmented parcels. This would seem to be at the expense of counting far more points than would be necessary using the direct method, but the short cut offered by the third method would appear to overcome this difficulty.

Emmot's Study of the Comparative Accuracy of Different Methods of Measurement Applied to Urban Land-Use Changes

The work was done on a site of area $70\,\text{km}^2$ occupied by the town of Preston. Photographs at a nominal scale of approximately 1:10,000 taken in the years 1946 and 1973 were used for the project, but it was found that the variations in scale were of the order of 1/10,500 to 1/12,000.

In order to determine the optimum method of area measurement, balancing accuracy against practicability, the following test procedure was adopted:

- A series of 1·00 ha dot grids were constructed to allow a close matching of grid scale to mean photoscale. Five grids were constructed ranging in scale from 1/12,000 to 1/10,500.
- By reference to corresponding map and photograph details a dot grid of the appropriate scale was placed in coincidence with the national grid transferred from map to photograph.
- Using the 1973 photography, the land-use category at each dot was recorded on a data sheet which indicated the position of each dot and the four-figure numeric land-use code.

- This procedure was then repeated with the 1946 photographs of the same place.

The advantages of this method are: first, that land use is recorded at identical points for each year; thus the changes noted are real rather than apparent, resulting from lack of correspondence between the dot positions. Secondly, land-use maps may be produced which are comparable point by point and which may be combined to cover the whole study area without gaps of overlaps. Thirdly, the land-use information is compatible with other data referenced to the national grid or the administrative units.

A major part of Emmott's work with respect to this was the investigation of comparable methods of measuring the very small land-use categories which characterise the urban land use of an old manufacturing town. With this object in mind, Emmott investigated

- derivation of a *standard* against which to compare the grid systems;
- selection of the type of grid to be used;
- selection of area measurement base, i.e. map or air photograph;
- selection of the optimum size of unit cell or grid density.

Accuracy Datum or Standard

The interpreted land-use units for a test area (4 km^2) were transferred from the photographs on to the 1/10,000 scale OS maps. The area of each unit was then measured using a polar disc planimeter and a list of category areas compiled.

Like most other workers in this field, Emmott considered the planimeter measurements to be without error. Thus, any discrepancies between areas derived from sampling techniques and those derived from planimeter measurements were treated as being errors in the results obtained by sampling.

Grid Type

Emmott considered that it was not possible, on the evidence of the published work, to determine the relative merits of point-counting as compared with the method of squares. Consequently he carried out the following tests to investigate their accuracy. Category areas were calculated from grid squares and dot counts carried out on the land-use maps compiled for the planimeter measurements. In each case systematic point patterns, being square grids of unit cell 100 mm^2 ($d_1 = 10$ mm), 50 mm^2 ($d_1 = 7{\cdot}1$ mm) and 25 mm^2 ($d_1 = 5$ mm) were used, these being equivalent to ground areas of 1·00 ha, 0·50 ha, 0·25 ha at the 1/10,000 scale of the maps. The results were analysed using the multiple regression equation derived by Yuill (1971):

$$\sqrt{D} = A + B \ln N + C \ln z \qquad (19.12)$$

where D is the difference between the grid and planimeter area, and the

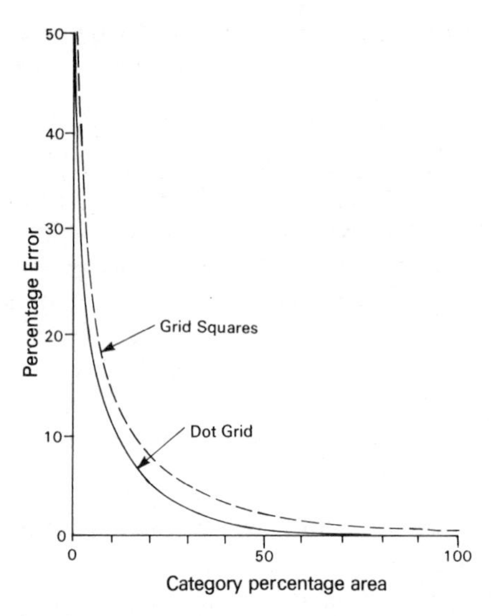

FIG. 19.4. Comparison between the results of measurements by means of grid squares and by dot grids. Percentage error plotted against category percentage area for a total count of 100. (Source: Emmott and Collins, 1980).

proportion z is expressed as a percentage. The Preston study yielded the following results:

Method of grid squares: $\sqrt{D} = 11.75 - 1.02 \ln N - 1.42 \ln z$

Dot grids: $\sqrt{D} = 11.62 - 1.01 \ln N - 1.57 \ln z$

Figure 19.4 illustrates the close agreement between both methods. In this, D is plotted against z. It may be seen that the dot grids produced somewhat better results, particularly for small category percentage areas. In addition, the following characteristics of the two grid types, noted during the tests, reinforce the choice of the dot grid as being most suitable.

The method of squares

- Categories occurring in small or narrow linear units rarely if ever form the dominant use within a square, and thus will tend to be omitted or systematically underestimated.
- The process of judging dominant use within a square is frequently difficult and always subjective even on a simple chorochromatic map; this difficulty would be very much greater when combining the three processes of photointerpretation, recognition of land-use boundaries and judgement of dominant land-use within a grid cell.

Dot grids

- The relationship of dot count to category depends on the proportion of the total area occupied by each category, not on the size or shape of individual land-use units; thus there is no tendency towards systematic under-estimation of categories occurring in small units.

- No estimations of proportions or dominance are required.

Measurement Base

Category area measurements were then carried out directly on photo overlays depicting the land-use units before transcription upon the base map. The areas

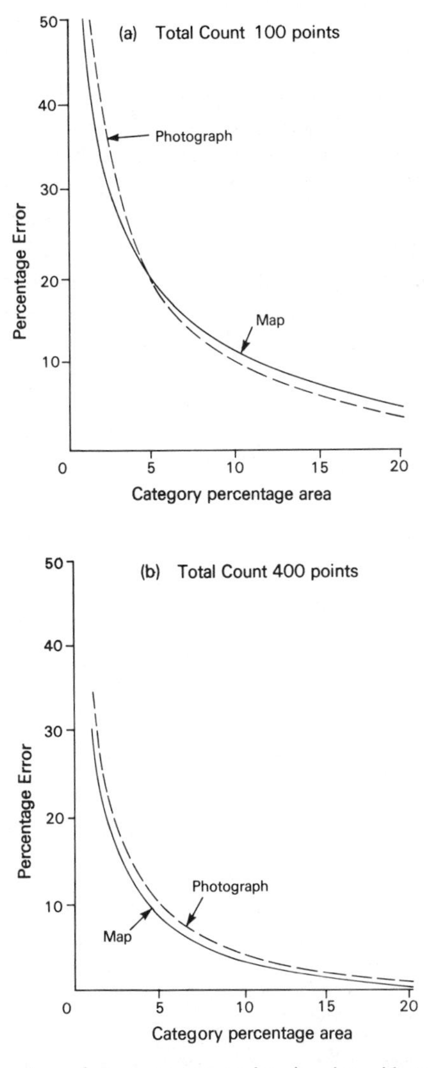

FIG. 19.5. A comparison of measurements made using dot grids to measure area on aerial photographs and maps. Plot of percentage error against category percentage area for (a) total count = 100 and (b) total count = 400. (Source: Emmott, 1981).

so derived were compared with the planimeter values and the results subjected to multiple regression analysis. The equation obtained was

$$\sqrt{D} = 11{\cdot}36 - 0{\cdot}93 \ln n - 1{\cdot}66 \ln z$$

As may be seen in Fig. 19.5, areas obtained directly from the photographs were very nearly as accurate as those taken from the map. Certainly the very small reduction in error resulting from transferring the land-use boundaries from the photographs to the map does not justify the effort involved.

Grid Density

From the plotted curves relating percentage error and total dot count it is apparent that for a category percentage area of 1% there is a marked increase in the rate of change of percentage error against total dot counts, for counts of less than 400 dots. This characteristic persists for category percentage areas up to 10%, but for areas of 20% the critical value is in the region of 200 dots and falls to 100 dots for a category percentage area of 50%.

These factors, considered with the sizes of the administrative districts within the study area, suggested that the optimum dot grid for the survey ought to be a ground area of 1·00 ha (a grid cell of approximately 10 × 10 mm on the photographs). This would result in dot counts of from less than 100 in the small town-centre wards to over 400 in the larger wards in the outer areas. This would mean that higher levels of accuracy would be expected in the areas of greatest land-use change.

20

Area Determined by Linear Measurements

It was a Sapper who invented it. Alas, his name is lost, yet it is recorded that he drew sixpence a day for the rest of his service as a reward.

(H. S. L. Winterbotham, *The National Plans*, 1934)

We have already seen in Chapter 4 that an alternative to the use of the method of squares or point-counting is to consider the parcel to have been subdivided into a series of contiguous rectangular strips, or intersected by a series of straight lines. The lengths of the rectangular strips or the lengths of the lines contained within a parcel are measured and the area of the figure determined from these data. Therefore the method of complete measurement is commonly known as the *method of strips*, and the sampling methods are often described as *transect methods*. The transect methods have been popular for use in forest evaluation, particularly in North America where much work has been done directly from aerial photographs rather than maps. They have also been much used in microscopy ever since this technique was first described by Rosiwall in 1898 for petrographic analysis.

Much of the work to be described in this chapter is now of historical interest. The most prolific period of contributions in this field of study were the years 1941–1942 when most of the references based on American forestry practice were published. The majority of these demonstrate that, first, the method is faster than the other methods of area measurement in common use at that time; secondly, the determination of area by measuring transects was only a little less accurate than the use of a polar planimeter and often proved more reliable than measurement by dot grid. Although the methods are less often used today than they were 30 or more years ago, this does not necessarily mean that they are either slow or inefficient. It is rather that they have become unfashionable compared with counting points. This is because of the ease with which durable and precise dot grids can now be constructed on transparent plastic. In the nineteenth century, when a measuring grid comprised a network of fine threads tensioned within a wooden frame (as are the strings of a tennis racket), a finely divided wooden or metal scale was the more reliable instrument.

Figure 20.1 indicates the various possibilities which will be considered and

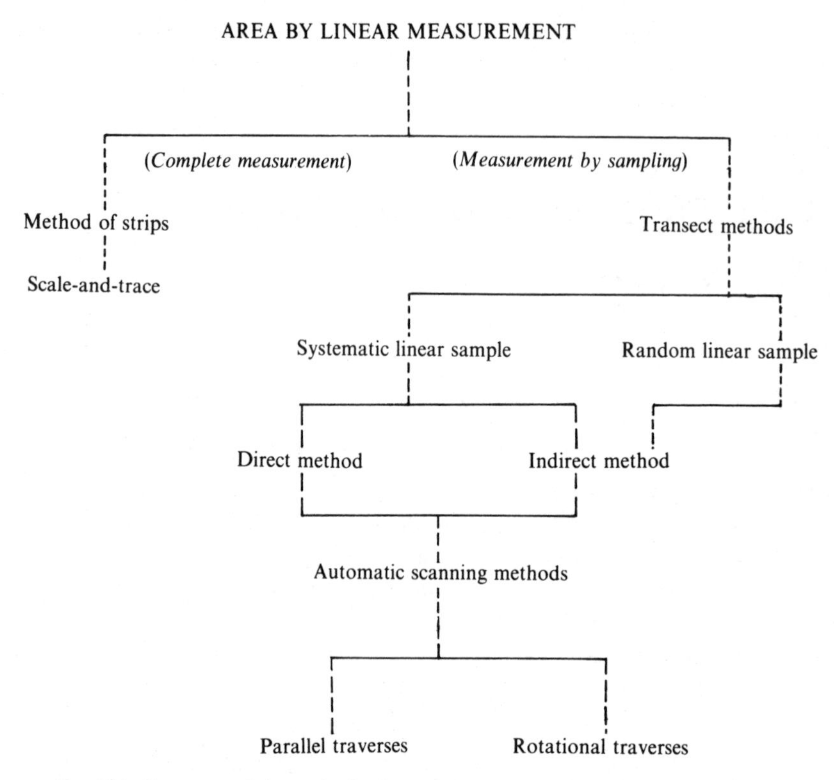

FIG. 20.1. Summary of the methods of area measurement using linear measurements.

therefore corresponds to Fig. 18.1, page 395, which did the same for counting squares and points.

The pattern of lines might be systematic, as illustrated in Fig. 8.6, p. 129, or random, is illustrated in Fig. 8.3(a). As in point-counting, it is evident that both direct and indirect methods may be used with a systematic pattern of lines, but only the indirect method may be used with a random sampling strategy.

THE METHOD OF STRIPS AND ITS APPLICATION IN THE SCALE-AND-TRACE METHOD OF AREA MEASUREMENT

The area of a parcel measured by the method of strips is the sum of the areas of the individual component rectangles. This was expressed algebraically by equation (4.8), page 61.

The best instrumental method of defining a strip and measuring its length was that adopted by the Ordnance Survey and known as the *scale-and-trace method*. It was so reliable in skilled hands that it remained the only method of measuring areas on the basic scale Ordnance Survey maps from its introduction before 1850, to its replacement more than a century later by the Stanley–

Cintel automatically recording polar disc planimeter. Two pieces of equipment were needed to make the measurements: *the computing* or *measuring scale* and *the trace*.

The computing scale comprised a wooden scale having a metal cursor set on one side of it. This could be moved along the length of the ruler but in no other direction. The centre of the cursor bracket contained an open window across which a fine wire was stretched and held by two screws. The surface of the scale was engraved with two linear scales, one numbered in the opposite direction to the other along the sides of a rectangular groove into which the cursor was mounted. That part of the cursor which lay in this groove was engraved with a datum mark and vernier, one for each scale and therefore engraved in opposite directions. As the cursor was moved along the scale so a linear distance was indicated by the scale reading.

The trace was a grid reproduced on transparent film. Although a single family of lines would have been sufficient, traces produced in the last years of its use were full square grids, each square representing an area of 0·1 acres at the scale of the map. Thus for use with 1/2,500 scale maps the sides of the grid squares each measured 8·05 mm.

Measurement Technique

The trace was placed on the map. It was important that it did not move during measurement so it was held in position with lead weights. This was better than using drafting tape because the trace had to be realigned for each parcel. The orientation of the trace is theoretically of no importance, but in practice it was found best to align one family of the grid lines with the longer axis of the parcel so that the parcel boundaries forming the ends of the strips to be measured were aligned at about 45° to the trace. Since the computing scale was to be aligned parallel with their direction, we shall call these *horizontal lines*, as in Fig. 20.2. The scale was laid upon the trace and the cursor was set to the zero reading of the vernier. The top portion of the parcel was framed within the cursor window and the computing scale was moved bodily until the cursor

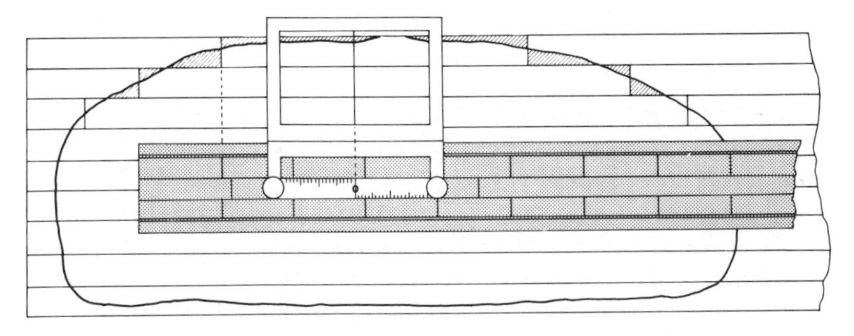

FIG. 20.2. The computing scale aligned over a parcel.

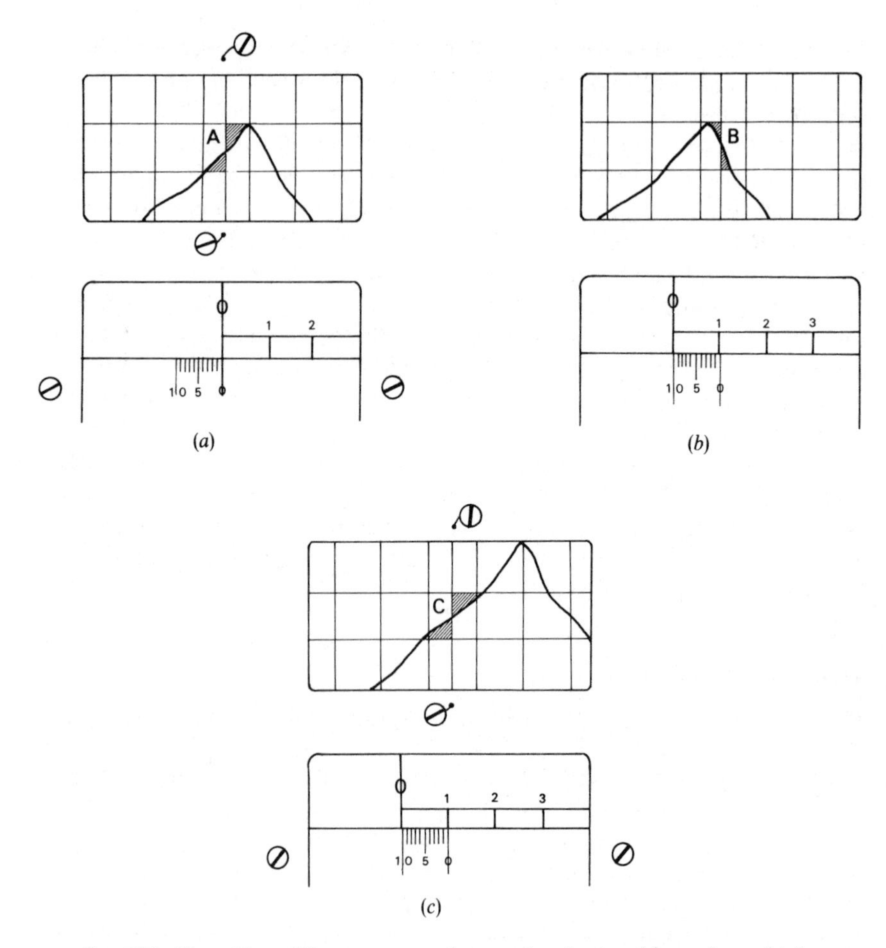

FIG. 20.3. The setting of the cursor over the parcel perimeter of form give-and-take lines. (*a*) The cursor wire is aligned through a point *A* in the first strip which has been identified by balancing the areas inside and outside the parcel perimeter which are shaded. The scale is set to zero. (*b*) The cursor is moved to the right until the point *B* has been identified as the give-and-take line for that end of the first strip. The scale reading has changed to 1·03. (*c*) Without altering the scale reading, the computing scale is moved to the next strip where the left-hand give-and-take line passing through *C* is identified. The next step is to move the cursor to the right hand of that strip.

hair bisected part of the boundary, within a square on the trace to form a give-and-take line as illustrated in Fig. 20.3. Holding the computing scale firmly, the cursor was then moved along it to bisect the boundary at the other end of this strip, again creating a give-and-take line. The movement of the cursor along the computing scale changes the vernier reading, the difference between the two being the measured side of the rectangle forming the strip. However, this intermediate scale reading did not have to be booked separately. The next stage was to move the computing scale bodily to the next horizontal line of the trace and to the position where the cross-hair intersected the parcel boundary. Provided that the cursor was not accidentally moved during this operation, the next strip measurement began with the scale reading obtained at the end of the first strip. The cursor was moved to the other end of the second strip. This procedure was followed, first moving the computing scale to one end of the strip to be measured and secondly moving the cursor along the computing scale, for as many strips as occurred within the parcel. On completion of the measurement procedure the vernier reading gave the cumulative sum of all the strip lengths within that parcel. This value was multiplied by the width of each strip (0·1 acres) to give a measurement of the parcel area. If the parcel was too large for the whole measurement to be carried out in a single reading of the computing scale, the point reached at the maximum reading was marked, the vernier reset to zero and measurement began again from the marked point.

The use of the cursor wire to represent give-and-take lines, and thereby identify the ends of each strip, was critical to the use of the equipment. Where the parcel boundary intersected the horizontal trace lines at about 45° it was easy to estimate the position of the cursor which divided the figure into two equal portions, as illustrated in Fig. 20.3. Often, however, the parcel boundary did not behave so well and some strips contained much-elongated figures whose proportional areas were less easy to estimate. This was most difficult to do where the parcel boundary made a small angle (10–15°) with the horizontal lines, and particularly if this line was irregular. For example, the top strip in Fig. 20.2 illustrates a difficult interpretation. It follows that the measurement of such a strip depended upon the ability of the operator to make satisfactory identification of the terminal points of a strip. In the opinion of the late C. J. McKay, who had been superintendent of the Areas Section of the Large Scale Devision of the OS at the time of the change to semi-automatic methods, those operators who were prone to making errors used to produce results which were either consistently high or low. In short, the commonest operating errors were systematic.

McKay (1968) believed that the following sources of error affected the accuracy of measurements made by scale-and-trace:

Environment
Temperature
Humidity, affecting the plastic trace
Operator
Physical posture of operator
Mental concentration
Vision: Parallax of the bisecting wire
Exterior and interior lighting

He thought that viewing angle was responsible for the greatest variation between different operators, and since this depended upon posture, he considered this to be a major cause of persistent error.

TABLE 20.1. *Acceptable tolerances for area measured by means of the scale-and-trace method as used by the Ordnance Survey*

Parcel size		Tolerance	
Acres	Hectares	Acres	Hectares
0–2	0–0·8	0·02	0·008
2–8	0·8–3·24	0·04	0·016
8–18	3·24–7·28	0·06	0·024
18–32	7·28–12·95	0·08	0·032

It was standard OS practice to measure every parcel twice, this being done by different operators. If the two results agreed within a tolerance of

$$t = \sqrt{(2A)}/100 \qquad (20.1)$$

the mean of them was accepted as the area of the parcel. If the results were larger than the tolerance, additional measurements were made until two acceptable measurements had been made. Table 20.1 indicates the maximum tolerance acceptable at the sizes of small parcels. We have seen in Chapter 10 that an additional check may be obtained if all the parcels on a given map are to be measured, for the sum of their areas may be compared with the theoretical area covered by the map. In OS practice this was done because all the parcels on a map were measured as a matter of course. Then the results were considered to be satisfactory if the total of the means of the individual measurements lay within the range $\pm\sqrt{A}/32$, where A is the area contained within the neat lines of the map. For the 1 × 1 km format 1/2,500 maps this corresponds to 247·1 acres (100·0 ha) and for the 2 × 1 km format it is 494·21 acres or 200·0 ha. Then the acceptable tolerances were ±0·5 acres (0·2 ha) and ±0·7 acres (0·3 ha) respectively. When the sum of the measured parcels, A', met this tolerance, the acceptable error was distributed proportionately through the map. The reader will notice that no attempt was made to correct for the variation in area scale for maps in different parts of the country as was subsequently done in the reduction of planimeter measurements. This was briefly mentioned on page 192.

AREA MEASUREMENT BY TRANSECT SAMPLING

Direct Measurement

It was shown in Chapter 7 that the linear sampling strategy involved sampling along certain lines or at certain points along lines. A typical procedure is that described by Kramer and Sturgeon (1942) as follows:

The transect method of estimating ground areas from aerial photographs consists simply in taking linear measurements on the photographs along a series of uniformly spaced traverse lines and applying a converting factor based on the spacing of the traverse lines and the scale of the photograph. The equipment needed includes a transect guide, a triangular engineer's scale, and a suitable marking crayon. The guide used in this case was prepared from a sheet of celluloid 0·025 inch in thickness, cut to a width of 8 inches and a length of 20 inches. Parallel lines were scored on the celluloid at an angle of 60° with the side; this angle made it possible to reverse the direction of the transect lines by merely inverting the guide. At the ends of the scored lines notches were cut to accommodate the point of the marking crayon. Because the aerial photograph index sheets were on a scale of approximately 1 inch to 1 mile, the transect lines were spaced 1 inch apart.

The guide is placed on the sheet preferably with its long axis parallel to the long axis of the area to be measured, and the ends of the transect lines are marked on the sheet with a crayon. This, in effect, transfers the lines from the guide to the sheet. Theoretically it is advantageous to have the area crossed by as many transect lines as possible. The direction to be taken by the transect lines on any sheet may be decided by flipping a coin. The guide is removed and the engineer's scale is placed along a transect line defined by one pair of marks. The sheet should be turned so that the estimator looks straight at the surface of the scale. In this case the 1/10-inch or 10 scale was used. As the first operation, the total distance across the area along one of the transect lines was noted and recorded in tenths of an inch. Next, the 1/10 inches of the line that represented forest area were counted, any fraction as large as 1/20-inch being included in the count as a full 1/10-inch and any smaller fraction being excluded. The procedure was followed until measurements had been made along every one of the transect lines. The measurements were then totalled and the total converted into ground-area values on the basis of 1 square mile to 1 square inch. Forested area was then computed as a percentage of total area. Estimates for extensive forested tracts were arrived at simply by counting all areas of forested land that are continuous through a full inch along a transect line.

Until a thorough acquaintance with the method is acquired, it is best to count both the units representing forested area and those representing nonforested area and see that the sum of the two counts equals the total distance along the line.

Indirect Measurement

The last sentence indicates that Kramer and Sturgeon recommended making the necessary measurements to apply the indirect method; nevertheless they had failed to appreciate that this was so. The equation to be used for direct measurement has already appeared as equation (4.8) on page 61, the corresponding expression to be used for indirect measurement is

$$A_B/A_M = l/L \tag{20.2}$$

where; A_B = the area of the parcel or the variable to be measured;
A_M = the total area of the map, these abbreviations being used as in equation (19.1) and the following text;
l = the total length of all the portions of the line corresponding to the feature to be measured;
L = the total length of all the lines measured on the map.

Aldred (1964) argued that measurement of the sample transects is slow when the work is done with an ordinary scale and the record for each line has to be booked manually. Obviously the scale-and-trace equipment can be used for transect measurement as effectively as it had been used by OS to measure strips, but it is evident that American foresters were, at that time, unaware of the instruments and methods used in Britain. Aldred therefore adapted a chartwheel for transect measurement by attaching a pencil to the measurer to mark the progress of the work, and thereby avoid missing or double-scanning of transect lines, just as the Markounter pencil is used for this purpose in point-counting. He also constructed a simple form of brake on the measuring wheel by inserting a small piece of rubber making contact with the side of the wheel. This reduced the risk of errors introduced by overrunning a transect or inadvertent movement of the wheel between measurements. He called the instrument a *transect area-meter*. The technique of using it is simply to draw the measuring wheel along each transect line and record the cumulative distance measured on the dial.

Drawing a chartwheel or a cursor to-and-fro over transect lines represents a crude form of scanning, and it is now common to measure area by the linear measurements recorded by an automatic scanning system. Latham (1963) drew attention to the cartometric possibilities of the scanning equipment then available, but much of the equipment which became available about that time had, as now, little value for map work. For example, a photoelectric planimeter was designed by G. N. Wilkinson of C.S.I.R.O. in Australia during the 1960s. The statistical theory of this was examined by Moran (1968). This instrument allowed for an opaque and more-or-less plane object, such as a leaf, to be inserted between two plastic rollers. Here it passed between a light source and a linear array of 32 photoelectric cells which sensed the opacity of the space between the rollers and, in effect, converted the intervals of time when the light source was cut off by the object scanned into a linear distance transverse to the direction of rolling. Obviously the size of the parcel (or leaf) which could be measured was governed by the lengths of the rollers and the number of photoelectric cells in the array. This equipment has little value for cartometric work unless, as in measurement by weighing, the operator is willing to mount the maps on an opaque material and cut out each parcel to be measured as a separate object. We have already seen that this is inconvenient, time-consuming and wasteful.

EXAMPLES OF APPLICATIONS

The following examples of area measurement by means of strips or linear transects have been described in some detail.

- Polushkin (1925) describes the use of the method in metallurgy, exploiting the abilities of a microscope stage to scan a slide along a series of parallel lines through measurable distances. Polushkin compared the results of indirect measurements made on the slide with planimeter measurements made upon photographs of the same features.
- Trefethen (1935) applied the method to geographical reconnaissance. His conclusions agree with those of Rosiwal; namely that good results could be

obtained when the total length of the traverse lines exceeds 100 times the average intercept of the field types traversed.

- Kramer and Sturgeon (1942) described the method of linear (transect) sampling to be applied on aerial photographs, index sheets and mosaics. In order to test the work they made a series of test measurements for three different localities where the extent of forest cover differed. They employed three different operators to make all the measurements, and compared the transect method with planimeter measurements.
- Proudfoot (1942) has described a rapid reconnaissance technique employed for the analysis of land use by Tennessee Valley Authority to sample the quantity and distribution of various types of land.
- Osborne (1942) investigated various sampling strategies in the measurement of land-use and forest types on maps of two parts of the western U.S.A. It is a pioneer paper in two-dimensional sampling methods. He was one of the first people to show that the variance and standard errors cannot be determined from systematic samples without the introduction of some bias. Because his samples were so large (in excess of 600 measurements) he was able to investigate fitting polynomial expressions to the measured data and thereby demonstrate that it was possible to obtain dependable estimates of the sampling errors.
- Latham (1963) drew attention to the possibilities of high-speed electronic scanning equipment in the rapid integration of irregular parcels. Possibilities of the method were illustrated by a study of cropland in Pennsylvania in which the operation of the scanning lines of sampling traverses of the scanner was simulated by manual traverses. By changing direction, i.e. rotating the angle at which the traverses crossed the cropland, Latham was able to derive important additional information on the orientation of the geographical distributions which he was studying.
- Haggett and Board (1964) followed up Latham's work. They both considered the effects of parallel shifts – which is, in effect, increasing the sampling density by increasing the number of parallel lines sampled – and also investigated the effect of rotation of the scan lines, so that a parcel was sampled in different directions.
- Aldred (1964) made a comparative study to evaluate the speed and accuracy of a transect method of measurement which employed the instrument developed by the writer, called a transect area-meter.
- Yates (1949) describes the methods of sampling employed in making the 1942 Census of Woodlands for Britain.

YATES' MEASUREMENT OF WOODLAND AREAS FROM THE ONE-INCH MAP

This was a procedure adopted as part of the 1942 Census of Woodlands made by the Forestry Commission during World War II when the results were needed with great urgency and little manpower could be allocated to undertaking it. Most of the sampling methods used for this census, which have been described by Yates (1944), (1949), were done using field checking on a sample of Six-inch to One-mile maps. These do not concern us in the present context. However, part of this census determined the total area of woodland in each county by estimating the area of land coloured green on each One-inch OS map. The method used was to measure the total length of the *E–W* kilometre grid lines which coincided with green areas. These maps were either copies of the immediate pre-war 5th Edition of the One-inch to One-mile map, or a wartime revision of the Popular Edition, the grid being a military grid of that period and not the National Grid, which did not become the base for small-scale OS maps until after the end of World War II. On these earlier editions of the One-inch map the neat lines did not correspond in position or orientation with integer grid lines as they do no modern maps. Consequently there were small variations in the lengths of each grid line shown on the map; a complication which does not arise in the use of later sources. All measurements were made using a scale, and reading this to the nearest millimetre. The results for OS sheet No. 115 covering part of Kent are given in Table 20.2.

Using the indirect method of equation (20.1) it follows that $l/L = 92{\cdot}7/716{\cdot}4 = 0{\cdot}129397$, so that knowing A_M, we may determine A_B. In this example, however, a further difficulty arises because part of this map (of northeast Kent) shows sea. Therefore the area A_M does not represent the product of the neat line dimensions. Yates does not provide an alternative and accurate measure for the land area, but an approximation may be obtained

TABLE 20.2. *Woodland areas measured from line intercepts (from Yates, 1949)*

Grid line	Length of line (cm) L	Length coloured green (cm) l	Grid line	Length of line (cm) L	Length coloured green (cm) l
98	3·5	0·0	83	30·0	3·8
97	4·2	0·9	82	29·4	4·1
96	9·2	0·0	81	29·1	4·9
95	12·6	0·0	80	28·8	6·0
94	15·5	0·3	79	28·6	5·4
93	21·2	0·1	78	28·2	2·3
92	25·2	0·5	77	27·2	2·9
91	25·4	3·1	76	26·3	2·1
90	31·2	2·8	75	25·4	6·3
89	34·2	2·7	74	25·5	8·2
88	34·1	2·8	73	25·2	5·4
87	33·0	2·6	72	24·9	6·6
86	31·4	2·3	71	24·6	6·6
85	31·0	3·5	70	20·8	4·1
84	30·7	2·4			
				$\Sigma L = 716{\cdot}4$	$\Sigma l = 92{\cdot}7$

from knowledge of the fact that the grids lines are 1 km apart. Since $\sum L =$ 716.4 cm = 453.911 km, it follows that the land area is approximately 453·9 km^2. Consequently the area of woodland measured and calculated by this method is equal to 58·7 km^2 or 14,513 acres. We must argue that because the lines sampled are parallel and equidistant this represents a systematic sampling scheme. Since this represents a single application of the measuring process we cannot determine the precision of the results.

Using the direct method, a length corresponding to 1 km represents an area of 1 km^2. Consequently the conversion factor to be applied to the total length measured in cm in order to give the estimated area in acres is 63,360 × 247·11/100,000 = 156·57. The total area of woodland is therefore

$$156{\cdot}57 \times \sum(l) = 156{\cdot}57 \times 92{\cdot}7 = 14{,}514 \text{ acres or } 58{\cdot}7\,\text{km}^2$$

as before. The same procedure can be followed for the other maps covering the county. Although this example demonstrates the ways in which the measurements may be handled as an example of statistical sampling, it does not provide any commentrary upon the speed of the measurement process, nor upon the relationship of the results to measurements made by other ways. To study these subjects we must turn to the American work.

THE SPEED OF AREA MEASUREMENT BY TRANSECT SAMPLING

Aldred tested both speed and accuracy of transect sampling using his transect area-meter. Two different localities shown on forest maps of scale 1/15,840 were tested against two established methods of measurement: the dot grid, used to judge speed performance; and the polar planimeter, treated as the criterion of accuracy. The MK Area Calculator was also tested and compared with the transect method.

Two blocks were used as test area: one with relatively small parcels (average size 11 ha) and the second with relatively large parcels (68 ha). The block of small parcels consisted of 35 forest types in a total area of 382 ha and the parcel sizes ranged from 0·8 to 34 acres; the block of large parcels consisted of 20 forest types in a test area of 1360 ha and parcel sizes ranged from 4 to 243 ha. The tests were made as comparable as possible by using a dot grid having the spacing $d = 5$ mm (grid density of 25 dots per square inch), a transect spacing of 5 mm (1/5 inch) and every second line of the MK/100 grid.

Test of Measurement Speed

In Table 20.8, speed is expressed as the time taken to measure the two test blocks; relative speed is expressed as percentage of work done in a given period of time compared to the work output of the dot grid in the same interval. The table shows that the block of large parcels were measured in about one-half of the time using the transect area-meter than was accomplished by point-

TABLE 20.3. *Summary of the time needed to measure two blocks. Relative speed measured as a percentage of the time take to measure the blocks by dot grid. Data from Aldred (1964)*

	Block 1 (Small parcels)		Block 2 (Large parcels)	
Test area	382 ha (152·4 cm²)		1360 ha (522·6 cm²)	
No. of forest type	35		20	
Average parcel size	10·9 ha (4·35 cm²)		168 ha (27·1 cm²)	
Method:				
Polar planimeter	270·0 min	(−94%)	120·0 min	(−83%)
Dot grid	16·0	(0%)	20·0	(0%)
Transect area-meter	16·0	(0%)	13·2	(51%)
MK Area Calculator	23·5	(−32%)	17·7	(13%)

counting. Even the MK Area Calculator operated about 13% faster than the dot grid. On the other block the dot grid and the transect measurements were performed at the same speed, whilst the MK Area Calculator was about one-third slower. Aldred found that measurement by polar planimeter was extremely slow in both blocks.

The three fastest methods were tested on areas of various size to indicate the effect of parcel size upon the relative speed efficiency. The graphic presentation of results in Fig. 20.4 shows that the transect method was faster than all the other methods tested for areas larger than 14 ha (5·6 cm²), and that the MK Area Calculator was faster than the dot grid on areas larger than 53 ha (21 cm²).

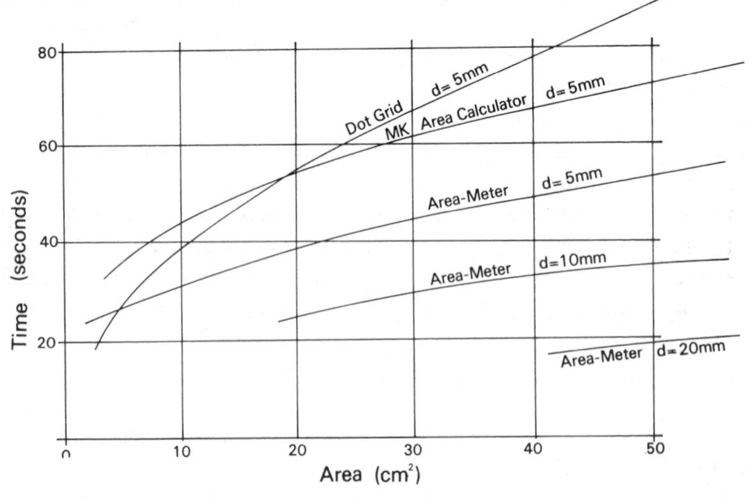

FIG. 20.4. Time required to measure parcels of different sizes by different methods. It should be noted that there is no evidence of operator stress causing a slackening of measurement speed such as was identified in Chapter 18. However, with the range of parcel sizes and separation of lines or dots used by Aldred, this is hardly surprising. For a dot separation of 5 mm it would need parcels of size 500 cm² or larger for the effects of fatigue to become evident. (Source: Aldred, 1964).

On average-size areas (about 6 ha $\approx$ 24·2 cm^2) the area-meter was about 45% faster than the dot grid. An additional reduction in time was achieved by the use of wider transect spacing. Doubling spacing increased speed a further 40%, and when spacing was quadrupled speed was increased by 80%. The results presented in Fig. 20.4 illustrate the speed relationships of the methods, and indicate which method is fastest for a particular size of parcel. A similar study was made by Kramer and Sturgeon (1942), who identified the need for a rapid and accurate method of estimating the extent of the areas having forest cover using aerial photograph index sheets as the source. We would call such a sheet a *print laydown.*

Like most other workers in this field, Kramer and Sturgeon considered measurement by planimeter to be a tedious and time-consuming. They also discarded the use of grids as also being unsuitable, although they offer no reasons why this would be so. Consequently they only compare measurements by planimeter with the transect method.

Their tests were made with three aerial photograph index sheets on which the proportions of forested areas were (1) low, (2) medium and (3) high. An elliptical area of approximately 100 in.^{2} was outlined on each of the sheets with coloured pencil and each of the three sample areas was measured both by planimeter and by transects, each of three operators (*A*–*C*). Therefore the total number of determinations was $2 \times 3 \times 3 \times 3 = 54$. In each of the three transect measurements made by each operator on any one sheet the lines were run in different directions. The time needed to set up the equipment and make each measurement was also recorded.

The results of the measurements are given in Table 20.4. These include, for each pair of average estimates, indices of relative efficiency of the two methods for that operator and the index sheet. This index is the standard error of percentage of sample area representing forested land computed on the basis of number of determinations per hour. The higher the index number, the lower is the efficiency of the method). In Table 20.5 the values given in Table 20.4 are summarised by method of measurement, by operator, and by index sheet. The average index values show that a relatively more precise estimate of forest area can be attained by transect than by planimeter within a standard length of time. The figures for average time required for an estimate strikingly show the greater rapidity of the transect method. The variation of estimate is, of course, less for total area than for forested area. The indices of efficiency presented in Tables 20.4 and 20.5, calculated on the basis of both time required for the estimates and dispersion of the estimates, constitute evidence that the transect method is the more efficient of the two.

Kramer and Sturgeon stress that, in estimating areas from aerial photographs, it is important that operators working on the same project reach a common opinion about classifying land-use. This is believed to be simplified by use of the transect method. Because the guide markings made by the first operator may be left on the sheet it is easy for another individual or group of

TABLE 20.4. *Results of tests made by Kramer and Sturgeon (1942) of comparative area measurements made on aerial photographic index sheets using planimeter and transect sampling. Each value tabulated, except in the final column, is the mean of three determinations*

Index sheet number	Operator	Method	Time required for one determination (min)	Total sample area (in^2)	Portion of Sampling area representing forested land (in.^{2})	(%)	Index of efficiency
1	*A*	Planimeter	44	84·5	18·5	21·9	1·013
1	*A*	Transect	10	83·2	17·3	20·8	0·417
1	*B*	Planimeter	37	84·6	17·7	20·9	2·111
1	*B*	Transect	10	84·2	20·2	22·7	0·943
1	*C*	Planimeter	29	84·9	15·6	18·4	1·230
1	*C*	Transect	8	83·4	17·4	20·8	0·658
2	*A*	Planimeter	67	101·5	49·2	48·5	5·582
2	*A*	Transect	11	101·5	46·9	46·2	1·709
2	*B*	Planimeter	52	102·6	48·6	47·4	1·170
2	*B*	Transect	12	101·5	46·1	45·4	1·098
2	*C*	Planimeter	25	101·3	47·0	46·4	0·854
2	*C*	Transect	15	101·9	43·7	42·9	0·535
3	*A*	Planimeter	62	89·9	73·2	81·4	2·458
3	*A*	Transect	15	88·9	73·3	82·4	0·276
3	*B*	Planimeter	40	89·1	74·5	83·6	1·297
3	*B*	Transect	13	88·1	76·1	86·4	1·094
3	*C*	Planimeter	20	89·8	73·5	81·8	2·021
3	*C*	Transect	12	90·2	76·2	84·4	0·424

TABLE 20.5. *Summary of results obtained by Kramer and Sturgeon (1942)*

Item	Time required for one determination (min)	Total sample (area) (cm^2)	Portion of sample area representing forested land (cm^2)	(%)	Index of efficiency* (%)
Method					
Planimeter	42	593·5	299·3	50·4	1·970
Transect	12	589·7	298·1	50·5	0·795
Operator					
A	35	591·0	299·3	50·6	1·909
B	27	591·6	303·2	51·2	1·285
C	18	592·9	294·2	49·6	0·954
Index sheet					
1	25	542·6	113·5	20·9	1·062
2	30	656·1	301·9	46·0	1·825
3	27	576·1	480·6	83·4	1·262
Total	27	591·6	290·7	50·5	1·383

*Arithmetic mean of the index of efficiency given in Table 20.4.

individuals to remeasure the area, and if care is taken in making the estimates any differences between them are entirely attributable to differences in defining land-use classifications. If a difference in estimate arises, by study of the area in question and of the tally sheets it can be narrowed down to individual transect lines, located and harmonised. Thereafter, such differences will be less likely to occur.

THE SPEED OF AREA MEASUREMENT UNDER CONDITIONS OF PRODUCTION CARTOGRAPHY

The foregoing examples provide some indication of the speed of measurement carried out by operators who are not involved every day with the measurement of area to the exclusion of all other work. Rather different results occur when we study the figures which have been derived from organisations such as the Areas Section of the Large Scales Division of Ordnance Survey. In the 6 years or so between 1961 and 1967 which spanned the introduction of the use of the automatic reading planimeter for this work in 1964, a fair amount of attention was paid to the output by both scale-and-trace and the method which followed it.

The comparison between the scale-and-trace method and that of area measurement using the automatic recording planimeter was described by Fortescue (1967) using a flow diagram to illustrate the improved speed of the automatic method. The times given in that diagram are summarised in Table 20.6. The left-hand column representing the times needed for the scale-and-trace method adds up to $19\frac{1}{2}$ hours but, because two independent measurements were made as part of the standard routine, the total time was effectively double this, or 39 hours. According to McKay (1968) the measurement rate over the 3 years before the introduction of planimeters was 15 parcels an hour; a reliable average obtained from approximately two million measurements. Average parcel content was 50 per km^2, i.e. a measurement rate of 3·33 hours per km^2 or $3{\cdot}33 + 3{\cdot}33 + 1{\cdot}00 = 7{\cdot}66$ per km^2 for at least two independent measurements. This may be converted into rates of 8·00 cm^2 per minute for a single measurement of an average-sized parcel or 3·48 cm^2 per minute if we include all the measurements needed to obtain acceptable results. However, before it was known whether a third check measurement was needed, a visual comparison of the two initial measures had to be made; this requiring a further $\frac{1}{2}$ hours per km^2. This has been included in the "manual calculation" stage in Table 20.6. This stage must be included as part of production process because satisfactory results cannot be obtained without making the check. So the total time per km^2 would be $7{\cdot}66 + 0{\cdot}5 = 8{\cdot}16$ hours per km^2 or 3·26 cm^2 per minute. In stating that the measuring rate is 8·00 cm^2 per minute it is important to add that this result has been obtained from average-sized parcels ($3{,}200/100 = 32$ cm^2); larger areas will usually be measured more quickly and smaller parcels more slowly. In an

TABLE 20.6. *A comparison between the time required (per 1 km^2) to measure and process the area measurements made by the Ordnance Survey on large-scale maps before and after the introduction of semi-automatic methods*

	Scale-and-trace method with manual calculation (hours)		Automatic reading planimeters and data processing (hours)
Preparation of original documents			
Selection of parcels	$4\frac{1}{2}$		$4\frac{1}{2}$
Preparation examination	$1\frac{1}{2}$		$1\frac{1}{2}$
Measurement			
Two independent measurements and partial third measurement	$7\frac{1}{2}$		$4\frac{1}{2}$
Subsequent processing			
Manual calculation	$3\frac{1}{2}$	Computer calculation	0·3
Monotype keyboard	0·1		0·1
Filmsetter	0·1		0·1
Data stickdown	$2\frac{1}{2}$		$2\frac{1}{2}$

evaluation of these figures, James McKay further commented that this is a rate to be expected from sustained conditions, and makes no allowance for tea and other natural breaks to continuity, which amount to a probable wastage of 20%. He wrote to me:

One of my experts some years ago measured a complete plan (2 km^2) of 80 parcels in 1 hour, but if an average of half this rate was enforced as a norm one would have a staff of gibbering idiots – and in all probability the third measurements would soar to 80% and nullify the time saving.

THE ACCURACY OF AREA MEASUREMENT DERIVED FROM LINEAR DISTANCES

As described in Chapter 17, Maslov (1955) made test measurements of a series of small parcels using both the method of squares and a transect method to determine their areas. The mean square error recorded in Table 18.6, page 410 for the method of squares corresponds within $\pm 0{\cdot}002$ cm of the corresponding linear measurements. In short, he found virtually no difference between the methods and considered that the expression (18.19)

$$s_p = 0{\cdot}03\sqrt{A}$$

represented the relationship between the mean square error and parcel area.

Aldred determined the standard errors obtained using the three methods of area measurement employed by him. These are given in Table 20.3 and the results are also illustrated by Fig. 20.5.

The percentage columns of Table 20.7 suggest that the transect area-meter was approximately 3 times more precise than the dot grid, and 1·5 times more precise than the MK Area Calculator for the block containing small parcels. In the block having larger parcels the area-meter was about 1·3 times more precise than the dot grid and 2 times more precise than the MK Area Calculator.

Aldred argues that the greater precision of the transect method compared with point-counting is that a dot is a discrete point which is used to sample

TABLE 20.7. *Standard error of the estimate of the methods of measurement made on two test blocks by Aldred (1964), using the results obtained by polar planimeter as standard*

	Block 1 (Small parcels)			Block 2 (Large parcels)		
Average parcel area:	11 ha			68 ha		
Standard error of estimate in absolute and percentage terms						
	Area		Percentage of area	Area		Percentage of area
Method	(ha)	(cm^2)		(ha)	(cm^2)	
Dot grid	±3·84	±0·086	±15·26	±2·42	±0·061	±2·13
Transect meter	±1·09	±0·027	±6·57	±2·64	±0·066	±1·66
MK Area Calculator	±1·41	±0·035	±8·49	±2·80	±0·070	±3·57

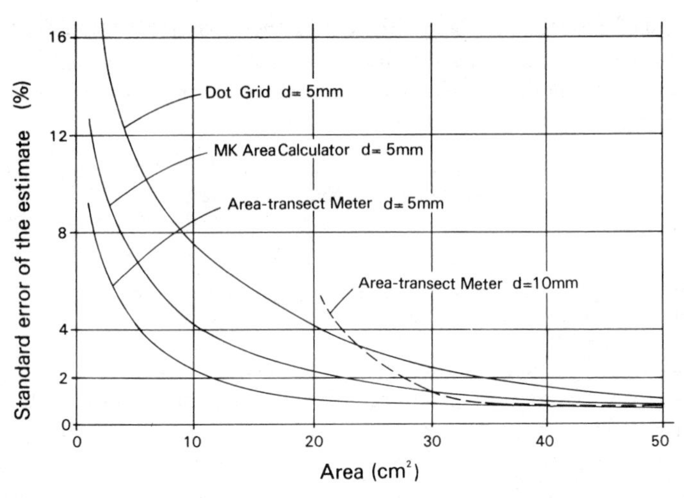

FIG. 20.6. The relative precision of transect sample measurements. (Source: Aldred, 1964).

both dimensions of the map. A transect, on the other hand, provides a complete measurement in one direction but is a sample in the other.

In addition to the sampling errors, which become proportionally less as the parcel size increases, there are instrument and operator errors to be considered. Assuming that some kind of calibration had been attempted, such as carefully checking the spacing of equidistant transect lines on the overlay, any systematic component may be removed so that the principal remaining source of error is the accuracy with which the scale may be read. We have seen that a standard error of $\pm 0{\cdot}08$ mm was employed by Maslov (1959), and this is consistent with most other work on the subject. However, Aldred states that his transect area-meter can only be read to $\pm 0{\cdot}05$ in., corresponding to $\pm 1{\cdot}3$ mm, or 16 times larger than the resolution of that to be expected with a simple scale. At the scale of 1/15,840 used for North American forestry maps this corresponds to $\pm 0{\cdot}4$ acres (or $\pm 0{\cdot}16$ ha). Operator error, on the other hand, was more difficult to define because it varied according to the skill and experience of the user. A test was made on several areas to find out how consistently an operator could remeasure an area. Aldred does not describe this work in detail, but claims that an operator could measure an area within $\pm 0{\cdot}1$ in.^{2} ($0{\cdot}65$ cm^2) 95% of the time. The variation was somewhat less for smaller areas.

The Influence of Sampling Intensity

Notwithstanding the conclusions of Hilliard and Cahn (1961), whose work in theoretical metallurgy was briefly mentioned on page 423, and who demonstrated that their novel views about intensity of point sampling applied equally to linear or Rosiwall sampling, we still expect greater precision of measurement to be obtained from using a systematic dot grid which is finely divided than from a coarse grid. Presumably the same rule applies in linear sampling so that we may expect more precise results from measuring a larger number of lines located closer together. This increases the time needed to make the measurements, so that, conversely, the increased speed resulting from wider spacing of dots or transects is gained at the expense of precision. For a given spacing larger areas were measured more precisely than smaller areas by dots and transects, because of the greater number of sampling units. Thus on large areas the spacing of dots or transects can be greater than on small areas for equal precision. This is, of course, a repetition of the conclusion reached by Frolov and Maling (1968) described on page 409. The problem is to define how much spacing can be increased or decreased. Aldred has argued that

> Theoretically, precision of measurement, as influenced by sampling error, is inversely proportional to the square of the transect interval or spacing between dots. Thus, if transect of dot spacing was doubled, precision declines four times; if spacing is quadrupled, the precision declines sixteen times.

As an example, let us assume that an area has to be measured to a precision of ±4·0% two-thirds of the time, with a transect spacing density of one line every 5 mm. Consulting Fig. 20.5, areas less than about 14 ha would not meet the standard. This is where the area-meter precision curve for 5 mm transect spacing crosses the horizontal ±4·0% line. Likewise, if the interval is doubled to 10 mm, the minimum area that could be measured is $2^2 \times 14$ or 56 ha to remain within the specified accuracy limit. This procedure could be used to define minimum area measurable for any transect interval and standard of accuracy.

The Rotational and Parallel Traverse Method in Sampling

Latham (1963) describes the *rotated parallel traverse method* and its operation. A transect line lying, for example, in a north–south direction on the map is systematically changed through successive bearings, such as 30°, 60°, 90°, 120° and 150°, and the measurement procedure is repeated for that line for each angular setting. This rotational method may be contrasted with the parallel traverse method in which a similar increase in the sampling intensity is achieved by increasing the number of lines measured. The latter technique was described in some detail by Osborne (1942). Figure 20.6 illustrates the difference between the two procedures. Differences between the two methods were tested over six comparable levels of sampling intensity. Sampling intensity is here expressed in "miles of traverse per square mile of survey area", so that for both methods tests were carried out by manual traverses at intensities in the range 0·6–3·7 km per 1 km^2.

Haggett and Board (1964) repeated Latham's experiments using a sample of 72 quadrats of approximate area 10 km^2 each taken from One-inch to One-

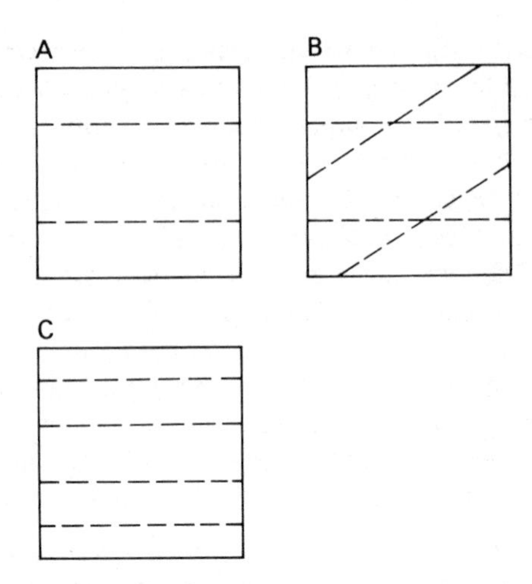

FIG. 20.6. Rotational and parallel sampling systems. Assuming that both systems being with (A) the sample lines running horizontally across the area, (B) the rotational system is intensified by rotating the lines; (C) the parallel system is intensified by increasing the number of lines all running in the same direction. (Source: Haggett and Board, 1964).

mile maps of southeast England. In each quadrat the area of woodland was measured by the indirect method. The accuracy of the two sampling methods at their various intensities was measured by comparing the estimated value given by the transect measurement with the "true" extent of woodland determined by planimeter. Table 20.7 provides an overall comparison for all 72 quadrats. The difference of 1·2% between the two values for mean accuracy is significant at the 95% confidence level.

Haggett and Board further demonstrated that this difference in accuracy is maintained at all the levels of sampling intensity tested. Figure 20.8 shows this association with the appropriate exponential regression lines for the two methods. Not only is the difference between the two methods clear, but the rapid increase in accuracy with higher levels of sampling intensity is shown. At the highest level of sampling intensity tested (3·7 km/km^2) the accuracy of the parallel method was nearly four times as great as that of the rotational method:

TABLE 20.8. *Difference in mean accuracy of parallel and rotational transects, as determined by Haggett and Board (1964)*

Sampling method	No. of quadrats	Mean accuracy compared with planimeter (%)	Standard deviation (%)
Rotational transect	72	6·1	±2·9
Parallel transect	72	4·9	±2·3

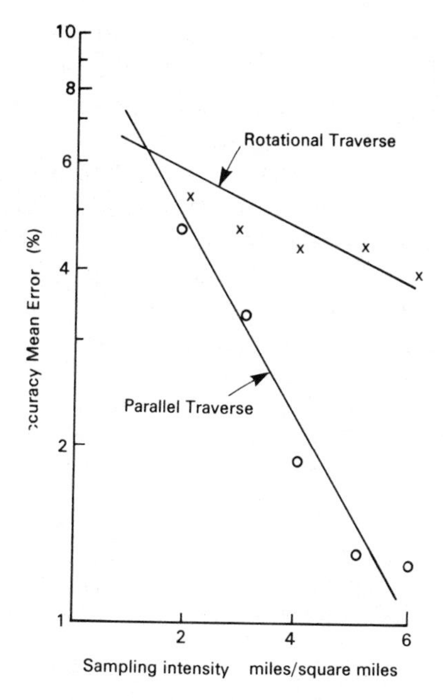

FIG. 20.7. Relationship of the accuracy of rotational and parallel sampling systems to the level of sampling intensity. (Source: Haggett and Board, 1964).

1·2% compared to 4·2% mean error. The implications of Table 20.8 and Fig. 20.7 are clear. Rotational transects give a generally less accurate estimate of area, and this difference appears to increase with intensity over the ranges tested.

The Effects of Other Characteristics

Haggett and Board also studied the possible effects of other characteristics of the distributions to be measured. These include:

- *Coverage*: the total area covered by woodland expressed as a percentage of total quadrat area.
- *Size*: the average area of each parcel of woodland expressed by the coverage value, divided by the number of such parcels.
- *Shape*: an estimate of fragment shape derived by dividing the coverage value by the number of intercepts made by the sampling transects in crossing and recrossing the parcel boundaries.
- *Orientation*: a measure of the orientation of the pattern of parcels derived by comparing the differences between the highest and lowest values obtained in the six rotational transect positions.

The differences between percentage deviation in measured area and sampling intensity were investigated for each of these pattern characteristics through regression analysis. In general the differences varied little in amount, and in no case exceeded 1%. Results obtained from these simulated traverses confirm Latham's conclusions about the general utility to be expected from rapid scanning of geographical patterns. Haggett and Board differ only in suggesting that the rotational traverse system advocated by Latham is less efficient than the conventional parallel scanning for the measurement of area. Where orientation is an important characteristic of the geographical pattern there is clearly no substitute for rotational traverses, but where this is not under separate investigation the simple parallel traverse appears to offer a more efficient alternative.

21

Cartometry and the Digital Computer

A student seeking training in cartography is opting for early obsolescence if he or she does not learn the basis of computer-assisted cartography..... One who specializes in current manual techniques will not be employable by the 1990's in the federal mapping agencies.

(Joel Morrison, *Computer Technology and Cartographic Change*, 1980)

The term *digital mapping* is often used to describe the processes involved in the production of maps by computer. It embraces conversion of cartographic or photographic images into data which can be processed by digital computer and thereby converted into a variety of different kinds of graphic output. Moreover, the subject is not confined to the output of paper maps or even the display of a map on the screen of a monitor or video display unit (VDU), for the stored spatial data may be used for many applications without having to go through the intermediate steps of producing a map. Cartometry is a major application of this sort. Indeed, Yoeli (1975) has referred to the solution of spatial problems by digital computer as "cartometrically oriented cartography" and "non-map cartography". Provided we have methods of accessing suitable spatial data, measurements of distance, area, angle or number may be obtained without having to use a map. Because microcomputers with quite modest specifications can handle the data, and because such microcomputers are now so readily available, it might be argued that digital methods of cartometry ought to supersede the conventional methods. For example, the BBC Advanced Interactive Video (AIV) system containing the geographical information system for Britain created for the 1985 Domesday Project can be used to measure distance and areas upon maps displayed upon the monitor screen. However, we must temper such enthusiasm for change with many qualifications about the availability, suitability and, above all, the cost of the data and the hardware needed to replace the paper map and conventional cartometric instruments.

The stages of digital mapping and digital cartometry are much the same as those for any other kind of automatic data processing. They may be listed as:

- digital data capture
- data storage
- data processing
- display of results.

Different degrees of emphasis are needed for different tasks. For example, in digital mapping the output, or display, stage is of primary importance, but in cartometric work it is only a minor consideration, for often the only required output is the single statement that a line is *L* kilometres long or that the area of a particular parcel measures *A* hectares.

In this chapter, much space is devoted to the subject of data capture in general, and vector digitising in particular, for this stage of digital cartometry is equivalent to the manipulation of dividers, chartwheel or planimeter by the classical methods. However, there is an important difference. Although we may assume that the user, who has access to the necessary hardware, but who only needs the data to make some cartometric measurements, may make the necessary measurements by digitising the coordinates of points along a line, it is possible that the necessary digital data have already been obtained for some other purpose and are already available on tape or disc. This means that the user may not have to make the measurements specially, but may rely upon existing data files. The case for using these is usually strong. Just as comparatively few of the scientists who use aerial photographs have access to a stereoplotter to make measurements on them, so unrestricted access to a digitising table of cartographic format and precision is a considerable luxury to the occasional user. On these grounds we might argue that "do-it-yourself" digitising for cartometric purposes is still primarily a research tool to investigate the potential of the methods, as has been done by Baugh and Boreham (1977) and by Maling, in Kretschmer (1977), for cartometry, or it has been directed towards other research interests such as the work of McMaster (1986) and Muller (1986, 1987) on cartographic generalisation. In many applications of digital mapping and cartometry we have to use extant files because we cannot collect the data ourselves. This is obviously true of most remote sensing applications, but in the final analysis it approximates to the statement that we do not normally have to produce the maps on which we intend to make the measurements. It is therefore reasonable to suppose that, if a digital database is available, it is far more convenient to use this rather than attempt to repeat the slow, exacting and error-prone operations of digitising part of a map and creating a new file. However, such a conclusion depends to a considerable extent on what meaning we read into the key work *available*.

A digital data file must be available in the obvious and simple sense that somebody has already collected the data. It must be available in a form which is compatible with the user's own hardware. It must be available at a cost which is acceptable to the user. If these desiderata cannot be fulfilled it is necessary either to employ "do-it-yourself" methods, or even abandon the intention of using digital methods and revert to the classical methods of cartometry. Sometimes these will provide better results. We shall refer to examples of digital databases which could be used for cartometric work but, because of their cost and hardware requirements, or the scale at which the original digitising was done, or the nature of the cartometric algorithm, digital cartometry offers little advantage to the user who can obtain better results using paper maps at more suitable scales with dividers, chartwheel and planimeter.

DIGITAL DATA CAPTURE

This is the process of collecting the original information, which are spatial data, into *machine-readable form.* It may be done manually, automatically or interactively, and in each case the process may be called *digitising.*

Manual digitising and some interactive digitising is carried out in the *vector mode.* This defines geographical entities by means of the coordinates of their characteristic points. For example the corners of a rectangular field or building may be digitised by moving the cursor to each of the four corner points and recording their coordinates. If the sides are straight this information is sufficient to describe the outline of the feature. If the boundaries are not straight additional points must also be measured, as many as may be needed for the linear elements joining successive points to approximate to an irregular feature such as a river bank or coastline. The result of digitising an irregular line on a map is a stream of coordinates which have been collected in a sequential string.

Reference has already been made in Chapter 3 (p. 40) to the possibility of operating in either of two modes. The first is the point mode, where each coordinate pair has to be recorded separately as the result of a decision by the operator who presses a switch to make each record. The second is the stream mode, in which the coordinate reading and recording process is controlled by means of a time switch or intervalometer which is set to read the coordinates of the cursor at a fixed interval in time. The processes are aptly called *line-*

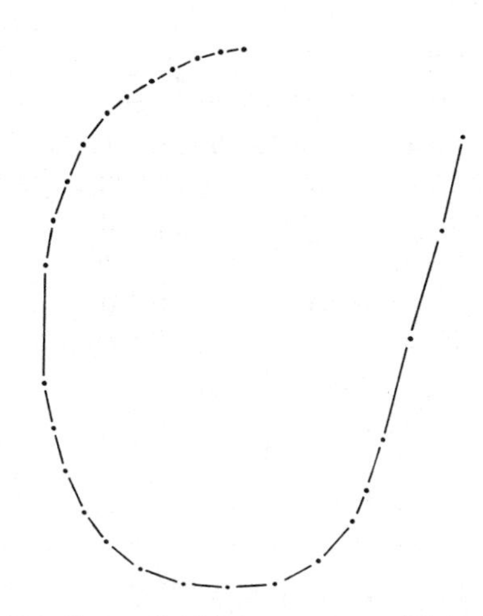

FIG. 21.1. Vector digitising as a means of data capture.

following, although in practice the term is often applied to all the manual digitising methods.

Automatic digitising is normally carried out in the *raster mode*, using grid cells for the definition of map detail. In the simplest case we imagine that a very finely ruled grid has been superimposed over the map. As the grid is scanned each square is allocated either the number 1 if it is occupied by an element of the map detail of interest to us, or 0 if it is not. The individual grid cell is known as a pixel, irrespective of whether this is the unit scanned from a map or from the ground by some kind of remote sensing device. The resulting information is held in store in an array or matrix whose lines and columns correspond to the grid. This is a faster method of data capture than vector digitising. Moreover,

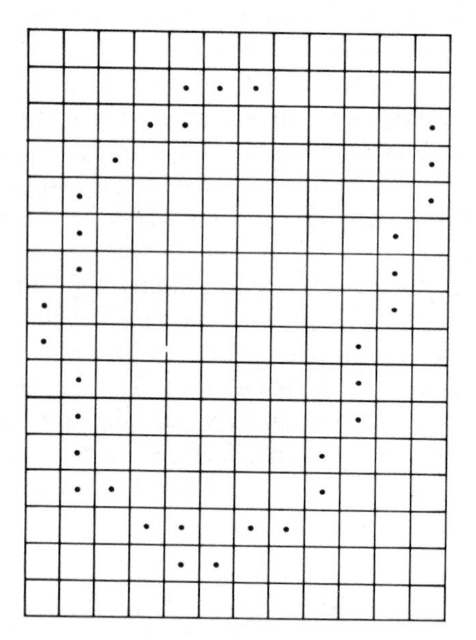

FIG. 21.2. Raster scanning as a method of data capture.

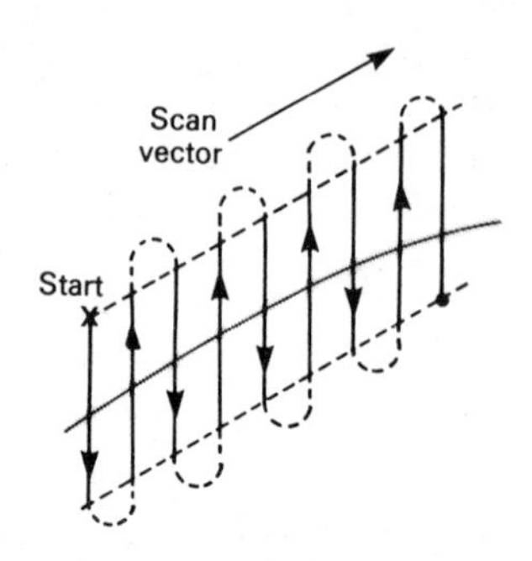

FIG. 21.3. Automatic line following as a method of data capture.

the process is entirely automatic, whereas vector digitising involves a considerable amount of careful manual work which is equivalent to draughting. An alternative to these methods is *interactive scanning*, which has also been called *automatic line-following*. The hardware required for this is a programmable scanner that can access the source in arbitrary patterns, and a means for the operator to follow and supervise this process. Figure 21.3 illustrates that the scanner operates at right angles to the line which is being followed.

Hardware Requirements

The principal hardware used for vector digitising is a table of the kind illustrated in Fig. 3.10, p. 40. The hardware requirements for automatic scanning are usually simple and costs are rapidly falling, since developments are underwritten by a market much greater than cartography. The instruments which are available either scan the document in a series of parallel lines to build up the matrix of pixels, or work like a television camera. The close relationship to television is further reinforced by the importance of this mode of data capture in remote sensing applications, particularly those images from unmanned satellites. For example, the data transmitted to earth from the multispectral scanning sensors on the Landsat satellites have been collected in the raster mode.

Early in the history of digital mapping it was realised that the high contrast of a conventional fair drawing in positive or negative form ought to make possible automatic line following using some kind of photoelectric cell device to follow the linework. The most important system which is currently available for interactive digitising is the Laserscan Fastrak, which uses a laser and tilting mirrors to achieve the result.

Data Handling

Hardware allows only serial or sequential storage of data. This is a handicap for later access for, in order to recover a particular piece of information from the file depicting a particular map, it may be necessary to copy, read and discard the mass of data which relates to features other than that required. This is a special disadvantage of searching for data stored on magnetic tape, for this may have to be read throughout in order to locate a particular item stored in the last inch of tape. Special hardware and software features such as random access to discs have been developed to increase speed of access by creating special directories, which locate where blocks of data have been stored on the tape or disc, but ultimately the directories themselves can only be searched sequentially.

In manual digitising the data collected may be limited to recording only the positions of points on the line to be measured. Interactive scanning also captures data selectively. In raster scanning, however, the whole source document must be scanned and processed, even if only a small amount of information is actually required. It follows that the data in raster form occupy significantly more storage space than vector data, and more time is also needed to load these data on line. Techniques exist for reducing the amount of data stored where, for example, there are large blank areas on a map, but inevitably the storage requirements for raster map data are much greater than for vector coordinates. Dale and DeSimone (1986) make the following comparison:

> Consider, for instance, a 1:2,500 scale map sheet covering an area of 1 km^2. If the resolution of the data is to be 0·1 m or one part in 10,000 of the area side lengths, there will be 100 million pixels each with a value of 1 or 0 depending on whether a line does or does not pass through that point. Given eight bits to a byte, that will require 12·5 million (or mega) bytes. Using efficient storage techniques, this requirement can be reduced by a factor of ten or more, depending upon the complexity of the map, but the amount of data to be stored still remains high. On the other hand it is possible that the data could be stored in less than 10,000 coordinate pairs and with five bytes per pair, this would require only 50,000 bytes. Although the storage of data is becoming easier and cheaper, the handling of the data still causes problems.

It seems, therefore, that each method of data capture has certain merits, which are usually a disadvantage of the other. Vector digitising involves low capital cost and is the better method if source documents require human interpretation. Slow speed and the difficulty of maintaining accuracy of output remain as fundamental disadvantages.

Raster digitising, on the other hand, has the advantage of capturing data very rapidly, for it requires only a few seconds to scan an entire map. Unfortunately, however, it also produces data for every speck of dust and blemish on the image to be digitised. Unless source documents are of a very high quality it is necessary to prepare the source document graphically before scanning, thereby defeating most of the advantages of speed in data capture. The grid format makes for straightforward manipulation of the data to the extent that many operations such as windowing can be hardware-based. Raster-digitised data are easily integrated with data from other sources such as satellite imagery, and the data records themselves can be stored in a straightforward way using direct access filestore.

It might appear that interactive digitising should possess most of the advantages with few of the disadvantages of the first two methods. Because the interactive operator can resolve such difficulties as they occur, and because high scanning resolutions can be used, interactive scanning can be less susceptible to the effects of defects or noise in the source document than automatic scanning. However, a major problem of interactive digitising is that the instrument requires operator guidance when completing one feature to start the next, or when it encounters a junction. For example, the basic scale Ordnance Survey maps contain innumerable junctions and short lines. This means that full-time operator attention is needed, which tends to slow down the work. In practice, therefore, interactive digitising is only marginally faster to use for large-scale work than is manual vector digitising. Moreover, the identification and coding of each feature still has to be done at a later stage. The advantages of automatic line following only become apparent in smaller-scale work where long and continuous lines, such as contours, need to be digitised.

Because no single method of data capture is wholly satisfactory for all cartographic purposes, it is clear that the ability to encode vector mode data into raster format or vice-versa would be valuable in any automated mapping system. In digital cartometry it allows for greater flexibility in the choice of method used to determine distance or area. Vector to raster conversions are in common use. For example, most digital plotters perform such a transformation when drawing vector data, but, as both Devereux (1986) and Dale and DeSimone (1986) have shown, the reverse operation of raster to vector encoding is far less straightforward.

This conclusion influences the techniques of digital cartometry. In vector mode it is sufficient to make use of coordinate difference formulae to obtain the appropriate measurements. For example, distance is measured by calculating the length of each straight line element joining adjacent digitised points by means of equation (3.5) and determining the length of the whole line from the sum of all such elements. The area of a parcel may be determined from the coordinates of points around its perimeter by any of the formulae (16.39), (16.40) and (16.45) which were originally described for the purpose of calculating the areas of simple geometrical parcels, but are equally appropriate for more complex figures. Cartometric techniques using raster data involve pixel counting such as has been described in Chapters 18 and 19 for measurement of area, or by counting boundary pixels as is described later in this chapter.

DATA STORAGE

Anyone who has worked in the field of digital mapping has soon become impressed by the fact that the map is an extremely compact way of storing a huge amount of spatial information, which has been difficult to match by means of any form of digital storage. Nevertheless, our ideas about data storage requirements have undergone a profound change during the past 20 years. At the beginning of this period, Howard (1968) made a statistical study of the information content of the Ordnance Survey One-inch to One-mile maps. She came to the conclusion that it would require about 1,300 reels of tape to accommodate all of the information contained on the 190 sheets of this series in digital form. Moreover, if the data were digitised at the scale of six inches to one mile the amount of store needed would rise to about 5,000 reels of tape. With improved hardware and a greater understanding of how to store the data in a more compact form this estimate has been much reduced. By 1971 the Experimental Cartography Unit and the Ordnance Survey had succeeded in producing a map of 100 km^2 of country in the vicinity of Bideford in Devon.

This involved digitising the map detail at 1/2,500 scale, which is the basic scale for the area, with contours extracted from the 1/10,000 scale maps, this being the largest scale on which contours are shown. According to Bickmore (1971), the data collected amounted to 1·25 million coordinate pairs, or about one-half reel of tape. This corresponds to about 15,000 coordinate pairs per square kilometre, or for England, Wales and Scotland about $3{\cdot}5 \times 10^9$ coordinate pairs on about 1,400 magnetic tapes. A decade later, Adams and Rhind (1981) estimated that it would need only 251 computer-compatible tapes to hold the digital database for the main urban areas, derived from 53,316 sheets of the 1/1,250 OS series and 331 computer-compatible tapes to hold the data digitised from the 164,461 sheets of the 1/2,500 basic mapping of Britain. At the smaller scales the storage requirements are much greater. Thus Rhind, in Visvalingam (1987) has noted that the 1/50,000 scale Girvan sheet (which is partly sea) occupies about 35 megabytes, and this implies that, even after data compression, the total storage required will be of the order of 3 to 5 gigabytes (i.e. 10^9 bytes). In these circumstances, single-feature national data sets may become particularly important and the contours, which amount to about half of the total line length, may be less useful than a regular grid of ground altitudes.

There is an unattributable statement (alleged to have been made by an American systems engineer) that if computers had been invented before photography we would now be looking upon photography as the great new way of storing information. The major breakthrough in storage of spatial data in the form of pictorial images did not come until the late 1970s when the first applications of video disc technology were tied to digital mapping. As early as 1980, Boyle hinted that this might well be the most likely direction for future development.

The video disc provides the advantage of holding large numbers of maps in the computer without overwhelming its storage capacity or without the substantial costs of manually digitising them. The contents of about 1,000 topographical maps may be stored on a single disc, so that, for example, the entire 1/24,000 U.S.G.S. basic mapping of the U.S.A., which exceeds 50,000 separate sheets, may be stored on only 50 discs. Playback units are simple and cheap, and can be fitted to a microcomputer. At present the most interesting and accessible application of this system of storage available in Britain is used in the BBC AIV system where the immense *Geographical Information System* created as the *BBC Domesday Project* and forming an archive for Great Britain in 1985. This is stored upon two compact discs, each side of which has a capacity of nearly 400 megabytes of digital data together with a further 54,000 analogue video frames.

However, a present disadvantage of this method of storage is its relatively poor resolution. Development work in this field indicates that approximately four-fold linear enlargement of the image is required for adequate image clarity when a map is viewed on a monitor screen. Thus, to show legible maps on the VDU, an area enlarged approximately four times has to be displayed. In the example of 1/50,000 scale maps this means that each picture can only measure 4 × 3 km, and it is shown at approximately 1/12,500 scale. This is the size of the module which has been used to store a great deal of the Domesday Project material. We shall see later that this creates problems in measuring features which do not lie conveniently within the single 4 × 3 km block.

AVAILABILITY OF DATA

In the early days of digital mapping, when it seemed that all things might be possible, much was written about the creation of *cartographic databanks* which would contain the digitised information of all the basic mapping of the continents, or even the whole world. This grand design was based upon the premise that since all maps are ultimately derived from the basic scales, it would be possible to create any maps of any country simply by accessing the appropriate files. It follows, therefore, that such a database would not have been affected by generalisation and the consequent loss in information content on the smaller-scale sources. In Ordnance Survey (1979), the term *scale-free database* was used to describe the concept.

When the truth dawned that a very long time would be needed to capture existing cartographic data, and that it would require a gigantic amount of storage, many of these ideas had to be shelved in favour of more modest aims. However, the importance of such national archives cannot be underestimated, and the evidence submitted to the OS Review Committee repeatedly stressed the need to persevere in its production. On the strength of this the Review Committee itself recommended that the OS should undertake, as core activity, the investigatory phase envisaged in a development strategy for a national topographical digital database.

Any question concerning the availability of digital data in the form of a scale-free database must take into consideration the fact that only intermittent cover will be available for many years to come. Proceeding from the estimate that it needs 13 man-hours to digitise one 1/1,250 scale map, evidently it will require 78 man-years to complete this work for all 53,000 maps at this scale, and produce a database for only the major towns and cities which are mapped at this scale. In order to produce the scale-free database for the whole of Britain involves the digitising of more than 220,000 maps. Table 21.1 indicates the state of affairs in the autumn of 1981. The current estimate by Proctor (1986) for completion of this undertaking is the year 2015. By the mid-1980s the principal effort has been the provision of blocks of continuous digital cover for certain limited areas, notably in Greater London and the West Midlands. Figure 21.4 shows the extent of the *designated digital areas*, which are the parts of Britain which will be completed first.

The urgent need for some kind of database derived from small-scale mapping has been clearly stated by different organisations. We quote here the

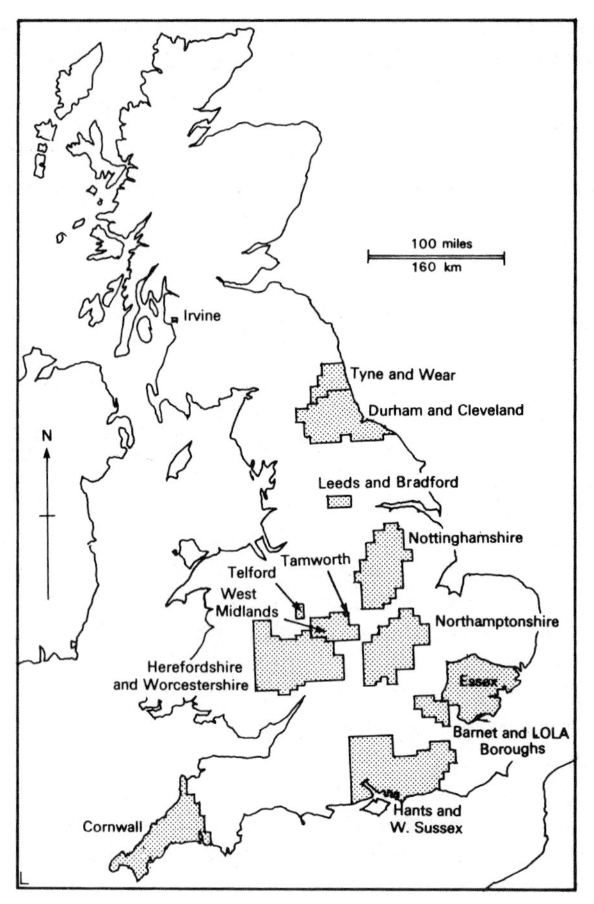

FIG. 21.4. The parts of the United Kingdom for which scale-free digitised data are intended. This map does not indicate the actual availability of digital tapes. The area covered in 1985 is substantially smaller.

views of the Institute of Terrestrial Ecology, as described by Bickmore and Witty (1980):

> We readily concede, of course, that it would be nice to have more detail than can be obtained at the 1:250,000 scale, but we do not want a situation in which the time for completion of the survey disappears over the horizon – excellent though the few maps produced may be. We would prefer a complete coverage with a limited amount of detail and precision.
>
> Our need for this digitisation is becoming urgent...

The early experiments by the NERC Experimental Cartography Unit culminated in 1970 in the Bideford experiment already mentioned. As a result of the case presented by the Institute of Terrestrial Ecology a small-scale geographical information system, known as *Ecobase Britain*, was based upon the detail digitised from OS maps by the Experimental Cartography Unit. The

TABLE 21.1. *The number of digital maps produced, and the time required to prepare both conventional and digital large-scale maps at the OS in 1981 (data from Fraser, 1982)*

			Time needed for preparation in man-hours	
Scale and format	Planned in series	Already available by digital methods	Conventional methods	Digital methods
1/1,250 ($\frac{1}{2} \times \frac{1}{2}$ km)	54,365	7,150	88	102
1/2,500 (1 × 1 km)	158,020	10,340	–	–
(2 × 1 km)	–	–	123	207
1/10,000 (Basic) (5 × 5 km)	3,659	3	–	–
1/10,000 (Derived) (5 × 5 km)	6,522	40	598	284
Total	222,566	17,533 (7·9%)		

greater part of Scotland was done at 1/50,000; the remainder of Britain at 1/250,000. Another important development was the substantial amount of digitising of contours and water features on the OS 1/50,000 scale maps of Britain carried out for the Directorate of Military Survey. From these, contractors have derived altitude grids or digital elevation models. Files for about half of the 204 map sheets covering Britain had been produced by the end of 1986. Ordnance Survey have agreed to act as marketing agents for these data.

In the meantime OS have concentrated almost entirely on the conversion of their large-scale maps into digital form so that only four "small-scale" maps are known to have been digitised – the coastline of Great Britain, some 1/50,000 scale mapping in the Dunfermline area for Rural Land-Use Information System project, Sheet 76 (Girvan) of the standard 1/50,000 scale maps and the 1/625,000 Route Planning Map.

In the terminology adopted by the Ordnance Survey Review Committee, a *scale-variable database* signifies that it has been prepared at a particular (and smaller) scale, usually for a specific purpose, and that it can only be used for mapping at smaller scales because of the known generalisation of the sources.

Since 1978, when the Royal Society and other organisations pressed the OS Review Committee to recommend that OS should digitise smaller-scale maps, there has been considerable external encouragement to OS to diversify their activities. Both the Review Committee Report and that more recently produced by the House of Lords Select Committee on Science and Technology accepted the case for providing such 'small-scale" data and urged governments and OS to take action. Mayes (1987) has examined much of the subsequent events from the OS point of view, especially their attempts to define the market for these data.

Because of the important place occupied by the 1/50,000 OS maps in many scientific studies, and the intention of the OS to meet this need, there is risk of some overlap of interest and therefore duplication of the work of data capture at the smaller scales, and this at a time when many users are waiting patiently for the scale-free database.

Provision of such services has to be paid for, but it is only fair to remark that the prices which have been proposed by the Ordnance Survey are somewhat unrealistic, especially to those working in branches of the environmental sciences which are traditionally poorly funded. Thus, when the OS circulated a questionnaire to potential users in 1984, it was suggested that the cost of the database in vector form would be £25,000 per sheet. Mayes (1987) has summarised some of the responses to this questionnaire, but we may leave the subject of costing to the diplomatically moderate response by Rhind.

> Whatever the final details, the demand for such data amongst scientific users seems high though their capacity to pay high prices is limited. It is to be hoped that OS and the appropriate scientific and university agencies can come to an early arrangement over the availability of small scale OS data for research and techning purposes.

By 1986 the Ordnance Survey database at 1/625,000 had been prepared, and this is now available as the following two datasets:

Dataset 1 – Coastline, rivers and canals, lakes and reservoirs, county boundaries and district boundaries. The cost (1986) is £2,500 + V.A.T.

Dataset 2 – Cities, towns and villages, built-up areas, roads. The cost (1986) is £5,000 + V.A.T.

There is, however, some uncertainty about the real value of a database at this scale, apart from the fact that this was easier to produce than anything at a larger scale and that it has national extent. Rhind, in Visvalingam (1987) has made the point that larger-scale datasets covering the whole country are an essential preliminary to testing such systems because the majority of those two wish to make use of them do not necessarily want to study distributions around Girvan. Nevertheless this opinion may not be widely held. An interesting response to the OS questionnaire, which is analysed in Mayes (1987), is that from a sample of 432 respondents, only 2% indicated a first-choice preference for a database at 1/625,000 compared with 43% for a dataset at 1/10,000, and 33% for the scale 1/50,000. Moreover, the present author finds that from the purely cartometric point of view there is really very little justification for the use of such a small-scale database. Enough has been said elsewhere in this book about the influence of map scale and the resulting generalisation at smaller scales upon measurement of distance to throw doubts upon measurements made at such a small scale. The reader can judge from the results presented in Chapter 4 whether measurement of the Yorkshire coast at 1/625,000 alone would provide an adequate measure of its length. However, we did see that this map scale is used to check the travelling expenses of Members of Parliament, so possibly there is a case for holding a copy of Dataset-2 in the Palace of Westminster.

Similarly the 1/625,000 maps series is too generalised to be of much use in area measurement. The only boundaries shown on this map are those for counties and districts, and the areas of these are readily accessible from other sources without the labour or expense of having to measure them independently. There remains the possibility of using these datasets as the bases for the production of other types of distribution map, for example maps of soil types, land-use, agricultural classification, river catchment basins and the like. Indeed some of these possibilities have already been published as paper maps at 1/625,000 scale. If additional software is available to create such distribution maps superimposed upon the topographical base provided by the OS datasets, then new boundaries are created and it is possible to measure the areas of the parcels thus defined. Since *Ecobase Britain* already exists at a larger scale, surely this would be the preferable base to use for such purposes.

A more sophisticated concept than the database or dataset is the *Geographical Information System* (*GIS*) which contains much more varied data than is normally found in the simple database of mapped information. For example, the GIS system currently being investigated and developed by the U.S. Geological Survey is intended to contain contributions from many different branches of the federal and state administration. Indeed the U.S.G.S./Connecticut G.I.S. Project, described by Nystrom (1986), combines data from 28 separate federal and state sources. The National Digital Cartographic Database, which provides the topographical component, depends upon digitising the basic 1/24,000 scale mapping. This is now being implemented through the system called MARK II (Starr, 1986). The production objectives are to complete digitising of the 1/100,000 scale database (1,800 maps), together with a proportion of the 1/24,000 basic mapping by 1990, attaining completion of this phase of data collection by the end of the century.

CARTOMETRY IN THE VECTOR MODE

In this section it is assumed that the user has access to the necessary hardware and therefore carries out the necessary digitising. Two specific examples of distance measurement are considered; that by the present author for the Yorkshire coast and that by Baugh and Boreham (1976) to determine the length of the mainland coast of Scotland.

Vector Digitising: Hardware and Operating Considerations

Most electromagnetic digitising tables correspond to the type of line follower described briefly in Chapter 3 and illustrated by Fig. 3.10, p. 40. Such instruments operate by precise timing of the delay between generating and receiving an electromagnetic impulse from a coil in the cursor by means of wires embedded beneath the table surface. During normal digitising the cursor coil is positioned by relating a graticule in the cursor to the map detail required.

In addition to this type of instrument there are other kinds of vector digitiser which are essentially based upon existing instruments for measuring or draughting, such as coordinatographs fitted with suitable encoders to convert the analogue (scale) readings into digital values. Some of these instruments are not really satisfactory to use for every kind of cartometric measurement, usually because they are awkward to operate and therefore tiring to use for prolonged periods. However, most of them can be pressed into service to do this work. Some instruments do not have the operating range for work on maps. We have already noted that the kind of planimeter provided by the instruments used in stereology, such as Quantimat, have been designed for use with photomicrographs and cannot accommodate the larger formats of maps (or even, in some instruments, the larger format of aerial photographs).

The Precision of Digitisers

Rollin (1986) has provided a valuable summary of the methods of accuracy testing of digitisers now used by the Ordnance Survey. As is customary in testing any kind of coordinatograph, this is done by obtaining the readout for points on a specially constructed grid which is more accurate than the reading resolution of the instrument to be tested. Testing the table is carried out by making a succession of pointings to the grid intersections. In OS practice four complete and independent measurements are made in one session by the same operator of all the points on the test grid.

The sets of coordinates are first checked for consistency by determining the mean and deviation of each pointing. This should not be greater than a tolerance of 0·127 mm (0·005 in.). Larger deviations are examined to determine whether they are the result of operator error, etc., or due to a more serious problem with the table. It is noteworthy that only one of the tables used for production work at the Ordnance Survey has been found to have seriously deteriorated since digitisers were first introduced in 1972.

Although the Freescan digitiser reads coordinates with a resolution of 0·001 mm, it is not possible for the operator to place the cursor over a point on a map with this precision. Experiments carried out by the author, using different operators to set the cursor over different kinds of symbol and on different parts of the table, suggest that in the point mode of operation a well-defined point such as a grid or graticule intersection can be determined with a standard error of $\pm 0{\cdot}15$ mm. On the other hand the precision of location of the terminal points of a line yields standard errors of approximately $\pm 0{\cdot}3$ mm. The author found that the attainable precision depends upon the location of the line to be measured on the table with respect to the operator. Repeated pointing to the terminal points of a stretch of coastline varied from $\pm 0{\cdot}25$ mm

TABLE 21.2. *The effect of introducing perturbations on digitised coordinates (from Baugh and Boreham, 1976)*

Standard deviation of perturbations metres/	2	4	6	8	10	20	30	40
millimetres	0·03	0·06	0·09	0·1	0·2	0·3	0·5	0·6
Variance of perturbation (m^2)	4	16	36	64	100	400	900	1600
Increase in apparent coastline length (km)	1·5	6·03	13·6	24·3	38·1	156·9	359·5	638·3
Increase in length variance $\times 10^5 km^{-1}$	3·77	3·77	3·78	3·79	3·81	3·92	3·99	3·98

at the end of the linear nearer the operator falling to ±0·53 mm at the end of the line situated near the working limit furthest from the operator.

Baugh and Boreham (1976) made their measurements on a d-mac table and claimed that the standard error of positioning of the pencil follower was ±0·1 mm, which is what the manufacturers state. They also suspected that random errors in digitising might have a systematic influence upon the calculated length, and therefore they tested this hypothesis as follows. They write:

> A computer program was written to simulate operator error by introducing random perturbations on each reading and to determine the apparent coastline length...
>
> The results illustrate that a random error in defining the coordinates introduces a systematic error in the length calculated from them; furthermore the increase is roughly proportional to the square of the standard error introduced (i.e. the variance).
>
> Therefore if a good estimate of the standard error of positioning can be made, the length can be corrected accordingly. This is not an easy exercise, as operator error is dependent on many factors and varies considerably. However, on the basis of control experiments the standard error of positioning was estimated to be 0·01 ± 0·05 mm (representing 6 ± 3 metres on the map) which is of the order of precision of the digitiser (which records to 0·1 mm).

The simulated operator errors calculated by Baugh and Boreham may be tabulated as shown in Table 21.2. Baugh and Boreham then consider an additional error which is the result of converting the individual digitised coordinates into their National Grid values before calculating the length of the coastline. This calculation should be simply an application of the *grid-on-grid transformation* which is used to transform from one Cartesian coordinate system into another. This calculation has been described in detail in Maling (1973). However, Baugh and Boreham appear to have used a somewhat less elegant method which, if anything, reduces the quality of their digitised result. In short, this error is wholly avoidable.

Practical Considerations in Digitising

Identification of Ground Features in a Data File

In addition to recording the coordinates of points, the digitising process must include a certain amount of additional data collection, or tagging so that the computer can be instructed to find the information relating to a particular feature, for example the high water line. At first sight it might appear that this should be done by manually inserting the additional information from a keyboard. However, the need for a keyboard, and the need to remember a series of code numbers, may be largely circumvented by using the *menu* technique. This was already being used for digital mapping in the late 1960s. Nowadays it has spread through interactive processing so much that it is quite difficult to find modern software for microcomputers which does not, at some stage, require the use of a menu to input a decision, preference or choice made by the operator. In digital mapping the rectangles or "boxes" representing the menu allow the operator to make distinctions between different ground features just as this is done by the legend of a map. This is a fast method of inputting a qualitative statement to the data file. Moreover, it is free from keying errors so that the operator may concentrate on line following without interruption.

Treatment of the Terminal Points of a Line

In order to measure distance by stream digitising at very small intervals of time it is desirable to investigate the manipulation errors which arise in the process. Experience in measuring the length of the Yorkshire coast has shown that the risk of introducing errors by accidentally straying away from the line during tracing is much less important than the creation of superfluous points beyond the terminal points, or failing to record any coordinates near them. These errors arise from pressing the key to start digitising too soon or too late. The discrepancy between the intended and actual terminal points appears to be greatest for digitising carried out with the timing device set to a very short interval in time. However, at very high speeds the size of the linear element derived from two pairs of successive coordinates is short. Consequently the following recommended procedure is partly cosmetic, but it is easy enough to introduce and it does have the additional advantage of providing a check against gross errors in setting the cursor over a line. The author recommends that measurement be carried out in two stages. First it is necessary to digitise the terminal points of the line using the point mode. This is repeated several times in order to obtain the mean coordinates for these points as well as a measure of pointing precision. Secondly the line itself is digitised several times in the stream mode. Finally the data file must be edited to ensure that the coordinate string begins and ends with the mean coordinate values measured at the terminal points.

Editing and Checking

Checking the result of digitising is extraordinarily simple if it is done visually, but the following procedure does require access to a graph plotter or similar peripheral to a mainframe computer. After completing a digitising job the file of output coordinates is used as data for a simple plotting program to copy the contents of the file just as it was collected. A comparison is made with the source map by placing the plot over the original map on a light table and examining the superimposed image for any departure from the digitised line. If the digitising has been done without error, the plotted line ought to correspond with the original. The fact that the plotted outline may have been produced on thin paper used with the graph plotter, and is therefore prone to changes in shape or size, is unimportant provided that some grid or graticule intersections have also been plotted so that the overlay may be fitted to the original with reference to this control, as described in Chapter 11. In any case, since this editorial check is to be made as soon as the graph plotter output is available, the amount of deformation of the plot is not as great as would occur after rolling up the copy and storing it for a prolonged period.

Blind Digitising and the Need for Supplementary Coordinate Displays

Some kind of visual display for the coordinates of the current position of the cursor is highly desirable, but not always provided. Systems engineers may argue that such information is not really necessary; that the information has been stored on tape or disc and that any corrections can easily be made at the initial editorial inspection of the data file. Such an argument indicates a touching degree of faith in the reliability of the hardware and also on-line access to the computer in order to identify a corrupt data file immediately. It is most undesirable to have to rely upon batch processing or any other kind of time-wasting procedure before verifying that the data are satisfactory. In the event of a malfunction it is essential that the operator should be aware of this at the first opportunity; otherwise time will be wasted digitising with faulty equipment and this unproductive work will continue until the malfunction has been recognised.

Experience of working with a Ferranti Freescan digitiser equipped with a visual display of the cursor coordinates has also shown that it can be a valuable adjunct for small jobs, for example to test a map sheet for paper deformation or to check the correctness of a map projection by measuring a few graticule intersections. Since the amount of output required is small, this can be read and booked manually and subsequently processed by pocket calculator. For this kind of job the results can be obtained far more quickly than the time needed to carry the magnetic tape over to the computer centre and submit it for batch processing, let alone wait until the results come back. In short, a digitiser with a visual coordinate readout may be used in the same way as an ordinary coordinatograph, but this is only possible if coordinates can be read directly when the cursor is set over a required point.

Measurement of Coastline Length by Vector Digitising

The Yorkshire Coast

For the Yorkshire coast measurements made by the author, the equipment used was a Ferranti Freescan digitiser with facilities for stream digitising output to magnetic tape and processed by ICL 1904S mainframe computer.

A large number of individual measurements were made of the Yorkshire coast on maps of scale 1/1,000,000 and 1/250,000, the results of which were given in Table 5.4, p. 69. Because the author was preoccupied with finding out what effect tracing speed had upon the results, the data comprised repeated measurements made with the intervalometer set to the same time, as well as measurements made at different settings of the intervalometer. The first of these tests is represented by a sample of 18 measurements made at 1/250,000 scale.

In order to take into consideration the fact that the unit step length was a variable which depended upon the setting of the intervalometer, the study of how tracing speed affected the results has been made in terms of the number of steps measured, N, rather than metric distances. If N_1 is the number of steps digitised at the faster speed, and N_2 is the number of steps digitised at the slower speed, the ratio $N_1{:}N_2$ provides a measure of the difference. The average ratio is 1:1·9, so that the distance between points which were digitised deliberately slowly is approximately double that of the same line digitised deliberately quickly.

The length of the coastline was determined in each case from the sum of the rectilinear elements joining successive coordinates, using equation (3.5). Thus the length of a feature is the sum of N unequal steps, which is equivalent to using the card method described on pp. 38–40. We call the mean of N unequal steps the *mean linear element*, $\bar{l}$.

If stream digitising is employed and the interval between readings is small (<1 second), the mean linear element is small and the standard error is small, indicating that $\bar{l}$ is nearly constant. There is clearly a temptation to treat this as the equivalent to the dividers separation, d, rather than a version of measurement by unequal steps. However, the spacing between digitised points varies according to the nature of the line, being greater along simple regular lines where the cursor can be moved relatively quickly, and shorter for less regular lines. This is exemplified by the following data extracted for the two sheets of the 1/250,000 Ordnance Survey maps. The northern sheet, which covers that coast as far south as Bridlington, is irregular with many bays and handlands. The southern sheet covers the coast from Bridlington to Spurn Head, and represents the much more regular coastline of Holderness. Table 21.3 compares the length, number of steps, average linear element and its standard error for some of the measurements of the two sheets.

These results demonstrate several interesting characteristics. First, the total measured length decreases as the interval is increased. This corresponds to the variation between measured length and the spacing of dividers in classical measurement. At the same time, of course, the length of the mean linear element increases. However, the standard error of the single observation remains remarkably constant for lines digitised with a very small interval ($<0{\cdot}1$ second) between readings. For larger intervals the variation in standard error more or less matches that of the mean linear element. The difference in the character of the coastline between the two map sheets is abundantly clear from the larger values for the mean linear elements on the southern sheet for every time setting listed.

If we treat $\bar{l}$ as being equivalent to d it is desirable to estimate whether there is any effect, resulting from variation in $\bar{l}$. In order to do this we determine a series of theoretical values for the length of the line, L', from

$$L' = \bar{l} \cdot N \tag{21.1}$$

and compare the difference $L - L'$. For example in the first line of Table 21.3, we have values for $L' = 1031 \times 0{\cdot}42 = 433{\cdot}02$ mm and $361 \times 0{\cdot}81 = 292{\cdot}41$ mm for the northern and southern sheets respectively. Repeated for all the measurements made at 1/250,000 scale, we find that the mean differences between L and L' are 0·38 mm and 0·57 mm for the two sheets, so that the lengths of the coastline determined in the two ways differ by less than 0·1% on the northern sheet and 0·2% on the southern sheet. From experience in comparing dividers measurements this is an acceptable variation between the results. Consequently we may regard $\bar{l} \equiv d$, and treat stream digitised distances as if they had been made with dividers set to a constant separation.

If we treat with L, or L' and $\bar{l}$ in the ways outlined in Chapter 14 we may calculate the limiting distance for an irregular line measured on a particular

TABLE 21.3. *Comparison of the measured length, number of steps, average linear element and its standard error, for the two parts of the Yorkshire coast at 1/250,000*

	Northern sheet				Southern sheet			
Time (seconds)	Length (mm)	Number (N)	Mean linear element (mm)	s.e.s.o. (mm)	Length (mm)	Number (N)	Mean linear element (mm)	s.e.s.o. (mm)
0·01	438·03	1031	0·42	±0·27	293·16	361	0·81	±0·45
0·02	438·84	856	0·51	0·30	291·93	344	0·85	0·33
0·03	436·05	893	0·49	0·28	291·45	329	0·88	0·37
0·04	437·85	770	0·57	0·31	291·84	321	0·91	0·39
0·05	442·46	747	0·59	0·33	291·29	328	0·89	0·33
0·06	439·66	678	0·65	0·31	292·08	261	1·2	0·53
0·07	438·73	701	0·63	0·31	291·13	279	1·04	0·53
0·2	431·57	293	1·47	0·65	291·47	138	2·11	1·01
0·3	429·62	215	1·99	0·91	290·91	77	3·78	1·59
0·4	427·28	178	2·40	0·95	291·44	88	3·31	1·39
0·5	424·63	139	3·05	1·10	292·23	71	4·11	1·93
0·6	421·87	130	3·23	1·27	290·80	59	4·92	2·11
0·7	421·71	111	3·77	1·38	290·71	52	5·59	2·24
0·8	416·32	101	4·12	1·56	290·50	41	7·04	3·22

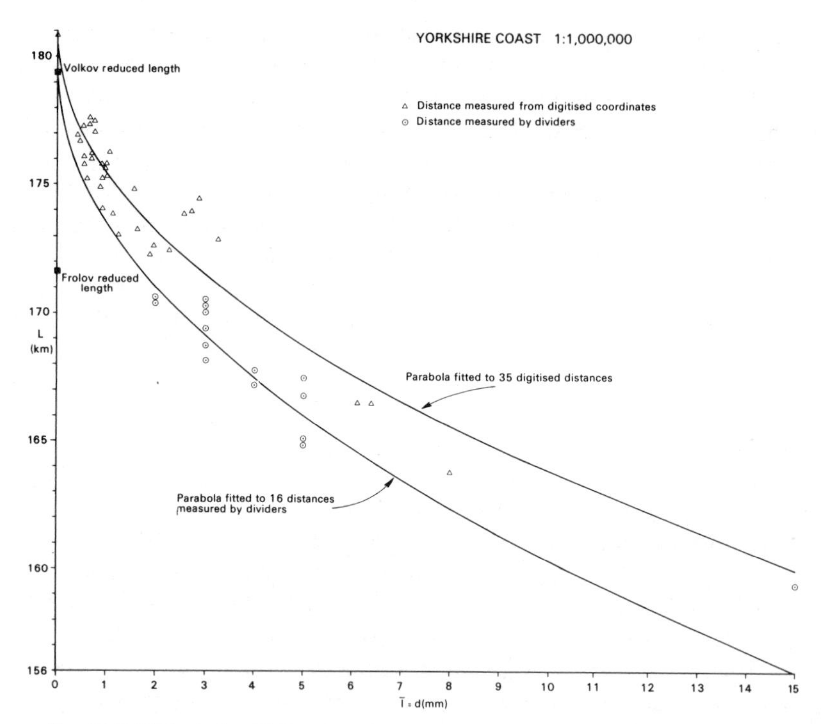

FIG. 21.5. The relationship between the mean distance digitised coordinate pairs and the total length of the Yorkshire coast measured on the *International Map of the World* at 1/1,000,000 scale, together with the corresponding dividers measurements.

map by digitiser. Figure 21.5 illustrates the relationship for the measurements at 1/1,000,000 scale and includes the dividers measurements given in Table 5.2, p. 68. One curve is fitted to the 16 measurements made by dividers; the other to the 35 measurements made by digitiser. We see that for comparable values for d and $\bar{l}$ the digitised distance is generally greater than that measured by dividers. Moreover a large number of the distances measured at very small intervals of time have given measurements which are even longer than the

TABLE 21.4. *Comparison of the limiting distances for the Yorkshire coast, measured at 1/1,000,000 scale by dividers and by digitiser*

Method	Limiting distance
Volkov reduction applied to 10 different combinations of d	$179{\cdot}369 \pm 0{\cdot}68$ mm
Least-squares fit of a parabola to the dividers measurements. Coefficient a in the expression $L = a + b\sqrt{d}$	176·653 mm
Least-squares fit of a parabola to the digitiser measurements. Coefficient a in the expression $L = a + b\sqrt{l}$	$180{\cdot}802 \pm 0{\cdot}7$ mm

limiting distance obtained from applying the Frolov reduction formula (14.25)–(14.30) to dividers measurements. A parabola fits these data better than the other functions tested. The three limiting distances based upon the parabola are sufficiently close to one another to express some confidence in this method of reduction.

The curve-fitting procedure can also be used to overcome a further difficulty in using a digitiser to measure distance. This is to find some way of choosing which measurements of portions of a line which occur on different map sheets should be combined to give the total length. In reducing measurements made by dividers there is no real problem in dealing with several part measurements. For example, all the portions measured on different map sheets with $d = 2$ mm may be added together to determine one estimate of the length of the line; the same is done using the measurements made with $d = 3$ mm and so on. However, the mean linear element is influenced by the nature of the line, as demonstrated in Table 21.3, so that it may be inappropriate to add together the results of part measurements made on different sheets even if these were all made at the same setting of the intervalometer. For example when the two parts of the Yorkshire coast at 1/250,000 were digitised at the same timing ($t = 0{\cdot}01$ second) the values on the top line of Table 21.3 are $\bar{l} = 0{\cdot}81 \pm 0{\cdot}45$ mm for the southern sheet and $\bar{l} = 0{\cdot}42 \pm 0{\cdot}27$ mm for the northern sheet. What justification have we in combining these results and, if we do, what is the appropriate value for $\bar{l}$?

A possible solution arose from the investigation made into the treatment of dividers measurements. This was to study whether there was any significant difference between

1 The length of the coastline measured and reduced sheet by sheet.
2 The application of the reduction formulae to the total distance obtained by adding together the crude results for each sheet first.

Application of the t-test for the difference between two means (7.21)–(7.24), p. 121, to these data indicated that there was no significant difference between the results at any of the customary levels of significance. Consequently the author used the reduction of individual sheet measurements by means of standard curve-fitting procedures and adding the components derived from the regression equations. For example, the results of fitting a parabola to 35 measurements on each of the 1/250,000 maps gives rise to the regression equations:

$$\text{northern sheet: } L'' = 451{\cdot}687 - 16{\cdot}768\sqrt{\bar{l}}$$
$$\text{southern sheet: } L'' = 293{\cdot}064 - 1{\cdot}0137\sqrt{\bar{l}}$$

By solving each regression equation for a given value of $\bar{l}$ we find the limiting distance for each portion of the coast and add the results. For example, if we substitute $\bar{l} = 1{\cdot}0$ mm in these equations we find the distances $L'' = 434{\cdot}92$ mm and $292{\cdot}05$ mm respectively, or a total distance $726{\cdot}97$ mm corresponding to

181·74 km. Usually we want the limiting distance for each part which we need to know for the final determination of the reduced length. This is simply obtained by summing the first (or *a*) coefficients, or

$$451{\cdot}687 + 293{\cdot}064 = 744{\cdot}751 \text{ mm or } 186{\cdot}19 \text{ km}.$$

This result agrees well with the Volkov limiting distance based upon 10 sets of dividers measurements (186·16 km) which is given in Table 14.11, p. 299.

The Mainland Coast of Scotland

The work of Baugh and Boreham was carried out by digitising the coastline at a scale of 1/63,360, using a d-mac digitising table. They state that the average distance between coordinated points was 140 m on the ground, corresponding to 2·21 mm on the map. Since the calculated length of the coastline, based upon the sum of rectilinear elements, was 5,285 km, as total of approximately 37,750 points were recorded in the digitising process. In addition to measuring the entire coastline at this scale, they also made a series of measurements of short lengths of the coast in sample areas on maps at 1/10,560, 1/25,000 and 1/63,360 scales. These measurements were also repeated by chartwheel.

For some inexplicable reason they converted the table coordinates into the corresponding National Grid values before determining the length of the coast. This necessitated solving their version of the grid-on-grid transformation about 37,750 times. It would have been distinctly more economical to have made all the calculations in the units of measure employed by the digitiser (which was centimetres on the older d-mac tables) and left the conversion of the result into the required ground units until last of all.

The most interesting and original contribution by Baugh and Boreham was their use of different models to determine the length of the coastline.

1. Thus far we have only considered the calculation of rectilinear elements, each being the straight-line distance between adjacent pairs of coordinates, the total length of the line being the sum of these elements. This is the traditional cartometric method for the creation of rectilinear chords is implicit in measurements made by dividers or card. As far as the author is aware, all other attempts to measure irregular lines by vector methods are similarly based upon straight-line elements. Baugh and Boreham obtained a total length of 5,285 km for the mainland coast of Scotland using the method.
2. In order to reduce the differences between the arc and chord measurement for each element, Baugh and Boreham tried two other solutions. The first was to determine the length of the parabola which may be fitted through groups of three consecutive digitised points, symmetrical about the middle point of the three, and summing the resulting arc elements. Baugh and

TABLE 21.5. *Comparison of measurements of the coastline near Wick, made by Baugh and Boreham (1976) by different methods and at different scales*

Map scale	Chartwheel (km)	No. of measurements	Digitised line (km)	No. of measurements
1/63,360	4·92	10	5·05	5
1/25,000	5·04	10	5·45	5
1/10,560	5·72	10	6·84	5

Boreham obtained a length of 5,410 km from their data treated in this manner.

3 The third model used by them determined the arc lengths of parabolae of maximum arc length which can be fitted to each group of three consecutive points. The sum of these arc elements resulted in a length of 5,360 km.

The relative merits of different types of curve have not been considered in detail, but the difference in estimate between the various models is relatively small. Baugh and Boreham argue that it would seem prudent to adopt the last of these models as representing the best estimate; for this would always give a close approximation to the true line for straight lines and smooth curves, and is the more likely to under-estimate rather than over-estimate length by smoothing, where the curve between the three points is not smooth.

Their final modification to the calculated length of line was derived from consideration of the digitising errors evaluated earlier and contained in Table 21.2. We have already seen that they considered the random error in positioning of the digitising cursor to have a systematic effect upon the calculated length of line. They have therefore applied a correction using the data contained in that table. Thus for a digitising precision of $\pm 0{\cdot}11$ mm (approximately 7 m on the ground) the corrected length of the coastline should be 20 ± 15 km greater than the calculated value. Consequently their final estimate of the length of the mainland coast of Scotland was $5{,}340 \pm 20$ km.

The measurements made on the three selected short sections of coast have not been properly tabulated in their paper. Table 21.5, which refers to a short stretch of coastline south of Wick, is all that can be presented here.

We see that these measurements correspond to the variation in distance with scale which we have seen in the other results studied in the book.

CARTOMETRY IN THE RASTER MODE

In order to exemplify the methods of digital cartometry in the raster mode we describe some of the attempts which have been made to measure distance and area from Landsat Multispectral Scanner imagery. Since the data have been captured and stored in the form of a raster or grid, the area of a parcel may be determined by counting of the number of grid cells or pixels occupied by that

TABLE 21.6. *The dimensions of the unit pixel of the Landsat M.S.S. output expressed at various map scales*

Map scale	Horizontal dimension	Vertical dimension	Area of pixel
1/1	57·10 m	79·06 m	4514·3 m² = 0·4514 ha
1/10,000	5·71 mm	7·91 mm	45.14 mm²
1/50,000	1·14 mm	1·58 mm	1·81 mm²
1/100,000	0·57 mm	0·79 mm	0·45 mm²
1/250,000*	0·23 mm	0·32 mm	0·072 mm²
1/500,000*	0·11 mm	0·16 mm	0·018 mm²
1/1,000,000*	0·06 mm	0·08 mm	0·005 mm²

*Indicates the scales for which Landsat M.S.S. imagery is normally available as photographs.

feature. The length of a feature may be determined from counting the number of pixels which coincide with it.

The Landsat Multispectral Scanner Imagery

The Landsat M.S.S. frame comprises 2,340 scan lines of 3,240 picture cells each. Consequently there are 7,581,600 pixels on the single frame. The *horizontal dimension* of the single pixel, which is that in the direction of the scan lines, is 57·10 m on the ground; that of the *vertical dimension* is 79·06 m. Therefore the area of one pixel corresponds to a rectangle of 0·4512 ha. The dimensions of the unit pixel expressed in millimetres at some familiar map scales may be tabulated as shown in Table 21.6. For purposes of inspection and superficial interpretation the M.S.S. imagery is reproduced at scale 1/3,369,000 on 70 mm negative film. As hard copy this is normally enlarged to 1/1,000,000 scale. Further enlargements to 1/500,000 and 1/250,000 are common. In conventional interpretation, using a photographic print, the limiting resolution with only moderate ($< \times 10$ magnification) is approximately a block of dimensions $0{\cdot}1 \times 0{\cdot}2$ mm, which is only as large as a prick-mark on the paper, corresponding to a block of four adjacent pixels. It follows that the digital resolution of the Landsat M.S.S. images is better than the human eye can detect without the aid of additional magnification. Consequently interpretation and discrimination by computer is an exceptionally powerful tool and, in practice, the more sophisticated kinds of interpretation are done this way, not with paper prints and magnifying glass. Interpretation, discrimination and measurement may be done by examining individual pixels or clusters of them, and comparing the spectral reflectance.

It is at this stage that the treatment of raster data extracted by scanning a map and that collected by remote sensing needs to be distinguished. In raster data derived from a map each pixel represents either an element of a mapped line or symbol (1) or it does not (0). In other words each pixel has a definite and unequivocal meaning. In remote sensing, however, the reflectance of the ground may be expressed by any of 64 equal grey scale steps for each of four wavebands, or a total of 256 steps. This is gradation in tone which, again, is far more sensitive than either photographic film or the human eye can detect. Moreover, the individual picture element may contain reflected images of different ground features which have been reflected differently. The single pixel may contain the reflected images of more than one category of land cover, such as land and water, or two different categories of land-use, and is sometimes called a *mixel*. Such pixels contain the boundary, or *interface*, between the categories. From the point of view of mapping, these boundaries are lines which need to be plotted; in cartometric applications these are the boundaries of the parcels whose areas are required.

The automatic interpretation process is accomplished in two stages. First it is necessary to allocate a spectral signature to each pixel. This is done by measuring the selective reflectivity of ground features at different wavelengths in the visible and near infra-red parts of the spectrum and the establishment of characteristic spectral signatures for each category of land cover. It leads to the establishment of *training sets* used to instruct the computer to recognise the characteristic spectral response of the target categories. The classification process can be *supervised*, which comprises preliminary delineation of spectral and textural classes by the human interpreter, followed by machine searching for similar classes, or it is *unsupervised*. In each case it is hoped that the derived spectral and textural signatures of distinctive land covers may be sufficiently dissimilar to avoid confusion.

The second stage is location of the interfaces using pattern recognition techniques which are now applied in other kinds of automatic data processing for the interpretation of various types of imagery depicting different categories of ground cover. By applying discriminant analysis to the spectral signatures at the different wavelength bands used in the multispectral scanner it is possible to establish which of them will facilitate automatic recognition of the

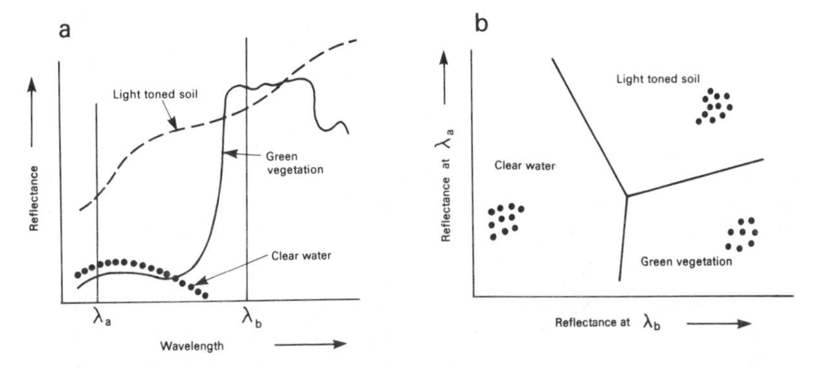

FIG. 21.6. Discriminant analysis applied to the spectral signatures of Landsat M.S.S. imagery: (a) measured at two different wavelengths, (λ_A and λ_B, using these data in (b) to define clusters for classifying additional pixels. (Source: Monmonier, 1982).

major categories of ground cover for a particular purpose. Figure 21.5 illustrates the principles which are adopted with M.S.S. imagery.

Procedures for Measurement from Landsat M.S.S. Data

Most of the literature concerned with the methods of analysing the output of the multispectral scanner is concerned with these methods of detection and discrimination. If it is possible to determine these interfaces, it is also possible to map them and measure them. In the first instance, the Landsat image has been converted into a map which may be measured using the conventional techniques described in other chapters. Here, however, we must consider the second group of automatic methods in cartometry to be carried out as part of the analysis of the M.S.S. imagery. We consider two contributions to this study; that of Faller (1977) concerning measurement of coastlines, and that of Crapper (1980) concerned with the theoretical accuracy of area measurement.

Linear Measurement

Faller's report is entitled *A procedure for detection and measurement of interfaces in remotely acquired data using a digital computer.* It is primarily concerned with the problem of detecting, locating and measuring shorelines of the Gulf Coast of the U.S.A., on the M.S.S. imagery produced by the early Landsat-1. Much of the report is naturally concerned with the methods used to locate the interface between land and sea from the scanned data, and make the appropriate geometrical corrections of these data which might affect the location of the interface. This is of interest in the context of Chapter 13 because it indicates the processing steps which are needed for such applications. Indeed the cartometric work itself might almost be considered to be an optional extra provided at the end of the sequence of processing stages, rather than an end in itself.

A shoreline has to be located as the interface between the different categories of image representing land and those representing water. By combining different categories of vegetation and land cover on the one hand to characterise land, and the different categories of water image to depict water, the numerous and varied images making up the M.S.S. pictures of the ground are simplified into two superclasses. The interface between these is the required boundary between land and water to be mapped and measured. The practical method applied in the computer with Landsat M.S.S. tapes as the input data, is to examine each set of images through a window of format 2 × 2 pixels. As the data are examined, two scan lines at a time, each pixel appears in four different positions in the window, allowing for the possiblity of comparing it with the adjacent pixels. If any two are the same there is no interface between them, but if they belong to different superclasses the interface, or boundary, has been identified. Such pixels are allocated a numerical value and a special data tape, comprising this information only, is prepared. Faller describes five main stages in the work which may be listed as:

- data selection,
- classification of images,
- isolation of the shoreline,
- scaling of the image,
- analysis and measurement.

Much of the report is concerned with the need to apply a correction to allow for the rotation of the earth beneath the Landsat M.S.S. scanner during the 25 seconds (for Landsat-2) to cover the ground area of one scene, the ground track drifts westwards one pixel in every 0·124 seconds. Faller makes use of the correction program SKEWCOR.

The first stage in measurement of a shoreline is essentially a matter of identifying those mixels containing land and sea and counting them; the second is to choose the distances corresponding to the length of the boundary in each of them. The metric contribution of the single mixel to the total interface measurement depends upon the orientation of the boundary to the sides of the picture element and therefore the scan direction. Some units will measure the distance H (≈ 57 m) and others will measure V (≈ 79 m) because the interface is respectively parallel to the scan line or perpendicular to it.

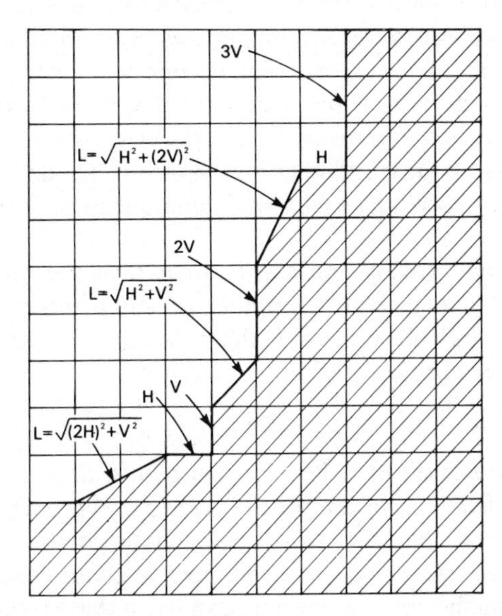

FIG. 21.7. Diagonal interfaces. Three types of diagonal interface elements are considered in the analysis, each with a unique contribution to the total length L. The cross-track pixel dimensions is H along the scan line; the other dimension is V. (Source: Faller, 1977).

However the element of the coastline is more commonly oriented at some other angle to the scan direction. Then the interface appears as a jagged-line constructed of alternating horizontal and vertical elements. As a means of smoothing, and for simplification of the measurement process, Faller recognised three possible configurations of the diagonal interface as it appears in the examination window and, because the examination window measures 2×2 pixels, only three possibilities need to be studied. The contribution of each type of diagonal to the interface length is

$$L_1 = [H^2 + (2V)^2]^{1/2} = 168\,\text{m} \quad (21.2)$$

$$L_2 = [H^2 + V^2]^{1/2} = 97\,\text{m} \quad (21.3)$$

$$L_3 = [(2H)^2 + V^2]^{1/2} = 139\,\text{m} \quad (21.4)$$

respectively. The numerical values are based upon the nominal values for H and V and apply to the single pixel.

Scaling of the Image

Although the values for the size of the unit pixel given in Table 21.6 are recorded to the nearest centimetre on the ground, it must be understood that they are approximate values and may vary slightly as the position of the satellite may depart slightly from that predicted from theory. If we are to use pixel-counting techniques to determine area and, moreover, use *direct* determination of area, in the sense adopted in this book, then a small change in pixel size will introduce a systematic error into the measurements. Likewise if pixel count is to be used to measure length, the error being systematic will be cumulative. Therefore the true dimensions of the pixel must be determined by means of a scaling procedure, which is tantamount to calibration of the unit pixel.

The procedure adopted by Faller is the time-honoured method used with aerial photography, namely to compare measurements made on a map between suitable control points and the coordinate distance measured in pixels on the image. The expression

$$d = \{(S_v \cdot \Delta_v)^2 + [S_H \cdot (\Delta_H - S_K \cdot \Delta_H]^2\}^{1/2} \quad (21.5)$$

is used where d is the distance between the points, S_v is the "vertical scale factor", i.e. the dimension of the picture element perpendicular to the scan direction, Δ_v is the number of scan lines between the points in the image, S_H is the "horizontal scale factor", Δ_H is the number of picture elements along the scan line separating the points in the image and S_K is the skew correction factor needed to take rotation of the earth into consideration.

Several scale determinations are needed just as in similar work with conventional aerial photography. Thus, in the study of the length of the coastline of Alabama, Faller used eight control points and measured five distances in order to scale the images.

Since these are the values for H and V which will be multiplied by very large numbers (the pixel counts), a small systematic error arising from inaccurate scaling will be greatly magnified in the resulting measurement. Enough has been written elsewhere in this book about the errors in making conventional measurements to suggest that it is important to determine an accurate value for d. However the measure of uncertainty in recognising terminal points which can be positively located in terms of single pixels suggests that there is a potential source for scaling errors here.

THE LENGTH OF THE SHORELINE OF ALABAMA AND RELATED MEASUREMENTS

Faller chose two sets of Landsat M.S.S. imagery, one dating from December 1972 the other from December 1973. Winter images were chosen so that any influence of vegetation upon representation of the shoreline in marshy areas should be reduced to a minimum. From the scaling procedure he found that the pixel dimensions were 57·34 × 80·80 m for the 1972 data and 57·94 × 80·04 m for the 1973 data.

The total amount of time required to complete the five stages and obtain the length of the shoreline of Alabama may be listed as follows:

Data selection	4 man-hours
Classification	39
Isolation	43
Scaling	6
Analysis	18
Total	100 man-hours

Admittedly Faller claims that some of the work (e.g. of classification) can now be done more quickly than his figures suggest, but it seems that we would be hard-pressed to justify the method of measurement as being any faster than

TABLE 21.7. *The length of the coastline of Alabama. A comparison of different methods and results (data from Faller, 1976 and Ringold and Clark, 1980)*

Faller, from Landsat M.S.S. data,	December 1972	1,452·5 km
	December 1973	1,162·6 km
	Mean	1,307·5 km
General shoreline (NOAA, 1975)		85·3 km
Tidal shoreline (NOAA)		976·9 km
National shoreline (U.S. Army, 1971)		566·5 km

TABLE 21.8. *Faller's analysis of the repeatability and accuracy of distance measurements made by Landsat*

Locality	No. of datasets	Mean shore-line length (km)	Sample deviation (km)	Sample deviation (%)	Correlation with state of tide	Chartwheel measurement at 1/24,000 (km)	Difference Map–Landsat (km)	Difference Map–Landsat (%)
Deer I.	8	15·786	0·313	1·98	−0·11	14·77	−1·02	6·48
Broadwater	7	12·676	0·343	2·70	0·16	12·01	−0·67	5·25
Pass Christian	4	22·226	1·047	4·71	−0·04	21·04	−11·19	5·34
Long Beach	6	4·621	0·219	4·75	0·73	4·71	0·90	1·93
Half Moon I.	4	12·717	0·731	5·75	0·14	12·22	−1·55	3·91
Petit Bois I.	4	17·155	0·336	1·96	0·44	15·61	−1·55	9·01
Dauphin I.	5	80·861	2·052	2·54	0·53	81·36	0·51	0·61
			Mean	3·48				4·64

those of conventional cartometry. Moreover, it can hardly be claimed that this is an automatic method of measurement, in view of the number of man-hours required at different stages of the operations.

The comparison betwen Faller's results and those made by classical methods may be tabulated as shown in Table 21.7. The mean length of the shoreline of Alabama obtained in this fashion is the 1,307·5 km and the difference between the two determinations is 289·9 km, which is 22% of the mean length. Obviously it is desirable to attempt an explanation of this difference, although it is no greater than many of the discrepancies in linear measurement which we have encountered in other work. Indeed this difference is much smaller than those recorded in Table 21·7 between the various official estimates of the length of the coast of Alabama made by conventional methods (42–91%).

We may dismiss any differences in scale of the images as being comparatively unimportant for the following reasons. If we assume that each of the pixel dimensions, H, V, L_1, L_2 and L_3 occurs with the same frequency, the mean of these five measurements represents the contribution of one pixel to the measured length. Thus the mean obtained from the calibrated 1972 data is 78·62 m. The corresponding value for 1973 is 78·46 m. Assuming that the mean length is correct, the number of pixels comprising the images of the Alabama shoreline is 16,631 for the 1972 imagery and 16,665 using the 1973 data, or a difference of 34 pixels, or just over 2·6 km in total distance (0·2%). This is an estimate of the effect of the amount of correction introduced by Faller's method of calibration. We see, in fact, the difference in the measurements is more than 100 times larger, or 289·9 km, which corresponds to 3,693 pixels.

In order to explain these differences, Faller investigated a number of possible variables such as the influence that the state of the tide might have upon the results. He did this using repetitive analyses of the lengths of seven different coastal features (nearly all less than 20 km in length) in Louisiana, Mississippi and Alabama for which multiple sets of Landsat imagery were available. The same procedures for discrimination and correction for earth rotation, scaling, etc. used for the Alabama study were employed in this work. The lengths of the same features were also measured by chartwheel from 1/24,000 scale U.S.G.S. maps. Correlation between state of tide and shoreline length was, with one exception, very poor. However, in order to test the relationship between pixel dimensions and shoreline length still further, a series of tests were made for another coastal marshland area in Louisiana.

Effect of Size of Minimum Sampling Unit

For this test Faller used the multispectral scanner mounted in an aircraft. The data were collected from a flying height of 1,400 m, giving a square pixel dimension of 3·4 m. The data were then processed to give pixel sizes × 2 (6·8 m), × 4 (13·6 m), × 8 (27·2 m) and finally a minimum mapping unit with

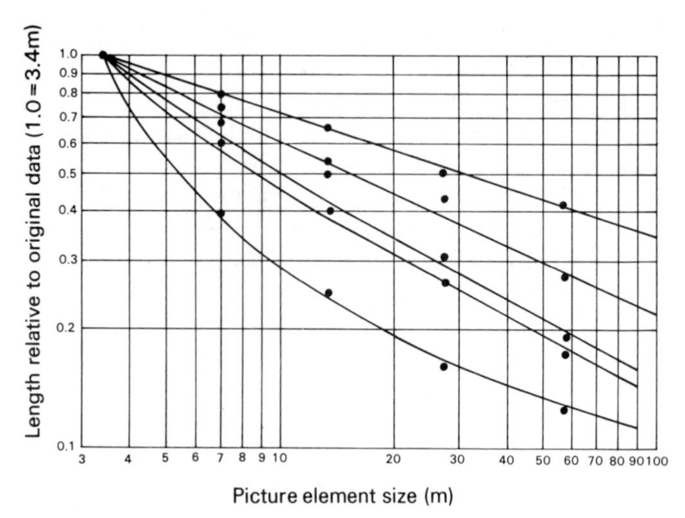

FIG. 21.8. Effect of minimum mapping unit size on the measured lengths of five different shorelines based upon M.S.S. data collected by conventional aircraft from 1,400 m flying height. (Source: Faller, 1977).

dimensions corresponding approximately to that of the Landsat pixel. By relating the measured distances as a fraction of the limiting distance, he obtained graphs of the form illustrated in Fig. 21.8. Unfortunately Faller did not indicate the location of the different lines tested, nor their character.

Application of this technique to repeated Landsat coverage of shoreline features demonstrates the reproducibility of shoreline length measurement. Tidal variation does not significantly affect the measurement which has been found to agree with manual measurements comparable to those made on small-scale maps, but the minimum mapping unit or pixel size greatly determines the complexity of an interface feature. Therefore Faller argues that no interface length measurement should be reported without a statement about the size of the unit pixel used in the determination corresponds to those by both Steinhaus (1954) and Perkal (1958) who insisted that the order of the measurement carried out by longimeter ought to be recorded with the result.

Area Measurement Using Landsat M.S.S. Imagery

Since the data have been captured and stored in raster mode, determination of the area of a parcel may be accomplished by counting of the number of grid cells or pixels occupied by that feature and multiplying by the area of a pixel.

Does this correspond to an application of the direct method of point-counting, or is it a version of the method of squares? If we were to treat is as a particular application of direct point-counting we would need to be able to repeat the count several times with the raster occupying different positions within the same scene or frame, just as we may shift the transparent grid about on the map. However, only one configuration of pixels is possible, so that an array of raster data cannot be treated exactly as if it were a separate grid. If we were to treat it as an application of the method of squares then we ought to make an estimate of the area contained in each of the marginal pixels through which a parcel boundary passes just as if it were a grid square, and apply a correction corresponding to that described in Chapter 4, p. 57. As in the conventional methods, the influence of the marginal cells is important. If we ignore the contribution of all the marginal squares in the method of squares the measured area is an under-estimate but if we include all of them the result is an over-estimate. We have seen that one way of overcoming the tedium of assessing the proportion of parcel area contained in each marginal grid square is to include a correction which is the addition of the area of one-half of all the marginal squares. Similarly, in counting mixels, one-half of these may be recorded as lying within the parcel and one-half as belonging to adjacent regions. In the former case an error of commission occurs in which the parcel area will be over-estimated, and in the latter case an error of omission has been introduced so that the parcel area will be under-estimated. In order to appreciate the magnitude of such errors we need to know what proportion of a given parcel is represented by single-category pixels; how many are mixels. Crapper (1980) argues that there are many more perimeter cells than one would intuitively expect. He quotes work showing that more than half the pixels are mixels. For example, in the parcel illustrated by Fig. 18.6, page. 404, the proportion of mixels to pixels is 45/56, or more than 80%, and argues that significant errors can occur in an area evaluation if no consideration is taken of the contribution of the marginal pixels.

If the individual squres depicted in Fig. 18.6 represented pixels, the parcel illustrated would measure $45 \times 0{\cdot}4514 = 21{\cdot}3$ ha, excluding all mixels, or 45·6 ha if all the mixels were included. We need to investigate the relationship between the percentage of mixels for parcels of different sizes.

The theoretical relationship between parcel size and the influence of the marginal cells, which we have seen, was studied in relation to point-counting methods of area measurement by Frolov and Maling (1969). The subject was subsequently studied by Lloyd (1976), by Goodchild and Moy (1976) and by Crapper (1980, 1981) with respect to counting pixels. Following the arguments presented by Crapper (1980), the number of perimeter pixels may be expressed as a function of the total area as follows:

$$L = 2K_1(\pi A)^{1/2} \qquad (21.6)$$

where L is the length of the perimeter of a parcel of area A units and K_1 depends upon the shape.

Frolov and Maling introduced the expression

$$k_1 = L/A^{1/2} \qquad (21.7)$$

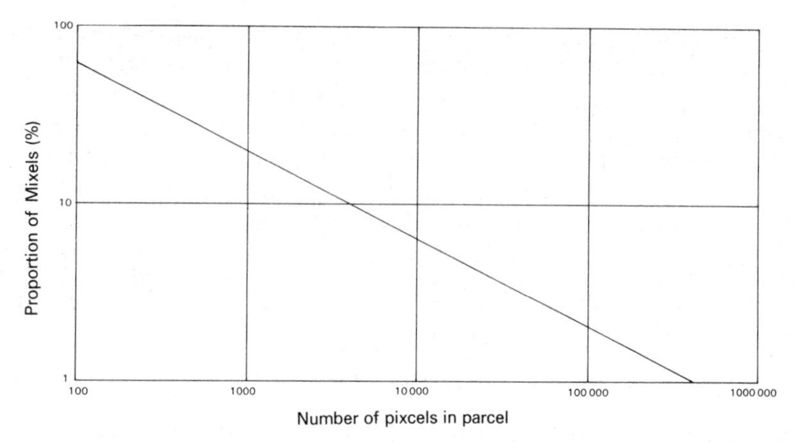

FIG. 21.9. The relationship between the number of mixels (expressed as a percentage) and the total number of pixels in a remotely sensed image of parcels for which $K_1 = 1{\cdot}82$.

as the shape factor in their equations, so that for a circle, $k_1 = 3{\cdot}545$ and for a square, $k_1 = 4{\cdot}00$.

The expression for shape in equation (21.7) has different meaning and the relationship between them is

$$K_1 = k_1 \cdot \sqrt{\pi}/2\pi \tag{21.8}$$

Therefore $K_1 = 1$ for a circle and $K_1 = 1{\cdot}128$ for a square. Crapper (1980) takes the average as being $K_1 = 1{\cdot}82$.

If we solve equation (21.6) using this value for K_1, we may determine the magnitude of the ratio of mixels to pixels for different sizes of parcel, expressed as N, the total number of pixels contained in them. The result is illustrated in Fig. 21.8, where we see that the ratio is less than 10% for $N > 4{,}000$ ($\approx 18\,\text{km}^2$) and less than 1% for $N > 400{,}000$ pixels ($\approx 1{,}800\,\text{km}^2$).

Error Variance for Area Measurement on Landsat M.S.S. Imagery

When a grid is placed over a parcel, the length of its perimeter, L, is given by

$$L = \sum_{i=1}^{i=N_b} l_i \tag{21.9}$$

where l_i is the length of the ith boundary pixel and N_b is the number of boundary pixels. The average length of the perimeter per boundary pixel l is given by

$$\begin{aligned} l &= L/N_b \\ &= K_2 \cdot L \end{aligned} \tag{21.10}$$

where K_2 is a measure of the average within-pixel distortion and L is the average length of a straight line laid across a rectangle. For Landsat pixel size, Crapper has determined $L = 53{\cdot}04$ m.

Thus the average number of perimeter pixels, N_b is given by

$$N_b = L/l$$
$$= [2K_1\sqrt{(\pi A)}]/K_2 \cdot L \tag{21.11}$$

The error variance for one pixel is equal to the mean square of the smaller area obtained when a randon straight line is laid across a pixel (A^2) and hence the overall error variance σ^2 is given by

$$\sigma^2 = N_b \cdot A^2 \tag{21.12}$$

The error variance is thus dependent on the number of pixels and the distortion parameters, K_1 and K_2 and is given by

$$\sigma^2 = \{[2K_1(\pi A)^2]/K_2\} \cdot [A^2/L] \tag{21.13}$$

The relative error is given by

$$\sigma/A = 0{\cdot}39A^{-3/4} \tag{21.14}$$

This curve is displayed in Fig. 21.10, from which it can be seen that for a relative error of 1%, $A = 132$ ha, for a relative error of 5%, $A = 15$ ha, and for a

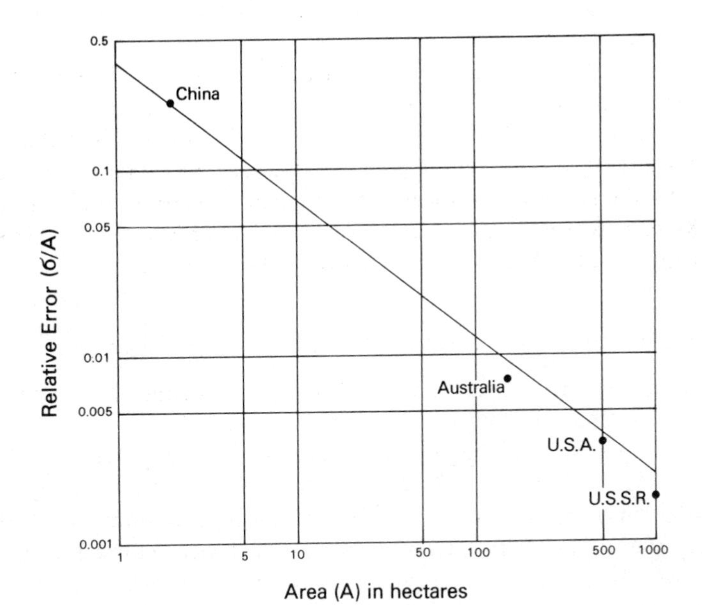

FIG. 21.10. The relationship between the relative error (σ/A) and parcel area, A. The relative errors for the area estimates of average-sized wheat fields in different countries (which are given in Table 21.9) are shown. (Source: Crapper, 1980).

TABLE 21.9. *The shape factor, typical area, and relative error in area estimate of wheat fields*

Country	Shape factor K_1	Typical area (ha)	Relative error (%)
U.S.A.	1·51	500	0·33
U.S.S.R.	1·13	1,000	0·17
Australia	1·13	150	0·72
China	1·82	2	23

relative error of 10%, $A = 6$ ha. An average value has been used for the shape factor (i.e. $K_1 = 1{\cdot}82$) in these calculations.

Crapper (1980) provides an example of how to make reasonably accurate estimates for the shape factor. He uses the comparison of wheat production in the U.S.A., U.S.S.R., Australia and China. Wheat fields in the United States are usually rectangular with an aspect ratio 5 and average area 500 ha, in the U.S.S.R. and Australia wheat fields are typically square with areas of 1,000 ha and 150 ha respectively, and in China wheat fields have irregular shape with typical areas of 2 ha. The shape factor can be calculated from equation (21.8) and after substituting into equation (21.13) the relative errors can be found. The shape factors, areas and relative errors in area estimates are shown in Table 21.9. For smaller areas, as typified by Chinese wheat fields, the relative errors are so large that great caution should be exercised when working with area estimates.

Sampling Procedures

Thus far we have assumed that all 7·5 million pixels which are contained within a single Landsat M.S.S. frame may be counted in order to measure the area of a feature which may occupy most of one frame, and possibly extend into adjacent frames. Such conditions may arise in the execution of certain kinds of study, for example the LACIE (Large Area Crop Inventory Experiment) to study world crop yields and food production. Sooner rather than later the C.P.U. time required to make the measurements becomes excessive, and it becomes necessary to select samples rather than struggle with the entire population of pixels.

The relative merits of different strategies of spatial sampling have already been examined in Chapter 8. Consequently there is no need to repeat the conclusions here. It is sufficient to refer the reader to a sample of the fairly extensive literature which now exists upon techniques to be used in studying Landsat M.S.S. imagery for land use, land cover, and agricultural purposes. The contributions by the LACIE team, for example Bauer, Hixson, Davis and Etheridge (1978), Hixson, Bauer and Davis (1979), Hixson, Davis and Bauer (1981), Hixson (1981) are all worthy of examination in this field.

CARTOMETRY AND THE BBC DOMESDAY INTERACTIVE VIDEODISC

In order to commemorate the 900th anniversary of the Domesday Survey of 1085, BBC Enterprises Ltd created and organised the venture known as the Domesday Project of 1985. It has been described by Goddard and Armstrong (1986), who treat with the organisation and execution of the project; Openshaw, Wymer and Charlton (1986) have described the geographical information and mapping systems which are employed, and a variety of other papers by Openshaw and Mounsey (1986), Rhind and Mounsey (1986), treat with different aspects of the content and uses which may be made of this remarkable venture. There is no doubt that this Geographical Information System will provide the data and the facilities for studying many variables within the environment, and that much more will be published about its potential and uses during the next few years.

Two of the facilities provided for interactive use is measurement of distance and of area. It is these cartometric applications which we consider here. The system operates with comparatively inexpensive hardware:

The BBC Master AIV Microcomputer
The BBC AIV Laser Vision Player.

Much of the operation of the microcomputer may be controlled by means of a tracker ball. The input is provided on a set of two videodiscs and is viewed on a colour monitor. Some textual information, but not the maps, may be printed as hard copy. At the end of 1986 the total cost of the complete system was approximately £4,500 including tax.

The two videodiscs are called the *Local* or *Community Disc* and the *National Disc*. The first of these contains the results of a survey of Great Britain in 1985 carried out by local communities, primarily by schools, but sometimes by other members of the public. This videodisc also has the facility of accessing Ordnance Survey maps, primarily based upon the 1/625,000 and 1/50,000 maps, with 1/10,000 cover for major urban centres. It is these maps upon which the cartometric work may be done.

We have already seen that approximately four-fold linear enlargement is required to represent the detail shown on Ordnance Survey maps with sufficient clarity on the monitor screen. Thus the part of a 1/50,000 scale map which may be shown at any one time comprises 4 × 3 km, and this, indeed, was the standard unit used for all contributions to the Local Disc. This corresponds to 50 × 30 km at the smaller scale of 1/625,000. In effect, therefore, Great Britain is divided into a succession of blocks of these dimensions, and any measurements which are to be made must be confined within these frames, which are as individual as the separate sheets of a map series. The immediate result is that any measurements of features which are larger than the individual frame, or which extend from one frame into the next, must be measured separately. In fact, the extent of the individual frames is further reduced by the superimposition of the menu bar along the top of each frame. This makes it impossible to make a measurement up to the grid line which represents the northern extremity of each frame. Hence there is some uncertainty about the point at which a measurement ends in one frame and starts in the next frame to the north.

Manipulation of the tracker ball creates a linear element, imaged in blue, which may be set upon the terminal point of the line to be measured. This line may be increased or decreased in length, and it may be rotated through an angle about the point at which it has been positioned. Using these variables it is possible to lay a rectilinear element upon a line on the map is such a way that it coincides with it just as the application of the card method illustrated in Fig. 39. In modern computing jargon this process is known as *rubber-banding*. Since the length of each rectilinear element may be chosen by trial-and-error, the result is exactly the same as the use of the card method, applied to a map image on a monitor rather than to the paper map. As each linear element is fitted to the irregular line, the menu space displays the measured distance, converted into statute miles or kilometres. This value is given to a variable number of decimals. Thus for lines which are shorter than 1 km the measurement is recorded to 0·001 km; for 1–10 km it is only to the nearest 0·01 km, and for lines greater than 10 km the answer is only recorded to the nearest 0·1 km. There is a limit of 50 such linear elements which can be measured in any single map frame.

It follows that the limitations upon obtaining satisfactory distance measurements using the BBC AIV system are:

- The comparatively poor resolution of the system characterised by the display in statute miles or kilometres.
- The limitation that only 50 steps are available to define a line on a particular frame inevitably means that extremely irregular lines are likely to be measured in rather large steps.
- The comparatively small size of the unit frame means that a measurement has to be made in a large number of separate entities. For example, an attempt to measure the length of the Yorkshire coast on the 1/625,000 source provided on the Local Disc, involves measurement on eight separate frames, compared with the two paper map sheets at this scale. At 1/50,000 scale the same measurement required the staggering total of 71 separate portions to be measured.
- The superimposition of the menu across the northern part of each frame prevents the operator from accessing the extreme northern edge of each frame. Consequently there is uncertainty whether the measurement of a feature is complete, or contains some unmeasured gaps.
- Although greater experience and practice in the use of the tracker ball might make it easier to place the measuring mark at the end of the line to be

measured, the author found it rather difficult to control its movements to the desired degree of precision.

These five limitations more or less rule out the BBC AIV system as a useful cartometric tool. It does not measure up to the order of precision which we have come to expect from dividers, chartwheel or possibly even a length of thread applied to the paper map. A solitary measurement of the length of the Yorkshire coast made by the author, using the 1/50,000 scale base maps for the area is given in Table 5.4. Compared with the other measurements obtained using conventional instruments the results seems to be acceptable, but there is only a single measurement done in this fashion and we have no idea how representative it might be. The fact that the whole measurement involved more than 70 separate 4×3 km displays, and the uncertain accuracy in the positioning of the index arrow where it intersected the grid lines bounding each of these frames, left the author with the firm conviction that there must be a better way of doing the work. Besides, the measurement of 71 frames took a whole day, which is far more time than any of the manual methods would have required.

22

Measurement on Mercator's Projection

If you keep reckoning according to *Mercator*, it will be requisite sometimes to sum up your reckonings past, namely so often as you make any noticeable alteration in your Course: And so this reckoning or any other may be set down almost as easily on Mercators Chart, the difference is, that here you must alter your Scale because the deg. of Latit. on this Chart are not equal but grow greater and greater towards the Poles. Now then the distance of two places is to be measured by that part of the Meridian which is intercepted between the Latitudes of those places; Or if both places lye in one and the same *Latitude*, their Distance is measured by a Degree or other lesser quantity, taken about that *Latitude*; namely halfe above, and halfe beneath.

(Richard Norwood, *The Seaman's Practice*, 1637)

Because Mercator's projection is so important as the mathematical base for most nautical charts and many marine distribution maps the special problems which arise in making and correcting measurements made upon them require special and detailed consideration. This chapter treats with these problems and offers some possible solutions.

Mercator's projection possesses two important properties which have made it indispensable for graphical dead-reckoning navigation ever since its introduction about the middle of the sixteenth century. First, the projection is conformal, so that all angles plotted or measured on the chart are equal to the corresponding angles on the earth's surface. Secondly, it is a cylindrical projection. Therefore equally spaced meridians are represented, in the normal aspect, by equidistant parallel straight lines. It follows that any straight line drawn obliquely across a Mercator chart intersects each meridian at the same angle and, because of its conformal property the line represents a constant direction on the earth's surface. Such a line is known as a *rhumb line* on the earth's surface. It is of importance in navigation because a ship or aircraft has to be steered with reference to a magnetic compass, directional gyro, or some combination of the two instruments known as a gyro-compass. This is easier to do if the vessel remains on the same heading, or *course*, for an appreciable period of time rather than changing direction continually. By virtue of maintaining a constant course (and allowing for any corrections which may have to be applied to the instruments used), the craft crosses each meridian at the same angle and therefore it is following a rhumb line.

MEASUREMENT OF DIRECTION

The technique of measuring direction on a Mercator chart relies upon these two properties and it is greatly simplified by the second of them. Because the meridians are parallel, a course or bearing may be measured anywhere upon the chart and transferred to another part without introducing any errors arising from convergence of the meridians as happens, for example, on conical projections. Therefore a Mercator chart has one or more compass roses printed upon it and since these are an integral part of the chart, they cannot be displaced accidentally, as might occur in the use of a separate protractor placed over the chart and oriented to the meridian before the required angle may be read. The marine navigator uses a parallel ruler which also serves as the straight edge for other graphical work. In order to identify and plot a particular course, track or position line, the parallel ruler is laid with an edge through the centre of the rose and in the specified direction indicated by the circular scale of degrees about this point. The line thus determined may be transferred as a parallel line to the required part of the the chart for plotting. Conversely, the direction of a line plotted on the chart may be measured by sliding or rolling the parallel ruler from that line to the nearest compass rose.

In the days when graphical D.R. was still commonplace in air navigation, a separate parallel ruler was seldom used because it had a tendency to roll off the chart table and be damaged. Direction was measured with a separate square *Douglas protractor*, specially designed for navigation, or by means of a *chart table plotter*, a straight edge attached by a parallel guidance mechanism to the table and an adjustable pivot which could be set to any bearing. Today there are few occasions when graphical D.R. is used in the air, and therefore Mercator charts are infrequently used for air navigation. The modern aeronautical chart is based upon the Lambert conformal conical projection, and because of the convergence of the meridians upon it, a separate protractor must be used, this being set with respect to the meridian through the point every time a bearing is measured.

MEASUREMENT OF DISTANCE

Notwithstanding the simplicity of measuring direction on Mercator's projection, measurements of both distance and area are much affected by the variations in particular scales with latitude. In order to satisfy the condition of conformality at the same time as maintaining equidistant meridians, it is necessary to increase the spacing between the parallels of latitude in the polewards directions. This corresponds to increases in the particular scales towards the poles.

The navigator has always measured distance on Mercator's projection with dividers, transferring the length of line spanned by them to the scale provided by the subdivisions of the marginal meridians forming two of the neat lines of

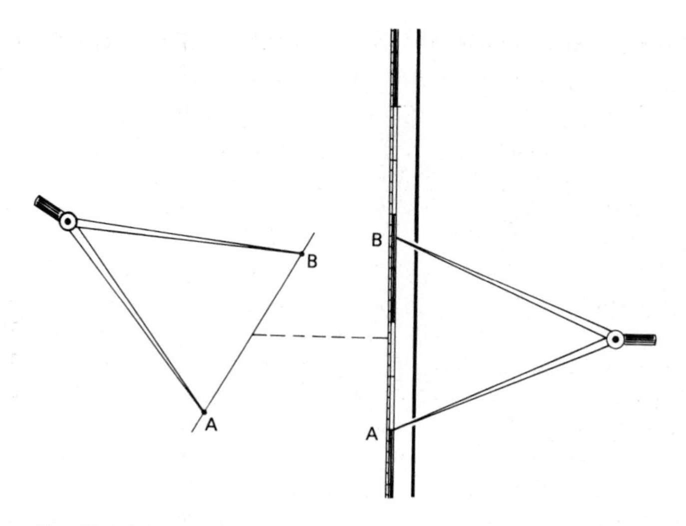

FIG. 22.1. Measurement of distance by dividers on a Mercator chart.

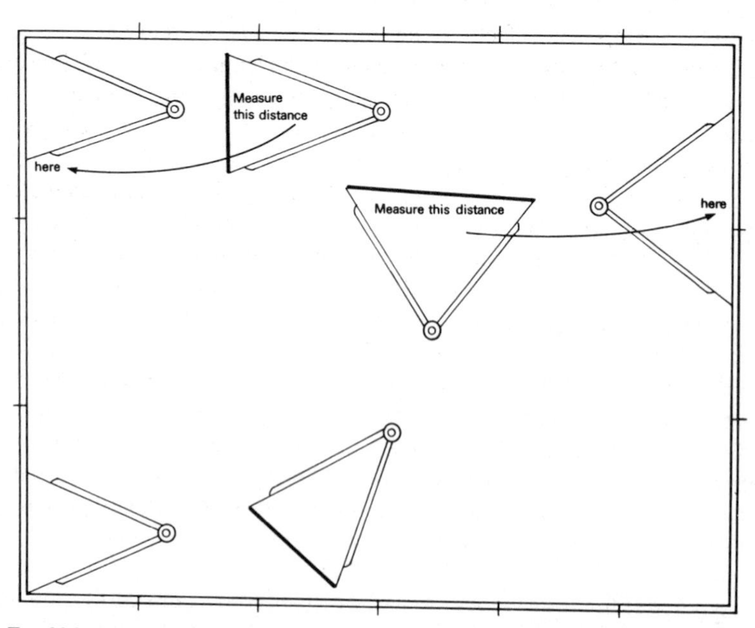

FIG. 22.2. Measurement of distance on a Mercator chart showing the location of the parts of the meridional scale to be used for measuring lines in different parts of it.

the chart, or, on aeronautical charts, to one of the meridians located within the body of the chart which have been specially subdivided into minutes of latitude for this purpose. Because of the increase in the particular scales with latitude, it is necessary to compare the separation of the dividers with that part of the meridional scale occurring in the sale latitude, as illustrated by Fig. 22.1. The rule to be followed is that the dividers are placed along the meridional scale symmetrically about the mean latitude (*halfe above, and halfe beneath*) of the terminal points of the line, as shown by Fig. 22.2. This procedure measures distance in units of minutes of arc. For many practical purposes it is sufficient to use the conversion 1 minute = 1 nautical mile = 1,852 metres, but the simplicity of recording distance in nautical miles (and calculating speed in knots) encourages the navigator to work directly in those units rather than convert distances into those more commonly used on land.

This method of measuring distance is, of course, an approximation, creating a balance between different parts of the line along which the scale is too large in one part of the chart but too small in another. Nevertheless it works well enough for D.R. navigation in open ocean, or in the air, where knowledge about position to the nearest nautical mile is usually adequate. Where greater accuracy of measurement is required it is better to calculate the course and distance, as described in a later section under the heading of *Mercator sailing.* For pilotage in enclosed waters the scales of the charts used are probably 1/100,000 and may be larger than 1/50,000. At such large scales the charts may be regarded as having constant scale everywhere, so that just as on topographical maps, the effects of the projection may be ignored.

In addition to the uncertainties of measuring the length of a rhumb line on a chart, we must stress that, apart from its use in navigation, the rhumb line distance is not a quantity which we want to know. Generally we require the shortest distance between two points and this is the length of the great circle arc passing through them. This distance is most easily obtained by calculation, using equations (3.6) and (3.7) of spherical trigonometry. However, their use is based upon the assumption that the earth is truly spherical. If we employ formulae to determine the length of the geodesic, which will be demonstrated in the next chapter, the calculations are more laborious. The magnitude of the differences between great circle and geodesic distances is also investigated in Chapter 23, pp. 547–551. Over short distances the rhumb line and great circle may be regarded as being coincident and of equal length. Therefore the difficulties of precise definition need not arise in the measurement of a succession of short linear elements, such as the components of an irregular or sinuous line representing a coastline, unless the purpose of the measurement demands extremely high accuracy.

MEASUREMENT OF AREA

Since the area scale on Mercator's projection varies according to the square of the particular scale, it follows that areas are even more exaggerated than distances. For this reason it has been customary to warn the beginner against ever attempting to make area measurements on a Mercator chart. However, this prohibition denies all possibility of measuring the area of an oceanic feature or distribution, such as the size of the Angola Basin used as a worked example later, without first compiling the required information upon an equal-area projection. This chapter described several techniques which have been developed in Russia to allow measurement of area to be made on nautical charts by applying suitable corrections.

THE GEOMETRY OF MERCATOR'S PROJECTION

In this chapter we are only concerned with the normal aspect of the projection, as illustrated by Fig. 22.4. This is because practically all nautical charts are based upon the normal aspect in which the parallels and meridians form a regular network of parallel straight lines. The reader will find superficial treatment of the transverse aspect, the transverse Mercator projection, which is used for the majority of modern topographical maps, in Chapter 10. The oblique aspect, or oblique Mercator projection, is also briefly mentioned with reference to mapping from satellite imagery in Chapter 13.

The following derivation of the projection is based upon the assumption that the earth is truly spherical. Frolov (1966) has investigated the different errors which arise in measuring areas on Mercator's projection. He concludes that area measurement is more sensitive to the various influences of the projection than to any of the other variables. Therefore no significant errors are likely to be introduced if we maintain the spherical assumption throughout the analysis.

Figure 22.3(a) represents an infinitely small quadrangle *ABCD* upon the spherical surface. Figure 22.3(b) depicts its plane representation as the infinitely small rectangle *A′B′C′D′*, on Mercator's projection. The points *A* and *D* both lie upon the same parallel of latitude, φ. The points *A* and *B* both lie in the same meridian λ. We denote the increment in latitude corresponding to the

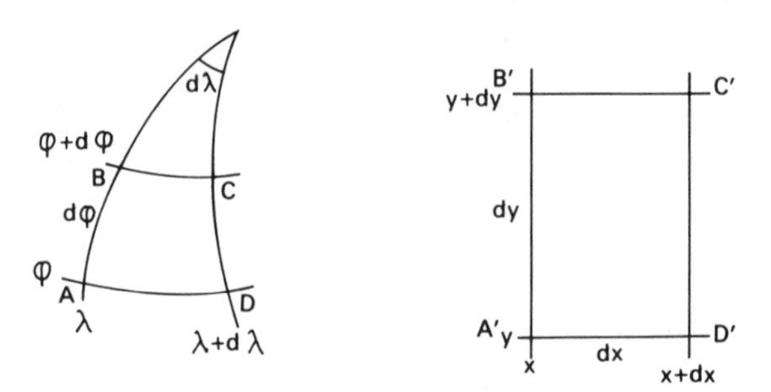

FIG. 22.3. The representation of (a) an infinitely small quadrangle, *ABCD*, on the spherical surface and (b), the corresponding rectangle *A′B′C′D′* on Mercator's projection. (Source: Maling, 1973).

distance AB as $d\varphi$ and the increment in longitude AD by $d\lambda$. On the chart we assign to the point A' the Cartesian coordinates (x, y), whose origin is the point where the equator intersects the meridian λ_0 and the y-axis coincides with that meridian. Within the figure $A'B'C'D'$ the increment in ordinate, dy, corresponds with the side $A'B'$ and the increment dx corresponds to $A'D'$.

At the point A' on the chart the meridional scale, h, is

$$h = dy/R \cdot d\varphi \tag{22.1}$$

where R is the radius of the earth.

Since the arc element of the parallel through A may be written

$$AD = R \cdot \cos\varphi \cdot d\lambda \tag{22.2}$$

it follows that the particular scale along the parallel is

$$k = dx/R \cdot \cos\varphi \cdot d\lambda \tag{22.3}$$

Since unmodified cylindrical projections have only one line of zero distortion, which is the equator (see Fig. 10.3(a), p. 182), the meridians must be correctly spaced along this line. In other words,

$$x = R \cdot \lambda \tag{22.4}$$

where λ is the longitude, expressed in radians, measured from the meridian λ_0 through the origin of the (x, y) coordinate system.

Equation (22.4) applies to all cylindrical projections, not only to Mercator. It follows that for an infinitely small increment in longitude, $d\lambda$, we have in equation (22.3)

$$k = R \cdot d\lambda/R \cdot \cos\varphi \cdot d\lambda \tag{22.5}$$

which simplifies to

$$\begin{aligned} k &= 1/\cos\varphi \\ &= \sec\varphi \end{aligned} \tag{22.6}$$

This, too, is true for all cylindrical projections. Because the normal aspect graticule of any cylindrical projection comprises perpendicular intersections of parallels and meridians, the principal directions coincide with these lines. Therefore following Maling (1973) and other books on map projections, $h = a$ and $k = b$ or vice-versa, depending upon the special properties of the projection. For a conformal projection, as in equation (7.6) we required that $a = b = h = k$. Thus for Mercator's projection we must make

$$dy/R \cdot d\theta = dx/R \cdot \cos\varphi \tag{22.7}$$

and for a sphere of unit radius,

$$dy/dx = d\varphi/\cos\varphi \tag{22.8}$$

Therefore

$$y = \int \sec\varphi \cdot d\varphi \tag{22.9}$$

TABLE 22.1. *Mercator's projection: particular scales and distortion characteristics for the unmodified version in which the line of zero distortion is the equator*

Latitude φ	Particular scales $\mu = h = k$	Area scale p	Maximum angular distortion $\omega°$
0	1·000	1·000	0
10	1·015	1·031	0
20	1·064	1·132	0
30	1·155	1·333	0
40	1·305	1·704	0
50	1·556	2·420	0
60	2·000	4·000	0
70	2·924	8·594	0
80	5·759	33·163	0

The solution of this integral is

$$y = \ln \tan(\pi/4 + \varphi/2) + C \tag{22.10}$$

where C is the integration constant. Because the origin of the plane (x, y) coordinates lies on the equator, where $\varphi = 0$, $y = 0$, we also have $C = 0$. Therefore the coordinates required to define and construct Mercator's projection for a sphere of unit radius are

$$x = \lambda$$

$$y = \ln \tan(\pi/4 + \varphi/2) \tag{22.11}$$

It follows that the particular scales

$$\mu = h = k = a = b \tag{22.12}$$

and

$$p = \sec^2 \varphi \tag{22.13}$$

Numerical values for the particular scales are given in Table 22.1. Figure 22.4 illustrates the development of the normal aspect of Mercator's projection for the whole world between latitude 80° north and south. This map also shows isograms for area scale, p.

Although the projection has been used in this way for world and atlas maps for a couple of centuries or more, it is unusual to find a chart like this being used for navigation and cartometric purposes. Normally the projection of a nautical chart of a limited sea area, such as the North Sea, has been modified so that the principal scale is preserved along some parallel other than the equator. In order to maintain this property the particular scale along the parallel φ_0 must satisfy the condition that

$$k_0 = x/R \cdot \cos \varphi_0 \cdot \lambda$$

$$= 1{\cdot}000 \tag{22.14}$$

or

$$x = R \cdot \cos \varphi_0 \cdot \lambda \tag{22.15}$$

In other words, the width of the east–west extent of the map has been reduced by the factor $\cos \varphi_0$. This is illustrated in Fig. 7.3(a). Elsewhere on the map the condition is that

$$k = R \cdot \cos \varphi_0 / R \cdot \cos \varphi \tag{22.16}$$

so that conformality is now satisfied by the condition that

$$\mathrm{d}y / R \cdot \mathrm{d}\varphi = R \cdot \cos \varphi_0 / R \cdot \cos \varphi \tag{22.17}$$

Using similar arguments to those advanced earlier we finally obtain as the coordinate expressions for the modified Mercator projection:

$$\begin{aligned} x &= \cos \varphi_0 \cdot \lambda \\ y &= \cos \varphi_0 \cdot \ln \tan(\pi/4 + \varphi/2) \end{aligned} \tag{22.18}$$

The Sailings

The term is used in marine navigation to describe the process of solving numerically the information about positions, courses and distances as an alternative to plotting and measuring the variables on a chart. We shall find in Chapter 23 that ordinary cartometric measurements may be inadequate as a means of locating the positions of maritime boundaries, and that numerical solutions are desirable. Here we review some of the relevant theory, but refer the reader to the standard textbooks on navigation, such as Admiralty (1954) for a more complete derivation of the equations which follow.

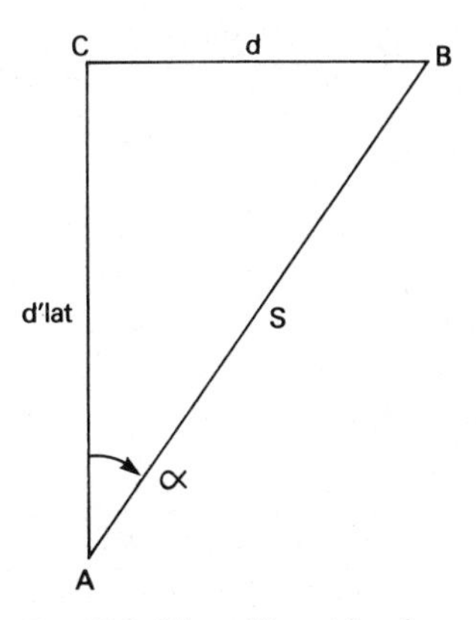

FIG. 22.5. The sailings triangle.

In navigation there are two typical problems to be solved from such calculations:

- When the positions of two points such as A and B are known, it is required to calculate the course and distance between them.
- When the position of one point, A, is known, together with the bearing and distance to another point B, it is required to determine the position of B. There are two solutions of interest to us. These are known as *plain sailing* and *Mercator sailing*. In both solutions we treat with the triangle ABC in Fig. 22.5 as if it were a plane triangle, solving unknown sides or angles by plane trigonometry. In the first case we assume that the earth is truly flat, which is useful enough for small working with distances in low latitudes. In the second case we simulate the conditions represented upon a Mercator chart but calculate the unknown values instead of plotting and measuring them by dividers or protractor.

In Fig. 22.5 the course for the rhumb line from A to B is given by the clockwise angle, α, measured from the meridian AC to the line AB. The distance AB along the rhumb line is designated s.

Proceeding along the rhumb line from A to B, a craft makes good the distance CB in the west–east direction. This is the difference in longitude $\lambda_b - \lambda_c = \delta\lambda$. The arc distance corresponding to CB is less than the corresponding along the equator because of the convergence of the meridians. We define the distance $CB = d$ as the *departure* and, for the parallel of latitude φ_B it may be expressed as

$$d = \delta\lambda \cdot \cos\varphi_B \qquad (22.19)$$

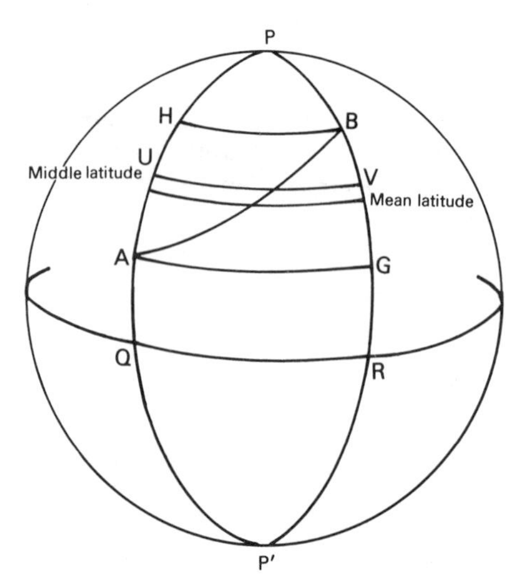

FIG. 22.6. The nature of middle latitude.

Then, treating the triangle ABP is a plane figure,

$$d = s \cdot \sin \alpha \tag{22.20}$$

The third side of each small triangle is an arc of the meridian, AC. Since the point C lies in latitude φ_B, this arc corresponds to the meridional arc $\delta\varphi = \varphi_B - \varphi_A$ which is, of course, the difference in latitude, $\delta\varphi$, also known as *d'lat*.

In the plane triangle ABC

$$\delta\varphi = s \cdot \cos \alpha \tag{22.21}$$

Because the arc length of a parallel varies with latitude, some degree of approximation is required in making some of the calculations. For example, an approximation is required to choose a value for φ which will provide an average value of $\cos \varphi$ to be used in equation (22.19). If the difference in latitude between A and B is not large, the mean latitude, or $\varphi_M = \frac{1}{2}(\varphi_A + \varphi_B)$, may suffice so that (22.19) takes the form

$$d = \delta\lambda \cdot \cos \varphi_M \tag{22.22}$$

but this expression is only strictly correct where $\varphi_A = \varphi_B = \varphi_M$.

With increasing difference in latitude, the error introduced by using the mean latitude also increases, so that a different solution must be sought. Since the departure is greater than the arc HB, and is less than the arc AG in Fig. 22.6, it must be equal to the length of some intermediate parallel such as UV. This is called the *middle latitude*, φ', and it has the property that

$$QR = UV \cdot \sec \varphi' \tag{22.23}$$

or

$$\delta\lambda = d \cdot \sec \varphi' \tag{22.24}$$

This is an accurate formula, but φ' is unknown and it is awkward to determine directly. The traditional method used in navigation is to apply a correction which varies with both mean latitude and $\delta\varphi$. It is available in published volumes of navigation tables where it is called *Workman's tables* or the *mid-latitude correction.*

The way in which the correction may be determined is described in the more advanced textbooks on navigation, e.g. Admiralty (1954) where the following equation is derived

$$\sec \varphi' = 1/(\varphi_B - \varphi_A)[\ln \tan(\pi/4 + \varphi_B/2) - \ln \tan(\pi/4 + \varphi_A/2)] \tag{22.25}$$

We shall see that the term in square brackets has especial importance, and is known as the *difference in meridional parts.* Sadler (1956) has shown that for small differences in latitude

$$\varphi' = \varphi_M + 0{\cdot}043\,\delta\varphi^2(2 \tan \varphi_M + \cot \varphi_M) \tag{22.26}$$

where $\delta\varphi$ is expressed in degrees and the result is given in minutes of arc. The correction represented by the right-hand term in equation (22.25) is always positive. It varies from 0′ for $\delta\varphi < 2°$ and $\varphi_M < 45°$ to $> 120'$ for $\delta\varphi > 20°$ in latitudes greater than $\varphi_M > 70°$.

Plane Sailing

Figure 22.6 represent the rhumb line AB as the hypoteneuse of a right-angled triangle. Clearly the course and distance solution may be obtained by calculating

$$d = \delta\lambda \cdot \cos\varphi' \tag{22.27}$$

$$\tan\alpha = d/\delta\varphi \tag{22.28}$$

$$s = d \cdot \operatorname{cosec}\alpha \quad \text{or} \quad \delta\varphi \cdot \sec\alpha \tag{22.29}$$

The coordinates of an unknown point may be found from

$$d = s \cdot \sin\alpha$$

$$\delta\varphi = s \cdot \cos\alpha$$

The plane sailing solution is so called because the assumption is made that the earth is flat and the solution is by plane trigonometry. It is generally considered to be adequate, for short distances in low latitudes where the errors are less than the accuracy of graphical D.R. navigation. It is often considered that use of the plane sailing solution is limited to lines of length < 600 n.m. and $\varphi_M < 60°$.

Mercator Sailing

In order to avoid such gross approximations so that it is applicable both to longer lines and in higher latitudes it is necessary to develop the argument a stage further and imagine how the plane triangle of Fig. 22.6 would appear on a Mercator chart. This is shown in Fig. 22.7. The principal difference between this and the preceding figure is that the distance AC, corresponding to $\delta\varphi$ on the earth, is now based upon the difference between the two ordinate values $y_B - y_A$ on the Mercator projection. These ordinates may be determined from equation (22.10) for each of the points A and B. Expressed in minutes of arc, these values are known as the *meridional parts* or *Mercatorial parts*, and designated algebraically by M. For example, if $\varphi = 40°$, the solution of (22.10) is

$$\begin{aligned} y &= \ln\tan(45° + 20°) \\ &= 0{\cdot}762\,9096 \text{ (radians)} \end{aligned}$$

and

$$\begin{aligned} M_A &= 0{\cdot}762\,9096 \times 3{,}437{\cdot}75 \\ &= 2{,}622{\cdot}9 \end{aligned}$$

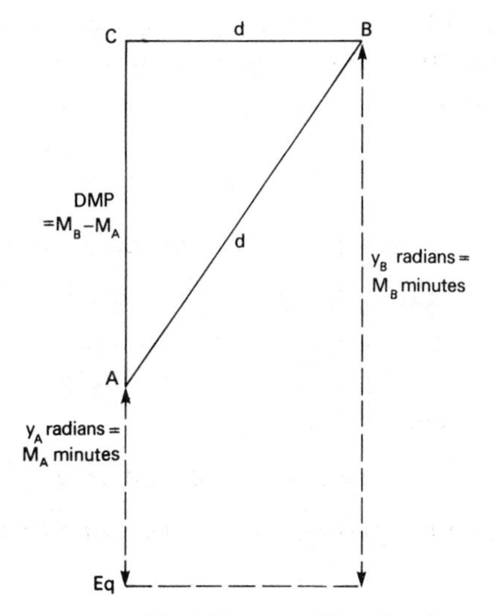

FIG. 22.7. The Mercator sailing triangle.

The quantity 3,437·75 is the factor needed to convert from radians into minutes of arc. A similar solution may be obtained for M_B. The side AC of the triangle in Fig. 22.7 is

$$D = (M_B - M_A)$$

or DMP, the *difference in meridional parts*

Tables of meridional parts have appeared in all navigation tables ever since the first were published by Wright in 1599. However, the solution of the ordinate is so easy to execute on a pocket calculator having hard-wired trigonometric and logarithmic functions that there is no longer any need to have access to tables.

For the course and distance solution, Fig. 22.7 gives

$$\tan\alpha = \delta\lambda/(M_B - M_A) \tag{22.30}$$

so that, as in (22.29),

$$s = \delta\varphi \cdot \sec\alpha$$

it should be observed that the use of $\delta\lambda$ with the difference in meridional parts in (22.30) avoids having to determine of departure with the additional complication of finding the middle latitude.

For the position-finding solution,

$$\delta\lambda = (M_B - M_A)\tan\alpha \tag{22.31}$$

and, as in (22.21)

$$\delta\varphi = s \cdot \cos\alpha$$

In order to demonstrate the use of these equations, we employ them to determine the course and distance betweeen two points of known position:

$$A\text{:}\quad 58°00'\,\text{N},\quad 0°00'$$

$$B\text{:}\quad 62°00'\,\text{N},\quad 10°00'\,\text{E}.$$

It follows that $\delta\varphi = 4° = 240'$ and $\delta\lambda = 10° = 600'$

(1) PLANE SAILING SOLUTION. From Inman's tables, under the heading *Mid. Lat. Corrn.* we find that for $\varphi_M = 60°$ and $\delta\varphi = 4°$, the correction is 3′. Therefore $\varphi' = 60°03'$. More accurately from Sadler's formula (22.26) it is 60°02′.78.

From (22.27),

$$d = 600 \times \cos 60°03'$$
$$= 299'{\cdot}55$$

Then in (22.28),

$$\tan\alpha = 299{\cdot}55/240$$
$$= 1{\cdot}248\,1102$$

or

$$\alpha = 51°\,2979$$

Finally, from (22.29),

$$s = 299{\cdot}546 \times 1{\cdot}281\,3819$$
$$= 383'{\cdot}83$$

or

$$s = 240 \times 1{\cdot}599\,3058$$
$$= 383'{\cdot}83$$

For navigation purposes it is generally sufficient to determine s to 1 decimal place.

(2) MERCATOR SAILING. From tables or by solution of (22.10) we obtain the meridional parts

$$A\text{:}\quad \varphi_A = 58°\,\text{N};\quad M_A = 4{,}294{\cdot}3$$
$$B\text{:}\quad \varphi_B = 62°\,\text{N};\quad M_B = 4{,}775{\cdot}0$$

Therefore D.M.P. = 480′·7. From (22.31),

$$\tan\alpha = 600/480{\cdot}7$$
$$= 1{\cdot}248\,2138$$

or

$$\alpha = 51°{\cdot}3002$$

Then from (22.29),

$$s = 240 \times 1{\cdot}599\,3864$$
$$= 383'{\cdot}85$$

It can be seen that, in this example, the distance s is only slightly longer than that obtained from the plane sailing solution, for this example has been chosen to represent a line shorter than 600 n.m., but is at the limiting latitude, $\varphi_M = 60°$ N. We shall return to the same example in Chapter 23, in order to demonstrate the errors likely to be encountered in the determination of the positions of maritime boundaries by different methods.

Measurement of Irregular Lines on Mercator's Projection

So far we have only considered the problem of measuring the rhumb line distance betweeen two points which may be several hundred miles apart. It is now desirable to consider some suitable method of measuring the length of a sinuous or irregular line on a nautical chart.

If we differentiate equation (22.6) for μ with respect to φ, we obtain

$$\mathrm{d}\mu/\mathrm{d}\varphi = \sin\varphi \cdot \sec^2\varphi \tag{22.32}$$

or

$$\mathrm{d}\mu = \sin\varphi \cdot \sec^2\varphi \cdot \mathrm{d}\varphi \tag{22.33}$$

Because we usually measure an irregular line in short rectilinear elements by dividers, we may investigate the behaviour of (22.33) with reference to such elements. If we define the latitude and particular scale at one end of such an element as φ, μ; the other end being φ_1 and μ_1 we may assume that φ and φ_1 do not differ much. Consequently there is also little difference between μ and μ_1. Therefore we may write

$$\mathrm{d}\mu/\mu_1 = (\sin\varphi \cdot \sec^2\varphi/\sec\varphi_1) \cdot \mathrm{d}\varphi \tag{22.34}$$

and

$$\mathrm{d}\mu/\mu_1 = \tan\varphi_1 \cdot \mathrm{d}\varphi \tag{22.35}$$

We know from earlier chapters that a practical limit to the precision of dividers measurements is about 1/300. Therefore we put $\mathrm{d}\mu/\mu_1 \approx 1/300$, and use this to

TABLE 22.2. *North–south extent of zones on Mercator's projection within which the particular scales do not vary by more than 1/300*

Latitude φ	Width of zone $(\varphi - \varphi_1)$	Latitude φ	Width of zone $(\varphi - \varphi_1)$
5	2° 12′	45	11′
10	1° 05′	50	9′·6
15	43′	55	8′·0
20	31′	60	6′·6
25	25′	65	5′·3
30	20′	70	4′·2
35	16′	75	3′·1
40	14′	80	2′·0

obtain a value for $d\varphi$. After some rearrangement we have

$$d\varphi = 0{\cdot}00333 \cot \varphi \qquad (22.36)$$

where $d\varphi = \varphi - \varphi_1$ is expressed in radians. Values of (22.36) are given in Table 22.2. If an irregular line lies wholly within a zone correspond to one of the tabulated values for $(\varphi - \varphi_1)$, its length may be measured just as if it were shown on a large-scale map or chart unaffected by variations in particular scale. Often, however, the line to be measured extends through several of these zones. Then it is necessary to measure the line in parts, each corresponding to a strip of width $d\varphi$. The following example demonstrates the use of this technique.

- We wish to measure the length of the coastline of the island of Timor from an Admiralty chart. The island lies between latitudes 8°25′ S and 10°20′ S. So close the equator, the projection in use is the version of Mercator having the principal scale preserved along it. Inspection of Table 22.2 indicates that $d\varphi$ is greater than 1° in these latitudes. Consequently it should suffice to measure the length of the coast in two separate zones. The boundary between them may be determined from either of the following expressions which give slightly different results:

$$\varphi'' = \varphi_N + 0{\cdot}00333 \cot 8°25' = 9°42' \text{ S.}$$
$$= \varphi_S - 0{\cdot}00333 \cot 10°20' = 9°18' \text{ S.}$$

  The mean of these values is 9°30′ S, which would be a suitable boundary to choose between the zones.

- Compare this with the problem of measuring the length of the coastline of South Georgia. This island lies between latitudes 53°55′ S and 55° S, so that $\delta\varphi = 65'$. From Table 22.3 it appears that for $\varphi = 55°$, $(\varphi - \varphi_1) = 8'$ so that it would evidently be necessary to subdivide the island into eight latitude zones and reduce the measured lengths in each of these sections separately. However Table 22.2 refers only to the unmodified version of Mercator's projection only a very small scale chart covering virtually the whole of the South Atlantic Ocean would show both the equator and South Georgia on the same sheet of paper, so that, on the grounds of scale

TABLE 22.3. *North–south extent of zones on Mercator's projection within which the particular scales do not vary by more than 1/300. Principal scale is preserved in latitude 60°*

Latitude φ	Width of zone $(\varphi - \varphi_1)$	Latitude φ	Width of zone $(\varphi - \varphi_1)$
45°	22′·9	60°	13′·2
50	19′·2	65	10′·7
55	16′·0	70	8′·3

alone, such a chart would not be suitable for measuring the length of the coastline of South Georgia.

It is therefore necessary to consider how the proposed theory applies to charts on which the principal scale is preserved along some parallel, φ_0, other than the equator. From consideration of equations (22.16) to (22.20) we may write

$$d\varphi = 0{\cdot}000333/\cos\varphi_0 \cdot \tan\varphi \qquad (22.37)$$

Thus, for a chart in which $\varphi_0 = 60°$ we have the following revised values for $d\varphi$. Using the same kind of argument applied to the first example we now see that using a chart of the Scotia Sea meeting this specification it would only be necessary to divide the coast of South Georgia into four different latitude zones of width 16′–18′ for measurement purposes.

Area Measurement on Mercator's Projection

In order to determine the corrected area, A_0, of a parcel from its measured value, A, it is necessary to correct for variations in the area scale. The simplest method is to use the equation

$$A_0 = A/p \qquad (22.38)$$

or, putting the reciprocal of p as the correction K,

$$A_0 = A \cdot K \qquad (22.39)$$

However, for this correction to be of any real use we must assume that p (or K) remains constant. A constant value for K can only occur in an equal-area projection, for which $p = K = 1$ so that there is no point in making any correction. In the normal aspect of Mercator's projection the area scale varies with latitude, but because p varies with φ according to $\sec^2\varphi$, the rate of change is rapid, particularly in high latitudes.

The simplest form of correction for a parcel lying between two parallels φ_N and φ_S is to determine the mean latitude of the parcel, φ_M, and calculate a coefficient K_M for this. We may also obtain a relative correction, δ_M, which has the form

$$\delta_M = (A_0 - A)/A \qquad (22.40)$$

from which it follows that

$$A_0 = A(1 + \delta_M) \qquad (22.41)$$

We introduce this apparently trivial result because this form of equation will be useful later. The suffix "$_M$" has been used in δ_M to indicate that the correction has been based upon the initial choice of a mean parallel φ_M. Similar relationships hold good for other methods of choosing a single value for purposes of calculation. For example we might determine the latitude φ_g

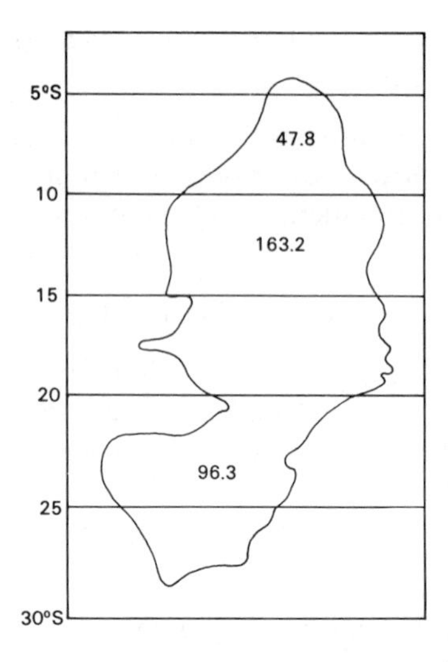

FIG. 22.8. The outline of the Angola Basin, traced from the *Morskoy Atlas* and showing the subdivision of it into three zones. The areas of the basin contained in each of these zones have been measured by planimeter and are expressed in vernier units. (Source: Ginzburg, 1958).

passing through the centre of gravity of the parcel to be measured. Then δ_g, p_g and K_g refer to this solution. We may also determine the mean value for p, either by calculating the average of the extreme values for area scale, or by determining a mean value of p from several equidistantly spaced parallels passing through the parcel. We shall see that each of these procedures gives rise to a different result.

The following example showing the effects of different crude corrections was originally used by Ginzburg (1958). The feature measured was the Angola Basin, defined as being bounded by the 5,000 metre isobath shown on a map of the South Atlantic Ocean at scale 1/3,000,000 in the *Morskoy Atlas* (1950). The outline of this parcel is shown in Fig. 22.8. This map is based upon Mercator's projection with the principal scale preserved along the parallel 30° S. The area was measured by planimeter, giving a mean result of 307·3 vernier units. Calibration of the planimeter indicated that 1 v.u. = 6,114 km². Making the assumption that there is no correction to be applied, i.e. that the projection has no influence upon the measurement, the calculated area was

$$A_1 = 307{\cdot}3 \times 6{,}114$$
$$= 1{,}879{,}000\ \text{km}^2 \qquad \text{(a)}$$

However, Fig. 22.8 indicates that the parcel extends from about 4° S through 28° S. Consequently the following area scales and correction coefficients apply in different parts of the map. Since the mean latitude is 16° S, the second result may be obtained by applying a correction to take this into account. Thus, from

TABLE 22.4. *Mercator's projection with $\varphi_0 = 30°$ S; particular scales, area scale and correction coefficients*

Latitude φ	Particular scale μ	Area scale p	Correction coefficient K
4	0·8681	0·7537	1·3268
5	0·8693	0·7537	1·3232
10	0·8794	0·7733	1·2931
15	0·8966	0·8038	1·2440
16 (φ_M)	0·9009	0·8117	1·2320
20	0·9216	0·8495	1·1774
25	0·9556	0·9131	1·0952
28	0·9808	0·9620	1·0395
30	1·0000	1·0000	1·0000

Table 22.4,

$$A_2 = 307{\cdot}3 \times 6{,}114 \times 1{\cdot}2320$$
$$= 2{,}315{,}000 \text{ km}^2 \qquad \text{(b)}$$

This is nearly 19% larger than the area determined in (a). If, however, we obtain the mean area scale from its value for the two limiting parallels (4° and 28°), we find $p = 0{\cdot}8579$ so that $K = 1{\cdot}1657$ and

$$A_3 = 307{\cdot}3 \times 6{,}114 \times 1{\cdot}1657$$
$$= 2{,}190{,}000 \text{ km}^2 \qquad \text{(c)}$$

which is 5% smaller than (b). We may also extract values for p at regular intervals of latitude through the parcel. Using those for p in latitudes 5, 10, 15, 20 and 25° in Table 22.4, we find the mean value of p to be 0·8492 so that $K = 1{\cdot}1775$ and

$$A_4 = 307{\cdot}3 \times 6{,}114 \times 1{\cdot}1775$$
$$= 2{,}212{,}000 \text{ km}^2 \qquad \text{(d)}$$

This is 1% larger than (c) but 4% smaller than (b).

Ginzburg considered that it was necessary to subdivide the parcel into the three zones illustrated in Fig. 22.8. Planimeter measurements of the area of each zone gave the following results.

$$A_5 = 2{,}273{,}000 \text{ km}^2 \qquad \text{(e)}$$

This is 2% smaller than (b). In Table 22.5 the values for the mean latitude of each zone were estimated, by the author, from Fig. 22.9. However Ginzburg used latitude 8° S in zone 1 and 23°30′ S in zone 3. This altered the areas of zones 1 and 3 so that the total area was increased to

$$A_6 = 2{,}284{,}000 \text{ km}^2 \qquad \text{(f)}$$

TABLE 22.5. *The area of the Angola Basin; measurement by zones, (data from Ginzburg, 1958)*

Zone no.	Planimeter reading (v.u.)	Mean latitude	Calculation	Area in km^2
1	47·8	5° S	47·8 × 6114 × 1·3232	386·70
2	163·2	15	163·2 × 6114 × 1·2440	1,241·269
3	96·3	25	96·3 × 6114 × 1·0952	644·830
Sum	307·3			2,272·803

The differences between the first determination (a) and all the others indicate that there is some wisdom in warning the beginner not to measure area on a Mercator chart. Moreover, comparison of (e) and (f) demonstrates that a very small change in the interpretation of the variables introduces a difference of 0·5% into the results. Consequently it is desirable to devise some method of correction which is not wholly based upon estimation.

In order to reduce such uncertainty it is necessary to study the changes in area scale, dp, which may occur between some parallel φ_1 and another parallel of latitude taken as datum. Using the northern hemisphere model, and to avoid problems of sign, comparison is made between the area scale p_1 in latitude φ_1 and that of p_s in latitude φ_s which lies nearer the equator.

For Mercator's equation we may write

$$dp/p_s = [2\sec^2\varphi\cdot\tan\varphi\cdot d\varphi]/\sec^2\varphi_s \tag{22.42}$$

If φ_1 and φ_s lie sufficiently close to one another, we may assume that

$$\sec\varphi_1 \approx \sec\varphi_s$$

so that equation (22.42) simplifies to

$$dp/p = 2\tan\varphi_1\cdot d\varphi \tag{22.43}$$

We now assume that we are able to measure the area of a parcel to 0·1%. This is a reasonable assumption, justified in Chapter 17, if we make use of a planimeter having one vernier unit representing 1/1,000 of the main scale to make the measurements. Then we may specify that the area scale within the measured zone between φ_s and φ_1 should not vary by more than 0·1% owing to the projection. In other words we have specified that

$$2\tan\varphi\cdot d\varphi = 1/1{,}000 \tag{22.44}$$

Solving this expression for dφ we obtain

$$d\varphi = 3{,}437{\cdot}75/1{,}000\tan\varphi = 3{\cdot}43775\cot\varphi \tag{22.45}$$

whereas the acceptable width of the zone within which measurements can be made is double this amout or

$$(\varphi_N - \varphi_s)' = 6{\cdot}8755\cot\varphi \tag{22.46}$$

TABLE 22.6. *Width of the zone $(\varphi_N - \varphi_s)'$ within which the variations in area scale do not exceed 1/1000*

Latitude	$d\varphi$	$(\varphi_N - \varphi_s)'$	Latitude	$d\varphi$	$(\varphi_N - \varphi_s)'$	Latitude	$d\varphi$	$(\varphi_N - \varphi_s)'$
5	39′·4	1° 18′·8	35	4′·9	9′·8	65	1′·6	3′·2
10	19′·5	39′·0	40	4′·1	8′·2	70	1′·3	2′·6
15	12′·8	25′·6	45	3′·4	6′·8	75	0′·9	1′·8
20	9′·4	18′·8	50	2′·9	5′·8	80	0′·6	1′·2
25	7′·4	14′·8	55	2′·4	4′·8	85	0′·3	0′·6
30	6′·0	12′·0	60	2′·0	4′·0			

As elsewhere, the constant 3,437·75 is the conversion factor from radians to minutes of arc. Table 22.6 gives comparative values for $(\varphi_N - \varphi_s)'$. Clearly the zones which satisfy these conditions are so narrow, particularly in high latitudes, that there is no obvious practical way in which these results can be used to control measurement of area. A solution must be sought which applies a correction for somewhat wider latitude zones if this is to have any practical value.

Volkov's Correction

In an early investigation into some practical method of making corrections, Volkov (1950b) describes a method which appears to work well at comparatively large scales, even in high latitudes. We start from the relationships of the area scales along two parallels, φ_s, which is the southern parallel of a parcel nearest the equator, where the area scale is $p_s = \sec^2 \varphi_s$ and some other parallel, φ_1 within the parcel, where the area scale is $p_1 = \sec^2 \varphi_1$. From the ratio of these area scales we may derive a relative correction, Q, where

$$Q = 1 - [\sec \varphi_s / \sec \varphi_1]^2 \qquad (22.47)$$

Then, from equations (22.42)–(22.45) the corrected area may be determined from

$$\begin{aligned} A_0 &= A(1 - Q) \\ &= A - Q \cdot A \end{aligned} \qquad (22.48)$$

Volkov devised a nomogram to determine Q, the construction and use of which we describe here, although given the power of modern pocket calculators it is easier to solve (22.47) directly as required, and there is no need to construct the nomogram.

In the nomogram the abscissa represents the difference in latitude $(\varphi_1 - \varphi_s)$ between some parallel of latitude passing through a given a quadrangle and φ_s which is the southern limit of that quadrangle. The ordinate represents Q in equation (22.47). For any latitude φ_1 there exists a sloping line relating $\delta\varphi$ to Q. For example, in Fig. 22.9, for $\varphi_1 = 64°30'$ and $\varphi_s = 64°20'$ it follows that $\delta\varphi = 10°$.

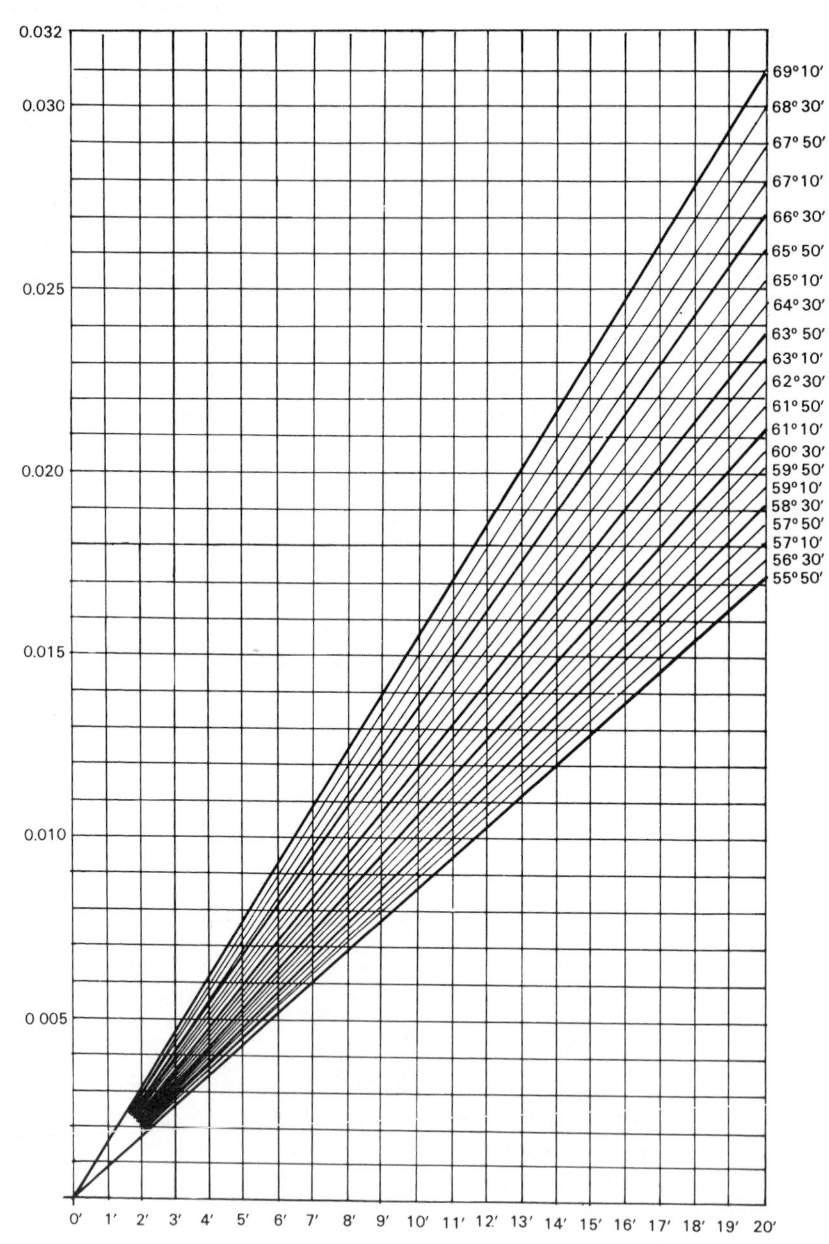

FIG. 22.9. Nomogram for correcting area measurement on Mercator's projection. (Source: Volkov, 1950b).

Moreover

$$\mu_1 = \sec \varphi_1 = 2{\cdot}32282$$

and

$$\mu_s = \sec \varphi_s = 2{\cdot}30875$$

From (22.47)

$$Q = 1 - 0{\cdot}9939\,452$$
$$= 0{\cdot}012$$

The use of this correction can be demonstrated by the example illustrated in Fig. 22.10. This shows a rectangle corresponding to a 20 × 30′ quadrangle on a Mercator chart and which is traversed by two boundaries dividing it into the three separate parcels labelled 1, 2 and 3. The areas of these parcels, together with the area of the whole quadrangle have been measured by planimeter and

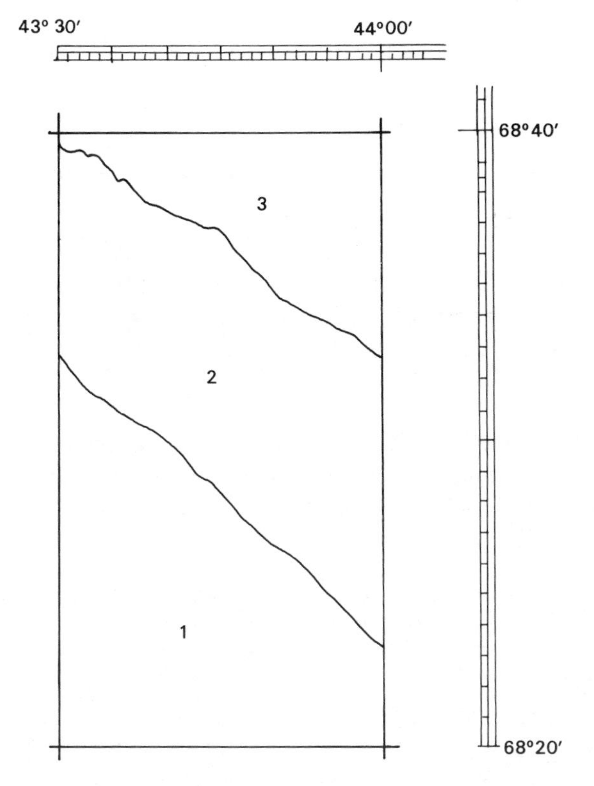

FIG. 22.10. An example of correcting area measurements in high latitudes on Mercator's projection. (Source: Volkov, 1950b).

the results of these measurements, expressed in vernier units, are

Parcel no. 1		528·5 vernier units
Parcel no. 2		537·8 vernier units
Parcel no. 3		247·2 vernier units
	Total:	1,313·5 vernier units

Independent measurement of the whole quadrangle: 1,313·5 vernier units

Use of the Nomogram

Stage 1: In order to determine the correction Q_M to be applied to the whole quadrangle, we must first find φ_M, the mean latitude. Then we determine $\delta\varphi = \varphi_M - \varphi_s$. In this case $\varphi_M = 68°30'$ and $\delta\varphi = 10'$.

On the nomogram select the line corresponding to the latitude 68°30′ and follow this to the intersection with the line $\delta\varphi = 10'$. Read the ordinate Q_M which gives a correction coefficient of 0·015

Stage 2: Multiply the planimeter reading for the whole quadrangle by this coefficient:

$$A \cdot Q_M = 1{,}313{\cdot}5 \times 0{\cdot}015 = 19{\cdot}7 \text{ v.u.}$$

Subtract this value from the planimeter reading for the whole quadrangle:

$$A - A \cdot Q_M = 1{,}313{\cdot}5 - 19{\cdot}7 = 1{,}293{\cdot}8 \text{ v.u.}$$

Stage 3: Estimate the positions of the mean parallels through each of the parcels. Using, in turn, the sloping lines corresponding to each of the values for φ_1 and the abscissa $\delta\varphi_1$, we obtain the corrections Q_1. These are listed in the fourth column of Table 22.7.

Stage 4: The corrections are applied to each of the parcels in the same way as they were to the planimeter readings for the whole quadrangle. The results are given in columns 5 and 6 of Table 22.7.

TABLE 22.7. *Corrections using the Volkov method to the areas of parcels measured by planimeter on Mercator's projection*

Parcel no.	φ_1	$(\varphi_1 - \varphi_s)$	Q_1	Correction to planimeter	Corrected readings	Area in km²
1	68° 24′	4′	0·006	3·2	525·3	306·7
2	68° 32′	12′	0·018	9·7	528·1	308·3
3	68° 38′	18′	0·027	6·7	240·5	140·4
			Totals	19·6	1,293·9	755·4

Stage 5: Calculate the area for the quadrangle measuring $20 \times 30'$ on the spherical surface. This may be done using equations (11.00) and (11.00). In this example the result is 755·29 km². Therefore 1 v.u. = 755·29/1,293·9 = 0·584 km². Multiplying the corrected readings in column (6) for each parcel by this conversion factor we obtain the areas in square kilometres which are recorded in the final column. The sum of the last three columns serves as checks upon the arithmetic, and upon the reliability of the estimates made of φ_1 for each parcel. If we have made a poor estimate of the mean latitude, then the sum of column (5) will not agree with the corresponding correction applied to the whole quadrangle in Stage 2.

We have seen that Volkov proposed using the area scale on the mean parallel to represent the average area scale for the whole quadrangle. We have already demonstrated that this is not necessarily the correct procedure because the mean of the area scale is not the same as the area scale on the mean parallel. We saw this in the study of measurement of the area of the Angola Basin where $p_M = 0{\cdot}8117$, but the average value for p is 0·8492, the last being the mean derived from the area scale on six different parallels at intervals of 5° in latitude. In that particular example there was a 4% difference between the areas computed from these figures. Frolov (1966) has shown that the substitution of p_M by p also alters the value of Q by

$$Q_M - Q_P = [\delta\varphi^2/12]\cdot\sec^2\varphi_s \tag{22.49}$$

where Q_p is the correction based upon the use of p for the quadrangle and Q_M is that determined in Stage 1 above.

Frolov's Method for Use on Small-Scale Charts

Volkov's correction is satisfactory for use with large-scale charts. This is indicated by both the small size of the unit quadrangle and the evident simplicity of the parcel boundaries in Fig. 22.10. At a smaller scale a larger portion of the earth's surface is contained within the unit quadrangle and the shape of the parcel is usually much more complicated so that the simple example illustrated in Fig. 22.10 may no longer be applicable.

In a detailed analysis of the various factors which may affect Volkov's method, Frolov (1966) proposed a number of different procedures which may be used to overcome the difficulties of measuring areas on small-scale charts. His main contribution was to introduce the concept of a triangular control figure formed by dividing the unit quadrangle by its diagonal. Since this diagonal is a rhumb line on the Mercator chart, he adopted the alternative and somewhat archaic word *loxodrome*, sometimes used to describe a rhumb line, and called the resulting figure a *loxodromic triangle*. This is indeed the same triangle illustrated in Fig. 22.7, which we used to describe Mercator sailing. Figure 22.12 represents the same triangle modified to show the different terminology used in this context. Obviously the two triangles in this figure

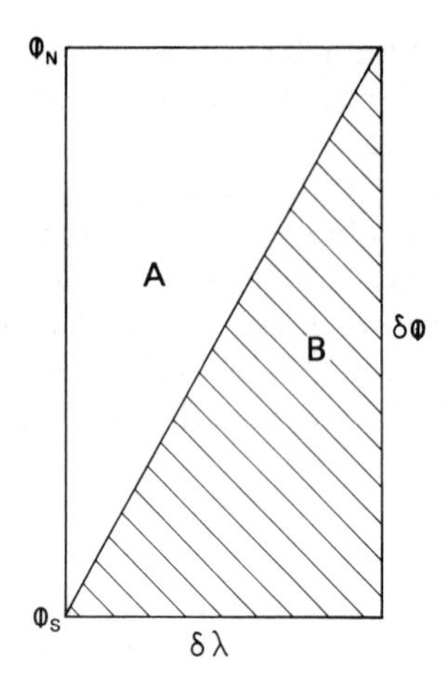

FIG. 22.11. Two forms of loxodromic triangle. (Source: Frolov, 1966).

differ in area, and, from the analysis of their geometry Frolov derived expressions relating these areas as well as the form of correction coefficient to be used with each.

The *type A triangle* is that having its base on the northern parallel of the quadrangle. He showed that its area could be expressed as

$$A_A = \tfrac{1}{2}A(1 - f) \tag{22.50}$$

The *type B triangle* is that having its base on the southern parallel and has area

$$A_B = \tfrac{1}{2}A(1 + f) \tag{22.51}$$

In both of these equations A is the area of the quadrangle of which they are components and

$$f = \delta\varphi/3 \cdot \tan\varphi_s(1 - \delta\varphi \cdot \operatorname{cosec} 2\varphi_s) \tag{22.52}$$

The corresponding correction coefficients were also determined by him as

$$Q_A = \tfrac{4}{3}\delta\varphi \cdot \tan\varphi_s + \tfrac{1}{2}\delta\varphi^2(1 - \tfrac{2}{3} \cdot \tan^2\varphi_s) \tag{22.53}$$

and

$$Q_B = \tfrac{2}{3}\delta\varphi \cdot \tan\varphi_s + \delta\varphi^2/6 \tag{22.54}$$

Table 22.8 represents Frolov's summary for values of Q, A, f, Q_A, Q_B and A_B for 5° intervals of $\delta\varphi$. Associated with the introduction of the loxodromic triangle as an alternative control figure to replace the quadrangle where the boundary of the parcel to be measured approximates to the diagonal of the quadrangle, is the need to seek a more realistic value than the mean latitude, φ_M, to be used in the initial calculations corresponding to Stage 2 of Volkov's method. The preferred parallel would be that passing through the centre of gravity of the parcel which, in effect, divides it into two portions of equal area

TABLE 22.8. *Values for Q and A for spherical quadrangles and loxodromic triangles determined by Frolov (1966); areas are expressed in hundreds of square kilometres*

	Quadrangle			Loxodromic triangle			
Latitude	Q (%)	Area (A)	f (%)	Q_A (%)	Area (A_A)	Q_B (%)	Area (A_B)
0°							
	0·2	3,074	0·127	0·4	1,535	0·1	1,539
5							
	1·0	3,051	0·383	1·4	1,520	0·6	1,531
10							
	1·8	3,005	0·644	2·4	1,493	1·2	1,512
15							
	2·6	2,938	0·915	3·5	1,456	1·7	1,482
20							
	3·4	2,848	1·203	4·6	1,407	2·2	1,441
25							
	4·3	2,737	1·511	5·7	1,348	2·8	1,389
30							
	5·2	2,605	1·849	7·0	1,278	3·5	1,327
35							
	6·3	2,453	2·226	8·4	1,199	4·2	1,254
40							
	7·5	2,282	2·657	10·0	1,111	5·0	1,171
45							
	8·8	2,094	3·163	11·8	1,014	6·0	1,080
50							
	10·5	1,889	3·774	13·9	909	7·1	980
55							
	12·4	1,669	4·193	16·6	800	8·4	869
60							
	15·0	1,436	5·546	19·8	678	10·2	758
65							
	18·4	1,191	6·949	24·2	554	12·6	637
70							

like a give-and-take line. Figure 22.12 illustrates the nature of the problem by means of four geometrical figures regarded as imaginary parcels.

In Fig. 22.12(a) the parcel to be measured approximates to a parallelogram so that the proportion of its area remains constant throughout the quadrangle bounded by the parallels φ_N and φ_S. For this configuration the mean parallel and that passing through the centre of gravity, g, are virtually identical so that the choice of φ_M is appropriate.

Although there are well-known methods of determining the centre of gravity of a plane figure or a flat disc, which are used in other branches of applied mathematics, the methods of determining this point for cartometric purposes have been wholly based upon visual estimation. Frolov studied the magnitude of the error which may arise in making such an estimation, and has shown that the error in the correction coefficient, δk, may be expressed as

$$\delta k = 2 \cdot \sin \varphi_g \cdot \delta\varphi \qquad (22.55)$$

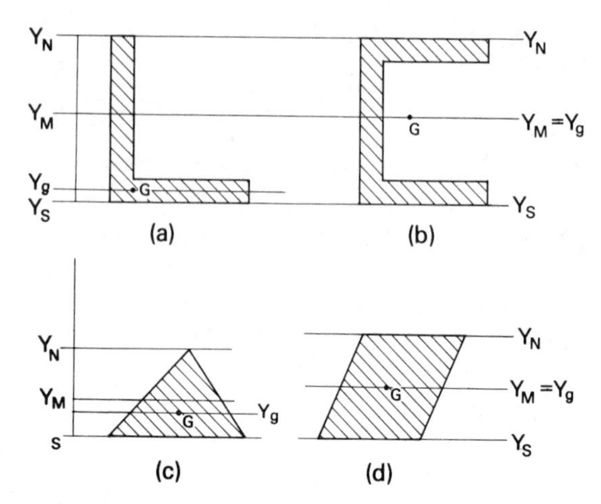

FIG. 22.12. Four extreme examples of parcels showing the relationship between the mean latitude and the centre of gravity. (Source: Isakova and Pavlov, 1958).

where φ_g is the latitude of the centre of gravity and $\delta\varphi$ is the error in estimating its position on a chart. For example, for a chart of scale 1/5,000,000 in which the estimated latitude of the centre of gravity is 60° N and the error in estimation amounts to 2 mm on the chart, the effect is a change. δk, of 0·14‰.

At the time when this work was done there were far fewer computing aids available to the scientist than today. There are, for example, many subroutines available which may be used to determine the position of the centre of gravity of a flat disc, given a succession of coordinates around its perimeter, such as might be determined by digitiser. Use of such techniques would much reduce the uncertainty in this part of the work.

23

The Cartometric Definition, Demarcation and Recovery of Maritime Boundaries

I am by no means a legal expert but a mere hydrographer engaged in the task of trying to fathom out the mathematical realities from a complex legal regime based on various shades of political philosophy. However, we hydrographers frequently find ourselves in a position where we have to translate broad philosophical concepts embodied in international law into such mundane terms as geographical coordinates, geodetic datums and low waterlines. It is therefore our hope that the emerging law will provide us with some appropriate direction for the future, particularly in view of the fact that demarcation problems will have to be resolved speedily if the international sea-bed authority is to function effectively. Hence from a technical point of view it is, indeed, important that the provisions contained in the universal treaty can be interpreted in terms of easily definable and recoverable boundaries so as to enable coastal states and the international sea-bed authority to establish permanent areas of jurisdiction and to exercise effective control over them.

(Rear Admiral D. C. Kapoor, *Marine Policy and Management*, 1977)

Reference was made in Chapter 1 to the connection between cartometry and the description of maritime boundaries for the purposes of international law. We returned to this subject in Chapter 12 where we developed the subject of the definition of features to be measured with particular reference to the sea coast. Chapter 22 treated with measurements made on Mercator's projection, this being the mathematical base of nautical charts which are the principal documents used to demarcate, illustrate and recover the positions of maritime boundaries. It is now desirable to consider the problems of definition and measurement of the various maritime boundaries which have been proposed by the Third United Nations Conference on the Law of the Sea, for the way in which this is done may have far-reaching political and economic consequences. We need to use three words having distinct and separate meanings:

Delimitation refers to boundary definition and is primarily a legal and political concept.

Demarcation refers to the technology, observing and measuring techniques and data processing involved in the establishment of the boundaries. This comprises mathematical concepts in geodesy and hydrographic surveying.

Recovery refers to the techniques which are available to return to and locate a boundary which has already been demarcated. This relates to the methods of navigation and hydrographic surveying.

We preface the study of this aspect of cartometry with the observation that some maritime boundaries are declared unilaterally, but others are the result of negotiation and treaty between pairs or groups of nations. The first of these categories comprise the boundaries of the maritime zones which lie successively further offshore and, as it were, surround a country like the skins of an onion. These have to be defined on the initiative of that country and will have to be presented to the United Nations in a form which has been specified in the 1982 *Draft Convention of the Law of the Sea.* This type of delimitation applies where the high seas lie beyond the outermost of these zones.

There is also need for a second group of boundaries which create or extend seawards the international frontiers on land. Smith (1982) has identified 376 possible international maritime boundaries, only one-quarter of which have so far been negotiated. They occur where two or more countries face one another across a stretch of sea or share the same coastline, and still normally apply only to countries sharing parts of the same continental shelf; not to countries lying on opposite sides of an ocean. Thus it is necessary to delimit a boundary between Britain and Norway within the comparatively narrow, and shallow North Sea, as yet there seems little point in determining some mid-Atlantic boundary between Britain and Canada or Britain and the U.S.A. Nevertheless we must reflect that 40 or 50 years ago the definition of a median line between Britain and Norway appeared to be equally pointless.

We emphasise the word *negotiation* because the only satisfactory boundary between two countries is one which has been agreed by both of them. Agreement of maritime boundaries has usually been straightforward enough where there have been no political or economic pressures caused by knowledge of the existence of valuable seabed resources, but if the presence of such resources is known or suspected, delimitation of the boundary becomes more difficult, for every proposed alteration to its position and alignment is likely to be hotly contested. The fact that such boundaries have been freely negotiated, and are the subject of treaties, must be borne in mind when we consider the technical problems of locating a series of points at sea which are supposed to be equidistant from both countries. Once their positions have been agreed, no matter how arbitrary were the original precepts, the boundary has a precise location. Because the boundary was not defined more accurately when it was originally negotiated, and its position was subsequently shown to give greater economic advantages to one state rather than the other, does not invalidate its legality. For example, when the North Sea was subdivided in the 1960s, most of the countries involved in the negotiations accepted that the boundaries with their neighbours should be settled on the basis of a *median line* equidistant from the respective coasts. However, at that time the exact distance across the North Sea could not be measured accurately because the differences between Ordnance Survey Datum, used in the United Kingdom, and the European Datum used by the other countries, had not been established. The agreed boundaries are now known to be incorrectly located but, they are still accepted. It is the expectation that this might happen again elsewhere which serves to sharpen the wits of those now intending to negotiate the positions of their own boundaries. This means that greater and greater accuracy is going to be needed to define and measure them; this in spite of the fact that some sea areas may never have much economic value. Irrespective of how the boundaries are defined, the negotiations and the treaty text should include enough technical information to avoid later disputes.

A huge amount has been written about different legal and geographical aspects of the Law of the Sea. That relating to the definition of maritime zones and their boundaries is generally preoccupied with the interpretation of the various articles of the convention in real or imaginary test cases. In this field we single out as examples only a handful of British contributions, such as those of Brown (1978, 1979, 1981), Prescott (1975), Couper (1983, 1985), McMillan (1985) and Barston and Birnie (1980). Our particular interest lies in the practical determination of the positions of boundaries by measurement. Far less has been written about this subject, but the contributions by Beazley (1971, 1978, 1982, 1986), by Thamsborg (1971, 1974) and by Hodgson and Cooper (1976), are all important.

MARITIME ZONES

In the early days of the Law of the Sea Conferences there was disagreement about the ideal breadth of each maritime zone. Indeed, when the first United Nations Conference was convened in 1958, about the only measure of agreement between the participating nations was that the old 3-mile limit was no larger adequate. The subject dominated the early conferences but it has now been resolved so that the following distances could be written into the Draft Convention of 1982.

We shall see that in the North Sea and other sea areas where there is intense economic activity, much of the discussion concerning the accuracy of position-fixing, the differences in position referred to more than one geodetic datum and distance measurements in general are expressed in metric units. Consequently the corresponding metric values are also given. In order to conform to modern international usage, the *international nautical mile*, of length 1852·00 metres, has been the unit employed for the conversion. The so-called *Admiralty nautical mile* of 6,080 feet (1853·167 m) is no longer acceptable for international use. It should be noted that the difference between the two units (1·167 m)

TABLE 23.1 *The breadth of maritime zones proposed by UNCLOS III*

Designation	Distance from baseline	Article No.
Territorial sea	12 nautical miles = 22,224 metres	3
Continguous zone	24 nautical miles = 44,448 metres	33
Exclusive economic zone	200 nautical miles = 370,400 metres	57

would produce an error of 233·4 m in the location of a boundary on the Exclusive Economic Zone (E.E.Z) if there were any uncertainty about which version of the nautical mile has been used. Notwithstanding the evident authority of an internationally agreed length for the nautical mile, we shall see later that this is not necessarily a suitable conversion factor to be used to obtain the precision needed in boundary demarcation. Indeed the use of this conversion factor can produce misleading results, as we shall show at the end of the chapter.

The limit of the E.E.Z. also serves as one definition for the breadth of the continental shelf. However, Article 76 of the Convention provides for other methods of defining this somewhat elusive feature of the seabed. The fact that there are five or six ways of defining its outer limit contained within the same Article is going to keep international lawyers in the manner to which they are

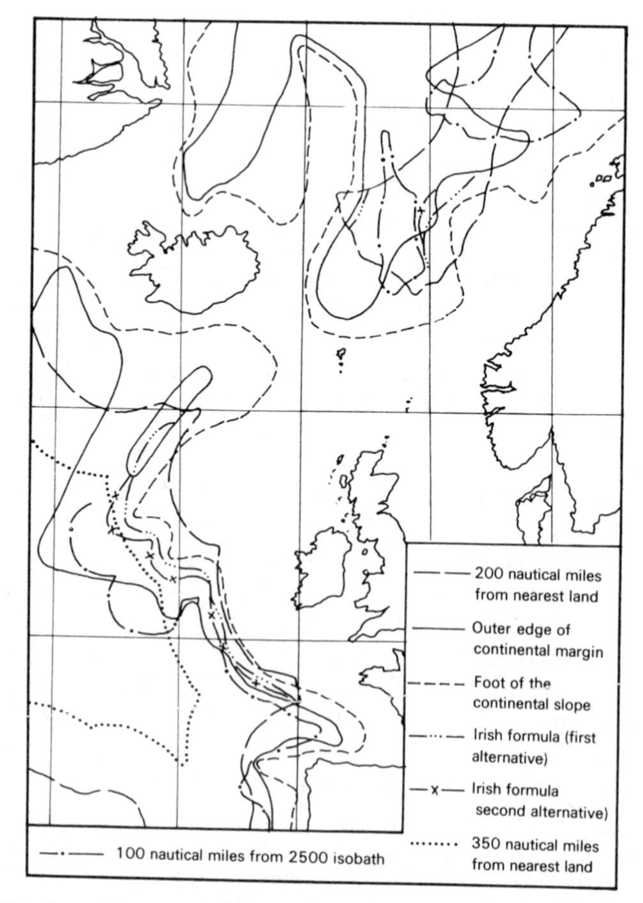

FIG. 23.1. Some different versions of how the limit of the exclusive economic zone may be delineated in the north-eastern Atlantic Ocean. (Source: Couper, 1986).

accustomed for many decades. Figure 23.1, which is derived from Couper (1986), illustrates some of the different possibilities to be deduced from Article 76 of the Convention when these are applied to the north-eastern part of the Atlantic Ocean. The reader is particularly recommended to refer to Couper (1983), *The Times Atlas of the Oceans*, for a valuable series of maps and diagrams which illustrate the methods of delimitation proposed by UNCLOS III. There is also a detailed textual commentary upon the Draft Convention of 1982 on pp. 241–247 of this atlas.

BASELINES AND EQUIDISTANCE LINES

There are three aspects of the definition of baselines and boundaries which need to be considered:

- establishment of *normal baselines,*
- the use of *straight baselines,*
- equitable subdivision of the seabed by means of *equidistance lines.*

The Normal Baseline

The breadth of each maritime zone is described as being measured from a baseline. It follows that accurate location of this is most important, for an error in the location of the baseline introduces a corresponding displacement of every other boundary referred to it, irrespective of other errors of measurement of definition which may occur.

Where the coast is uncomplicated by inlets, bays and offshore rocks or islands there is little legal uncertainty about how the boundary between land and the territorial sea should be drawn. According to the Draft Convention (1982):

Article 5
Normal Baseline

Except where otherwise provided in this Convention, the normal baseline for measuring the breadth of the territorial sea is the low-water line along the coast as marked on large-scale charts officially recognised by the Coastal State.

Bays are defined by Article 10 of the Convention through application of the "semi-circular rule", described in Chapter 12, pp. 236–237, to distinguish between shallow indentations of the coastline and "true" bays. Moreover, all bays and inlets which measure less than 24 nautical miles across are automatically regarded as lying within the internal waters of a State. This is equivalent to the old 10-mile rule (p. 236) used to define the U.S.A. but extended to correspond with the 12-mile limit for the territorial sea.

We saw in Chapter 12 that a low-water line on a nautical chart corresponds to chart datum. This may correspond to the level of the lowest astronomical tide or even lower (p. 240), but there are some oceans and enclosed seas where other definitions are required. We also saw that the delineation of the low-water line is essentially an office compilation, for it can only be drawn after the soundings made close inshore have been reduced to the local chart datum. Consequently it is not a line which has ever been seen or surveyed from the shore, although it is theoretically possible to map it using infra-red photography provided that other conditions are suitable for aerial photography on the one occasion in every 19 years when L.A.T. is experienced. Various other factors may influence the correct positioning of the low-water line. An example of how variation in the height of tidal datum may be reflected in offsetting of the planimetric position of chart datum was referred to in Chapter 12 and illustrated by Fig. 12.4. Theoretically each step, representing a local change in chart datum, should also give rise to an horizontal displacement in the position of the low-water line. In practice, however, those places where it might be recognised and even portrayed on large-scale charts usually lie some distance from the open sea. Thus Fig. 12.4 illustrated an example of what may occur near the head of the Bristol Channel, where an enormous tidal range accentuates the variation in local chart datum. However, the extreme example illustrated here does not influence the location of any baselines, for the width of the Bristol Channel is already less than 24 nautical miles between Worms Head and Bull Point (which is more than 60 miles further west than the area illustrated), and it is there that the baseline is represented by the straight line joining these headlands. The point has been made here to emphasise that, whatever the simplicity of the legal definition, the low-water line depicted on a nautical chart is neither easily recognisable nor is it a well-defined feature of the landscape. Indeed, it may be argued that really it is a thoroughly unsatisfactory datum from which to start the delineation of maritime boundaries.

The Straight Baseline

According to the first part of Article 7 of the Convention:

> In localities where the coastline is deeply indented and cut into, or if there is a fringe of islands along the coast in its immediate vicinity, the method of straight baselines joining appropriate points may be employed in drawing the baseline from which the breadth of the territorial sea is measured...

The classic example of use of this technique, which in 1935 was the first straight baseline to be defined, was that for northern Norway where, as shown in Fig. 23.2, a series of straight baselines connect the outermost parts of the offshore chain of the Lofoten Islands and their neighbours. These straight baselines enclose the whole of the comparatively sheltered waters inshore of the skaergård, making this part of the internal waters of the country. Many other straight baselines have now been declared, so many and so indiscriminately that their justification must sometimes be questioned. Prescott (1975)

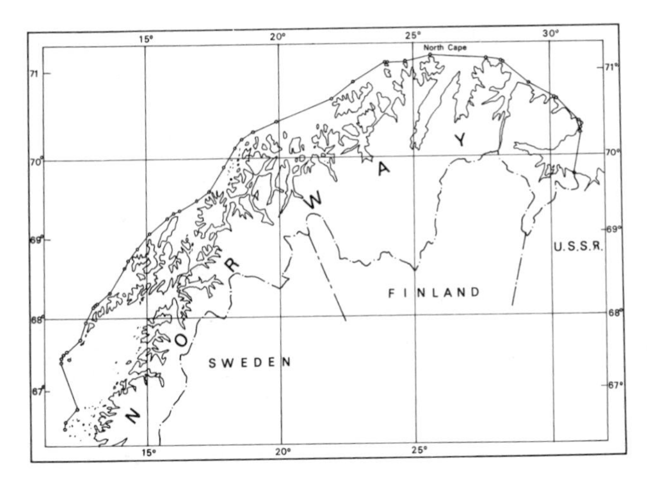

FIG. 23.2. The use of straight baselines to delineate the coastline of northern Norway. All sea inshore of the straight baselines is deemed to be part of the internal waters of the country.

has provided a detailed commentary on the geographical validity of some of them.

Maritime Boundaries Between States

We should distinguish between boundaries to be drawn where the States face one another across a sea or strait, as Britain and Norway face one another across the northern part of the North Sea, or the boundary may be the prolongation of the land frontier between two countries sharing the same coast, such as those between the Netherlands, Federal German Republic and Denmark. According to Article 15:

Article 15
Delimitation of the territorial sea between States
with opposite and adjacent coasts.

Where the coasts of two states are opposite or adjacent to one another, neither of the two States is entitled, failing agreement between them to the contrary, to extend its territorial sea beyond the median line every point of which is equidistant from the nearest points on the baselines from which the breadth of the territorial sea of each of the two States is measured. The above provision does not apply, however, where it is necessary by reason of historic title or other special circumstances to delimit the sea of the two States in a way which is at variance therewith.

Langeraar (1984) considers that the term *median line* should be confined to the description of the boundary between two States with opposing coasts, using the word *equidistance* to describe the offshore prolongation of the frontier between two adjacent States. However there is no consistent agreement about this usage. The principles are illustrated in their most elementary form in Figs 23.3–23.5.

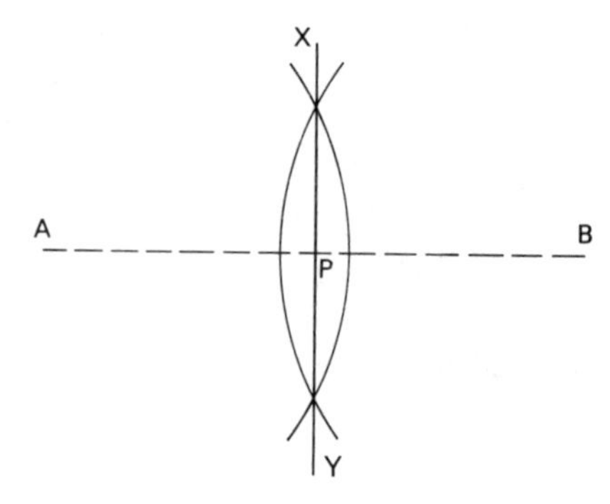

FIG. 23.3. The determination of the point, *P*, which is equidistant between two fixed points, *A* and *B*.

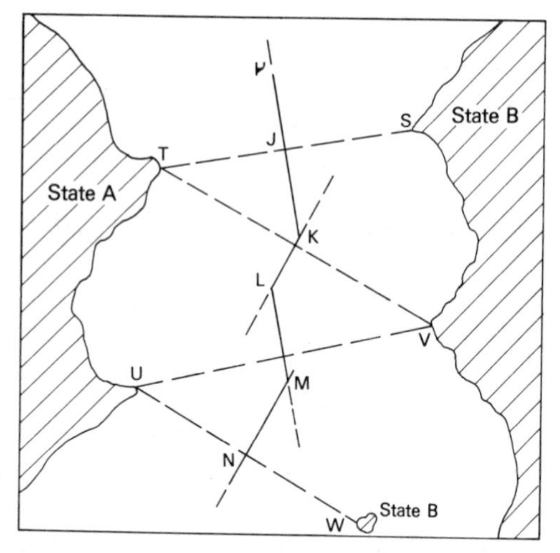

FIG. 23.4. The determination of the median line passing between five points, *S*, *T*, *U*, *V*, *W*. (Source: Hodgson and Cooper, 1976).

In plane geometry the locus of a point which is equidistant from any two fixed points, such as *A* and *B* in Fig. 23.3, lies on the perpendicular bisector of the straight line joining these points, i.e. at *C*, *D* and *E*, etc. Applying the same principle repeatedly to a succession of fixed points, we obtain the more complicated example illustrated in Fig. 23.4, where the median line *H*, *J*, *K*, *L*, *M*, *N* passes through a succession of points such as *S*, *T*, *U*, *V*, *W*, each of which has been located by similar construction. Figure 23.5 illustrates the corresponding construction for equidistance from a single coastline.

The reader will appreciate that these diagrams illustrate conditions when there is little doubt about the choice of the coastal points from which equidistant measurements are to be made. It should be appreciated that usually the choice of points is less obvious. Then the positions and alignment

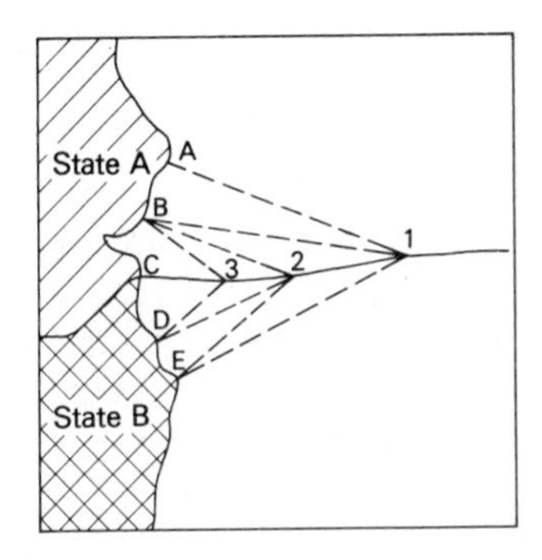

FIG. 23.5. The determination of an equidistant line, *c*, 3, 2, 1, which is the seaward prolongation of an international frontier on land.

of parts of the boundaries are more likely to be open to doubt. This is particularly true of equidistance lines constructed from features on the same coastline, where a small change in the interpretation of the alignment of the boundary may create larger positional changes further offshore. Brown (1978) has demonstrated this with reference the offshore boundary between England and Scotland in the North Sea.

Beazley (1978) has illustrated many different kinds of graphical construction which may be used where islands, offshore rocks and other features complicate these simple examples. He has also shown, in Beazley (1979, 1986), how islands lying some distance offshore may further complicate matters, and has offered a series of additional geometrical rules which may be used to facilitate negotiations. For example, determination of the median line between the United Kingdom and France in the Celtic Sea and English Channel, depends to no small measure upon how much weight should be attached to the baseline drawn round the Scilly Islands.

Another kind of empirical rule which may be applied in some difficult cases is that of *proportionality of coastline length to maritime area.* As has been described by McDorman *et al.* (1985), this technique was accepted by the International Court of Justice in their decision in 1984 which resolved the dispute between Canada and the U.S.A. over the delineation of their boundary through the Gulf of Maine. Enough has been said about the problems of measuring coastline length elsewhere in this book to warn the reader about the perils of straying into this particular minefield.

PRACTICAL LOCATION OF BOUNDARIES AT SEA AND ON THE CHART

The Draft Convention and other documents concerning the Law of the Sea make or imply two fundamental assumptions. First, it is supposed that it is possible to demarcate and recover the positions of maritime boundaries to the same order of accuracy as this can be done on land. The second assumption is that the breadths of maritime zones, or the lengths of lines measured from shore to median line, can be measured accurately on a chart by those methods of plane geometry employed in navigation. In fact, the demarcation of a boundary line can only be as reliable and accurate as the charts and maps used for this purpose. This, in turn, implies a third assumption; that charts of suitable scale are already available for a particular sea area.

The first assumption may now be satisfied in certain sea areas, for example, parts of the North Sea, Persian Gulf and Gulf of Mexico, which are particularly well endowed with electronic position-fixing equipment. Ultimately it will become possible everywhere in the world with the use of increasingly more sophisticated satellite navigation methods. However, it should be appreciated that this is a comparatively recent development, and that until now legal ideas about the accuracy of boundary delimitation have outstripped the ability of the hydrographic surveyor to locate them. The second assumption must obviously be qualified by considerations of chart scale, and should be clear from earlier chapters.

Boundaries are intended to separate what belongs to one State from what belongs to another State. Moreover, where there is active exploration and exploitation of hydrocarbons, a boundary between two licence blocks, which separates the activities of rival companies, may seem to be almost as important as an international boundary. Since offshore boundaries cannot be demarcated at the surface without resorting to the use of fairly sophisticated equipment, it follows that the nations or companies sharing a common boundary must both be able to recover it with sufficient accuracy in order to avoid disagreement, or to avoid conflict by giving the actual boundary a comparatively wide berth. In the hydrocarbon industry it has been, and still is, usual practice for government licensing authorities to prohibit activities within a certain distance of the boundaries of an operating area, unless special permission to work closer to the boundary has been granted. In the North Sea this distance is 125 metres in British blocks, but so great is the value of a few square metres of the seabed in productive oil or gas fields that where a licence boundary intersects a proven field there are strong pressures to reduce the width of this *cordon sanitaire* by invoking the special permission clause.

If the place where the boundary is located is visible from land, it is easy enough to fix the position of a craft or buoy using conventional methods of field survey from control points on land. A variety of E.D.M. instruments have been developed especially for maritime use, and under ideal conditions ship-to-shore equipment can now measure distances up to 80 km with an accuracy of ±1 metre or better. However many boundaries lie much further out to sea. We have seen that the outer edge of the E.E.Z. is 200 nautical miles from a baseline. Obviously this is far beyond the range of E.D.M. and conventional surveying methods.

During World War II there was a revolution in navigation methods following the development of the early hyperbolic navigation systems, such as

Shoran, Loran and Decca, by which positions could be fixed by measuring the time taken for different high-frequency radio signals to reach a ship by different routes. By these means it became possible to locate position at sea within a few hundred metres in areas of optimum reception. However, such areas were comparatively limited in extent, and generally the accuracy of fixing was less than ± 1 nautical mile. Consequently these aids were of smaller value to contemporary hydrographic surveying than has often been claimed. Nevertheless, they were still the most accurate methods of fixing position far offshore at the time of the first U.N. Conference of 1958 (UNCLOS I) and the drafting of its so-called Geneva Conventions. Many of the articles relating to the definition of maritime boundaries used in the Draft Convention of 1982 were originally drafted for the Geneva Conference in the knowledge of what was then practically possible at sea. Hence the implied acceptance of a cartometric solution to most of the problems of boundary demarcation has tended to survive to the present with only minor modification, and this despite the fact that graphical methods should now give way to digital solutions.

Later technological progress created a variety of related systems offering higher orders of accuracy. For example, in the North Sea where distances seldom exceed 150 km from land, *Hi-Fix 6*, which is a much-improved version of the old Decca equipment, can be used to fix position with a repeatability of 25 metres. Two other systems, known as *Pulse/8* and *Argo*, are only slightly less accurate. However, these systems are only available where it has been considered worthwhile to install the land stations.

True world-wide capability for fixing position is now provided by the navigation satellites whose principal mode is to measure distances between a cluster of artificial satellites and the ground station by determining the Doppler shifts in signal frequencies. The notable system in this field is the U.S. Navy Navigation Satellite System, known as TRANSIT, which first became available to civilian users in the early 1970s. This allows position to be fixed with an accuracy of ± 80 to ± 100 metres anywhere. The NAVSTAR-GPS system, which represents the next generation of satellite-based Doppler systems, and which is due to become available for civilian use in the late 1980s, is expected to fix position anywhere within ± 10 metres. In ordinary navigation practice the movement of the ship reduces the precision of position fixing by these methods, but for an anchored vessel or a platform from which a long-continued series of observations may be made from the same fixed point, even better results may be obtained. An example of the use of this technique was the *Frigg positioning project*, in the oil field of that name, located adjacent to the median line of the North Sea, where the platform CDP 1 was located by different methods and evidently to geodetic accuracy. The reader is referred to the work of Blankenburgh and co-workers (1978, 1980), Leppard (1980), Bakkelid and Rekkedal (1983), Hedge (1985) and Seeber (1985) for more detailed statements concerning the standards of positional accuracy which are now possible in marine geodesy.

MEASUREMENT OF BREADTH AND EQUIDISTANCE

There are three stages in the process of defining the maritime zones of a country. The first is that of determination of the baselines; the second is measurement of the breadth of each specified zone; the third is the presentation of the information in the form required by international law. The last of these stages is covered by articles referring to each of the boundaries to be defined. For example, in the determination of the boundary of the territorial sea:

Article 16

Charts and lists of geographical coordinates

1. The baselines for measuring the breadth of the territorial sea determined in accordance with articles 7, 9 and 10, or the limits derived therefrom, and the lines of delimitation drawn in accordance with articles 12 and 15, shall be shown on charts of a scale or scales adequate for determining them. Alternatively, a list of geographical co-ordinates of points, specifying the geodetic datum, may be substituted.

2. The coastal State shall give due publicity to such charts or lists of geographical co-ordinates and shall deposit a copy of each chart or list with the Secretary-General of the United Nations.

A similar form of words is found in Article 47 with reference to Archipelagic States; in Article 75 referring to definition of the E.E.Z. and in Article 84 concerning the definition of the Continental Shelf.

Availability of Charts

The use of the word *chart* here, and in all the other relevant articles, indicates the intention to use nautical charts rather than land maps for these purposes. We have seen that it is necessary to know the position of the low-water line to define normal baselines. We have seen, moreover, that the low-water line corresponding to chart datum is shown on charts but it does not necessarily appear on land maps. In some cases, such as the basic scale mapping of England and Wales by the Ordnance Survey, the low-water line which is depicted does not correspond to that line on nautical charts of England and Wales because a different stage of the tide is used to define its position. Also, Article 13 of the Convention permits the use of *low-tide elevations* as points on a normal baseline. A low-tide elevation is a naturally formed area of land which is surrounded by and above water at low tide but is submerged at high tide. Such features do not necessarily appear on land maps, but because they are hazards to navigation they are always shown on nautical charts.

Unlike topographical map series, which provide uniform cover of a country by scale, nautical charts have been traditionally designed to portray a particular sea area upon a standard size of paper which can be accommodated without folding on a chart table. This may be a smaller-scale chart to cover the feature in its entirety or a larger-scale chart which depicts the sea area between two ports, or capes which are commonly the destination of vessels, or the points of departure and landfalls employed in route planning and the execution of D.R. navigation. Until the 1960s the scale of the chart was nearly always selected on the basis of fitting its neat lines to standard paper sizes, giving rise to such curiosities as 1/9,130, 1/16,300, 1/24,200 or 1/91,200, but since graphical D.R. navigation invariably involved distance measurement by comparison with the meridional subdivisions of the chart, most navigators were hardly aware of the fact. During the last quarter-century there has been an international attempt to standardise the scales used for nautical charts into the following categories:

Ocean charts	1/20,000,000
Ocean passage charts	1/3,500,000 or 1/10,000,000
Landfall indentification charts	1/1,000,000
Coasting charts	1/300,000
Port approach charts	1/100,000
Harbour charts	1/50,000 or larger.

Because the two largest scales obviously provide extremely intermittent cover of the coastline of a country, it is probable that the largest scale of chart which is likely to be available to offer anything like continuous cover is at 1/300,000. In fact there are still enormous stretches of the world's coastlines which are not charted at such large scales. Figure 23.6 indicates the state of hydrographic

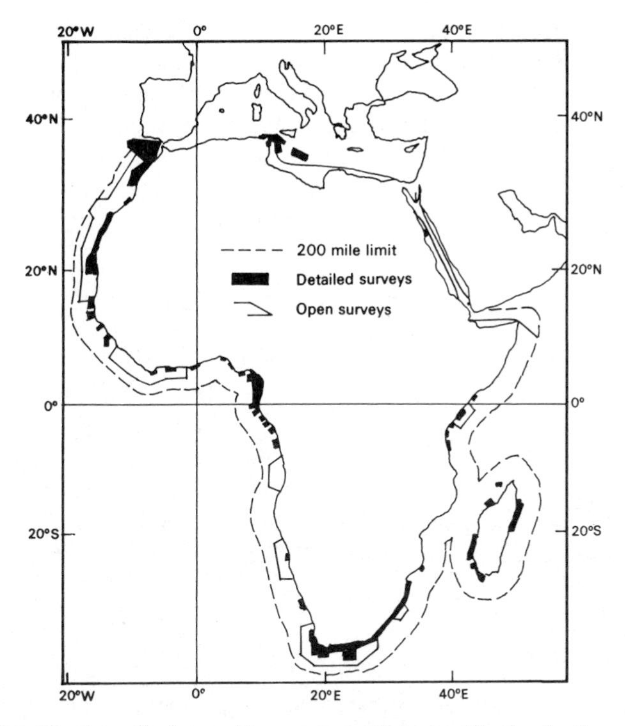

FIG. 23.6. The state of hydrographic surveying and chart publication for the coasts of Africa in 1975. (Source: Haslam, 1983).

surveying and the availability of Admiralty charts around the coasts of Africa in 1975.

Even in British waters the extent of hydrographic knowledge is by no means complete. Thus Haslam (1985) has noted that:

> at the end of December 1981, only some 14,445 sq. miles, or about 8·4% of the U.K.'s coastal waters, had been fully surveyed to modern standards... and there were some 116,500 sq m, or about 67·3% of our coastal waters which had never been surveyed at all or only by hand lead-line many years ago.

The concept of E.E.Z., and all that goes with the creation of these, has revolutionised the need for much additional work to be done in a zone 200 nautical miles wide round every nation-state with direct access to the sea. Kapoor (1981) and others have stressed the huge increase in the economic potential of otherwise ill-provided states, such as Barbados and Mauritius created by their E.E.Z.s.

Since there are evident gaps in our hydrographic knowledge of the continental shelf areas of the world, and since it is possible, on grounds of scale alone, that existing nautical charts may be inadequate for the representation of maritime boundaries in the amount of detail and with the precision required for boundary negotiation, we must attempt to answer the following questions:

- Is it possible to bring the cover of existing nautical charts sufficiently up to date and, where possible, produce large-scale charts for this kind of use?
- Is it preferable to create new kinds of bathymetric maps for the continental shelf areas of the world, the prime purpose of which should be for use in the exploration and exploitation of the seabed resources?
- Can some of the problems be resolved by dispensing with conventional charts altogether, and simulating the geometrical problems by digital computer?

An answer to the first two questions must refer to the availability of suitably skilled personnel. We do not attempt to treat with this subject here, but refer the reader to the numerous warnings given by Ritchie (1975, 1981, 1982, 1986), Kapoor (1977, 1981), Haslam (1977, 1985) and Myres (1986) about the urgent need for an army (or should it be a navy?) of suitably qualified surveyors to undertake this massive task of investigating the vast new territories, for providing national nautical charts for these areas and for delimiting the national maritime boundaries upon the charts. The amount of labour involved in doing this, and the time that it is going to take, should not be underestimated.

Bathymetric Maps

The future exploitation and subsequent management of marine resources within E.E.Z.s produces a need for thematic or bathymetric maps of these offshore areas containing *all* rather than a skilled *selection* of the available data, which is all that is shown on a navigation chart. It is therefore desirable to see what other kinds of map may be available or may be specially produced for these purposes. Those which already exist are only intended to provide cover of national claims to the continental shelf, such as those being currently produced in the U.S.A. and Australia.

The bathymetric maps of the U.S.A. are now being prepared and published jointly by the N.O.S. and the U.S.G.S. as the Bathymetric Mapping Program. This series has been described by Banks (1983). The Program was initiated in 1966 to produce a uniform series of maps at scale 1/250,000 for the continental shelf areas of the United States, but is still incomplete. In addition, work on larger-scale series at 1/100,000 and 1/24,000 began in 1977, but obviously production of these maps will take even longer to complete because of the enormous amount of detailed offshore surveying which has to be done to complete such an ambitious project. There are, for example, more than 1,800 individual sheets of the 1/24,000 scale series to be prepared in order to complete this cover.

In Australia, where much of the vast continental shelf area is yet to be charted, the National Mapping Authority has been carrying out bathymetric surveys on the scale 1/150,000. The zone covered by these maps is roughly from the 20-metre isobath out to the 300-metre isobath. The intention is to provide a base map of the continental shelf for investigation of resources.

Chart Scale

What is the meaning of *adequate scale* mentioned in Article 16?

There are three reasons why it is desirable to use the largest possible scale for determination of baselines. First, the information content at the largest scale is greater than for any smaller scale. As on land maps, the detail shown on charts must be increasingly simplified or generalised with reduction in scale in order to preserve legibility. Secondly the smaller scale chart may not be as completely up to date as the largest scale chart of that part of the sea. This is because it is axiomatic that the largest scale chart of an area is that used for navigation, so that any significant amendments are incorporated upon it first. Depending upon the significance of any new discoveries or changes in the seabed to navigation, these may not be incorporated into the smaller-scale charts until these are taken up for a major revision. This may not happen for several years or even decades. Phillips (1971) suggests that the scale of 1/20,000 ought to be adequate for the purpose, but thinks that in some areas it may be necessary to work with charts of scale 1/10,000 to determine the low-water line with sufficient accuracy. However, he fails to mention that more than 99·9% of the world's coasts have never been charted at such large scales.

The third reason why the largest-scale chart should be used for any graphical work relating to the location of baselines and boundaries is that errors are introduced in the measurement process. This applies equally to the determination of the breadth of a zone or the distance between the baseline and the median line as it does to other kinds of cartometric work. If the geographical coordinate of points along a median line are determined by scaling them from a chart by dividers, we know that the precision of such work should be about $\pm 0{\cdot}4$ mm. If this is done on a chart of scale 1/500,000 the precision of the measurements corresponds, therefore, to ± 200 metres, which is about $\pm 6''$ in angular measurement. Because charts of scale 1/500,000 were used during the treaty negotiations between the different North Sea countries, this explains why the positions of points along the median line are recorded in multiples of 6″ of arc.

In determining the positions of the offshore maritime boundaries graphically, we shall see that there are a series of cartometric constraints which more or less restrict us to the use of rather small scales.

Construction of the Envelope

Each of the articles defining a maritime zone refers to the breadth of that zone measured from the baseline which we assume to have been plotted already. For want of any other information this must evidently be regarded as the *straight-line distance measured on the chart at right angles to the baseline.* Consequently it would seem that there are two measurements to be made; one linear, the other angular. The angular measurement would be the more difficult to plot precisely, for we must assume that the right angle is measured from the tangent to the baseline at each point on it. If so, a small error in the determination of the orientation of the tangent is equivalent to an error in measuring the right angle. This error would be exaggerated into a large displacement of the line defining the breadth of the zone. Obviously the greater the distance, the larger the displacement. For example, an angular error of only $\frac{1}{2}°$ at the baseline becomes a displacement of more than 3 km on the 200 nm boundary. However, it is not necessary to make specific angular measurements because the usual method of locating a boundary by graphical construction on a chart is to construct an *envelope* formed by intersecting circular arcs. Strictly speaking the boundary must be defined by an infinity of such arc elements but, in practice, the envelope is constructed from a comparatively small number of them. Beazley (1978) has recommended spacing the centres from which the arc are drawn at $\frac{1}{4}$-mile intervals along the baseline but obviously a major consideration determining this interval must be the scale of the chart on which plotting is done.

The traditional method of measuring distance on a nautical chart is by dividers, as described in Chapter 22, pp. 499–501. In order to determine the breadth of a maritime zone and define the envelope by drawing short circular arcs, the dividers must be set to the appropriate spacing, e.g. 12 nautical miles, placed at a point on the baseline and the arc is plotted over the sea. Using dividers of the size and style commonly employed in navigation, it is generally

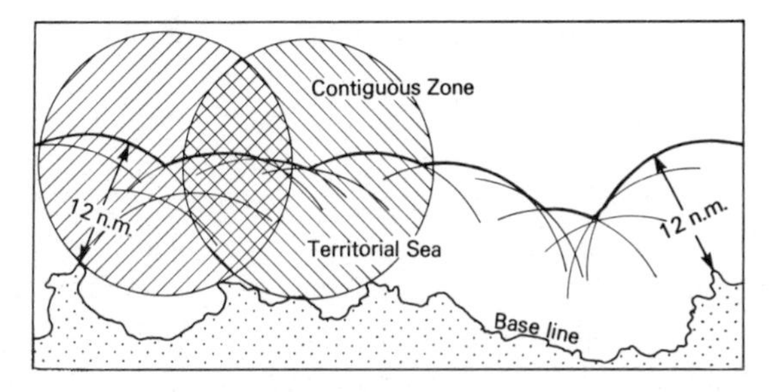

FIG. 23.7. The geometrical construction of an envelope line to represent a maritime boundary. (Source: Shalowitz, 1962).

possible to measure a distance of 180 mm or thereabouts in a single step. As indicated in Chapter 3, if the span is made larger, the angles at which the points meet the plane of the chart are small and there is consequential loss of accuracy resulting from the oblique penetration of the points. We therefore take 180 mm to be the maximum distance which can be measured easily by dividers.

We stress that the method of construction of the envelope by sweeping circular arcs has to be done with the dividers set to the total required distance. It cannot be done by stepping off the distances. In order to measure the distance from the baseline to the 12 nautical mile limit within these limitations, we must work on a chart whose scale is not larger than

$$0{\cdot}18/22224 = 1/123{,}466$$
$$\approx 1/125{,}000 \text{ in round figures.}$$

In this expression, 0·18 is the maximum spread of the dividers expressed in metres and, as we have already seen, 22,224 metres is the breadth of the territorial sea. Using analogous reasoning, we find that to plot the edge of the contiguous zone, this must be done at a scale of 1/250,000 or smaller. For the

TABLE 23.2. *The standard errors to be expected in using different graphical techniques to construct the envelope to define offshore boundaries by graphical methods*

	Dividers		Beam Compass	
Outer boundary of	Largest practicable scale	Standard error	Largest practicable scale	Standard error
Territorial sea	1/125,000	± 50 metres	1/25,000	± 5 metres
Contiguous zone	1/250,000	± 100 metres	1/50,000	± 10 metres
Exclusive economic zone	1/2,000,000	± 800 metres	1/500,000	± 100 metres

200 nm limit at the edge of the E.E.Z. or the continental shelf, the largest practicable chart scale has decreased to 1/2,000,000.

In Chapter 3 we described the use of the beam compass for plotting and measurement (p. 34). This instrument is particularly important in this field of geometrical construction because the points are always perpendicular to the plane of the chart and an arc of greater radius may be described. It has been suggested that the limiting working distance of a beam compass is about 800 mm and that it has greater precision, about $\pm 0{\cdot}2$ mm. Substituting these values, the corresponding limiting scales are now, $0{\cdot}8/22224 = 1/27{,}780 \approx 1/25{,}000$ for the 12 nm limit, 1/50,000 for the seaward limit of the contiguous zone and 1/500,000 for the outer edge of the E.E.Z. Table 23.2 indicates the ground distances corresponding to these standard errors at the largest practicable scales which may be employed to construct the envelope.

THE NATURE OF THE LINE USED TO PLOT BOUNDARIES AND MEDIAN LINES

Most maritime boundary agreements, including the Draft Convention of 1982, include a statement that "straight lines" shall be used to connect a set of geographical coordinates. However the nature of the straight line is not usually defined. This is yet another omission which can lead to differing interpretations. From our point of view the distinction needs to be made between the geodesic, great circle arc and the rhumb line. Similarly we need to know the meaning of the word "distance" and "breadth". To the navigator "distance" is the rhumb-line distance, but to the surveyor it is the length of the geodesic. The straight line of the early "treaties" was normally interpreted by mariners as a rhumb line, for the simple reason that this was represented by a straight line on the ordinary nautical chart. Later, for example in the negotiation of the North Sea Continental Shelf Agreements, it was specified that the lines should be great circle arcs. Still later, for example in the negotiation of certain boundaries between dependencies of the U.S.A. and other countries (e.g. U.S.A. and New Zealand, which is the median line between American Samoa on the one hand and the Cook Islands and Tokelau on the other), the geodesic is taken to be the line.

Because of the importance of making the correct choice, it is desirable to consider the ways in which the result depends upon the correct choice of line.

Since a chart is based upon a projection, the nature of a straight line measured on the chart depends upon the mathematical properties of the projection. We saw in Chapter 22 that a straight line on Mercator's projection is a rhumb line. Therefore a measured distance obtained on a nautical chart by lying a straight edge alongside this line, or the span of dividers between its terminal points, or the radius of a circular arc made to define the envelope corresponding to a particular boundary, is the rhumb line distance. We have already seen that this is not the shortest distance between two points which is, of course, the great circle arc or the geodesic passing through them. There is only one map projection which represents all great circles as truly straight lines. This is the *gnomonic projection*, which is now seldom used for any practical purposes save as a name (which, strictly speaking, is incorrectly used) on Admiralty charts of scale 1/50,000 and larger. Except near the centre of the chart the measurement of distance on the gnomonic projection tends to be unreliable because the particular scales vary rapidly from point to point. Special graphic scales may be calculated, and were sometimes printed on gnomonic charts intended for air navigation when such existed, but even these provide only an approximate distance for long lines. In general the length of a line measured on the gnomonic projection will be determined with even less accuracy than that of a line of similar length measured on Mercator's projection.

We have seen that on the Lambert conformal conical projection, described in Chapter 10, p. 194, a great circle arc approximates to a straight line, and this is sufficiently close for the assumption to be acceptable in navigation. Moreover the relatively small variations in particular scales make it possible to measure the lengths of these lines by dividers, or even by placing a straight edge along the line.

The rhumb line and the great circle are the only two important types of line which need to be considered in the determination of maritime boundaries on maps or charts, for the small scale of them more or less precludes the

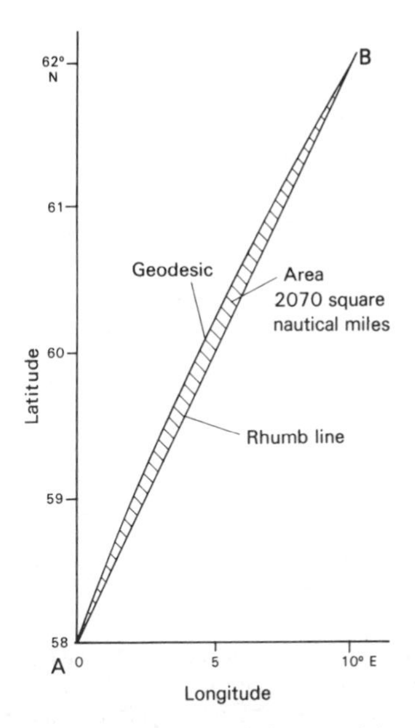

FIG. 23.8. The discrepancy between the great circle arc and rhumb line for the example used in the text. (Source: Thamsborg, 1977).

consideration of any differences between the great circle and the geodesic. The difference in the measured length depends upon the overall length of the line and upon its orientation. Truly north–south lines, being meridional arcs, are, at the same time, both great circles and rhumb lines. In all other directions the two lines diverge and it is greatest in an east–west line. In this case the rhumb line coincides with the arc of the parallel (which is a small circle) and the great circle arc lies polewards of this parallel. This discrepancy is illustrated in Fig. 23.8, and is based upon a specific example used by Thamsborg (1977), and later in this chapter, to demonstrate the differences which occur between the two arcs as these might appear plotted on a Mercator chart. Even for comparatively short distances (713 km in this example) the zone formed between the two lines has an area of 2,070 nm^2 (7,100 km^2).

The kind of displacement which may be encountered may be illustrated by Fig. 23.9, from Hodgson and Cooper (1976). The figure illustrates the difference between the same equidistance line located by graphical construction on both Mercator's projection and the Lambert conformal conical projection. In addition to the displacement of one of the boundaries with respect to the other, the area contained between them, and bounded by the 200 nm line offshore, containing the points *s* and *s'* in the figure, measures in

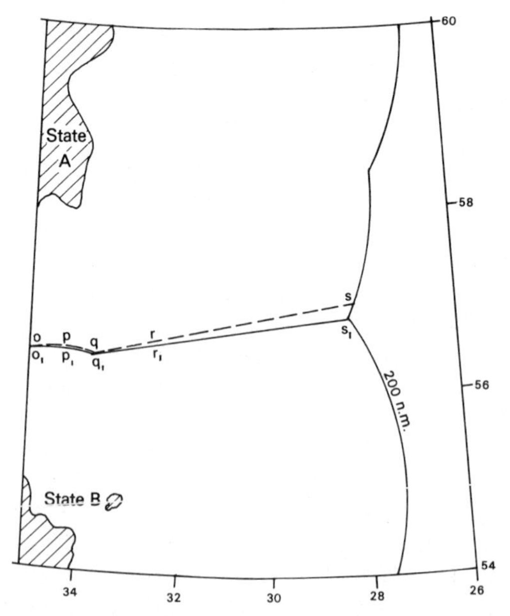

FIG. 23.9. The displacement in position owing to using the Lambert conformal conical and Mercator projections to plot the same point at 200 nautical miles from the baseline. The line o_1, p_1, q_1, r_1, s_1 is that determined by plotting the points on the Lambert Conformal Conical Projection. The line *o*, *p*, *q*, *r*, *s* is that transferred from the corresponding plot on a Mercator chart. (Source: Hodgson and Cooper, 1976).

this case more than 700 sq nautical miles (2,338 km^2). This area would, presumably, be lost to State A if the equidistant boundary had been constructed on a Mercator chart. Hodgson and Cooper argue that, in middle and high latitudes, the Lambert conformal conical projection is a more accurate chart for manual construction of equidistant boundaries.

BOUNDARY DETERMINATION BY CALCULATION

The lack of suitable charts of scale larger than 1/300,000 or thereabouts, and the impossibility, therefore, or putting limits closer than about 130–150 metres on the seabed suggests that, in fact, graphical cartometric methods and the presentation of the results on suitable charts are unsuitable to locate maritime boundaries wherever it is likely that there may be future mining or drilling activities. The case has therefore been made for determining the positions of points along the boundaries by calculation, and describing them in the form of coordinate lists. We have seen that according to Article 17 this is an acceptable alternative to drawing the boundaries on a chart.

Strictly speaking, therefore, the preparation of coordinate lists from calculation of the latitudes and longitudes of points which are precisely 12, 24 or 200 nautical miles from a point on a baseline is not a cartometric technique. However, we cannot just abandon the subject at this point because it is still desirable to emphasise some of the pitfalls which may arise if dividers are superseded by a computer. If these are not appreciated it is possible that the results may be inferior to graphical construction even when this has been carried out on a small-scale chart.

There is no doubt that the distances between pairs of points of known position may be calculated to a higher degree of accuracy than could ever be attained by cartometric methods, but in order to proceed it is necessary to specify certain variables which are involved in the computations. The specification of some of them are covered by the term *geodetic datum* contained in Article 16 and elsewhere.

We must therefore provide a brief explanation of the object and purpose for this requirement. First, it is necessary to decide about a suitable model of the earth for the job in hand, whether it is sufficient to assume that the earth's surface is part of the curved surface of a sphere, or that it is part of the curved surface of a reference spheroid. If the earth is regarded as spherical it is necessary to choose a suitable radius for the body; if it is a spheroid, then both the size and shape must be taken into account. Secondly, it is necessary to specify the nature of the line to be measured, as indicated in the preceding section. Third, it is necessary to specify whether the distance to be used is that along the geodesic, great circle or rhumb line. Fourth, it is necessary to specify the units of measurement. If an unsuitable conversion factor is used for the calculations, such as the metric definition of the nautical mile mentioned on page 527, this has the effect of introducing systematic errors into the results

Geodetic Datum

Any national or international control survey must be based upon some origin of assumed or known position together with an accepted value for the figure of the earth used in the computation of the results. The positions of points within a trigonometric system are then computed sequentially with reference to these data, so that the location of a station which is remote from the origin is referred indirectly to the origin through the positions of many intervening stations.

Before World War II, geodesy was primarily concerned with the creation of national control surveys, and each national survey had its own geodetic datum. The origin was usually a station whose position has been determined after decades of repeated astronomical observations at a major observatory such as Greenwich, Potsdam, Pulkova or Paris. The choice of the figure of the earth had often been for historical or political reasons as much as for geodetic suitability. During the first half of the twentieth century there were almost as many different geodetic systems in use in Europe as there were nation-states. Consequently it was not surprising that the positions of points which were common to different national surveys sometimes differed by several metres. Resulting from the discrepancies between different networks, and therefore the difficulty of defining position consistently throughout Europe, the various geodetic systems of central and mainland Western Europe were readjusted and recomputed by the U.S. Army Map Service. It is of historical interest that this formidable task, carried out in the immediate postwar period, was one of the first geodetic applications undertaken by digital computer. The new system was based upon the origin at the observatory at Potsdam, which had already been employed as the origin of several central European systems, and made use of the international spheroid. This system is now described as ED50, or the European Datum of 1950. Neither the geodetic systems of Norway nor the United Kingdom were included in this readjustment. The Ordnance Survey had reobserved and recomputed the primary triangulation of Britain during the 1930s and the datum is often referred to as OSGB36. This system had its origin at the site of the Royal Observatory at Herstmonceux in Sussex, and is computed on the Airy spheroid, which had also been used for the computation of the original triangulation of the country in the middle of the nineteenth century. Following the measurement of the lengths of many sides of the network during the quarter-century since the introduction of E.D.M., the network was readjusted to create a new version known as OSGB70.

Although the British network was not included in the adjustment to ED50, a few lines had been observed across the Straits of Dover, sufficient to give a geodetic connection between England and France, allowing the relationship between the two systems to be determined.

In southern Norway the old primary network which had also been observed during the nineteenth century had its astronomical origin at Oslo Observatory, and employed a version of the Bessel figure of the earth which differs from the figure of that name used in Germany and other parts of central Europe. The Norwegian datum is known as NGO48. The system was not part of ED50 but was added to it later, in 1969, when development of the North Sea hydrocarbon industry made it imperative to use a consistent geodetic datum throughout the whole area. In the 1965 Agreement on the North Sea, the actual median lines were used only as the basis for negotiation, and the agreed boundary was taken as a series of points defined by their geographical coordinates on ED50 and joined by arcs of great circles. The European Datum was chosen because of the existing connection between OSGB 36 and ED50 across the Straits of Dover.

With the introduction of position fixing from artifical satellites by Doppler methods, yet another form of geodetic datum enters the argument. Each of the extent systems depends upon slightly different values for the earth's figure, and particular versions have to be used with particular satellites. Thus we have the spheroid derived by the Naval Weapons Laboratory known as NWL-9D which is used with the TRANSIT system and the World Geodetic System of 1972, or WGS 72, for NAVSTAR GPS.

COMPARISON OF DIFFERENT CALCULATED DISTANCES

In order to demonstrate the differences in distance calculated by various methods we take the example of a line across the North Sea between the points, $A = 58°, 0°$ and $B = 62°$ north, $10°$ east. This line extends well into the mountains of central Norway, and is therefore much longer than those needed to locate the median line between Britain and Norway, but it is the example

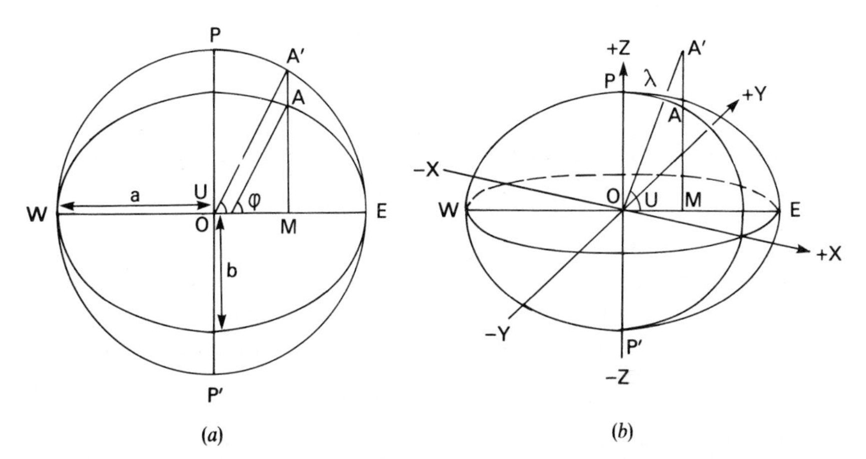

FIG. 23.10. The parameters used to determine the three-dimensional Cartesian coordinates of a point on the surface of a spheroid as a preliminary to the determination of the length of the geodesic: (a) definition of the *reduced latitude, u*; (b) the three-dimensional Cartesian coordinate system employed to calculate the (X, Y, Z) coordinates of A.

used by Thamsborg (1974) and also the line illustrated in Fig. 23.8. It is also, as we have seen in Chapter 22, a line which corresponds to the approximate limits of length and mean latitude for use of the plane-sailing assumption.

The Length of the Geodesic

The length of the geodesic between two points which are defined by their geographical coordinates may be determined in a variety of ways. The method used here is that described in Admiralty (1964), which is based upon the determination of the straight line chord distance between the points, to which a chord-to-arc correction is applied to find the length of the geodesic. The stages in the determination are as follows:

1 Find the reduced latitude, u, of each point from equation (23.5)

$$\tan u = (b/a)\tan\varphi \qquad (23.1)$$

2 A point on the spheroidal surface may be expressed in three-dimensional Cartesian coordinates as

$$\begin{aligned} X &= a\cdot\cos u\cdot\cos\delta\lambda \\ Y &= a\cdot\cos u\cdot\sin\delta\lambda \\ Z &= b\cdot\sin u \end{aligned} \qquad (23.2)$$

Expressing the coordinates of the point A as X_A, Y_A, and Z_A and those of the point B as X_B, Y_B and Z_B respectively, the coordinate differences may be written

$$\begin{aligned} \delta X &= X_A - X_B \\ \delta Y &= Y_A - Y_B \\ \delta Z &= Z_A - Z_B \end{aligned} \qquad (23.3)$$

3 The chord distance, K between these points may be determined from

$$K^2 = \delta X^2 + \delta Y^2 + \delta Z^2 \qquad (23.4)$$

4 The correction $D - K$ to convert from chord into arc distance varies with the length and the direction of the line. This is the most awkward part of the computation and, because it is not an exact solution, a small degree of approximation is introduced at this stage. The required expression is

$$(D - K) = K^3/24R^2 + 3K^5/640R^4 \qquad (23.5)$$

where R is the mean radius of the spheroidal arc. This has to be computed from

$$R = \rho v/(\rho\sin^2\alpha + v\cos^2\alpha) \qquad (23.6)$$

and α is the bearing, or azimuth of the line from A to B. This, in turn, has to

TABLE 23.3. *The length of the geodesic between point A ($\varphi_A = 58°$ N, $\lambda_A = 0°$) and point B ($\varphi_B = 62°$ N, $\lambda_B = 10°$ E) based upon different figures of the earth (distances in metres)*

Spheroid	*a*	*b*	Distance
Airy	6,377,563·4	6,356,256·9	712,748
International	6,378,388	6,356,911·9	712,851
Bessel	6,377,397·2	6,356,079·0	712,731
Bassel NG048	6,377,492	6,356,173·3	712,741
WGS 72	6,378,135	6,356,750·52	712,816
GRS 80	6,378,137	6,356,752·3	712,818
NWL-9D	6,378,145	6,356,759·8	712,818
I.A.U.	6,378,160	6,356,774·5	712,820

be found from the equation

$$\tan\alpha = \sin\delta\lambda/\{\sin\varphi_a[\cos\delta\lambda - (\cos u_a/\cos u_b)(1 - b^2/a^2)] - b^2/a^2 \tan\varphi_b \cos\varphi_a\} \quad (23.7)$$

This solution for R is based on Euler's theorem.

For lines which are less than 100 km in length the correction $(D - K)$ is less than 1 m (or $< 0{\cdot}001\%$); at 350 km, which approximates to the 200 nm limit, the correction is about 44 m ($\approx 0{\cdot}013\%$). For a line of length 1000 km the correction is of the order of 1,030 m, which is about 0·1% of the length of that line. The calculated lengths of the geodesic are tabulated in Table 23.3 for those figures of the earth most commonly in use in the northern North Sea. The range in the calculated distance is 120 metres. This variation results wholly from the difference between the shapes and sizes of the earth's figure.

In order to establish the position of an equidistant boundary based upon two or three fixed points on an ellipsoid, it is necessary to employ more complicated algorithms. We do not deal with this problem here, but refer the reader to the paper by Carrera (1987) which describes a solution adopted by the Canadian Hydrographic service.

The Great Circle Arc

The method of calculating the length of a great circle arc was described by equations (3.6) and (3.7) on p. 45. Using (3.6) it is easy to determine that $z = 0{\cdot}1115523$ radians, and only one solution for z is possible from the geographical coordinates of A and B. In order to convert z from angular to linear measure we use equation (3.7) with a suitable value for R, the radius of the sphere. We must therefore decide what is the suitable value from our knowledge about the various parameters of the spheroid, a, b, ρ and ν.

The problem is not confined to the present study, but arises fairly often in the study of geodesy and map projections, whenever it is considered

TABLE 23.4. *The values of the length of the great circle arc AB determined for different spherical radii and compared with the length of the geodesic through the same pair of points calculated for the International Spheroid*

Definition of R	Spherical radius (metres)	Length of the arc AB (metres)	Difference from geodesic (metres)
Major axis, a	6,378,388	711,523	−1328
Arithmetic mean of a and b	6,367,650	710,325	−2526
Geometric mean of a and b	6,367,641	710,324	−2527
Arithmetic mean of three axes $(2a+b)/3$	6,371,229	710,725	−2126
Geometric mean of three axes $(2a \cdot b)^{1/3}$	6,371,221	710,724	−2127
ρ for $\varphi = 45°$	6,367,586	710,318	−2533
ν for $\varphi = 45°$	6,389,135	712,722	
g.m. for $\varphi = 45°$	6,378,351	711,519	−1332
ρ for φ_M	6,383,727	712,119	−732
ν for φ_M	6,394,529	713,324	+473
Frolov's equation	6,389,009	712,708	−143
Sphere of equal volume	6,371,221	710,724	−2127
Authalic sphere	6,371,228	710,725	−2126
Rectifying sphere	6,367,655	710,326	−2525
Radius from Euler's theorem	6,391,462	712,982	+131
Radius implicit in the use of the conversion factor 1 nm = 1852 m.	6,366,706	710,221	−2630

appropriate to simplify the algebra or the calculations, for example in the derivation of certain versions of the Transverse Mercator projection. The spherical body, which is to be regarded as tangential to the spheroid at a particular point, is, as we saw in Chapter 11, also referred to as an auxiliary sphere. We saw in Chapter 11 how the dimensions of $1 \times 1°$ quadrangles vary according to what radius has been chosen for the reference sphere. We demonstrate in Table 23.4 similar conclusions for the calculated lengths of the great circle arc AB, which is based upon the same value for z but have been determined using different spherical radii. The range in these tabulated values is more than 3·1 km. Moreover, none of the determinations agree with the length of the geodesic within 100 metres, and all but two of the arcs are shorter than the length of the geodesic. Many of these determinations indicate discrepancies of the order of 1 part in 250. This is the order of relative precision to be expected in chain surveying. Assuming a graphical standard error in location of a point on a chart of $\pm 0{\cdot}4$ mm, these discrepancies are equivalent to what might occur if the measurement and plotting had been done on a chart of scale 1/6,225,000. In other words it is far too crude for the intended purpose.

Plane Sailing and Rhumb Line Distances

For comparison with these results we give the corresponding plane and Mercator sailing distances as these were determined in Chapter 22, for this line was used as the data for the worked examples on page 510.

Plane Sailing

We have $\delta\varphi = 240'$, $\delta\lambda = 600'$ and $\varphi_m = 60°$ north. Then, from equation (22.30), we find the middle latitude to be $\varphi' = 60°\ 02'\ 47''$. From equation (22.31) the departure is $299'{\cdot}58$. It follows from equation (22.33) that the distance, $s = 383'{\cdot}859$. Using the conversion 1 nm = 1,852 m we find that this corresponds to the distance of 710,907 m. However, comparison of this distance with those already determined for the geodesic and great circle shows that the plane sailing distance is evidently less than what we know from theory to be theoretically the shortest distance between two points. In fact it is the use of an unsuitable conversion factor which is to blame for this impossible result. If we take as the value for $R = 6{,}391{,}462$ m which, in Table 23.5, is the spherical radius determined from Euler's theorem, we obtain the plane sailing linear distance to be $s = 713{,}672$ m, which is more realistic.

Mercator Sailing

If we assume the earth to be spherical, the meridional parts for the points A and B may be calculated as 4,294·3019 and 4,774·9865 respectively. Then the difference in meridional parts is 480·6846 and the rhumb line distance is $383'{\cdot}8538$.

Using the spheroidal approximation which employs the reduced latitudes u_A and u_B, the meridional parts are reduced to 4,284·4739 and 4,764·753 respectively, giving a difference in meridional parts of 480·2791. It follows that the rhumb line distance is $384'{\cdot}0513$.

A summary of the determinations relating to the International Spheroid are listed in Table 23.6. The range between the values listed in Table 23.5 is 2,178 m. This is the measure of uncertainty to be expected in calculating the breadths of zones if the type of line to be used has not been properly specified, even if there is no doubt about the figure of the earth. If, in addition, we add other measures of uncertainty about the geodetic datum or the radius of the auxiliary sphere, the possibility of error in the calculated distance increases to more than 3 km.

It follows that unless the enumerated variables have all been carefully specified, the calculated distance is no more reliable than one which has been scaled from a small-scale chart.

TABLE 23.5. *The length of the line between point A ($\varphi_A = 58°$ N, $\lambda_A = 0°$), and B ($\varphi_B = 62°$ N, $\lambda_B = 10°$ N) (in metres)*

Geodesic on the International Spheroid	712,851
Great circle, using $R = 6{,}391{,}462$ m	712,982
Plane sailing distance	713,672
Rhumb line distance (sphere with radius $R = 6{,}391{,}462$ m)	713,412
Rhumb line distance (International Spheroid)	714,029

References

Aangeenbrug, R. T. (Ed), (1980): *Proceedings of the International Symposium on Cartography* and *Computing: Applications in Health and Environment. Auto Carto IV*, Falls Church, Va., American Congress on Surveying and Mapping and American Society of Photogrammetry, 2 volumes: Vol. 1, 623 pp., Vol. 2, 479 pp.

Abell, C. A. (1939): A method of estimating area in irregularly-shaped and broken figures. *J. For.*, **37**, 344–5.

ACIC (1962): *Map Accuracy Evaluation. Part 1. Evaluation of Horizontal Map Information.* Aeronautical Charting and Information Center Reference Publication No. 2. St. Louis, 46 pp.

ACIC (1963): *Map Accuracy Evaluation. Part 2. Evaluation of Vertical Map Information.* Aeronautical Charting and Information Centre Reference Publication No. 2, St. Louis, 46 pp.

ACSM (1976): *Proceedings of the International Conference on Automation in Cartography. "Auto Carto I"*. Falls Church, Va., American Congress on Surveying and Mapping, 318 pp.

ACSM (1979): *Proceedings of the International Conference on Computer-Assisted Cartography. Auto-Carto III.* Falls Church, Va., American Congress on Surveying and Mapping, 519 pp.

Adams, T. A. and Rhind, D. (1981): The projected characteristics of a national topographic digital databank. *Geogr. J.*, **147**, 38–53.

Admiralty (1948): *Admiralty Manual of Hydrographic Surveying.* London, Hydrographic Department, 1938; 2nd edn, 1948, 572 pp.

Admiralty (1954): *Admiralty Manual of Navigation, Vol. III.* London, H.M.S.O., 437 pp.

Admiralty (1964): *Admiralty Manual of Hydrographic Surveying.* London, The Hydrographer of the Admiralty, Vol. 1, 1964, 671 pp.

Admiralty (1973): *Admiralty Manual of Hydrographic Surveying.* London, The Hydrographer of the Admiralty, Vol. 2, 1973 onwards (published as looseleaf chapters).

Aherne, W. A. and Dunnill, M. S. (1982): *Morphometry.* London, Arnold, 225 pp.

Ahmed, Fouad, A. (1982): Rigorous Determination of Irregular Areas. *Can. Surveyor*, **35**, 103–7.

Aldred, A. H. (1964): Evaluation of the transect area-meter method of measuring map areas. Dept. Forestry, Canada, Forest Research Branch Contribution No. 588, *For. Chron.*, 175–83.

Alford, M., Tuley, P., Hailstone, E. and Hailstone, J. (1974): The measurement and mapping of land-resource data by point sampling on aerial photographs. In: Barrett, E. C. and Curtis, L. F. (Eds), *Environmental Remote Sensing: applications and achievements.* London, Arnold, 113 pp.

Amiran, D. H. K. and Schick, A. P. (1961): *Geographical Conversion Tables.* Zürich, International Geographical Union, 315 pp.

Aronoff, S. (1982a): Classification accuracy: a user approach. *Photogram. Eng. & Remote Sens.*, **48**, 1299–1307.

Aronoff, S. (1982b): The map accuracy report: a user's view. *Photogram. Eng. & Remote Sens.*, **48**, 1309–1312.

Atlas of Britain (1963): See Bickmore, D. P. and Shaw, M. A. (Eds) (1963).

Baer, H. (1937): Genauigkeitsuntersuchungen am Polarplanimeter. *Z. InstrumKde.*, **57** Jahrgang, Fünftes Heft, 177–89.

Bakkelid, S. and Rekkedal, S. (1983): Analysis of the positioning problems in the North Sea. *Marine Geodesy*, **6**, 117–37.

Balandin, V. N. (1985): Vychislenie ploshchadi sferoidicheskoy trapetsii (Calculation of the area of a quadrangle on the spheroid), *Geodeziya i Kartografiya*, **9**, 18–19.

Banks, N. E. (1983): National Ocean Survey. Bathymetric Mapping Program. *Int. Hydrogr. Rev.*, **60**, 105–16.

Barford, N. C. (1967): *Experimental Measurements: precision, error and truth.* London, Addison–Wesley, 143 pp.
Barrett, E. C. and Curtis, L. F. (Eds) (1974): *Environmental Remote Sensing: applications and achievements.* London, Arnold, 1974, 309 pp.
Barrett, J. P. and Philbrook, J. S. (1970): Dot grid area estimates: precision by repeated trials. *J. For.*, **68**, 149–51.
Barry, A. B. (1964): *Engineering Measurements.* New York, Wiley, 1964, 136 pp.
Barston, R. P. and Birnie, P. (Eds) (1980): *The Maritime Dimension.* London, Allen & Unwin, 1980, 194 pp.
Bauer, M. E., Hixson, M. M., Davis, B. J. and Etheridge, J. B. (1978): *Area estimation of crops by digital analysis of Landsat data. Photogram. Eng. & Remote Sens.*, **44**, 1033–43.
Baugh, I. D. H. and Boreham, J. R. (1977): Measuring the coastline from maps: a study of the Scottish mainland. *Cartogr. J.*, **13**, 167–71.
Beazley, P. B. (1971): Territorial sea baselines. *Int. Hydrogr. Rev.*, **48**(1), 143–54.
Beazley, P. B. (1978): *Maritime Limits and Baselines.* The Hydrographic Society, Special Publication No. 2, 2nd edn, 1978, 41 pp.
Beazley, P. B. (1979): Half-effect applied to equidistance lines. *Int. Hydrogr. Rev.*, **56**, 153–60.
Beazley, P. B. (1982): Maritime boundaries. *Int. Hydrogr. Rev.*, **59**, 149–59.
Beazley, P. B. (1986): Developments in maritime delimitation. *Hydrographic J.*, **39**, 5–9.
Beckett, P. H. T. (1977): Cartographic generalisation. *Cartogr. J.*, **14**, 49–50.
Bellhouse, D. R. (1981): Area estimation by point-counting techniques. *Biometrics*, **37**, 303–12.
Berry, B. J. L. (1962): *Sampling, coding and storing flood plain data.* Agric. Handbook No. 237. Washington, Dept. of Agriculture, Farm Economics Division.
Berry, B. J. L. and Baker, A. M. (1968): Geographic sampling. In: Berry, B. J. L. and Marble D. F. (Eds), *Spatial Analysis*, pp. 91–100.
Berry, B. J. L. and Marble, D. F. (Eds) (1968): *Spatial Analysis.* New York, Prentice-Hall, 512 pp.
Bibby, J. (1972): Infinite rivers and Steinhaus' Paradox. *Area*, **4**, 214.
Bickmore, D. P. and Shaw, M. A. (1963): *Atlas of Britain and Northern Ireland.* Oxford, Clarendon Press, 222 pp.
Bickmore, D. P. (1971): Experimental maps of the Bideford area. Conference of Commonwealth Survey Officers, 1971, *Report of Proceedings, Part I*, Paper No. E1. London, O.D.A., pp. 217–223.
Bickmore, D. P. and Witty, G. J. (1980): Ecobase of Britain: status report on a digital data base of Britain. In Aangeenbrug, R. T. (Ed). (1980), pp.
Bird, E. C. F. (1984): *Coasts.* Oxford, Blackwell, 3rd edn, 320 pp.
Blachut, T. J. (Ed.) (1964b): Compendium of photogrammetric contouring. X International Congress of Photogrammetry, Commission IV. *Nachr. Karten-u. Vermessungsw.*, R. V. Sonderheft Nr. 9, 1964.
Blachut, T. J., Tewinkel, G. and Finsterwalder, R., (1960): Second International Mapping Experiment. Renfrew Test Area. *Can. Surveyor*, 138–70.
Blakemore, M. (Ed.) (1986): *Proceedings Auto Carto London*, 2 volumes. Vol. 1: *Hardware, Data Capture and Management Techniques.* Vol. 2: *Digital Mapping and Spatial Information Systems.* London, Auto Carto London Ltd, for I.C.A.
Blankenburgh, J. C., Fossum, B. A., Osterholt, A. and Torsen, H. O. (1978): Norwegian marine geodetic projects. *Marine Geodesy*, **1**, 125–45.
Blankenburgh, J. C. (1980): Doppler–European Datum transformation parameters for the North Sea. *Phil. Trans. R. Soc. Lond. A*, **294**, 277–88.
Bocharov, M. K. (1953): Zavisimost' mezhdu rasstoyaniem i chislom tochek na dannoy ploshchadi. (The relationship between distance and the number of points in a given area). *Vestnik VIA*, 66.
Boggs, S. W. (1930): Delimitation of the territorial sea. *Amer. Journ. Int. Law*, **24**, 541–55.
Bomford, G. (1950): *Geodesy*, Oxford, Oxford University Press, 1st edn, 561 pp.
Bonnor, G. M. (1975): The error of area estimates from dot grids. *Can. J. For. Res.*, **5**, 10–17.
Boyle, A. R. (1979): Digital Hardware: Mass Digitization. Introductory remarks. In A.C.S.M. (1979), p. 237.
Brazier, H. H. (1947): A skew orthomorphic projection with particular reference to Malaya. *Conf. Commonwealth Survey Officers, Report of Proceeding.* London, H.M.S.O., 1951, pp. 42–62.

Briarty, L. G. (1975): Stereology: methods for quantitative light and electron microscopy. *Sci. Prog. Oxf.*, **62**, 1–32.
Brown, E. D. (1978): "It's Scotland's oil"? *Marine Policy*, **2**(1), 3–21.
Brown, E. D. (1979): Rockall and the limits of national jurisdiction of the U.K. *Marine Policy*,
Brown, E. D. (1981): Delimitation of offshore areas. *Marine Policy*, **5**, 172–84.
Bryan, M. M. (1943): Area determination with the modified acreage grid. *J. For.*, **41**, 764–6.
Bunge, W. (1962): Theoretical geography. *Lund Stud. Geogr.*, Ser. C, General and Mathematical Geography No. 1, 2nd edn, 1966, 289 pp.
Burnside, C. D. (1979): *Mapping from Aerial Photographs.* London, Granada, 1979, 304 pp.

Card, D. H. (1982): Using known map category marginal frequencies to improve estimates of thematic map accuracy. *Photogram. Eng. & Remote Sens.*, **48**, 431–9.
Carrera, G. (1987): A method for the delimitation of an equidistant boundary between coastal states on the surface of a geodetic ellipsoid. *Int. Hydrogr. Rev.*, **64**, 147–59.
Carrier, A. (1947): *Traité de Topographie Générale.* Paris, Girard, Barrère et Thomas, 757 pp.
Chatfield, C. (1970): *Statistics for Technology.* Harmondsworth, Penguin, 1970, 359 pp.
Chayes, F. (1956): *Petrographic Modal Analysis.* New York, Wiley, 1956, 113 pp.
Chernyaeva, F. A. (1958): K voprosu opredeleniya dlin izvilistykh liniy po kartam s pomoshchya sirkulya. *Kartografiya. Uchenye zapiski Leningradskogo Gosudarstvennogo Universiteta*, Nr 226. Seriya Geogr. Nauk, Bd 12,182–200. (On the question of obtaining the lengths of sinuous lines on maps using dividers).
Chernyaeva, F. A. (1967): Sravnenie metodov planimetrirovaniya i vzreshivaniya pri opredelenii ploshchadey po kartam. *Novye probl. i metody kartogr.*, M, Nauka, 1967. (A comparison of the methods of planimeter and weighing for the determination of area on a map).
Chernyaeva, F. A. and Molchanova, Z. P. (1963): Primenenie metodov planimetrirovaniya i vzveshivaniya dlya priblizennych opredeleniy ploshchadey po kartam. *Vestnik Leningradsk. Univ.*, 12, 132–135. (The use of planimeter methods and weighing for preliminary determination of areas on maps).
Chorley, R. J. (Ed.) (1972): *Spatial Analysis in Geomorphology.* Edited for the British Geomorphological Research Group. London, Methuen, 393 pp.
Chorley, R. J. and Haggett, P. (Eds) (1967): *Models in Geography.* London, Methuen, 816 pp.
Cliff, A. D. and Ord, J. K. (1981): *Spatial Processes: Models and Applications.* London, Pion, 266 pp.
Cliff, A. D., Haggett, P., Ord, K. A., Bassett, K. A. and Davies, R. B. (1975): *Elements of Spatial Structure.* Cambridge, Cambridge University Press, 258 pp.
Cochran, W. G. (1946): Relative accuracy of systematic and stratified random samples for a certain class of population. *Ann. Math. Statist.*, **17**, 164–77.
Cochran, W. G. (1977): *Sampling Techniques.* New York, Wiley, 3rd edn, 428 pp.
Cochran, W. G., Mosteller, F. and Tukey, J. W. (1954): Principles of sampling. *J. Amer. Statist. Assoc.*, **49**, 13–35.
Cole, J. P. and King, C. A. M. (1968): *Quantitative Geography.* London, Wiley, 692 pp.
Coleman, A., Isbell, J. E. and Sinclair, G. (1974): The comparative statics approach to British land-use trends. *Cartogr. J.*, **11**, 34–41.
Colvocoresses, A. P. (1972): ERTS-A Satellite Imagery. *Photogram. Eng.*, **38**, 555–60.
Colvocoresses, A. P. (1974): Space oblique Mercator. *Photogram. Eng.*, **40**, 921–6.
Colvocoresses, A. P. and MacEwan, R. B. (1973): EROS Cartographic Progress. *Photogram. Eng.*, **39**, 1303–9.
Colwell, R. N. (Ed.) (1960): *Manual of Photo Interpretation.* Washington, Amer. Soc. Photogrammetry, 868 pp.
Colwell, R. N. (Ed.) (1983): *Manual of Remote Sensing.* Vol. 1: *Theory, Instruments and Techniques.*, Falls Chruch, Va., Amer. Soc. Photogrammetry, 1232 pp.
Congalton, R. G., Oderwald, R. G. and Mead, R. A. (1983): Assessing Landsat classification accuracy using discrete multivariate analysis statistical techniques. *Photogram. Eng. & Remote Sens.*, **49**, 1671–8.
Copland, D. R. (1977): Correspondence. *Cartogr. J.*, **14**, 6.
Coppock, J. T. and Johnson, J. H. (1962): Measurement in human geography. *Econ. Geogr.*, **38**, 130–7.

Cormack, R. M. and Ord, J. K. (1979): *Spatial and Temporal Analysis in Ecology*, Fairland, Int. Co-op. Publishing House, 356 pp.
Countryside Commission (1968): *The Coasts of England and Wales. Measurements of Use, Protection and Development*. London, H.M.S.O.
Couper, A. D. (1978): The Law of the Sea. London, Macmillan, 28 pp.
Couper, A. D. (Ed.) (1983): *The Times Atlas of the Oceans*. London, Times Books, 272 pp.
Couper, A. D. (1985): The Marine Boundaries of the United Kingdom and the Law of the Sea. *Geogr. J.*, **151**, 228–36.
Crapper, P. F. (1980): Errors incurred in estimating an area of uniform land cover using Landsat. *Photogram. Eng. & Remote Sens.*, **10**, 1295–1301.
Crapper, P. F. (1981): Geometric properties of regions with homogeneous biophysical character-to grid cells. *Photogram. Eng. & Remote Sens.*, **50**, 1497–1503.
Crapper, P. F. (1984): An estimate of the number of bounding cells in a mapped landscape coded to grid cells. *Photogram. Eng. & Remote SEns.*, **50**, 1497–1503.
Crofton, M. W. (1868): On the theory of local probability applied to straight lines drawn at random in a plane; the methods used being also extended to the proof of certain new theorems in the integral calculus. *Phil. Trans. R. Soc. Lond.*, **158**, 181–99.
Crone, D. R. (1953): The accuracy of topographical maps. *Emp. Surv. Rev.*, **12**, 88, 64–70.
Czartoryski, J. (1987): Accuracy of cartographic processes in the construction of nautical charts. *Int. Hydrogr. Rev.*, **64**, 79–90.

Dale, P. (1976): *Cadastral Surveys within the Commonwealth*. Overseas Research Publication No 23. London, H.M.S.O., 281 pp.
Dale, P. and DeSimone, M. (1986): Vector–raster–vector conversions. *Land. Min. Surv.*, **4**, 126–33.
Davis, J. C. and McCullagh, M. J. (Eds) (1975): *Display and Analysis of Spatial Data*. London, Wiley, 378 pp.
De Hoff, R. T. and Rhines, F. N. (Eds) (1968): *Quantitative Microscopy*. New York, McGraw-Hill, 422 pp.
Demina, V. V. (1960): Otsenka tochnosti izmereniy po aerosnimkam i fotoskheman pri geologo-geograficheskikh issledovaniyakh. *Trudy Laboratorii Aerometodov* Tom IX. Moscow and Leningrad, A. N. SSSR, 244–259. (An appraisal of the accuracy of measurement on aerial photographs for geological and geographical research).
Devereux, B. J. (1986): The integration of cartographic data stored in raster and vector formats. In Blakemore, M. (Ed.): (1986), Vol. 1, pp. 257–66.
Diaconis, P. (1976): Buffon's problem with a long needle. *J. Appl. Prob.*, **13**, 614–18.
Dickinson, G. C. (1969): *Maps and Air Photographs*. London, Arnold, 286 pp.
Doornkamp, J. C. and King, C. A. M. (1971): *Numerical Analysis in Geomorphology*. London, Arnold, 372 pp.
Dowman, I. J. (*et al.*) (1985): Photography from space. *Photogram. Rec.*, **11**, 667–727.
Doyle, F. J. (1985): The large format camera on shuttle mission 41-G. *Photogram. Eng. & Remote Sens.*, **51**, 200–1.
Dozier, J. and Strahler, A. H. (1983): Verification and accuracy assessment for thematic maps. In Colwell, R. N. (Ed.), *Manual of Remote Sensing*.
Dumitrescu, V. (1966): Perspectives cosmographiques – un système utile de projections azimutales. *Int. Jahrb. Kartograph.*, **6**, 25–32.

Eckert, Max (1921): *Die Kartenwissenschaft*. Berlin and Leipzig, de Gruyter, Vol. 1, 640 pp.
Ellis, M. Y. (Ed.) (1978): *Coastal Mapping Handbook*. Washington, U.S. Govt. Printing Office, 200 pp.
Emmott, C. (1981): Computer aided photointerpretation for land use. *Surveying and Mapping 1981*. Paper No. F5., London, R.I.C.S., 1981.
Emmott, C. and Collins, W. G. (1980): Air photo interpretation for the mesurement of changes in urban land use. *Int. Arch. Photogramm.*, **XXIII**, B7, 281–90.

Faller, K. (1977): *A procedure for detection and measurement of interfaces in remotely acquired data using a digital computer*. Technical Report TR-P-472, N.A.S.A., N.S.T.L. Station, Mississippi.
Fisher, R. A. and Yates, F. (1948): *Statistical Tables for Biological, Agricultural and Medical Research*. Edinburgh, Oliver and Boyd, 3rd edn, 112 pp.

Fitzherbert, A. (1598): *The Boke of Husbandry.* Facsimile reprint as No. 926 of The English Experience. Amsterdam, Elsevier, 1979, 207 pp.
Fitzpatrick, K. A. (1977): The strategy and methods for determining accuracy of small and intermediate scale land-use and land-cover maps. *Proc. 2nd W.T. Pecora Symposium,* A.S.P., pp. 339–61.
Fitzpatrick-Lins, K. A. (1978): Accuracy and consistency comparisons of land-use and land-cover maps made from high altitude photographs and Landsat multispectral imagery. *U.S.G.S. J. Res.,* **6**, 23–40.
Fitzpatrick-Lins, K. A. (1978): An evaluation of errors in mapping land-use changes for the Central Atlantic Regional Ecological Test Site. *U.S.G.S. J. Res.,* **6**, 339–46.
Fitzpatrick-Lins, K. A. (1981): Comparison of sampling procedures and data analysis for a land-use and land-cover map. *Photogram. Eng. & Remote Sens.,* **47**, 343–51.
Förstner, R. (1957): Schichtlinienfehler. *Z. VermessWes.,* 12.
Förstner, R. and Schmidt-Falkenberg, H. (1964): The accuracy of photogrammetrically plotted contour lines. *Int. Arch. Photogramm.,* **XV-6**, 1965, 29–42.
Fortescue, F. S. (1967): The latest developments in cartographic and printing techniques in the Ordnance Survey. Conf. Commonwealth Survey Officers 1967, *Report of Proceedings,* Vol. II, 525–529. London, HMSO.
Fraser, S. E. G. (1982): Digital cartography at the Ordnance Survey. In Rhind, and Adams, (Eds): *Computers in Cartography,* pp. 29–34.
Frazier, B. E. and Shovic, H. F. (1980): Statistical methods for determining land-use change with aerial photographs. *Photogram. Eng. & Remote Sens.,* **46**, 1067–77.
Frolov, Y. S. (1964): Analiticheskie formuly dlya opredeleniya privedennykh znacheniy dlin liniy. *Vestnik Leningradskogo Universiteta.* No 18. Seriya Geologii i Geografii, Vyp 3. (An analytical formula for the determination of the reduced lengths of lines).
Frolov, Y. S. (1966): Izmerenie ploshchadey po melkomasshtabnym kartam v merkatorskoy proektsii. *Vest. Leningr. Universiteta.,* **6**, 121–134. (Measurement of area on small scale charts on Mercator's Projection).
Frolov. Y. S. and Maling, D. H. (1969): The accuracy of area measurement by point counting techniques. *Cartogr. J.,* **6**, 21–35.

Galloway, R. W. and Bahr, M. E. (1979): What is the length of the Australian coast? *Austral. Geogr.,* **14**, 244–7.
Gamble, J. K. (1974): *Global Marine Attributes.* Ballinger.
Gamble, J. K. and Pontecorvo, G. (Eds) (1974): *Law of the Sea: Emerging Regime of the Oceans.*
Gannett, H. (1881): *The Areas of the United States, the several states and territories and their counties.* Extra Census Bulletin. Washington, Government Printing Office, 22 pp.
Gardiner, V. (1977): On generalisations concerning generalisation. *Cartogr. J.,* **14**, 135–6.
Gerke, K. (1951): Gewichte von Flächenberechnungen mit Umfahrungsplanimetern. *Z. VermessWes.,* **4**, 103–10.
Gierhart, J. W. (1954): Evaluation of methods of area measurement. *Surv. and Mapp.,* **14**, 460–5.
Ginevan, M. E. (1979): Testing land-use map accuracy: another look. *Photogram. Eng. & Remote Sens.,* **45**, 1371–7.
Ginzburg, G. A. (1958): *Posobie po izmereniyam na melkomasshtabnykh kartakh.* Trudy TsNIIGAiK, Vyp 119, 136pp. (Text-book of measurement on small-scale maps).
Gladman, T. and Woodhead, J. H. (1960): The accuracy of point counting in metallographic investigations. *J. Iron Steel Inst.,* **194**, 189–93.
Glusic, A. M. (1961): *The Positional Accuracy of Maps.* Army Map Service Technical Report No. 35. Washington, U.S. Army, Corps of Engineers, 40 pp.
Goddard, J. and Armstrong, P. (1986): The 1986 Domesday Project. *Trans. Inst. Brit. Geogr.,* N.S. **11**, 290–5.
Goodchild, M. F. (1980): Fractals and the accuracy of geographical measures. *Math. Geol.,* **12**, 85–98.
Goodchild, M. F. and Moy, W. S. (1977): Estimation from grid data: the map as a stochastic process. In: Tomlinson, (Ed.): Proc. Commission on Data Sensing and Processing.
Goodier, R. (Ed.) (1971): *The Application of Aerial Photography to the Work of the Nature Conservancy.* Edinburgh, Nature Conservancy, 185 pp.

Gordon, D. L. (1968): Rationalisation of chart datum in the British Isles. *Int. Hydr. Rev.*, **45**, 55–65.
Grattan-Bellew, P. E. Quinn, E. G. and Sereda, P. J. (1978): Reliability of scanning electron microscopy information. *Cement and Concrete Res.*, **8**, 333–42.
Greenwalt, C. R. and Schultz, M. E. (1962): *Principles of Error Theory and Cartographic Applications.* U.S. Air Force, A.C.I.C. Tech. Rep. No. 96, St. Louis, 90 pp.
Gregory, S. (1963): *Statistical Methods and the Geographer.* London, Longmans, 240 pp.
Greig-Smith, P. (1964): *Quantitative Plant Ecology.* London, Butterworth, 2nd edn, 256 pp.; Oxford, Blackwell, 1983, 3rd edn, 359 pp.

Haggett, P. (1963): Regional and local components in land-use sampling: a case-study from the Brazilian Triangulo. *Erdkunde*, Band XVII, 108–14.
Haggett, P. and Board, C. (1964): Rotation and parallel traverses in the rapid integration of geographic areas. *Ann. Assoc. Amer. Geogr.*, **54**, 406–10.
Håkonson, L. (1978): The length of closed geomorphic lines. *J. Int. Assoc. Math. Geol.*, **10**, 141–67.
Hallert, B. (1960): *Photogrammetry.* New York, McGraw-Hill, 340 pp.
Harley, J. B. (1975): *Ordnance Survey Maps: a descriptive manual.* Southampton, Ordnance Survey, 200 pp.
Harris, L. J. (1959): Present position and trends in applied cartography. *Conf. Commonwealth Survey Officers 1959, Report of Proceedins*, Paper No. 20b, London, HMSO.
Hart, C. A. (1940): *Air Photography Applied to Surveying.* London, Longmans, 366 pp.
Harvey, D. (1959): *Explanation in Geography.* London, Arnold, 521 pp.
Haslam, D. W. (1977): Hydrographic Services, *Marine Policy*, **1**, 3–15.
Haslam, D. W. (1983): Why a hydrographic Office? *Conference of Commonwealth Surveyors, Report of Proceedings*, Vol. 1, Paper A4. London, O.D.A.
Haslam, D. W. (1985): Contract hydrographic surveys. *Int. Hydrogr. Rev.*, **62**, 7–19.
Hay, A. M. (1979): Sampling designs to test land-use map accuracy. *Photogram. Eng. & Remote Sens.*, **45**, 529–33.
Hedge, A. R. (1985): Navigation and positioning requirements for marine survey operations. *Int. Hydrogr. Rev.*, **62**, 95–104.
Henrici, O. (1894): Report on planimeters. *Report of the British Association for the Advancement of Science, 1894.* London, John Murray, pp. 496–523.
Hicks, S. D. (1986): Tidal datums and their uses – a summary. *Hydrographic J.*, **39**, 17–20.
Hilliard, J. E. and Cahn, J. W. (1961): An evaluation of procedures in quantitative metallography for volume-fraction analysis. *Trans. Metal. Soc. A.I.M.E.*, pp. 344–52.
Hiwatashi, M. (1968): A study of the comparison between the efficiency of the dot-grid method and planimeter method in the area measurement. *For. Exp. Stn. Tokyo, Japan, Bull.* **214**, 111–125. (In Japanese).
Hixson, M. M. (1981): Techniques for evaluation of area estimates. *1981 Machine Processing of Remotely Sensed Data Symposium*, pp. 84–90.
Hixson, M. M., Bauer, M. E. and Davis, B. J. (1979): Sampling for area estimation: a comparison of full-frame sampling with the sample segment approach. *1979 Machine Processing of Remotely Sensed Data Symposium*, pp. 97–103.
Hixson, M. M., Davis, S. M. and Bauer, M. E. (1981): Evaluation of a segment-based Landsat full-frame approach to crop area estimation. *1981 Machine Processing of Remotely Sensed Data Symposium.*
Hodgson, R. D. and Cooper, E. D. (1976): The technical delimitation of a modern equidistance boundary. *Ocean Dev. Int. Law*, **3**, 361.
Hoiz, H. (1957): Vorschlag für die einheitliche Deutung der Schichtlinienfehler. *Z. VermessWes.*, **7**.
Holdgate, M. W. (1979): *A Perspective of Environmental Pollution.* Cambridge, Cambridge University Press, 278 pp.
Hord, R. M. and Brooner, W. (1976): Land-use maps accuracy criteria. *Photogram. Eng. & Remote Sens.*, **42**, 671–7.
Hotine, M. (1931): *Surveying from Air Photographs.* London, Constable, 250 pp.
Howard, S. M. (1968): A cartographic data bank for Ordnance Survey maps. *Cartogr. J.*, **5**, 48–53.
Husch, B. (1971): *Planning a Forest Inventory.* FAO Forestry and Forest Products Studies, No. 17. Rome, FAO, 121 pp.

I. C. A. (1973): *Multilingual Dictionary of Technical Terms in Cartography*, International cartographic Association, Commission II. Wiesbaden, Franz Steiner Verlag, 573 pp.
I. C. A. (1984): *Basic Cartography*. International Cartographic Association, Vol. 1, 206 pp.
Idler, R. (1951): Genauigkeit von Flächenberechnungen mit Umfahrungsplanimeteru. *Z. VermessWes.*, **76**, 216–17.
Imhof, E. (1965): *Kartographische Geländarstellung*. Berlin, de Gruyter, 1965.
Imhof, E. and H. J. Steward (1982): *Cartographic Relief Presentation*. Berlin, de Gruyter.
Institut Géographique National (1958): Standard cartographic practices recommended for international use for land maps. Presented to the 2nd Regional Cartographic Conference of the United Nations for Asia and the Far East, Tokyo. *World Cartogr.*, **6**, 19–46.
Isakov, I. S. and Demin, L. A. (Eds) (1950–53): *Morskoy atlas*, Vol. 1, 1950, Vol. 2., 1953. (Atlas of the Oceans).
Isakova, L. N. and Pavlov, A. A. (1958): Izmerenie Ploshchadey po kartam v Merkatorskoy Proeksii. *Ucheniye Zapiski L.G.U.* Nr 226, Seriya Geograficheskikh Nauk, Vyp 12, 201–212. (Measurement of area on maps on Mercator's Projection).
I. T. C. (1979): Instructions for handling and measuring aerial photographs. Chapter 7: Linear and Areal Measurement. International Training Centre, *Textbook of Photointerpretation*, Enschede, VII 6, 18–21.

Jeffers, J. N. R. (Ed.) (1972): *Mathematical Models in Ecology*. Oxford, Blackwell, 398 pp.
Johnsson, O. (1968a): Drawing accuracy. *R.A.K. Medd.*, NR **A37**, pp. 61–62.
Johansson, O. (1968b): Orthophotomaps as a basis for the economic map of Sweden at the scale of 1/10,000. *R.A.K. Medd.*, NR **A37**, pp. 53–60.
Jones, A. (1979): Land measurement in England, 1150–1350. *Agric. Hist. Rev.*, **27**, 10–18.
Jordan, W. and Eggert, O. (1914): *Handbuch der Vermessungskunde*.

Kahan, B. C. (1961): A practical demonstration of a needle experiment designed to give a number of concurrent estimates of π. *J. Roy. Statist. Soc.*, A, **124**, 227–39.
Kapoor, D. C. (1977): The delimitation of exclusive economic zones. *Martime Policy and Management*, **4**, 255–63.
Kapoor D. C. (1981): Hydrographic surveys to aid exploration and exploitation in the exclusive economic zone. *Int. Hydrogr. Rev.*, **58**, 7.
Karo, H. A. (1956): World coastline measurements. *Int. Hydrogr. Rev.*, **33**(1), 131–40.
Kenady, R. M. (1961): Electronic dot counter. *For. Sci.*, **7**, 146–7.
Kishimoto, Haruko, (1968): *Cartometric Measurements*. Geographisches Institut, Universität Zürich, 143 pp.
Kneissl, E. H. Max (1963): *Handbuch der Vermessungskunde, Band II, Feld- und Landmessung Absteckungsarbeiten*. Stuttgart, Metzlersche Verlagsbuchhandlung, 10th edn, 816 pp.
Koppé, C. (1902): Über die zweeckentsprechende Genauigkeit det Höhendarstellung in topographischer Plänen und Karten für allgemeine technische Vorarbeiten. *Z. VermessWes.*, **34**.
Köppke, H. (1967): Flächenberechnung mit Punktraster. *Vermessungstechnik*, 15 Jg. Heft 3, 111–112.
Kopylov, A. P. (1980): Poluavtomaticheskiy kurvimetr. *Geodeziya i Kartografiya*, 1980. 42–44. (A semiautomatic curvimeter).
Kramer, P. R. and Sturgeon, E. E. (1942): Transect method of estimating area from aerial photography index sheets. *J. For.*, **40**, 693–6.
Kretschmer, I, (Ed.) (1977): *Beiträge zur Theoretischen Kartographie*, Festschrift für Erik Arnberger, Vienna, Franz Deuticke, 303 pp.
Kuhlmann, H. (1951): Das Planimeter in der Vermessungstechnik, *Vermess. Rdsch.*, **13**, 230–235.

Langeraar, W. (1984): *Surveying and Charting of the Seas*. Amsterdam, Elsevier, 612 pp.
Latham, J. P. (1963): Methodology for an instrumented geographic analysis. *Ann. Assoc. Amer. Geogr.*, **53**, 194–209.
Lawrence, G. R. P. (1979): *Cartographic Methods*. London, Methuen, 2nd end, 153 pp.
LeBlond, P. H. (1981): The insensitivity of fortnightly water-level modulations to tidal bores in rivers. *Marine Geodesy*, **5**, 35–41.
Leppard, N. A. G. (1980): Satellite Doppler fixation and international boundaries. *Phil. Trans. R. Soc. Lond.*, A, **294**, 289–98.

Libaut, A. (1961): *Les Measures sur les Cartes et leur incertitude*, Paris, 231 pp.
Lindig G. (1956): Neue Methoden der Schichtlinienprüfung. *Z. VermessWes.*, **81**(7), 244–51; **81**(8), 296–303.
Lins. H. T. (1976): Land-use mapping from Skylab S-190B photography. *Photogram. Eng. & Remote Sens.*, **62**, 301–7.
Lloyd, P. R. (1976): Quantisation error in area measurement. *Cartogr. J.*, **13**, 22–5.
Loetsch, F. and Haller, K. E. (1965): *Forest inventory*. Bayerischen Landwirtschaftsverlag, München, Vol. 1, 436 pp.
Loetsch, F., Zöhrer, F. and Haller, K. E. (1973): *Forest inventory*. Bayerischen Landwirtschaftsverlag, München, Vol. 2.
Lüdemann, K. (1927): Die Leistungsfähigkeit det Kompensations-Polarplanimeter von G. Coradi und A. Ott und des Kompensationsplanimeters mit Kugellagerung von J. Schnöckel. *Z. VermessWes.*, **LVI**, Heft 10, 305–11.
Lyon, A. J. (1970): *Dealing with Data*. Oxford, Pergamon Press, 392 pp.

McCaw, G. T. (1942): Terrestial superficies. *Emp. Surv. Rev.*, **6**, 70–4.
Macdonald, R. B. and Hall, F. G. (1980): Global crop forcasting, *Science*, **208**, 670–9.
McDorman, T. L., Saunders, P. M. and VanderZwaag, P. L. (1985): The Gulf of Maine Boundary. *Marine Policy*, **9**, 90–117.
McKay, C. J. (1966): Automation applied to area measurement. *Cartogr. J.*, **3**, 22.
McKay, C. J. (1967): The pattern map: a reliability standard for sampling methods. *Cartogr. J.*, **4**, 114.
McKay, C. J. (1968): Personal communications.
McMaster, R. B. (1986): A statistical analysis of mathematical measures for linear simplification. *Amer. Cartogr.*, **13**, 103–16.
McMillan, D. J. (1985): The extent of the continental shelf: factors affecting the accuracy of a continental margin boundary. *Marine Policy*, **9**, 148–56.
Maling, D. H. (1962): Quelques ideés sur la généralisation cartographique au point de vue quantitatif. *Bull. Com. franc. cartogr., Fasc.*, No. 171–183.
Maling, D. H. (1963): Some quantitative ideas about cartographic generalisation. *Nachr. Karten-Vermessungsw.* R.V. Sonderhefte Nr 5. Frankfurt, 1963.
Maling, D. H. (1967): The terminology of map projections. *Int. Jahrb. Kartogr.*, **7**, 1–56.
Maling, D. H. (1968): How long is a piece of string? *Cartogr. J.*, **5**, 147–56.
Maling, D. H. (1973): *Coordinate Systems and Map Projections*. London, George Philip, 1973, 255 pp.
Maling, D. H. (1973): Cartometry – the neglected discipline? In Kretschmer, (Ed.) (1977), pp. 229–46.
Maling, D. H. (1984): Mathematical cartography. In I.C.A.: (1984), pp. 32–78.
Malovichko, A. K. (1958): Po povodu izmereniya na kartakh krivykh liniy. *Geodeziya i Kartografiya*, **8**. (Concerning the measurement of sinuous lines on maps).
Mandelbrot, B. (1967): How long is the coast of Britain? *Science*, 156, 636–8.
Mandelbrot, B. (1977): *Fractals: form, chance and dimensions*. San Francisco, Freeman, pp.
Mantel, N. (1953): An extension of the Buffon needle problem. *Ann. Math. Statist.*, **24**, 674–7.
Marsden, L. E. (1960): How the National map accuracy standards were developed. *Surv. Mapp.*, **20**, 427–39.
Maslov, A. V. (1955): *Sposobi i tochnosti opredeleniya ploshchadey* M. Geodezizdat, 1955, 227 pp. (Methods and accuracies in the determination of area).
Maslov, A. V., Larchenko, E. G., Gordeev, A. V. and Aleksandrov, N. N. (1958): *Geodeziya, Chast' I*, Moscow, Geodezizdat, 511 pp. (Geodesy, Vol. I).
Maslov, A. V. and Gorokhov, G. I. (1959): *Geodeziya Chast' III*, Moscow, Geodezizdat, 172 pp. (Geodesy, Vol. III).
Mátern, B. (1947): Metoder att uppskatta noggrannheten vid linje-och provytetaxering (Methods of estimating the accuracy of line and sample plot surveys). *Medd. fr. Statens Skogsforsknings Institut*, **36**, 1–138.
Mátern, B. (1960): Spatial variation. *Medd. fr. Statens Skogsforsknings Institut*, **49**(5), 144 pp.
Matthews, A. E. H. (1976): Revision of 1/2,500 scale topographic maps. *Photogramm. Rec.*, **8**(48), 794–805.

Maxim, L. D. and Harrington, L. (1983): The application of pseudo-bayesian estimators to remote sensing data: ideas and examples. *Photogram. Eng. & Remote Sens.*, **49**, 649–58.

Mayes. M. H., (1987): OS small-scales digital development – past, present and future. In: Visvalingam, M. (Ed.) (1987), pp. 2–15.

Milne, A. (1959): The centric systematic area-sample treated as a random sample. *Biometrics*, **15**, 270–97.

Ministry of Defence (1973): *Manual of Map Reading, 1973*. London, H.M.S.O., 142 pp.

Monkhouse, F. J. and Wilkinson, H. R. (1952): *Maps and Diagrams*. London, Methuen, 330 pp.

Monmonier, M. S. (1982): *Computer-assisted Cartography: Principles and Prospects*. Englewood Cliffs, N. J., Prentice-Hall, 214 pp.

Montigel, R. (1926): Genauigkeitsuntersuchungen üder Flächenbestimmung mit dem Planimeter. *Z. VermessWes.* LV, Heft **9**, 257–64.

Moran, P. A. P. (1968): Statistical theory of a high-speed photoelectric planimeter. *Biometrika*, **55**, 419–422.

Morskoy Atlas (1950–53): See Isakov, I. S. and Demin, L. A. (Eds) (1950–53).

Muehrcke, P. (1978): *Map Use, Reading, Analysis and Interpretation*. Madison, JP Publications, 474 pp.

Muller, J. C. (1986): Fractal dimensions and inconsistencies in cartographic line representations. *Cartogr. J.*, **23**, 123–30.

Muller, J. C. (1987): Fractal and automated line generalization. *Cartogr. J.*, **24**, 27–34.

Myres, J. A. L. (1975): Surveys in support of the energy programme. *Conf. Commonwealth Survey Officers, 1975, Report of Proceedings*. Paper A4, Part 1. London, Ministry of Overseas Development, 1975.

Myres, J. A. L. (1977): Energy surveys and the Hydrographic Department. *Hydrographic J.*, **3**, 22–30.

Myres, J. A. L. (1986): Survey tasks arising from the United Nations Convention on the Law of the Sea. *Int. Hydrogr. Rev.*, **63**, 65–76.

Nash, A. J. (1949): A photoelectric planimeter. *Surv. and Mapp.*, **8**, 64–9.

Naylor, R. (1956): The determination of area by weight: forest service experience. *N.Z.J. For.*, pp. 109–111.

Nystrom, D. A. (1986): Geographic information system developments within the U.S. Geological Survey. In: Blakemore, M. (Ed.) (1986), Vol. 1, pp. 33–42.

Nystuen, J. D. (1966): Effects of boundary shape and the concept of local convexity. *Michigan Inter-University Community of Mathematical Geographers*. Discussion Paper No. 10. Ann Arbor, 22 pp.

Openshaw, S. and Mounsey, H. (1986): Geographic Information Systems and the BBC's Interactive Videodisk. In Blakemore, M. (Ed.) (1986), Vol. 2, pp. 539–46.

Openshaw, S., Wymer, C. and Charlton, M. (1986): A geographical information and mapping system for the BBC Domesday optical discs. *Trans. Inst. Brit. Geogr.*, N.S., **11**, 296–304 (20).

Open University (1978): *Law of the Sea*, S334, Oceanography, Units 15 and 16, Milton Keynes, The Open University, 86 pp.

Ordnance Survey (1932): *Instructions to Field Revisers 1/2,500 scale, 1932*. London, H.M.S.O., 80 pp.

Ordnance Survey (1971): *Annual Report of the Ordnance Survey*. London, H.M.S.O.

Ordnance Survey (1972): *The Overhaul of the 1:2,500 County Series Maps*. O.S. Professional Paper, N.S., No. 25. Southampton, 30 pp.

Ordnance Survey (1979): *Report of the Ordnance Survey Review Committee*. London, H.M.S.O., 185 pp.

Orlóci, L. (1975): *Multivariate Analysis in Vegetation Research*. Den Haag, Junk, 276 pp.

Osborne, J. G. (1942): Sampling errors of systematic and random surveys of cover-type areas. *J. Amer. Statist. Assoc.*, **37**, 256–64.

Penck, A. (1894): *Morphologie der Erdoberfläche*. Stuttgart, 1894.

Perkal, J. (1958a): O Dlugosci Krzwych Empirycznych. *Zast. Mat.* **3**, 258–284. (transl. Jackowski, R.) (1966): On the length of empirical curves. *Michigan Inter-University Community of Mathematical Geographers*. Discussion Paper No. 10. Ann Arbor, December, 34 pp.

Perkal, J. (1958b): Próba obiekytywnej generalizacji. *Geodezja Kartogr.* VII.2.130–142. (transl. Jackowski, R.) (1966): An attempt at objective generalisation. *Michigan Inter-University Community of Mathematical Geographers.* Discussion Paper No. 10. Ann Arbor, December, 1966, 18 pp.
Phillips, J. O. (1971): Coastal boundary surveys. *Int. Hydrogr. Rev.*, **48**, 129–41.
Pielou, E. C. (1974): *Population and Community Ecology.* New York, Gordon & Breach, 424 pp.
Polushkin, E. P. (1925): Determination of structural composition of alloys by a metallographic planimeter. *Trans. AIMME*, **71**, 669–90.
Prescott, J. R. V. (1975): *The Political Geography of the Oceans.* Newton Abbott, David & Charless, pp.
Proctor, D. W. (1986): The capture of survey data. In Blakemore, M. (Ed.) (1966), Vol. 1, pp. 227–36.
Proudfoot, M. J. (1942): Sampling with transverse traverse lines. *J. Am. Statist. Assoc.*, **37**, 265–70.
Proudfoot, M. J. (1946): *Measurement of Geographic Area.* Sixteenth Census of the U.S., 1940. U.S. Dept. of Commerce, Bureau of the Census. Washington, Govt. Printing Office, 120 pp.

Quenouille, M. H. (1949): Problems in plane sampling. *Ann. Math. Statist.*, **20**, 355–75.

Reignier, F. (1957): *Les Systèmes de Projection et leurs applications...*, Vol. I, Paris, I.G.N., 312 pp.
Rhind, D. (1987): Introduction. In Visvaligam, M. (Ed): (1987), p. 1.
Rhind, D. and Mounsey, H. (1986): The land and people of Britain: a Domesday record, 1986. *Trans. Inst. Brit. Geogr.* N.S., pp. 315–25.
Richardson, L. F. (1961): The problem of contiguity: an appendix to statistics of deadly quarrels. *General Systems Yearbook*, **VI**, 139–87.
Richardus, P. and Adler, R. K. (1972): *Map Projections.* Amsterdam, North-Holland, 174 pp.
R.I.C.S. (1981): *U.K. National Land Survey and Mapping Conference, at the University of Reading, 30th March–3rd April, 1981.*
Rideout, E. E. (1962): Measurement by the MK area calculator, *Dept. of Forestry, Canada, Forest Research Branch Report*, 62–19.
Ringold, P. L. and Clark, J. (1980): *The Coastal Almanac for 1980 – The Year of the Coast.* San Francisco, Freeman.
Ritchie, G. S. (1975): The work of the International Hydrographic Organisation. *Conf. Commonwealth Survey Officers, Report of Proceedings, 1975*, Vol. 1, Paper No. A1. London, O.D.M.
Ritchie, G. S. (1981): The new challenge facing the world's hydrographers. *Hydrographic J.*, 6–9.
Ritchie, G. S. (1982): Hydrography, yesterday, today and tomorrow. *Int. Hydr. Rev.*, **59**, 7–13.
Ritchie, G. S. (1986): The role of the hydrographic surveyor in offshore exploration and exploitation. *Int. Hydr. Rev.*, **63**, 29–35.
Rollin, J. (1986): A method of assessing the accuracy of cartographic digitising tables. *Cartogr. J.*, **23**, 144–6.
Rosenfield, G. H. (1981): Analysis of variance of thematic mapping experiment data. *Photogram. Eng. & Remote Sens.*, **47**, 1685–92.
Rosenfield, G. H. (1982a): Sample design for estimating change in land use and land cover. *Photogram. Eng. & Remote Sens.*, **48**, 793–801.
Rosenfield, G. H. (1982b): The analysis of areal data in thematic mapping experiments. *Photogram. Eng. & Remote Sens.*, **48**, 1455–1462.
Rosenfield, G. H. (1986): Analysis of thematic map classification error matrices. *Photogram. Eng. & Remote Sens.*, **52**, 681–6.
Rosenfield, G. H. and Fitzpatrick-Lins, K. (1981): Comparison of sampling procedures and data analysis for a land-use and land-cover map. *Photogram. Eng. & Remote Sens.*, **47**, 343–51.
Rosenfield, G. H. and Fitzpatrick-Lins, K. (1986): A coefficient of agreement as a measure of thematic classification accuracy. *Photogram. Eng. & Remote Sens.*, **52**, 223–7.
Rosenfield, G. H. and Melley, M. L. (1980): Applications of statistics to thematic mapping. *Photogram. Eng. & Remote Sens.*, **46**, 1287–94.
Rosenfield, G. H., Fitzpatrick-Lins, K. and Ling, H. S. (1982): Sampling for thematic map accuracy testing. *Photogram. Eng. & Remote Sens.*, **48**, 131–7.
Royal Society (1978): *The Future Role of the Ordance Survey. A Submission to the Ordnance Survey*

Review Committee by the Council of the Royal Society (Submitted 24 May 1978). London, The Royal Society, 60 pp.

Sadler D. H. (1956): Spheroidal sailing and the middle latitude. *J. Inst. Nav.*, **9**, 371–7.
Scheidegger A. E. (1970): *Theoretical Geomorphology*. London, George Allen & Unwin, 2nd edn, 435 pp.
Schmidt-Falkenberg, H. (1962): Grudlinien einer Theorie der Kartographie. *Nachr. Karten-Vermessungsw.* RII, Heft 22.
Schmiedeberg, W. (1906): Zur Geschichte der geographischen Flächenmessung bis zur Erfindung des Planimeters. *Z. Ges. Erdk. Berl.* Nr 3, 152–76; Nr 4, 233–56.
Schuster, E. F. (1974): Buffon's needle experiment. *Amer. Math. Monthly*, **81**, 26–9.
Seeber, G. (1985): Point positioning in marine geodesy. *Marine Geodesy*, **9**, 365–80.
Senkov, E. P. (1958): Izmerenie dliny rek s pomoshchu usovershenstvovannogo kurvimetra KS. *Meteorol. i Gidrologiya*, No. 4, 28–30. (Measurement of the lengths of rivers with the help of the improved KS curvimeter).
Seymour, W. A. (1963): Revision. *Conf. Commonwealth Survey Officers, 1963, Report of Proceedings*, Paper No. 33. London, H.M.S.O.
Shalowitz, A. L. (1962): *Shore and Sea Boundaries*, U.S. Dept. of Commerce, Coast & Geodetic Survey Publication 10–1. 2 vols. Washington, Government Printing Office; Vol. 1, 420 pp.; Vol. 2, 749 pp.
Sheppard, H. L. (1953): A note on chart distortions. *J. Inst. Nav.*, **6**, 159–60.
Shokal'skiy, U. M. (1930): *Dlina glavneyshikh rek aziatskoy chasti SSSR i sposob izmereniya dlin rek po kartam.* Moscow, Rechtransizdat, 1930. (The lengths of the principal rivers of the Asiatic part of the USSR and methods of measuring the lengths of rivers on maps).
Skellam, J. G. (1972): Some philosophical aspects of mathematical modelling in empirical science with special reference to ecology. In Jeffers, J. N. R. (Ed.) (1972), pp.
Smartt, P. F. M. and Grainger, J. E. A. (1974): Sampling for vegetation survey. *J. Biogeog.*, **1**, 193–206.
Smirnov, L. E. (1967): *Teoreticheskie osnovy i metody deshifrirovaniya aerosnimkov*, Lenningrad, Izd. Leningradskogo Universiteta, 1957, 241 pp. (Theoretical basis and methods of geographical photographic interpretation).
Smirnov, L. E. and Chernyaeva, F. A. (1964): Opyt izmereniya ploshchadey na aerosnimkakh. *Vestnik Leningradskogo Universiteta, Seriya Geologii i Geografii, Vyp. 2*, No. 12, Leningrad, 1964. (Experiments in measuring areas on aerial photographs).
Smith, C. S. and Guttman, L. (1953): Measurement of internal boundaries in three-dimensional structures by random sectioning. *Trans AIME, J. Metals*, 81–7.
Smith, R. W. (1982): A geographical primer to maritime boundary making. *Ocean Devel. Int. Law J.*, **12**, 1–22.
Smith, W. P. (1979): National mapping: a case for government responsibility. *Conf. Commonwealth Survey Officers, 1979. Report of Proceedings*, Part II, Paper M2. London, O.D.A.
Smith, W. P. (1980): The Ordnance Survey: a look to the future. *Cartogr. J.*, **17**, 75–82.
Snyder, J. P. (1978): The space oblique Mercator projection. *Photogram. Eng. & Remote Sens.*, **44**, 585–96.
Snyder, J. P. (1982): *Map Projections used by the U.S. Geological Survey.* Geological Survey Bulletin 1532. Washington, U.S. Government Printing Office, 313 pp.
Snyder, J. P. (1987): *Map Projections – A Working Manual*, U.S. Geol. Survey Professional Papers 1395. Washington, Govt. Printing Office, 383 pp.
Solomon, H. (1978): *Geometric Probability*. Philadelphia Society for Industrial and Applied Mathematics, 174 pp.
Starr, L. E. (1986): Mark II: The next step in digital systems development at the U.S. Geological Survey. In: Blakemore, M. (Ed.) (1986), Vol. 2, pp. 200–5.
Steinhaus, H. (1954): Length, shape and area, *Colloquium Mathematicum*, **III**(1), 1–13.
Steward, H. J. (1974): *Cartographic Generalisation – Some Concepts and Explanation*, Cartographica Monograph No. 10, 77 pp.
Stobbs, A. R. (1967): Recent trends in land resource investigation. *Conference Commonwealth Survey Officers, 1967. Report of Proceedings.* London, H.M.S.O.
Stobbs, A. R. (1968): Some problems of measuring land use in underdeveloped countries: the land use survey of Malawi. *Cartogr. J.*, **5**, 107–10.

Strahler, A. H. (1981): Stratification of natural vegetation for forest and rangeland inventory using Landsat digital imagery and collateral data. *Int. J. Remote Sensing*, **2**, 15–41.
Sukhov, V. I. (1957): *Sostavlenie i redaktirovanii obshchegeograficheskikh kart*. Moscow. Geodezizdat, 279 pp. (The compilation and editing of geographical maps).

Taylor, E. G. R. (1947): The surveyor. *Econ. Hist. Rev.*, **17**, 121–33.
Taylor, J. I. and Steingelin, R. W. (1969): Infrared imaging for water resources studies. *J. Hydraulics Div., Proc. Amer. Soc. of Civil Eng.*, **95**, No HY1, pp. 175–89.
Taylor, P. J. (1977): *Quantitative Methods in Geography*. Boston, Houghton Mifflin, 386 pp.
Thamsborg, M. (1971): Notes on hydrographic assistance to the solution of sea boundary problems. *Int. Hydrogr. Rev.*, **48**, 149.
Thamsborg, M. (1974): Geodetic hydrography as related to maritime boundary problems. *Int. Hydrogr. Rev.*, **51**, 157.
Thompson, M. M. (1956): How accurate is that map? *Surv. and Mapp.*, **16**, 164–73.
Thompson, M. M. and Rosenfield, G. H. (1971): On map accuracy specifications. *Surv. and Mapp.*, **31**, 57–64.
Tobler, W. R. (1966): Foreword to *Discussion Paper No 10. Michigan Inter-University Community of Mathematical Geographers.* Ann Arbor, 53 pp.
Tobler, W. R. (1970): A computer movie simulating urban growth in the Detroit region. *Econ. Geogr.* **46**, 234–40.
Tomasegovič, Z. (1968): Direct determination of area distribution based upon topographic features by means of the Wild B9 Aviograph, *Photogrammetria*, **23**, 123–37.
Tomkieff, S. I. (1945): Linear intercepts, areas and volumes. *Nature*, **155**, 24.
Tomlinson, R. F. (1970): Computer based geographical data handling. In Cox, (Ed): *New Possibilities and Techniques for Land Use and Related Surveys*, pp. 105–20.
Trefethen, J. M. (1935): A method for geographic surveying, *Amer. J. Sci.*, **32**, Series 5, 454–64.
Trorey, L. G. (1950): *Handbook of Aerial Mapping and Photogrammetry*. Cambridge, Cambridge University Press, 178 pp.
Tryon, T. C., Hale, G. A. and Young, H. E. (1955): Dot gridding air photos and maps, *Photogram. Eng.*, **21**, 737–8.
Turnbull, L. G. and Ellis, C. E. (1952): An examination of the estimation of tenths by different observers. *Surv. and Mapp.*, **12**, 270.

U.S. Army, (1955): *Guide to the Compilation and Revision of Maps*, TM 5–240. Washington, U.S. Govt. Printing Office, 166 pp.
U.S. Department of State (1969): *Sovereignty of the Sea*. Geographical Bulletin No. 3, 33 pp.

van Genderen, J. L. and Lock, B. F. (1977): Testing land-use map accuracy. *Photogram. Eng. & Remote Sens.*, **43**, 1135–7.
van Genderen, J. L., Lock, B. F. and Vass, P. A. (1978): Statistical testing of thematic map accuracy. *Remote Sensing of Environment*, **5**, 3–14.
Visvalingam, M. (ed), (1987): Research based on Ordnance Survey small-scales Digital data. *Institute of British Geographers, Cartographic Information Systems Research Group*, The University of Hull, 79 pp.
Vogel, T. J. (1981): Horizontal datums for nautical charts. *Int. Hydrogr. Rev.*. **58**, 53–64.
Volkov, N. M. (1950a): O Tochnosti Kart. *Trudy TsNIIGAiK*, Vup 52, 67–91. (About Map Accuracy).
Volkov, N. M. (1950b): *Printsipy i Metody Kartometrii*, Moscow and Leningrad, Izd. Akademii Nauk, 327 pp. (The Principles and Methods of Cartometry).
Volkov, N. M. (1949): Novyy sposob izmereniya dlin rek po kartam, *Izv. AN SSSR, Seriya Geografiya i Geofizicheskaya*, T. 13.2. 1949. (New methods of measuring the lengths of rivers on maps).
Volkov, N. M. (1964): Vneshnie perspektivnye proektsii s positivnym izobrazheniem poverkhnosti zemnogo. *Geodeziya i Kartografiya, 70–73.* (External perspective projections with positive representation of the surface of the Earth).
Vorob'ev, V. I. (1959): Dlina beregovoy linii morey SSSR. *Geograficheskii Sbornik*, Vyp XIII, 1959, 63–89. (The Length of the Coastline of the USSR).
Wagner, K.-H. (1949): *Kartographische Netzentwürfe*. Leipzig, Bibliographisches Institut, 263 pp.

War Office (1926): *Survey Computations: To be used "in conjunction with the Text Book of Topographical and Geographical Surveying.* London, H.M.S.O., 3rd edn, 179 pp.
War Office (1957): *Manual of Map Reading, Air Photo Reading and Field Sketching, Part I, Map Reading.* London, H.M.S.O., 1957, 132 pp.
Weibel, E. R. (1979): *Stereological Methods,* Vol. 1: *Practical Methods for Biological Morphometry.* London, Academic Press, 415 pp.
Weibel, E. R. (1980): *Stereological Methods,* Vol. 2: *Theoretical Foundations.* London, Academic Press, 340 pp.
Weightman, J. A. (1982): Satellite geodesy and offshore oil. *Int. Hydrogr. Rev.,* **59**, 31–57.
Werkmeister, P. (1935): Ein neuer Linienmesser von A. Ott, *Z. InstrumKde.,* **55**, 176–7.
Willers, Th. (1911): Zur Geschichte der Geographischen Flächenmessung seit einführung des Planimeters. *Pet. Mit. Erg.* Nr 70, 96 pp.
Wilson, R. C. (1949): The relief displacement factor in forest area estimates by dot templets on aerial photographs. *Photogram. Eng.,* 225–236.
Winkworth, A. V. (1956): A comparison of methods employed in measuring area. *University of British Columbia Research Note No. 15.*
Winterbotham, H. S. L. (1934): *The National Plans.* Ordnance Survey Professional Papers, New Series No. 16. London, H.M.S.O., 112 pp.
Wood, W. F. (1954): The dot planimeter. *Prof. Geographer,* **6**(1),
Wray, T. and Weiss, C. C. (1982): Auxiliary spheres of common spheroids. *Can Surveyor,* **36**, 191–6.

Yates, F. (1944): Methods and purposes of agricultural surveys. *J. Roy. Soc. Arts,* **91**(4640), 367–79.
Yates, F. (1949): *Sampling Methods for Censuses and Surveys.* London, Griffin, 1st edn; 4th edn, 1981, 458 pp.
Yoeli, P. (1975): Methodology of computer-assisted cartography. In: Davis, J. C. and McCullagh, M. J. (Eds): (1975), pp. 130–6.
Yuill, R. S. (1971): Areal measurement error with a dot planimeter: some experimental estimates. *U.S. Geological Survey. Interagency Report No. 213.* Washington, U.S.G.S., 9 pp.
Yule, G. U. (1927): On reading a scale. *J. Roy. Statist. Soc.,* **90**, 570–87.

Zeimetz, K. A., Dillon, E., Hardy, E. E. and Otte, R. C. (1976): Using area point sampling and air photos to estimate land use change. *Agric. Econ. Res.,* **28**(2), 65–74.
Zill, W. (1955): Zur Genauigkeit der Flächenbestimmung mit Umfahrungsplanimetern, *Wiss.Z. Tech. Univ., Dresden,* **4**, 146–54.
Znamenshchikov, G. I. (1963): *Izmerenie po kartam protyazhemosti ob'ektov, izobrazhushchikhsya kriv'mi liniyami.* Avtoreferat, MIIGAiK, M, 1963, 38 pp. (The measurement on maps of the lengths of objects depicted by sinuous lines).
Zöhrer, F. (1978): *On the Precision of Dot Grid Estimates.* Resource Inventory Notes. U.S.D.I. Bureau of Land Management, Denver, pp. 1–6.

Index